Lecture Notes in Computer Science 9502

More information about this series at http://www.springer.com/series/7409

Ching-Hsien Hsu · Feng Xia
Xingang Liu · Shangguang Wang (Eds.)

Internet of Vehicles – Safe and Intelligent Mobility

Second International Conference, IOV 2015
Chengdu, China, December 19–21, 2015
Proceedings

Editors
Ching-Hsien Hsu
Department of Computer Science
Chung Hua University
Hsinchu, Taiwan
Taiwan

Feng Xia
School of Software
Dalian University of Technology
Dalian
China

Xingang Liu
School of Electronic Engineering
University of Electronic Science
and Technology of China
Chengdu
China

Shangguang Wang
State Key Laboratory of Networking
and Switching Technology
Beijing University of Posts
and Telecommunications
Beijing
China

ISSN 0302-9743 ISSN 1611-3349 (electronic)
Lecture Notes in Computer Science
ISBN 978-3-319-27292-4 ISBN 978-3-319-27293-1 (eBook)
DOI 10.1007/978-3-319-27293-1

Library of Congress Control Number: 2015955877

LNCS Sublibrary: SL3 – Information Systems and Applications, incl. Internet/Web, and HCI

Springer Cham Heidelberg New York Dordrecht London

Printed on acid-free paper

Springer International Publishing AG Switzerland is part of Springer Science+Business Media
(www.springer.com)

Preface

This volume contains the proceedings of IOV 2015, the International Conference on Internet of Vehicles, which was held in Chengdu, China, during December 19–21, 2015.

The Internet of Vehicles (IOV) is emerging as an important part of the smart or intelligent cities being proposed and developed around the world. IOV is a complex integrated network system that interconnects people within and around vehicles, intelligent systems on board vehicles, and various cyber-physical systems in urban environments. IOV goes beyond telematics, vehicle ad hoc networks, and intelligent transportation, by integrating vehicles, sensors, and mobile devices into a global network to enable various services to be delivered to vehicular and transportation systems, and to people on board and around vehicles, using contemporary computing techniques such as swarm intelligent computing, crowd sensing and crowd sourcing, social computing, and cloud computing. IOV 2015 intended to play an important role for researchers and industry practitioners, offering them the chance to exchange information regarding advancements in the state of the art and practice of IOV architectures, protocols, services, and applications; it also aimed to identify emerging research topics and to define the future directions of IOV.

This year, the technical program of IOV 2015 received submissions from 15 countries: China, India, Japan, Korea, New Zealand, Pakistan, Saudi Arabia, Sudan, Taiwan, Tunisia, Uganda, UAE, UK, and USA. All papers submitted underwent a rigorous peer-review process with at least three reviewers, coordinated by the international Program Committee. The Program Committee finally accepted 41 papers for inclusion in the conference program and proceedings. The accepted papers provide research contributions in a wide range of research topics including telematics, wireless communication networks, services and applications, social information systems, swarm intelligence, economics for IOV, modeling and simulation, as well as cloud computing and big data. We believe that this volume not only presents novel and interesting ideas but will also stimulate future research in the area of IOV.

The organization of conferences requires a lot of hard work and dedication from many people. We would like to take this opportunity to thank the numerous individuals whose time and efforts made this conference possible and ensured its high quality. We wish to thank the authors for submitting high-quality papers that contributed to the conference technical program. We wish to express our deepest gratitude to all Program Committee members and external reviewers for their excellent job in the paper review process. Without their help, this program would not be possible. Special thanks go to the entire local Arrangements Committee for their help in making the conference a wonderful success. We take this opportunity to thank all the presenters, session chairs, and participants for their presence at the conference, many of whom traveled long distances to attend this conference and make their valuable contributions. Last but not least, we would like to express our gratitude to all the organizations that supported our efforts to bring the conference to fruition. We are grateful to Springer for publishing these proceedings.

We hope that you enjoy the IOV 2015 proceedings.

December 2015

Sajal Das
Hsiao-Hwa Chen
Athanasios V. Vasilakos
Feng Xia

Organization

Executive Committee

2015 International Conference on Internet of Vehicles (IOV 2015)

Honorary Chair

Fangchun Yang	Beijing University of Posts and Telecommunications, China

General Co-chairs

Sajal Das	Missouri University of Science and Technology, USA
Hsiao-Hwa Chen	National Cheng Kung University, Taiwan

Program Committee Chairs

Athanasios V. Vasilakos	Lulea University of Technology, Sweden
Feng Xia	Dalian University of Technology, China

Executive Chair

Xingang Liu	University of Electronic Science and Technology of China, China

Steering Committee

Robert C.H. Hsu	Chung Hua University, Taiwan (Chair)
Shangguang Wang	BUPT, China (Vice Chair)
Victor C.M. Leung	The University of British Columbia, Canada
Chung-Ming Huang	NCKU, Taiwan

Awards Chair

Chu-Sing Yang	National Cheng Kung University, Taiwan

Publicity Chairs

Yan Zhang	Simula Research Laboratory and University of Oslo, Norway
Der-Jiunn Deng	National Changhua University of Education, Taiwan
Atilla Elci	Aksaray University, Turkey

International Liaison Chairs

Shawkat Ali	The University of Fiji, Fiji
Daming Wei	Tohoku University, Japan
Yang Xiang	Deakin University, Australia

Publication Chair

Yao-Chung Chang	National Tai Tung University, Taiwan

Technical Program Committee

Abdelmadjid Bouabdallah	Université de Technologie de Compiègne, France
Adão Silva	University of Aveiro/Instituto de Telecomunicações, Portugal
Alexandre Santos	University of Minho, Portugal
Anthony Lo	Delft University of Technology, The Netherlands
Aravind Kailas	LLC, USA
Benoît Parrein	University of Nantes, French
Cailian Chen	Shanghai Jiao Tong University, China
Carlos Calafate	Universidad Politecnica de Valencia, Spain
Carsten Röcker	Fraunhofer IOSB-INA, Germany
Cathryn Peoples	University of Ulster, UK
Chau Yuen	Singapore University of Technology and Design, Singapore
Christian Prehofer	LMU München, Germany
Chuan-Ming Liu	National Taipei University of Technology, Taiwan
Constandinos Mavromoustakis	University of Nicosia, Republic of Cyprus
Constantine Kotropoulos	Aristotle University of Thessaloniki, Greece
Daqiang Zhang	Tongji University, China
Daxin Tian	Beihang Unviersity, China
Dimitrios Koukopoulos	University of Western Greece, Greece
Domenico Ciuonzo	Second University of Naples, Italy
Donghyun Kim	North Carolina Central University, USA
Dongliang Duan	University of Wyoming, USA
Enzo Mingozzi	University of Pisa, Italy

Esa Hyytiä	Helsinki University of Technology, Finland
Georgios Kambourakis	University of the Aegean, Greece
Gerard Parr	University of Ulster, UK
Ghalem Boudour	LIG Laboratory, France
Giacomo Verticale	Politecnico di Milano, Italy
Hai Jin	Huazhong University of Science and Technology, China
Haitao Xia	LSI Corporation (Acquired by Avago Tech), USA
Han-Shin Jo	Hanbat National University, Korea
Hung-Min Sun	National Tsing Hua University, Taiwan
Hung-Yu Wei	National Taiwan University, Taiwan
Ignacio Soto	UC3M, Spain
Ing-Chau Chang	National Changhua University of Education, Taiwan
Jenq-Haur Wang	National Taipei University of Technology, Taiwan
Jeremy Blum	Penn State University, USA
Jiannong Cao	Hong Kong Polytechnic University, Hong Kong, SAR China
Jingon Joung	Institute for Infocomm Research, Singapore
John Mcgregor	Clemson University, USA
Juan-Carlos Cano	Technical University of Valencia, Spain
Kan Zheng	BUPT, China
Khalil Ibrahimi	University IBN Tofail, Faculty of Sciences, Morocco
Luca Caviglione	CNR-ISSIA, Italy
Marco Listanti	DIET, Italy
Massimiliano Comisso	University of Trieste, Italy
Mehmet Celenk	Ohio University, USA
Miguel López-Benítez	University of Liverpool, UK
Minseok Kwon	Rochester Institute of Technology, USA
Momin Uppal	LUMS School of Science and Engineering, Pakistan
Mujdat Soyturk	Marmara University, Turkey
Mu-Song Chen	Dayeh University, Taiwan
Na Li	Northwest Missouri State University, USA
Naixue Xiong	Colorado Technical University, USA
Nary Subramanian	University of Texas at Tyler, USA
Natarajan Meghanathan	Jackson State University, USA
Nikolaos Papandreou	IBM Research - Zurich, Switzerland
Oscar Esparza	Universitat Politècnica de Catalunya, Spain
Paolo Carbone	University of Perugia, Italy
Pascal Lorenz	University of Haute Alsace, France
Rachid Outbib	Aix-Marseille Université, France
Ramin Yahyapour	GWDG - University of Göttingen, Germany
Razvan Stanica	INSA Lyon, France
Rui J. Lopes	ISCTE-IUL Lisbon University Institute/Instituto de Telecomunicações IT-IUL, Portugal
Shin-Feng Lin	National Dong Hwa University, Taiwan
Shiqiang Wang	Imperial College London, UK
Shujun Li	University of Surrey, UK

Sokratis Katsikas	University of Piraeus, Greece
Thinagaran Perumal	Universiti Putra Malaysia, Malaysia
Thomas Lagkas	The University of Sheffield International Faculty, UK
Tzung-Shi Chen	National University of Tainan, Taiwan
Uei-Ren Chen	Hsiuping University of Science and Technology, Taiwan
Vincenzo Piuri	University of Milan, Italy
Winston Seah	Victoria University of Wellington, New Zealand
Woong Cho	Jungwon University, Korea
Xavier Masip	UPC, China
Yao-Chung Chang	National Taitung University, Taiwan
Yiming Ji	University of South Carolina Beaufort, USA
You-Chiun Wang	National Sun Yat-sen University, Taiwan
Yu Cai	Michigan Technological University, USA
Yuan-Cheng Lai	National Taiwan University of Science and Technology, Taiwan
Yufeng Wang	University of South Florida, USA
Yunchuan Sun	Beijing Normal University, China
Zbigniew Dziong	École de technologie supérieures, Canada
Zhe Chen	Northeastern University, China
Zheng Yan	Xidian University, China

Contents

IOV Architectures and Applications

Intelligent Mobility

V2V and M2M Communications

Modeling and Simulations

IOV Architectures and Applications

Vehicle Cardinality Estimation in VANETs by Using RFID Tag Estimator

Jinhua Song[1], Ching-Hsien Hsu[2], Mianxiong Dong[3], and Daqiang Zhang[1](✉)

[1] School of Software Engineering, Tongji University, Shanghai, China
dqzhang@tongji.edu.cn
[2] Department of Computer Science and Information Engineering, Chung Hua University, Hsinchu, Taiwan
[3] Muroran Institute of Technology, Hokkaido, Japan

Abstract. Nowadays, many vehicles equipped with RFID-enabled chipsets traverse the Electronic Toll Collection (ETC) systems. Here, we present a scheme to estimate the vehicle cardinality with high accuracy and efficiency. A unique RFID tag is attached to a vehicle, so we can identify vehicles through RFID tags. With RFID signal, the location of vehicles can be detected remotely. Our scheme makes vehicle cardinality estimation based on the location distance between the first vehicle and second vehicle. Specifically, it derives the relationship between the distance and number of vehicles. Then, it deduces the optimal parameter settings used in the estimation model under certain requirement. According to the actual estimated traffic flow, we put forward a mechanism to improve the estimation efficiency. Conducting extensive experiments, the presented scheme is proven to be outstanding in two aspects. One is the deviation rate of our model is 50 % of FNEB algorithm, which is the classical scheme. The other is our efficiency is 1.5 times higher than that of FNEB algorithm.

Keywords: Vehicle estimation · VANETs · RFID tag · Privacy preservation

1 Introduction

The automobile popularity provides much convenience for people, together with significant serious traffic problems [1]. In this situation, Intelligent Transportation System (ITS) [2] is the direction and goal for traffic management, where many vehicles are attached with a RFID-enabled module for wireless communication. ITS alleviates traffic pressure and reduces traffic accident occurrence frequency. In ITS, Internet of Vehicle (IOV) [3] achieves the two kinds of communication between vehicles and vehicles, vehicles and roads, drivers and managers. Based on the communication between the Road-Side Units (RSU) and On-Board Units (OBU) [4], traffic managers can evaluate the real-time traffic condition through estimating the number of vehicles and thus make more effective traffic management. Meanwhile, the drivers can access current traffic situation to adjust more effective transportation plan in time. Therefore, it is important to create a good method for vehicle cardinality estimation, and that is the purpose of this paper.

C.-H. Hsu et al. (Eds.): IOV 2015, LNCS 9502, pp. 3–15, 2015.
DOI: 10.1007/978-3-319-27293-1_1

In the past few years, researches on RFID tag estimator have developed very quickly, and many of them can achieve high performance when making tag cardinality estimation. As we can see in Fig. 1, there are correspondent relationship between IOV and RFID system [5], in which RSU corresponds to a RFID reader and OBU corresponds to an independent tag. That means, estimating vehicle cardinality in IOV equals to estimating tag cardinality in RFID system, so we can use the RFID tag estimator to estimate the number of vehicles.

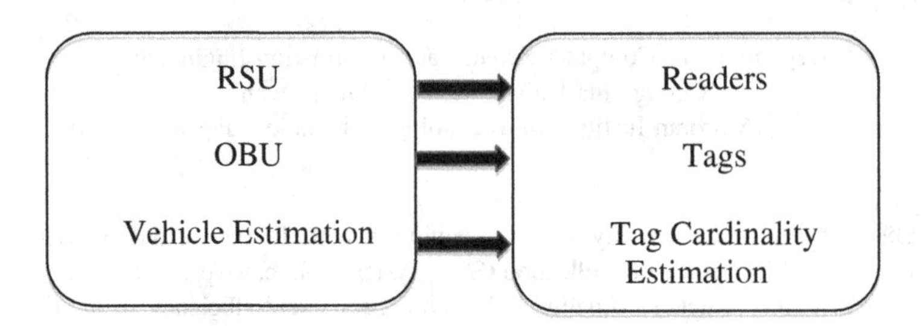

Fig. 1. Correspondent relationship between IOV and RFID system

To this end, we propose a scheme for vehicle cardinality estimation by using RFID tag estimator, which leverages the location distance between the first vehicle and the second vehicle to make estimation. It achieves high performance both in accuracy and time efficiency without privacy leakage [6]. Specifically, our scheme derives the relationship between the distance and number of vehicles. Then, it deduces the optimal parameter settings used in the estimation model under certain requirement. Finally, according to the actual estimated traffic flow, we put forward a mechanism to improve the estimation efficiency. We only need identify two vehicles in a round, which helps to protect privacy in the estimation process.

The major contributions of our scheme are three-fold.

(1) Our scheme has high accuracy while estimating vehicle cardinality, which can be seem from that the deviation rate of our model is 50 % of FNEB algorithm that is the classical scheme.
(2) Our scheme has high time efficiency while estimating vehicle cardinality. Compared with FNEB algorithm, our efficiency is 1.5 times higher than that of it.
(3) Our scheme avoid privacy leakage, because it can estimate vehicle cardinality without identifying vehicles one by one.

The following sections of the paper is as follows. Section 2 presents the related work. Section 3 gives the problem description and system overview. Section 4 describes the prosed scheme in detail. Section 5 reports the experiment results. Section 6 contains the final conclusion of our work.

2 Related Work

In recent years, tag cardinality estimation has attracted much attention from the research community. Based on the different ways for tag coding, these schemes are divided into two approaches: uniformly distribution of hash function based and geometric distribution of hash function based.

In the first approach, Kodialam proposed a tag cardinality estimator based on probability analysis with Anonymity, which broke with the tradition of using tag identification protocol to estimate tag cardinality. However, it had two drawbacks, one was the reader must read all tags in a single time round. The other was the scale of a tag set should be a known quantity. Because of these disadvantages, a FNEB model was presented [7], which makes tag cardinality estimation using the first slot chosen by tags. It avoided reading all tags in a frame.

Uniformly distribution of hash function for tag coding is widely used in tag cardinality estimation, but it is not the only choice. Lottery of Frame (LoF) estimator was proposed [8], which utilized the average run size of 1 to estimate tag cardinality. It used the geometric distribution of hash function for tag coding, so enlarged the estimation range.

At present, research on RFID tag cardinality estimation is still ongoing, such as Zero-One Estimator [9], Simple RFID Counting estimator [10] and Average-run-based estimator. In this paper, we compare the classical FNEB algorithm with our new proposed scheme when used in VANETs, and the result shows that our scheme achieves better performance in both time cost and accuracy.

3 Problem Description and System Overview

This section presents the problem description and system introduction. The aim is to estimate the vehicle cardinality accurately and quickly without identifying each vehicle individually. Firstly, we introduce the frame-slotted ALOHA model [11] and its application in our paper briefly. Then, we describe the way in which vehicles choose road segment locations and present the communication protocol [12] between RSU and OBU.

3.1 Problem Description

Assuming the OBUs are all in the communication range of RSUs, and vehicles keep static during the estimation phase. The problem is how to estimate the vehicle cardinality accurately and quickly without reading each vehicle individually. In our scheme, we introduce two variables, accuracy probability and confidence interval, to define the estimation requirement. Their relationship can be described as follows. With vehicle cardinality t_0, accuracy probability γ and confidence interval β, our scheme returns an estimation value t_1, which satisfies the formula $P\left[\frac{|t_1-t_0|}{t_0} \leq \beta\right] \geq \gamma$. For example, if $t_0 = 3000$, $\gamma = 90\ \%$ and $\beta = 10\ \%$, the probability of our result between 2700 and 3300 is above 90 %. Table 1 introduces the symbols appeared in our paper.

Table 1. Symbol description used in the design the proposed scheme

Symbol	Description	Symbol	Description
β	Confidence interval	t_{max}	Upper bound
γ	Accuracy probability	l	Road size
t_0	Vehicle cardinality	s	Random seed
t_1	Estimated value of vehicle cardinality	ρ	Load factor
F	Location number of the first vehicle	n	Times of cycles
S	Location number of the second vehicle	$T(\cdot)$	Estimation time
W	Location number distance of the first and second vehicle	$h(\cdot)$	Hash function

3.2 System Overview

Our scheme references the framed-slotted ALOHA protocol, whose idea is to unify user's data transmission through clock signal. By dividing time into discrete time slice, the user can only send data in the start of any time slice, which avoids data transmission casually and reduces the probability of data conflict. In our scheme, we will divide a road length into discrete road segments. All road segments are numbered uniquely and sequentially, then a vehicle chooses one of them.

The RSU and OBU communicate with each other through multiple roads, and every road is divided into multiple road segments. Here, a single road corresponds to a collection cycle. Firstly, the RSU transmits a random seed s and a road size l to OBU. If the road size is l, there are l road segments numbered by l consecutive integers that can be chosen in a road. Then, the vehicles within reception range choose any road segment in the road, which decides the location of a vehicle. The selection method is OBU of a vehicle uses road size l, random number s, OBU *ID* and uniformly distributed hash function $h(f, R, ID)$ to decide which road segment to choose in this collection cycle. In essence, each vehicle selects a road segment number from the uniformly distributed integers between 1 and l randomly.

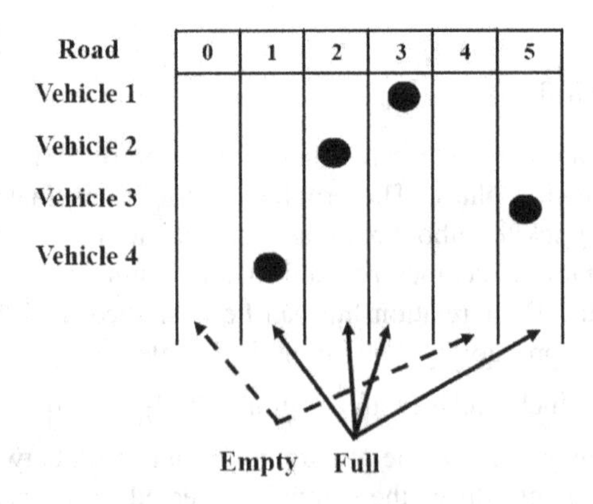

Fig. 2. Process of vehicles choosing road segments

As each vehicle choose a road segment independently, there may be road segments without any vehicle choosing or multiple vehicles choosing. However, the road here are all single-lane road, so we do not consider the collision problem. According to different chosen conditions, road segments can be divided into two categories: empty and full. This process can be seen in Fig. 2. After executing the query of a road, the RSU will get a binary sequence with 0 and 1, such as the Fig. 2 results to a sequence of 011101, in which, 1 represents full road segment and 0 represents empty road segment.

4 Vehicle Cardinality Estimation in VANETs

Our vehicle cardinality estimation scheme includes four parts. The first part is to derive the estimation formula, and that is to get the mathematical relationship between expectation of location distance and number of vehicles, where the location distance is the location number difference between the second vehicle and first vehicle. The second part is to determine the times of cycles according to the estimation requirement. The third part is to decide the road size by means of minimizing estimation time. The fourth part is to adjust the upper bound of vehicle cardinality according to the actual estimated traffic flow. The whole function of our estimator is illustrated in Fig. 3.

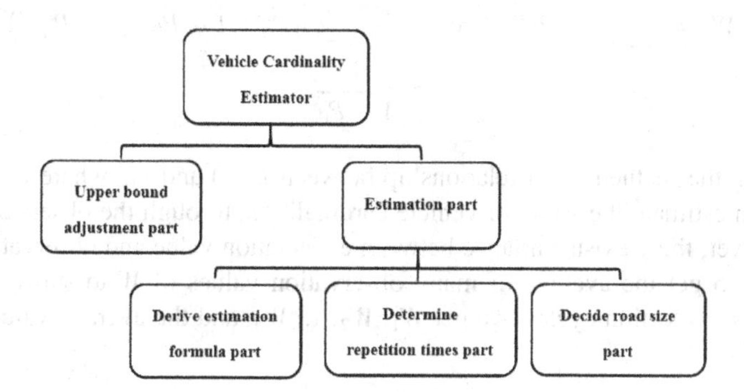

Fig. 3. Overall framework of our estimator

4.1 Vehicle Cardinality Estimator

Each vehicle chooses a road segment number in a road randomly and independently, which means the probability of any road segment being selected by any vehicle equals to each other. That is to say, the probability of any road segment being empty or full is the same. We define the probability of any road segment being empty as P_0,

$$P_0 = \left(1 - \frac{1}{l}\right)^{t_0} \quad (1)$$

When the road size l is large, P_0 can be simplified to

$$P_0 \approx e^{-\rho}, \text{ where } \rho = \frac{t_0}{l} \tag{2}$$

The location number of the first vehicle is defined as F and the location number of the second vehicle is defined as S respectively, by the probability formula of independent events, we can get

$$P[F = u] = P_0^{u-1}(1 - P_0) \tag{3}$$

$$P[S = v] = (v - 1)P_0^{v-2}(1 - P_0)^2 \tag{4}$$

The location number difference between the first vehicle and second vehicle is defined as W, by the probability formula of discrete random variable, we can get

$$P[W = w] = P_0^{w-1}(1 - P_0)\left(1 - P_0^{l-w}\right) \tag{5}$$

The expectation of W is

$$\begin{aligned} E(W) = \sum\nolimits_{w=1}^{l-1} wP(W = w) &= \sum\nolimits_{w=1}^{l-1} wP_0^{l-1}(1 - P_0)\left(1 - P_0^{l-w}\right) \\ &\approx \frac{1}{1 - P_0} \end{aligned} \tag{6}$$

Based on the mathematical relationship between $E(W)$ and P_0, where P_0 is related to t_0, we can estimate the value of vehicle cardinality t_0 through the observation value of W. However, there exists variance between expectation value and observation value, so we need to get the average of many observation values of W to substitute $E(W)$. Conducting n collection cycles, we get W_1, W_2..... W_n, and the average value, defined as V, is

$$V = \sum\nolimits_{i=1}^{n} \frac{W_i}{n} \tag{7}$$

Where W_i is the ith observation value of W.

According to $E(W_i) = E(W)$, and $W_1 \sim W_n$ is not correlated mutually, we can get

$$E(V) = \frac{\sum_{i=1}^{n} W_i}{n} = E(W) = \frac{1}{1 - P_0} = \frac{e^{\rho}}{e^{\rho} - 1}, \text{ where } \rho = \frac{t_0}{l} \tag{8}$$

That is

$$t_0 = l \cdot ln \frac{E(V)}{E(V) - 1} \tag{9}$$

Simplifying $E(V)$ to V, we can derive the vehicle cardinality estimation value t_1 is

$$t_1 = l \cdot ln \frac{V}{V-1} \tag{10}$$

There is a special case when $V = 1$, which means the second vehicle is next to the first vehicle. In that circumstance, we can draw a conclusion that the real-time traffic condition can't be worse without estimating vehicle cardinality.

4.2 Determining Repetition Times *n*

Collecting data process should be conducted repeatedly, because there exists variance between $E(V)$ and V. We can determine the times of cycles n through the variance of $E(V)$.

Based on the calculation formula of variance [13] we can get

$$Var(W) = E(W^2) - E^2(W) = \frac{1 + P_0}{(1 - P_0)^2} - \left(\frac{1}{1 - P_0}\right)^2 = \frac{P_0}{(1 - P_0)^2} \tag{11}$$

As $Var(W_i) = Var(W)$ and $W_1 \sim W_n$ is not correlated mutually,

$$Var(V) = \frac{Var\left(\sum_{i=1}^{n} W_i\right)}{n^2} = \frac{Var(W)}{n} \tag{12}$$

The expectation of V and standard deviation are defined as ε and τ respectively,

$$\varepsilon = E(V) \tag{13}$$

$$\tau = (Var(V))^{\frac{1}{2}} = (Var(W)/n)^{\frac{1}{2}} \tag{14}$$

According to mathematical theorem, we can get

$$T = \frac{V - \varepsilon}{\tau} \tag{15}$$

Where the distribution of parameter T is normal and standardized, we can get the cumulative distribution function [14] of T as

$$\emptyset(x) = \frac{1}{\sqrt{2\pi}} \int_{-\infty}^{x} e^{-\frac{u^2}{2}} du \tag{16}$$

A constant h can be found to make

$$P[|Z| \leq h] = erf\left(h/\sqrt{2}\right) = \gamma \tag{17}$$

Where we can figure out h through error function $erf()$. In the error function, if $\gamma = 95\ \%$, we can get $h = 1.96$ correspondingly.

The requirement defined by β and γ can be described as

$$\begin{aligned} P[|t_1 - t_0| \leq \beta t] &= P[(1-\beta)t_0 \leq t_1 \leq (1+\beta)t_0] \\ &= P\left[(1-\beta)t_0 \leq l \cdot ln\frac{V}{V-1} \leq (1+\beta)t_0\right] \\ &= P\left[\frac{e^{\rho(1+\beta)}}{e^{\rho(1+\beta)}-1} \leq V \leq \frac{e^{\rho(1-\beta)}}{e^{\rho(1-\beta)}-1}\right] \end{aligned} \tag{18}$$

Combine the formula (16), (17) and (18), we need to satisfy the following two conditions

$$\frac{\frac{e^{\rho(1+\beta)}}{e^{\rho(1+\beta)-1}} - \mu}{\tau} \leq -h \text{ and } \frac{\frac{e^{\rho(1-\beta)}}{e^{\rho(1-\beta)}-1} - \mu}{\tau} \geq h$$

Then we can get the times of cycles n

$$n = \frac{h^2 e^{t_{max}/l}\left(e^{\beta \cdot t_{max}/l} - e^{-t_{max}/l}\right)^2}{\left(1 - e^{\beta \cdot t_{max}/l}\right)^2} \tag{19}$$

4.3 Determining Road Size l

The total estimation time is decided by collection cycle times and each cycle's execution time. Our scheme requires the RSU identify the first vehicle and the second vehicle, so we use the location number of the second vehicle, defined as S, to measure each cycle's execution time. The total estimation time can be simplified to the product of n and $E(S)$.

The expectation of S is

$$\begin{aligned} E(S) &= \sum\nolimits_{v=2}^{l} v \cdot (v-1)P_0^{v-2}(1-P_0)^2 \\ &= \frac{2}{1-P_0} = \frac{2e^{\rho}}{e^{\rho}-1} \end{aligned} \tag{20}$$

Based on the formula (19) and (20), we can get the calculation formula of estimation time as

$$T(t_0, l) = n \cdot E(S) = \frac{2h^2\left[e^{(1+\beta)\cdot\rho} - 1\right]^2}{(1-e^{\beta\rho})^2 \cdot (e^{\rho}-1)} \tag{21}$$

Where h calculated by error function $erf()$ [15] and β defined as confidence interval are all known quantities.

So the estimation time $T(t_0, l)$ is only dependent on the load factor ρ. We minimize $T(t_0, f)$ to find the optimized load factor ρ using the matlab tool. Then, as given above $\rho = \frac{t_0}{l}$, we can get the optimized road size l through ρ and t_{max}.

4.4 Adjusting Upper Bound t_{max}

When the given upper bound of vehicle cardinality is too large, there will be more empty road segments in the road. It causes the location of the first vehicle and second vehicle backward, which means the RSU needs to identify more road segments, that is, more time cost. We utilize the location number of the first vehicle F to judge whether the original upper bound t_{max} is too large for the current traffic flow. If so, we will shrink the upper bound t_{max} adaptively. For the randomness, we need to collect F many times and get the average $\overline{F}$. Through several experiments, we find $\overline{F}$ can get a stable value when the data collections reach 50 times. Hence, we use the average value $\overline{F}$ of 50 collected F_i to adjust the upper bound t_{max} in the simulation. The flow chart of the adjustment process is in Fig. 4.

F is defined as the location number of the first vehicle. In the ith cycle, we will get a F_i. Firstly, we calculate the average $\overline{F}$. Then the whole cycle begins, we decrease

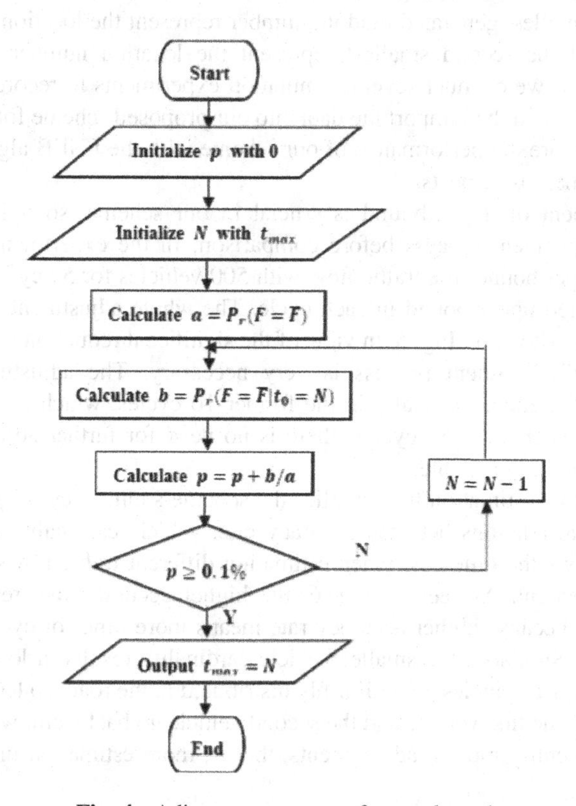

Fig. 4. Adjustment process of upper bound

N from t_{max} to 1 to traverse all possible vehicle cardinality. Every time, we calculate $P_r(t_0 = N|F = \overline{F})$, which is the probability of vehicle cardinality being N when the location number of the first vehicle is $\overline{F}$. With the decrement of N, we accumulate $P_r(t_0 = N|F = \overline{F})$ calculated in each cycle, defined as p. p is the probability of vehicle cardinality t_0 between N and t_{max}, written as $P_r[N \leq t_0 \leq t_{max}]$. When p is controlled in a limited range, such as 0.1 %, that is, the probability of vehicle cardinality larger than N is very low (0.1 %). In other words, there are high probability (99 %) of vehicle cardinality less than N. Therefore, we can shrink upper bound t_{max} to N.

5 Experiments

In order to evaluate the performance of our vehicle cardinality estimation scheme, we simulate estimating different traffic flows using the Matlab R2012a. The evaluation is conducted in two aspects, accuracy and time efficiency. We do not consider any collision and interference problems. In the simulation, we set the confidence interval as 5 %, accuracy probability as 99 %, vehicle cardinality from 500 to 5000, the original upper bound as 10000. The simulation experiment is conducted as followed. Firstly, in order to get the location number of the first vehicle and the second vehicle, we use the process of Matlab tool generating random number to represent vehicles choosing road segments. The smallest generated random number represent the location number of the first vehicle and the second smallest represent the location number of the second vehicle. Secondly, we conduct several simulation experiments to record and integrate experimental data, and then import the data into our proposed scheme for estimation. In the end, we compare the performance of our scheme with the FNEB algorithm in both accuracy and time cost aspects.

The adjustment of upper bound is crucial in our scheme, so it is necessary to simulate the adjustment process before comparison. In the experiment, we simulate adjusting the upper bound of a traffic flow with 500 vehicles for 50 cycles in total, and record the adjusted upper bound in each cycle. The whole adjustment process can be seen in Fig. 5. As shown in Fig. 5, in view of the significant reduction on upper bound, we can get the adjustment process is very necessary. The adjustment algorithm achieves high efficiency, especially in the former 10 cycles, which shrinks the upper bound sharply. After these 50 cycles, there is no need for further adjustment, as the upper bound tends to be stable.

It is known that estimation time implies the scheme's efficiency. Figure 6 gives the three-dimensional relations between accuracy rate, vehicle cardinality and estimation time, which shows the time cost when estimating different traffic flows with different accuracy requirement. As seen in Fig. 6, the higher accuracy rate results in longer estimation time, because higher accuracy rate means more times of cycles, thus, more estimation time. Similarly, the smaller vehicle cardinality results in longer estimation time. That's because vehicles are uniformly distributed in the road, so less vehicles lead to the location of the first vehicle and the second vehicle go backward, which means the RSU needs to identify more road segments, that is, more estimation time.

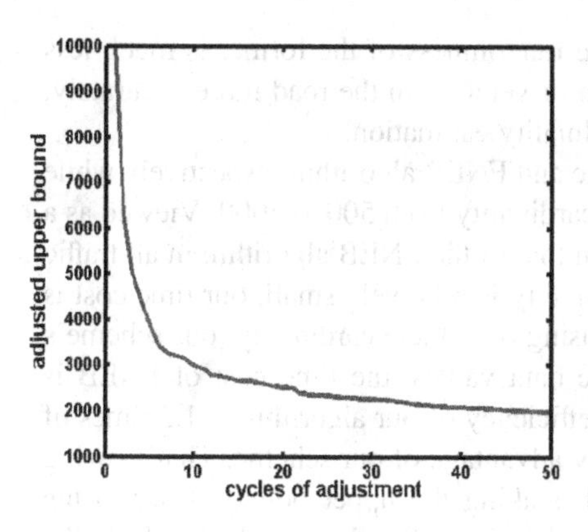

Fig. 5. Upper bound adjustment of vehicle cardinality

Fig. 6. Estimation time of our scheme

After that, our proposed scheme and FNEB algorithm are used to estimate different traffic flows respectively, whose cardinality is from 500 to 5000. Based on the result of experiments, we will compare these two tag estimators in aspects of deviation rate and time efficiency.

For comparison, we define deviation rate as d, which can be calculated with $d = |t_1 - t_0|/t_0$. Experimental results of deviation rate d are shown in Fig. 7. Viewed as a whole,the deviation rate of our scheme is stable and keeps lower than 0.05, however, the FNEB's arises more than 0.1, even close to 0.35. Observing the specific data values, the deviation rate of our scheme is at least 50 % lower than that of the FNEB algorithm, which proves superiority of our scheme in the aspect of accuracy. That's because our scheme leverages the location number difference between the second vehicle and the first vehicle to estimate vehicle cardinality, while FNEB algorithm only

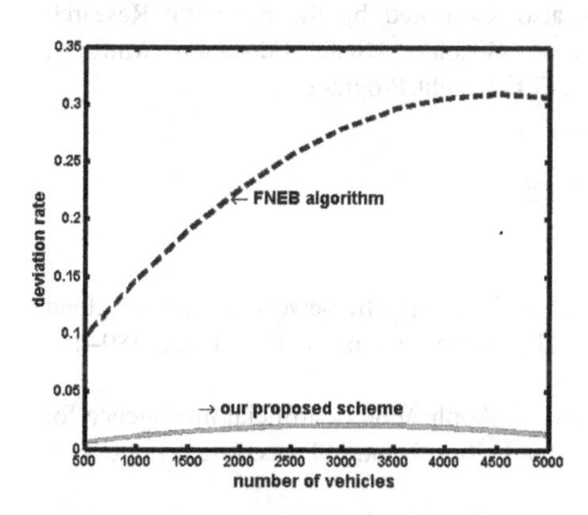

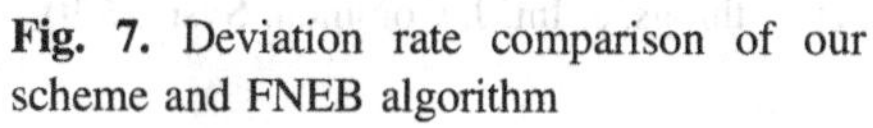
Fig. 7. Deviation rate comparison of our scheme and FNEB algorithm

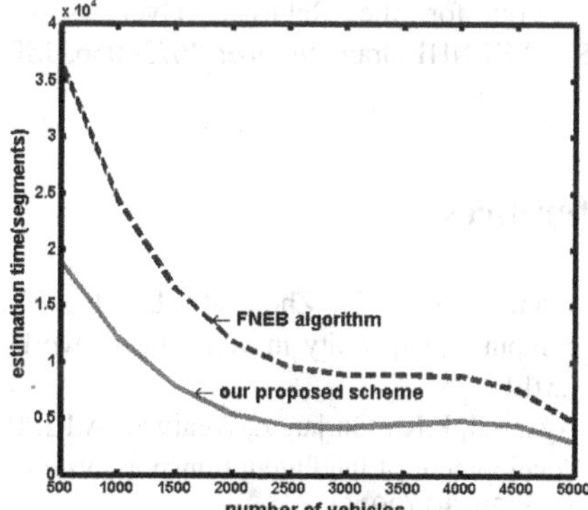

Fig. 8. Time efficiency comparison of our scheme and FNEB algorithm

leverages the location of the first vehicle. The randomness of the former is much less than the latter, so it can reflect the distribution of vehicles in the road more accurately, which means higher accuracy in vehicle cardinality estimation.

Figure 8 shows the time cost of our scheme and FNEB algorithm respectively while estimating different traffic flows with vehicle cardinality from 500 to 5000. Viewed as a whole, the time cost of our scheme is less than that of the FNEB algorithm in all traffic flows. Especially in the cases of vehicle cardinality is relatively small, our time cost is almost half of FNEB's. While with the increasing of vehicle cardinality, our scheme's gain in time cost shrinks. Observing specific data values, the time cost of FNEB is almost 1.5 times of our scheme, that is, time efficiency of our algorithm is 1.5 times of FNEB algorithm, which proves time efficiency advantage of our scheme. That's owing to the process of adjusting the upper bound, making the upper bound closer to the actual vehicle cardinality. The road size l and times of cycles n calculated by the adjusted upper bound will be more accurate, which contributes to higher time efficiency.

6 Conclusion

This paper has proposed a newly-fashioned scheme based on the location distance between the first vehicle and the second vehicle for cardinality estimation. It includes the process to optimize parameters' value and adjustment process of original upper bound, which contribute to its high performance applied in VANETs. To be specific, both theoretical analysis and extensive simulations demonstrate that our scheme achieves high accuracy and time efficiency. In addition, it helps protect privacy, because it makes estimation without identifying vehicles one by one.

Acknowledgement. This work is partially supported by the National Science Foundation of China (Grant No. 61472283), the Fok Ying-Tong Education Foundation, China (Grant No. 142006), the Fundamental Research Funds for the Central Universities (Grant No. 2013KJ034, 2100219043 and 1600219246). This project is also sponsored by the Scientific Research Foundation for the Returned Overseas Chinese Scholars, State Education Ministry, JSPS KAKENHI Grant Number 26730056, JSPS A3 Foresight Program.

References

1. Wan, J., Zou, C., Zhou, K., Lu, R., Li, D.: IoT sensing framework with inter-cloud computing capability in vehicular networking. J. Electron. Commer. Res. **14**(3), 389–416 (2014)
2. Agarwal, P.K., Gurjar, J., Agarwal, A.K., Birla, R.: Application of artificial intelligence for development of intelligent transport system in smart cities. J. Int. J. Trans. Eng. Traffic Syst. **1**(1), 20–30 (2015)
3. Xia, F., Yang, L.T., Wang, L., Vinel, A.: Internet of things. J. Int. J. Commun. Syst. **25**(9), 1101 (2012)

4. Chuang, M.C., Lee, J.F.: PPAS: a privacy preservation authentication scheme for vehicle-to-infrastructure communication networks. In: 2011 International Conference on CECNet, pp. 1509–1512. IEEE (2011)
5. Chen, M., Gonzalez, S., Zhang, Q., Leung, V.C.: Code-centric RFID system based on software agent intelligence. J. IEEE Intell. Syst. **2**, 12–19 (2010)
6. Krishnamurthy, B., Naryshkin, K., Wills, C.: Privacy leakage vs. protection measures: the growing disconnect. In: Proceedings of the Web, pp. 1–10 (2011)
7. Han, H., Sheng, B., Tan, C., et al.: Counting RFID tags efficiently and anonymously. In: IEEE INFOCOM 2010, pp. 1–9. IEEE (2010)
8. Qian, C., Ngan, H., Liu, Y., Ni, L.M.: Cardinality estimation for large-scale RFID systems. J. IEEE Trans. **22**(9), 1441–1454 (2011)
9. Zheng, Y., Li, M.: Towards more efficient cardinality estimation for large-scale RFID systems. J. IEEE Trans. Network. **22**(6), 1886–1896 (2014)
10. Jiang, W., Zhu, Y.: A Unified Approach for fast and accurate cardinality estimation in RFID systems. In: 11th International Conference on Mobile Ad Hoc and Sensor Systems, pp. 407–415. IEEE (2014)
11. Ning, H., Cong, Y., Xu, Z. Q., Hong, T., Zhao, J. C., Zhang, Y.: Performance evaluation of RFID anti-collision algorithm with FPGA implementation. In: AINAW 2007. In: 21st International Conference on Advanced Information Networking and Applications Workshops, pp. 153–158. IEEE (2007)
12. Guo, B., Sun, L., Zhang, D.: The architecture design of a cross-domain context management system. In: 2010 8th IEEE International Conference, pp. 499–504. IEEE (2010)
13. Markowitz, H.: Mean–variance approximations to expected utility. J. Eur. J. Oper. Res. **234**(2), 346–355 (2014)
14. Zhou, Z., Zhao, D., Shu, L., Tsang, K.F.: A Novel Two-tier cooperative caching mechanism for the optimization of multi-attribute periodic queries in wireless sensor networks. J. Sens. **15**(7), 15033–15066 (2015)
15. Ding, C., Kong, D.: Nonnegative matrix factorization using a robust error function. In: 2012 IEEE International Conference on Acoustics, Speech and Signal Processing, pp. 2033–2036. IEEE (2012)

An Efficient Transmission Scheme Based on Adaptive Demodulation in Wireless Multicast Systems

Mingming Li[1(✉)], Jiansheng Ma[1], Congcong Li[2], Tingting Xu[1], Xiaoliang Wang[1], and Wenbo Li[1]

[1] Department of Information and Communications, State Grid Weifang Power Supply Company, Weifang 261021, People's Republic of China
mmliboy@163.com

[2] Department of Power Distribution Network Operation and Maintenance, State Grid Dongying Power Supply Company, Dongying 257000, People's Republic of China

Abstract. To raise the transmission rate of wireless multicast system, an efficient transmission scheme based on adaptive demodulation (ADS) is proposed. Firstly, the working principle how ADS can exploit user's transmission ability better is described. Then, the bit mapping method of ADS is formulated. Finally, to analyze the transmission ability of ADS, the expression of transmission rate is derived. Numerical results show that the transmission rate of ADS is much higher than the existing schemes.

Keywords: Multicast scheduling · Transmission rate · Adaptive demodulation

1 Introduction

Multicast transmission is defined as a unidirectional point-to-multipoint bearer service in which data is transmitted from a single source entity to multiple recipients, which is a remedy to the inefficient resource usage [1–3]. Dynamic resource allocation schemes for multicast services have been extensively studied [4–8]. However, in these literatures, transmission rate would be determined by the worst channel gain, which is very low. Being motivated by recent advances in erasure codes and fountain codes [9, 10], opportunistic multicast scheduling (OMS) schemes have been proposed to improve the system throughput performance. The main idea of OMS is that during each transmission time interval (TTI), BS only transmits data to a group of users that have fine channel gains, and with the help of erasure and fountain codes, each user can recover the original message as long as a minimum set of encoded bits are received.

Some literatures that have focused on OMS should be emphasized. In [11, 12], the authors proposed to transmit data according to the user whose instantaneous signal-to-noise ratio (SNR) is the median of the ordered list of the users. The scheme in [13] predefines a transmission rate, and the data is only transmitted during the TTI that more than a defined number of users can support the predefined rate. In [14], the

C.-H. Hsu et al. (Eds.): IOV 2015, LNCS 9502, pp. 16–24, 2015.
DOI: 10.1007/978-3-319-27293-1_2

authors proposed to only transmit data to a ratio of users that have fine channel gains during each TTI. The scheme in [15] transmits data according to a selected SNR threshold and only the user with a SNR that is larger than the threshold can receive data. However, the mentioned OMSs are all fixed-rate schemes. In these schemes, once the bits are modulated and transmitted, according to the received SNR, each user just has two choices: if the received SNR is higher than the SNR requirement, the bits are demodulated and collected, otherwise, the bits are abandoned, which would restrict the system performance.

To overcome the problems in OMSs and raise the transmission rate of multicast system, we introduce a novel transmission scheme based on adaptive demodulation (ADS). In ADS, as the bits are modulated a high level modulation and coding scheme (MCS), each user can adaptively select one MCS level to demodulate bits according to the channel state information (CSI). For the users with fine channel condition, the bits are demodulated by the original MCS, and all the bits are received. For users with poor channel conditions which can't demodulate bits by the original MCS, the bits will be demodulated by a lower level MCS, and several bits can still be received.

The rest of the paper is organized as follows. In Sect. 2, the system model is introduced. In Sect. 3, for ADS, the working principle is described, the bit mapping method is formulated, and the expression of transmission rate is derived. Simulation results and comparisons are shown in Sect. 4. Finally, we draw our conclusion.

2 System Model

Consider a wireless point-to-multipoint downlink system supporting multicast service for a group of K users. Fading coefficients of the BS-user link remain constant within each transmission time interval (TTI), but may vary independently from one TTI to the next. At the BS side, the bits are processed by a rateless encoding scheme and turn into a continuous bits stream. Then, during each TTI, several encoded bits are mapped to the constellation of a selected MCS. At the user side, every user collects the bits demodulated in every TTI, and as L bits are collected, the original data can be recovered. If all the users have received L bits, transmission terminates.

The system supports a set of different MCSs $M = \{m_1, m_2, \cdots, m_{|M|}\}$. γ_{m_i} denotes the required SNR of MCS m_i and c_{m_1} denotes the corresponding number of bits carried in one TTI. For multiple phase shift keying (MPSK), γ_{m_i} can be given by

$$\gamma_{m_i} = \begin{cases} \frac{1}{2T}\left[Q^{-1}(p_e)\right]^2, & \text{if } c_{m_i} = 2 \\ \frac{1}{4Tc_{m_i}}\left[\frac{Q^{-1}(p_e c_{m_i}/2)}{\sin(\pi/2^{c_{m_i}+1})}\right]^2, & \text{if } c_{m_i} \geq 2 \end{cases} \quad (1)$$

where p_e is the required BER, T is the length of one TTI, and $Q(x) = \frac{1}{\sqrt{2\pi}}\int_x^{+\infty} e^{-t^2}dt$. To facilitate the following discussion, we assume $\gamma_{m_1} < \gamma_{m_2} < \cdots < \gamma_{m_{|M|}}$ and $c_{m_1} < c_{m_2} < \cdots < c_{m_{|M|}}$. We use $\gamma_k(n)$ to denote the received SNR of user k over TTI n. If $\gamma_{m_i} \leq \gamma_k(n) < \gamma_{m_{i+1}}$, m_i can be defined as the desired MCS level of user k in the $n-$th TTI, because with m_i, user k can get the maximum number of bits.

3 Transmission Based on Adaptive Demodulation

In OMS, during each TTI, BS selects a MCS m_i to modulate the bits. Then, at the user side, each user would compare the received SNR $\gamma_k(n)$ with the required SNR γ_{m_i}, each user has two choices: (1) if $\gamma_k(n) \geq \gamma_{m_i}$, all the bits would be correctly demodulated; (2) otherwise, none of the bits would be received. However, in this scenario, if the received SNR of user k satisfies $\gamma_{m_{i-1}} \leq \gamma_k(n) < \gamma_{m_i}$ ($i > 1$), user k has the ability to receive $c_{m_{i-1}}$ bits. Thus, OMS cannot fully exploit each user's transmission ability.

3.1 The Main Idea of ADS

To exploit each user's transmission ability better, we designs ADS for Phase Shift Keying (PSK). The main idea is that if the bits are modulated by a high level MCS, each user can adaptively demodulate bits with each user's desired MCS level. The number of bits received by each user during one TTI are determined by each user's channel condition, thus, each user can take full use of the channel condition. To demonstrate how ADS works, the transmission of multicast bits modulated by 8PSK is given as an example, which is shown in Fig. 1.

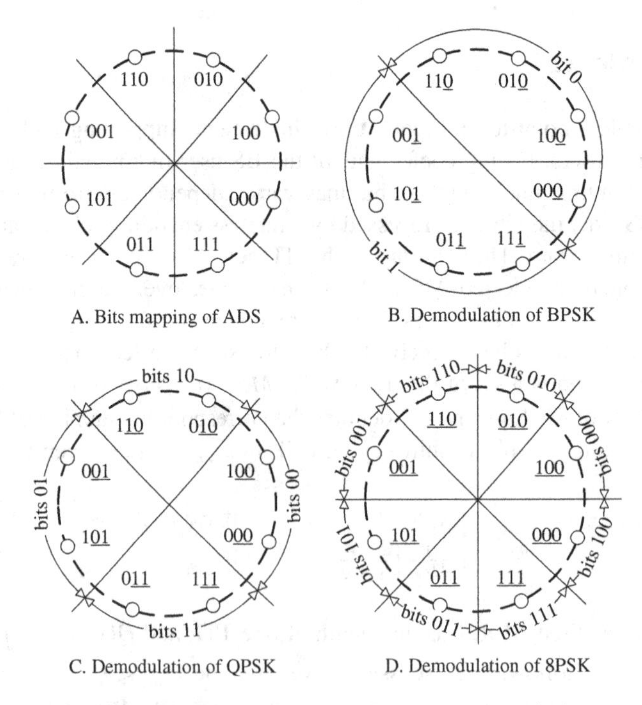

Fig. 1. Bits mapping and demodulation of ADS as $|M| = 3$

During each TTI, $|M| = 3$ bits are mapped to the constellation of 8PSK. According to the received SNR, each user selects the desired MCS level from $[m_1, m_2, m_3] =$ [BPSK,QPSK,8PSK] to decode data. For example, during TTI n, if $\gamma_{\text{BPSK}} \le \gamma_k(n) < \gamma_{\text{QPSK}}$, BPSK is the desired MCS level of user k, and according to Fig. 1B, user k can receive 1 bit. Similarly, user k would receive 2 bits or 3 bits, if $\gamma_{\text{QPSK}} \le \gamma_k(n) < \gamma_{\text{8PSK}}$ or $\gamma_k(n) \ge \gamma_{\text{8PSK}}$. Thus, each user can receive as many bits as the received SNR can support, though the bits are modulated by a high level MCS 8PSK.

3.2 The Bit Mapping Method of ADS

ADS is based on a special bit mapping scheme. Thus, in this section, we propose the bit mapping method according to PSK, which can be formulated as follows:

Algorithm: Bit mapping of ADS

```
Initialization: Ω = zeros(2^|M|, |M|);
for m = 1:|M|
    for i = 1:2^(m-1)
        down = 2^(|M|-m) · (2i - 1) + 1;
        upper = i · 2^(|M|+1-m);
        for j = down:upper
            Ω[j, |M| + 1 - m] = 1;
        end
    end
end
```

In ADS, $|M|$ is the number of bits transmitted during one TTI. Ω is a $2^{|M|} \times |M|$ matrix, and the bits in one row would be mapped to one constellation point. We can select one constellation point as the starting constellation point (SCP). On the transmitter side, the bits of $\underbrace{00\cdots0}_{|M|}$ in the first row are mapped to the SCP, and the bits in the n - th row are mapped to the n–th constellation point in counterclockwise direction. On the user side, if m_i is the desired MCS level to demodulate bits, all the constellation points are divided into 2^i sets, where SCP is the first point of the first set. With the process of ADS, the $2^{|M|-i}$ constellation points in each set would contain i bits identical data, thus, they can be treated as one demodulation point, and the i identical bits are the decoded data of each set.

3.3 The Transmission Rate of ADS

In this section, to analyze the transmission ability of ADS, we derive the expression of transmission rate.

In multicast system, transmission would terminate as long as every user collects L bits. Thus, the number of TTIs used to completed transmission should be expressed as

$$N = \min\{N : [\min_k \sum_{n=1}^{N} c_k(n)] \geq L\} \tag{2}$$

where $c_k(n)$ is the number of bits received by user k during TTI n.

The received SNR of user k during TTI n is

$$\gamma_k(n) = P|h_k(n)|^2/N_0, \ \forall k \tag{3}$$

where N_0 denotes the power density of additive white gaussian noise (AWGN), $h_k(n)$ is the instantaneous channel gain of user k during TTI n and P is the transmitting power.

According to $\gamma_k(n)$, the number of bits that each user can receive is

$$c_k(n) = \begin{cases} c_{m_{|M|}}, & \mathit{if}\ \gamma_k(n) \geq \gamma_{m_{|M|}} \\ c_{m_i}, & \mathit{if}\ \gamma_{m_i}(n) \leq \gamma_k < \gamma_{m_{i+1}} \\ 0, & \mathit{if}\ \gamma_k(n) < \gamma_{m_1} \end{cases} \tag{4}$$

where $m_i = \{m_1, m_2, \cdots, m_{|M|-1}\}$.

Assume that the channel gains of different users obey the same distributed function, and during each TTI, the probability that m_i is selected as the desired MCS is identical for all the users, which is denoted by p_{m_i}. With the help of Appendix, p_{m_i} is given by

$$p_{m_i} = \begin{cases} [F_{|h|^2}(\frac{N_0\gamma_{m_{i+1}}}{P}) - F_{|h|^2}(\frac{N_0\gamma_{m_i}}{P})], & \mathit{if}\ m_i \neq m_{|M|} \\ 1 - F_{|h|^2}(\frac{N_0\gamma_{m_i}}{P}), \ \mathit{if}\ m_i = m_{|M|} \end{cases} \tag{5}$$

where $F_{|h|^2}(\cdot)$ is the cumulative distribution function (CDF) of the channel gain.

Let $C_{k,N}$ be the number of bits received by user k during N TTIs. $C_{k,N}$ can be expressed as

$$C_{k,N} = \sum_{n=1}^{N} c_k(n) \tag{6}$$

By invoking the central limit theorem, for sufficiently large number of TTIs, we can approximate $C_{k,N}$ as a Gaussian distributed random variable such that

$$C_{k,N} \sim \mathrm{N}(N\mu, N\sigma^2) \tag{7}$$

$$\mu = \sum_{m=m_1}^{m_{|M|}} p_m c_m \tag{8}$$

$$\sigma^2 = \sum_{m=m_1}^{m_{|M|}} p_m [c_m - \sum_{m=m_1}^{m_{|M|}} p_m c_m]^2 \tag{9}$$

Let $C_{\mathrm{min,N}} = \min_k C_{k,N}$, and by lemma 2 in [14], we can get the mean value of $C_{\mathrm{min,N}}$

$$E[C_{\mathrm{min,N}}] \approx \sqrt{N}\sigma\lambda + N\mu \tag{10}$$

$$\lambda = (1-\varepsilon)\Phi^{-1}(\frac{1}{K}) + \varepsilon\Phi^{-1}(\frac{1}{Ke}) \tag{11}$$

where $\varepsilon \approx 0.5772$ is the Euler-Mascheroni constant, $e \approx 2.7183$ is the Euler's number, and $\Phi(\cdot)$ is the standard Gaussian CDF.

Transmission would terminate as long as every user receives L encoded bits. By setting $E[C_{\mathrm{min,N}}] = L$, we get

$$\mu N + \sigma\lambda\sqrt{N} - L = 0 \tag{12}$$

We find that (12) is a quadratic equation about $\sqrt{N}$, so the value of N can be obtained by the standard quadratic-root formula

$$N = (\frac{-\sigma\lambda + \sqrt{(\sigma\lambda)^2 + 4\mu L}}{2\mu})^2 \tag{13}$$

The transmission rate of ADS can be obtained by

$$R\_ADS = L/(N \cdot T) \tag{14}$$

where T is the length of every TTI.

4 Numerical Results

In this section, the performance of ADS is estimated. In simulation, the BS wants to transmit $L = 10000$ bits data to users, and transmission terminates as long as 10000 bits are received by every user. The BER requirement is $p_e = 10^{-4}$. The system supports six MCSs: BPSK, QPSK, 8PSK, 16PSK, 32PSK and 64PSK. The channel gain of BS-user link is $h = \sqrt{x^2 + y^2}$, where both x and y are Gaussian random variables with mean 0 and variance 0.5. The duration of one TTI is 1 ms. The power spectrum density of AWGN is -90 dB. The numerical results are averaged over 10000 channel realizations.

As comparison, the opportunistic multicast scheduling scheme (OMS) in [14] and the conventional scheme (CON) are simulated. In OMS, an optimized ratio of the users would be served and the MCS level would be determined by the worst channel gain of the selected users. In ADS, during each TTI, 6 encoded bits are mapped to the constellation of 64PSK.

Figure 2 shows the transmission rate according to the value of SNR. It can be seen that ADS performs better than other simulated schemes. For a given value of SNR,

OMS and CON would determine the MCS level according to the channel condition of a selected user, and only the users whose channel condition are better than the selected user can correctly demodulate the bits. Though some users with fine channel conditions can support high level MCS, they have to receive small number of bits determined by the selected user, which is a waste of resource. However, in ADS, the bits are modulated by high level MCS, but each user can decode bits according to the current channel condition. The users with fine channel condition can receive more bits and the users with bad channel condition receive fewer bits, which can take full use of each user's channel condition.

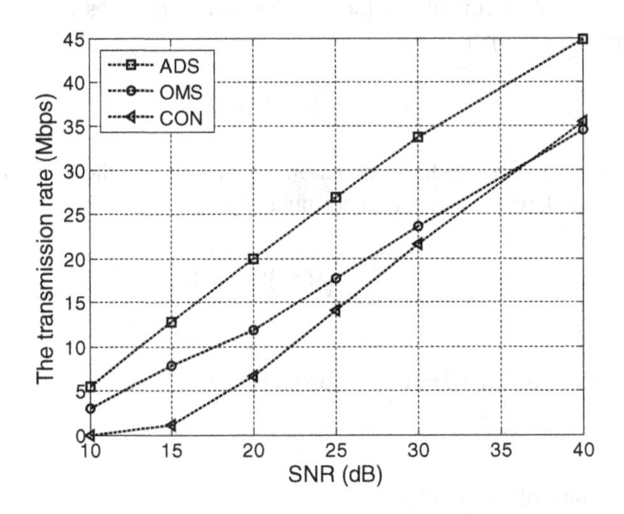

Fig. 2. Transmission rate comparison of ADS, OMS and CON as SNR increases

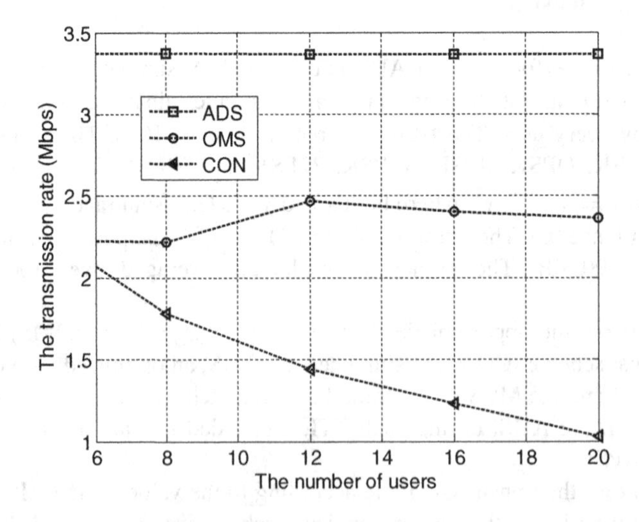

Fig. 3. Transmission rate comparison of ADS, OMS and CON as the number of users increases

As the number of users that subscribe to the multicast service increases, the transmission rates of ADS, OMS and CON are presented in Fig. 3. It is observed that the transmission rate of ADS is higher than OMS and CON. It is because ADS can exploit each user's transmission ability better. Besides, as the number of users increases, the transmission rate of ADS is a constant. It is because ADS modulates bits with the highest level MCS in each TTI, so the transmission rate doesn't decrease as the number of users increases.

5 Conclusions

This paper presents an adaptive demodulation mapping scheme (ADS) to raise the transmission rate of wireless multicast system. Different from the previous literatures, in ADS, every user can adaptively select a suitable modulation and coding scheme (MCS) to demodulate bits according to the channel state condition, which can utilize each user's channel condition better. Numerical results show that ADS can significantly raise the transmission rate for different number of users and for different values of SNR compared with the existing schemes.

References

1. Varshney, U.: Multicast over Wireless Networks. Commun. ACM **45**(12), 31–37 (2002)
2. Wang, J., Sinnarajah, R., Chen, T.: Broadcast and multicast services in cdma2000. IEEE Commun. Mag. **42**(2), 76–82 (2004)
3. Parkvall, S., Englund, E., Lundevall, M.: Evolving 3G mobile systems: broadband and broadcast services in WCDMA. IEEE Commun. Mag. **44**(2), 68–74 (2006)
4. Liu, J., Chen, W., Cao, Z.: Dynamic power and subcarrier allocation for OFSMA-based wireless multicast systems. In: IEEE International Conference on Communications, pp. 2607–2611. IEEE Press, New York (2008)
5. Kagan, B., Wu, M.Q., Liu H.: Adaptive resource allocation in multicast OFDMA systems. In: IEEE Wireless Communications and Networking Conference, pp. 1–6. IEEE Press, New York (2010)
6. Li, M.M., Wang, X.X., Zhang, H.T., Tang M.W.: Resource allocation with subcarrier cooperation in OFDM-based wireless multicast system. In: 73th IEEE Vehicular Technology Conference, pp. 1–5. IEEE Press, New York (2011)
7. Tang, M.W., Wang, X.X.: Resource allocation algorithm with limited feedback for multicast single frequency networks. J. Zhejiang Univ. Sci. C-Comput. Electron. **13**(2), 14–154 (2012)
8. Zhang, H.B., Wang, X.X., Li, F.: Channel-aware adaptive resource allocation for multicast and unicast services in orthogonal frequency division multiplexing systems. IET Commun. **6**(17), 3006–3014 (2012)
9. Mackay, D.: Fountain codes. IET Commun. **152**(6), 1062–1068 (2005)
10. Luby, M., Watson, M., Casiba, T.: Raptor codes for reliable download delivery in wireless broadcast systems. In: 3rd IEEE Consumer Communications and Networking Conferenc, pp. 192–197. IEEE Press, New York (2006)

11. Gopala, P., Hesham, E.: Opportunistic multicasting. In: 38th Asilomar Conference on Signals, Systems and Computers, pp. 84–849. IEEE Press, New York (2004)
12. Gopala, P., Hesham, E.: On the throughput-delay tradeoff in cellular multicast. In: International Conference on Wireless Networks, Communications and Mobile Computing, pp. 1401–1405. IEEE Press, New York (2005)
13. Ge, W.Y., Zhang, J.S., Shen, S.A.: A cross-layer design approach to multicast in wireless networks. IEEE Trans. Wirel. Commun. **6**(3), 1063–1071 (2007)
14. Low, T., Pun, M., Hong, Y.: Optimized opportunistic multicast scheduling (OMS) over wireless cellular networks. IEEE Trans. Wirel. Commun. **9**(2), 791–801 (2010)
15. Quang, L., Tho, L., Ho, Q.: Opportunistic multicast scheduling with erasure-correction coding over wireless channels. In: IEEE International Conference on Communications, pp. 1–5. IEEE Press, New York (2010)

A Full Service Model for Vehicle Scheduling in One-Way Electric Vehicle Car-Sharing Systems

Hongman Wang, Zhaohan Li(✉), Xiaolu Zhu, and Zhihan Liu

State Key Laboratory of Networking and Switching Technology,
Beijing University of Posts and Telecommunications, Beijing, China
lizhaohanbupt@126.com

Abstract. The optimization of vehicle scheduling problem in electric vehicle (EV) Car-Sharing systems has always been a hotspot with the rapid development of EVs these years. In this paper, a research and design of full service model based on vehicle recommendation and scheduling algorithm for EV Car-Sharing systems has been presented. It leads to efficient vehicle management when there are needs to control cars in the whole station network and recommend appropriate vehicle to customers. Based on the IOV (Internet of Vehicles) Platform for EV Operation and Service, a new algorithm of vehicle recommendation and scheduling strategy for Car-Sharing systems is proposed that this problem can be regarded as a multi-objective optimization model. Through the analysis of EV Car-Sharing system model, factors like time matching, station recommendation, vehicle utilization and customer satisfaction have been discussed and data sharing is also the superiority in our full service model, compared with the controlled service model and conditional service model. Meanwhile, the algorithm has been simulated and verified in the station network model of GreenGo Car-Sharing system, which is a fast developing electric Car-Sharing project in Beijing. The simulation results have demonstrated that the proposed algorithm is efficient.

Keywords: EV Car-Sharing systems · Vehicle scheduling · Multi-objective optimization

1 Introduction

In these years, electric vehicle has come into the public view as a new means of energy conveyance, which has unparalleled advantages in energy saving, green carbon, etc. Car-Sharing systems, which is a new business operation model of car rental where people rent cars for short periods of time, often by the hour, is an alternative to private vehicle ownership, especially for electric vehicles. The EV Car-Sharing system is attractive to customers who make only occasional use of a vehicle, as well as others who would like occasional access to a vehicle of a different type than they use day-to-day. The organization renting the cars may be a commercial business or the users may be organized as a company, public agency, cooperative, or ad hoc grouping. The concept of this system is to share a small number of vehicles among a large

C.-H. Hsu et al. (Eds.): IOV 2015, LNCS 9502, pp. 25–36, 2015.
DOI: 10.1007/978-3-319-27293-1_3

number of users. There are three contributions by the system: cost sharing, earth sharing, and road sharing [1, 2]. So then, Car-Sharing systems really have a broad prospect for development.

Usually the Car-Sharing system is classified into the one-way type and the round-trip type according to where users return a vehicle. In the round-trip type, a user has to return a vehicle at the station where he rents it. On the other hand, in the one-way, a user can return a vehicle at any station. Thus, users can use vehicles like a public bus or taxi. Consequently, the one-way type is more flexible and useful than the round-trip type. However, it is hard to keep distribution balance of parked vehicles among stations. If distribution balance is disrupted, a lack of vehicles is occurred at any stations. As a result, it is impossible to assign vehicles to new users immediately [3]. Therefore, the one-way type needs an effective vehicle scheduling and dispatching algorithm for balancing parked vehicles among stations. In this paper, we propose a full service model to meet users renting demands as more as possible, using data sharing method to dispatch vehicles and schedule vehicles more reasonably. The model keeps the station networks in a relatively balance condition and gives users better experience and service.

The contribution of this paper lies in the following three aspects:

- We propose a novel optimization model called full service model to solve the problem of vehicle recommendation and scheduling in EV Car-Sharing systems.
- Our simulation takes the IOV Platform for EV Operation and Service as the supporting platform and experiment precondition, which takes IOV and data sharing seriously.
- We evaluated our model using reliable data sources from GreenGo Car-Sharing system consisting of actual station networks and station data, which simulate the real situation as far as possible.

The remainder of the paper is organized in the following way. Section 2 reviews related works and introduces two existing service model of one-way Car-Sharing systems. In Sect. 3, firstly we analyses conditional service model and controlled service model in detail. Then, the optimization full service model based on vehicle recommendation strategy and scheduling algorithm is given. Section 4 presents the detailed optimization mathematical model which includes rental station network model, station model, vehicle model, charging model and other important parameters. Section 5 discusses the experimental results of our approach and gives the analysis of these results. Major conclusions and future work are then discussed in Sect. 6.

2 Related Works

The majority of researches focus on the one-way Car-Sharing system and there are mainly two existing service model, one is conditional service model which don't reject users' trip demands unless there are no vehicles available; the other is controlled service model which accepts or rejects demands completely by the income-maximization objective [3]. These two existing service model both have evident limitations, such as lower service response and single income objective, so there are still better optimization strategies or algorithms for vehicle assignment and scheduling problem which have a comprehensive consideration of all the influencing factors. In order to provide users a better humanized and convenient rental experience, a full service model is really in urgent need.

For the relocation problem in the EV Car-Sharing systems, [4] presents the first Mixed Integer Linear Programming formulation of the EV relocation problem by a method of manpower scheduling, where cars are moved by personnel of the service operator to keep the system balanced. [5] describes different relocation strategies for free-floating Car-Sharing Systems, which are grouped into two different approaches: user-based and operator-based approaches, as well as a comprehensively comparison of them. [6] proposes a method for optimization vehicle relocation according to distribution balance of parked vehicles by modeling service area, station, vehicle and user. [7] presents a stochastic optimization method based on Monte Carlo sampling to solve the Car-Sharing dynamic vehicle allocation problem, and a multistage stochastic linear integer model with recourse is formulated that can account for system uncertainties such as Car-Sharing demand variation. [8] analyzes the membership prediction, mode share, and vehicle allocation optimization methodologies in free-floating Car-Sharing systems based on real-time information of users. Many related works deal with the relocation problem by finding solutions from details, ignoring the significance of IOV and backstage data sharing in such Car-Sharing systems. At the same time, the current situation is far more complex, and real-time demands must be taken into consideration.

3 Problem Description

The main problem we address in this paper is how to assign and schedule cars in EV Car-Sharing system in order to satisfy demands of different kinds of users as more as possible, that is to enhance the service rate as well as vehicle utilization. Usually Car-Sharing systems can be classified with respect to organization goals, geographic scope, station location, and trip configuration. As regards the organization goals, some systems are run by volunteers and are non-income making, while others are commercial ventures run by international companies. Today, an increasing number of EV Car-Sharing projects and companies are being operated and in the long term, it must be a trend that Car-Sharing systems be a promising business model and popularized by users throughout the world.

Now in a Car-Sharing system, two existing schemes of model are common [3].

One is **conditional service model (S1)**, in which there is no obligation of satisfying all trips between existing stations, but rather, they can only be rejected if there are no vehicles available at the pick-up station. This means that, if there is a vehicle at a station and a request is made for a trip starting at this station, there is no possibility of rejecting that trip. This certainly makes sense in a situation where clients can act as walk-ins asking for a vehicle instantly at a station.

The other is **controlled service model (S2)**, in which the Car-Sharing organization has total control over the selection of trips from a list of demands made by clients. In this model, the organization or the company is free to decide whether accept trips or not in the period according to the income-maximization objective. This means that a vehicle is only allocated to a specific trip if this is advantageous from the perspective of income. Usually, only demands whose renting time is 2 h or more can be accepted. If this is not the case, the trip will not be satisfied even when there are vehicles available in the pick-up station. Such model can only be feasible if there is a central request management service which is able to compute income and give instructions on which vehicles should be reserved for later trips. However, this may lead to great discontent as some clients may find themselves in a situation where they know of the existence of vehicles at a desired pick-up station and still have their demands rejected. In addition to this, there is no guarantee that a decision to accept or reject a trip will lead to the maximization of expected incomes given that demand is not deterministic.

Obviously, both of the two service models above are useful to solve the vehicle assignment and scheduling problem in one-way Car-Sharing systems, but not that efficient. Because of the changes of real-time demands in all of the stations of a city network, data sharing and vehicle scheduling between stations must be taken into consideration. So in this paper, an optimized solution called **full service model (S3)** is presented. Full service model is a multi-objective optimization model that significant factors like time matching, station recommendation and vehicle scheduling, vehicle utilization and customer satisfaction are all paid much attention and whenever the pick-up station doesn't have electric vehicles available for users, vehicle optimization assignment strategy based on data sharing will start. More user demands can be satisfied because some vehicles will be returned to the pick-up station in that period of time and we can make full use of them. This really helps increase both service rate and vehicle utilization as well as the quality of experience (QoE). Additionally, if there is still no vehicle available in this condition, full service model will go to the last section that available vehicles in other station not far can be allocated to the pick-up station via a special algorithm for vehicle scheduling.

4 Optimization Model

In this section, we present the formulation of full service optimization model aimed at the best vehicle assignment and scheduling solution and trying to satisfy all demands of different kinds of users. In the one-way Car-Sharing system, there are at most three steps to allocate the most suitable vehicle to the user who has already asked for a trip request at one point.

1. Recommending the nearest station as the pick-up station based on user's location, and to see if there are vehicles available for the trip based on time matching strategy. If there is some vehicle available, allocate one to the user, if not, go to step 2.
2. Starting the vehicle optimization assignment strategy, attempt to make sure if there will be vehicles that should return to the pick-up station before pick-up time via data sharing and prediction activities. If there are some, allocate one to the user, if not, go to step 3;
3. Starting the vehicle scheduling algorithm, choosing the nearest station which has a vehicle available for the trip request and letting a staff drive the vehicle to the pick-up station for user.

As for this optimization model, we need a supporting platform, which is a data sharing and distribution platform called the IOV Platform for EV Operation and Service [9, 10]. In our paper, the most significant precondition is the internet of all the EVs, stations and users (user terminals), and this platform successfully provides many fundamental capabilities in the EV systems. Charging management system and billing system indicate the interactive relationship of vehicles and stations; Car-Sharing service system presents the information processing of users and stations; unified backstage data management system ensures the whole IOV, etc. Then we can use data sharing to solve the vehicle optimization assignment strategy and scheduling problems. S3 is applied to the administer terminal, user terminal and vehicular terminal, which are the service providing forms.

4.1 Model Construction

This model applies whenever trip demands are submitted to the Car-Sharing organizations or companies. We consider EV Car-Sharing systems in a metropolitan area network, and the optimization model is defined from the following four aspects.

Rental station network model.

- $G = \{V, E, d\}$: set of the whole rental station network which is an undirected weight graph;
- $V = \{1, 2, \ldots, n\}$: set of rental stations in the rental station network;
- $E = \{(i,j)i, j \in V\}$: set of arcs in the rental station network;
- d_{ij}: distance of arc (i,j);

Station model.

- $C_i = \{T, R, P\}$, $i = \{1, 2, \ldots, n\}$: set of vehicles at some rental station;
- $T = \{1, 2, \ldots, n\}$: type of vehicles in the rental station;
- $R \in N$: reservation quantity available at some rental station;
- $P(P > 0)$: price of vehicles in different types accumulated by hour;
- (unit: yuan / hour; less than an hour regard as an hour)

Vehicle model.

- $C_{itj} = \{e, b, s\}$: the vehicle numbered j in type t at station i;
- $e(0 < e < 1)$: percentage of the battery power. Suppose that an electric vehicle can run S_0 mileage when it is in full power condition, trip mileage is directly proportional to batter power, current percentage of the battery power is e, then the distance an EV can run is:

$$dis = eS_0 \tag{1}$$

- $B = \{1, 2, \ldots, 36\}$: time resource block. Time available for renting cars is from 6:00 a.m. to 24:00 p.m. Every half an hour can be regarded as a time resource block;
- $s = \{0, 1\}$: status code of the time resource block. Code 0 stands for that the vehicle has been booked or rented by other users; code 1 stands for that the vehicle is available currently;

Charging model.

- $ch = \{t_k, e_k\}$: charging parameter notation;
- t_k: upper limit time on the quick charge;
- e_k: percentage of the electric vehicle battery power by quick charging of t_k time. Therefore, a period of t time by quick charging can support a vehicle to run a mileage as follows:

$$mileage = \frac{t}{t_k} e_k S_0 \tag{2}$$

Other parameters.

- $return - t$: vehicle return time, which is used for data sharing in vehicle optimization assignment strategy;
- $return - s$: vehicle return station, which is used for data sharing in vehicle optimization assignment strategy;
- v: average speed of urban vehicles (for example vehicle's average speed in Beijing is about 50 km/h). If user's trip request for renting is a mileage of S_x(an estimated value from the pick-up station to the return station), then the percentage of battery power should satisfy the condition as follows when user taking the car.

$$e > \frac{S_x + d_{ij}}{S_0} \tag{3}$$

d_{ij} is the distance between station i and station j, $d_{ij} = 0$ when there is no need of vehicle scheduling from other stations.

4.2 Objective Functions

Four main objective functions are researched here in the vehicle assignment and scheduling problem, which are time matching function, station recommendation & vehicle scheduling function, vehicle utilization function and customer satisfaction function. In terms of the condition that user already has a certain trip time in a fixed location, we have the following thoughts:

- Time matching is to solve the problem of whether there are available vehicles in the nearby stations.
- Station recommendation & vehicle scheduling helps to make sure the best station among candidate stations.
- Vehicle utilization is a goal in higher level which is based on satisfying users' trip demands as possible as we can.
- Customer satisfaction is a consideration of QoE. Our solution not only provides users with available vehicles but also give them the best service, for example, taking the car nearby rather than going to a certain station.

In S3, we have worked out time matching function and station recommendation & vehicle scheduling function in detail, the formulation of these two functions will be presented as follows. At the same time the service rate and vehicle utilization can be assessment criteria for EV Car-Sharing systems in this paper.

Given conditions: user's position which is the coordinates of latitude and longitude, trip time ($B_x \sim B_y$), return time and return station of a vehicle on trip. The nearest station from user is defined as V_0, in this paper we use V_0 to mark user's position approximately.

Time Matching. Traverse vehicles numbered $\mathrm{C_{itj}}$ in station $\mathrm{V_i}(i = 0, 1, 2, \ldots, n)$, the vehicle must meet the conditions that there are enough battery power, available time resource blocks and a guarantee of flex time for pick-up and return cars. So we can define these as follows:

$$(e - \frac{2v}{S_0}) * \prod_{i=x-1}^{y+1} B_i > 0 \tag{4}$$

In formula (4), $x - 1$ and $y + 1$ are respectively the starting and ending time block, where actually the renting time is from time block x to y. Here we deal with the renting time flexibly and make sure that the time interval between two renting periods is sufficient enough. If there are some vehicle available, then assignment the vehicle on demand; if not, then go to the next step to have a optimization assignment based on data sharing, which have a good use of the order information with the service provided by the IOV Platform for EV Operation and Service.

Vehicle Optimization Assignment. When there is no vehicle available for user's demand at the current station, we go to the optimization assignment section. Depend upon the data such as return time and return station of every trip, we can have a thoughtful solution to the vehicle assignment problem. Suppose that we fail to assign a car for user now at the pick-up station, however, there may be some vehicles on their way to the current station as their destination station. As long as this return time can match that pick-up time, as well as a short period of flex time, then this will be a feasible optimization assignment.

Station Recommendation & Vehicle Scheduling. Nevertheless, some trip may still have no appropriate vehicle assignment, in order to provide the full service model for users and customers, station recommendation & vehicle scheduling is created. $\mathrm{V_0}$ is regarded as the approximate user site. When there are vehicles available in $\mathrm{V_0}$, accept the trip demand and serve the user; when there are not, start the vehicle optimization assignment module based on data sharing, which mainly has relations with $return - t$ and $return - s$ parameters; otherwise, go to the cross-station scheduling algorithm, just as shown in Fig. 1.

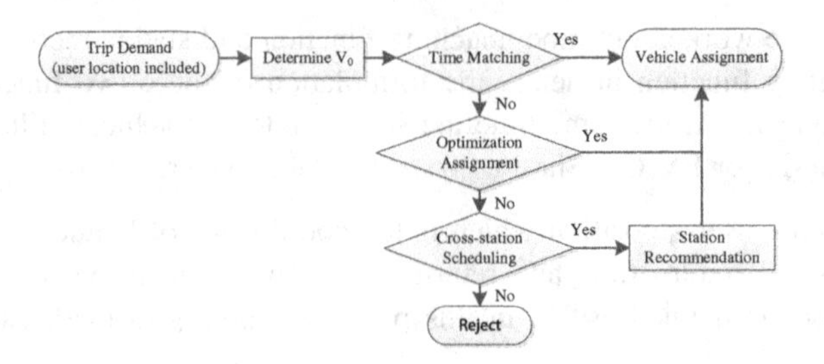

Fig. 1. Vehicle scheduling flow chart

In the cross-station scheduling module, we use the formula below to find the most appropriate station. Formula (5) is derived on the basis of (2) and (3).

$$F = \min\{\frac{d_{0i}}{v} + (S_x - e_{itj}S_0 + d_{0i})\frac{t_k}{e_k S_0}\} \tag{5}$$

In curly braces, the first part is scheduling time and the second part is charging time. The formula above is to find the minimum time of cross-station scheduling. e_{itj} is the percentage of battery power of the vehicle numbered j in type t in station i. After the traverse of vehicles in the station, the most suitable scheduling station and the appropriate vehicle can be confirmed by i and j.

Meanwhile, on the other hand, the influence on the following reservation and vehicle rentals due to the vehicle scheduling from another station cannot be neglected. Only when this scheduling has no effect on other vehicle assignment or situation can be solved, can the recommended station and scheduling be done.

5 Experiment and Results

5.1 Simulation Environment and Data Set

GreenGo EV Car-Sharing system serves six main districts of Beijing and the data set of it contains 25 rental stations as well as around 350 EVs. Figure 2 describes the station location network, which comes from the official website of GreenGo EV Car-Sharing system. Our simulation has collected real data set of GreenGo, Table 1 is the vehicle numbers of 25 stations, E150 and eQ two types included.

As for the simulation process, firstly, 25 rental stations of the station network are all in their real data set condition. And the user demands, including parameters such as user's current position, vehicle renting time, vehicle type, return time and station, are created randomly at the same time, ranging from 0 to 2500. Renting time has the same chance to be diffused in 36 time blocks. Around 350 vehicles and every vehicle has 36 time blocks, when the demands is 2500, the average using time for vehicles is 5 time blocks, namely 2.5 h. If there are some new demands, put them together in the demand pool and execute the service model once more. Finally, assign a suitable vehicle for user via S1, S2 and S3 three models.

Fig. 2. Station network of GreenGo

Table 1. Numbers of vehicle type E150 and eQ at every station

Staion	Location	E150	eQ	Staion	Location	E150	eQ
V1	116.374082,39.984315	10	5	V14	116.475578,39.956743	14	7
V2	116.332963,39.943897	9	4	V15	116.499902,39.99409	3	2
V3	116.318917,39.903663	4	2	V16	116.474864,40.008843	7	3
V4	116.344941,39.97813	12	6	V17	116.443869,39.89064	3	2
V5	116.359857,40.008125	13	6	V18	116.332477,39.885893	14	7
V6	116.284092,39.938006	8	4	V19	116.23425,39.911155	12	6
V7	116.318265,40.047255	18	9	V20	116.403735,39.861944	15	8
V8	116.241855,39.95991	8	4	V21	116.494608,39.804515	5	2
V9	116.46179,39.910477	10	5	V22	116.501841,39.806011	4	2
V10	116.459556,39.910429	6	3	V23	116.533639,39.796705	5	3
V11	116.459991,39.924284	10	5	V24	116.530212,39.794422	5	2
V12	116.50141,39.92122	8	4	V25	116.519702,39.788437	19	10
V13	116.516186,39.918675	9	4				

5.2 Simulation Method and Result

In order to show the excellent performance of S3, there are some comparisons with S1 and S2. And for the results and efficiency of the assignment and scheduling process, we can evaluate three models from multiple aspects. Obvisiously, the experiment environment, data set and initial conditions for three models are all the same, otherwise the comparison will be of no significance.

(1) *Service rate.*

Car-Sharing service as a novel trial, must focus on enhance the user experience, which need to increase service rate at first. With the amount of EVs (nearly 350) remains constant and the increase of demands ranging from 0 to 2500, service rate of three models has a rather distinct performance. In Fig. 3, for S1 and S2, service rate sharply reduces to below 90 % when demands are around 900 and 1600, while for S3, service rate is still above 95 % even when demands is around 2000 or more. Therefore, S3

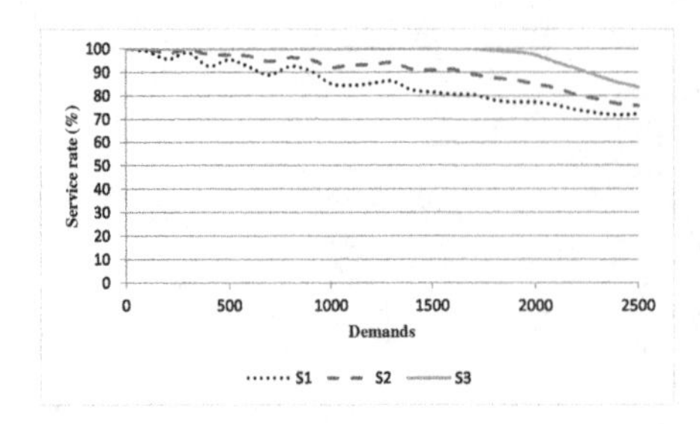

Fig. 3. Comparison of service rate

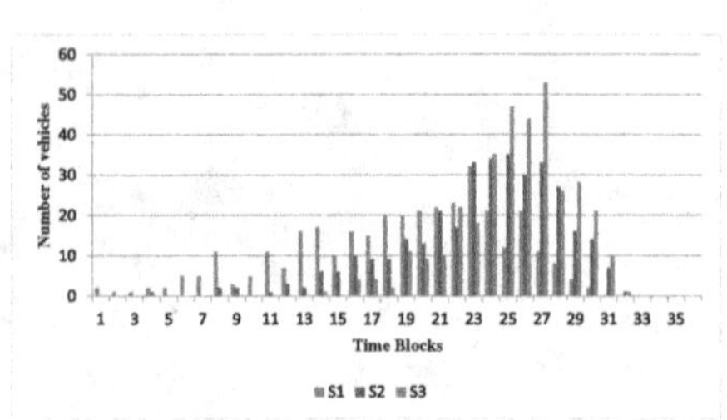

Fig. 4. Comparison of renting time distribution for vehicles every day

shows an outstanding result in service rate, which is the most important target of Car-Sharing systems.

Three curves all have big drop when demands are around 2000 $\sim$ 2500, the reason is that the total demands is beyond the capability of that existing vehicles can bear, which means more EVs are in crying need.

(2) *Total renting time distribution for vehicles every day.*

There are totally 18 h of renting time one day, which is divided into 36 time blocks. We can find that in Fig. 4, compared with S1 and S2, S3 has a more concentrated distribution. The majority of vehicles in S3 are in the range of 19 $\sim$ 31 time blocks, which means they run 10 $\sim$ 15 h per day, and almost very few vehicles run too little or too much. Therefore, S3 has a balanced use of vehicles, which prevents the situation of some vehicles over worn while others idle. The number of demands here is 2000, which is an appropriate choice. It is not only within the capability of existing vehicles, but also well reflects the average vehicle usage.

(3) *Incomes of EV Car-Sharing systems.*

For the EV Car-Sharing system, income is a quite indispensable target. In our simulation, the rent is 19 yuan/hour for both vehicle types. In Fig. 5, when demands are within 2500, S3 always have a distinct superiority to S2, which puts income-maximization objective as the first goal. That's because actually users' renting time below 2 h occupies a large proportion in the whole trip demands. In this case, as long as there is a reasonable scheduling strategy taking into account the short period of renting, high yield can be achieved smoothly. S3 not only has better QoE and users' satisfaction, but also has an advantage over S2 in income aspect.

(4) *Average scheduling distance.*

The trend of the two curves in Fig. 6 is roughly similar because of the similar scheduling strategy. S3 needs to respond to more trip demands, so with the increasing number of demands, there is an early appear of cross-station scheduling in S3. The average scheduling distance of S2 and S3 are mostly same, when the number of demands is ranging from 650 to 1500. S3 doesn't have any more vehicles used for scheduling, while satisfying more trip demands. When the number of trip demands grows even larger, the average scheduling distance of S3 is not that longer than S2. For

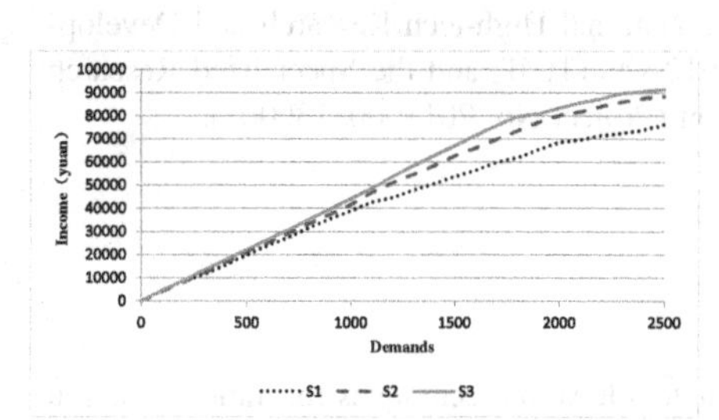

Fig. 5. Comparison of incomes

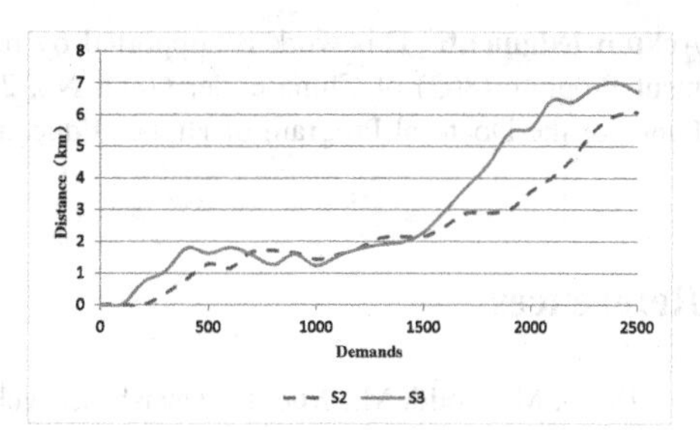

Fig. 6. Comparison of average scheduling distance

example, at the position of 2000 demands, the difference is only 2.5 km. And in the peak position, average distance is no more than 7 km. According to the vehicle's average speed of 50 km/h in Beijing, all the scheduling time is around 3 ~ 8 min. To sum up, S3 is reasonable and acceptable because it responds to more trip demands, keeps the scheduling distance in a limited range and the additional operating expenses are not much.

6 Conclusions

To optimize the vehicle recommendation and scheduling problem, we have presented an optimization model called full service model for EV Car-Sharing systems, which is based on the IOV Platform for EV Operation and Service. This platform provides charging management and fundamental capabilities in the EV systems, taking full advantage of IOV and data sharing, consequently having had a better scheduling even cross-operator scheduling. The vehicle demand in the future can use this model to have a better management of vehicle assignment and scheduling, in this way there will be a striking promotion of both service rate and vehicle utilization. The efficiency and accuracy of the full service model is evaluated by extensive experiments compared with existing maturity models, such as conditional service model and controlled service model. While owing to the constraint of real data of GreenGo Car-Sharing system, which is still at the initial stage of development, our simulation and experimental study are not that sufficient.

Our EV Car-Sharing system is built on the IOV Platform for EV Operation and Service and is also a successful expansibility of this supporting platform, together with the optimization vehicle scheduling model, EV Car-Sharing systems must have a more promising prospect. In the future, we plan to consider the programming of full service model which regards vehicle utilization and customer satisfaction as the parameters of our optimization and to have a further study of the other extensional services based on our platform.

Acknowledgment. This work is supported by the National High-tech Research and Development Program (863) of China under Grant No. 2012AA111601, and the Specialized Research Fund for the Doctoral Program of Higher Education (Project No. 20110005130001).

References

1. Barth, M., Todd, M., Xue, L.: User-based vehicle relocation techniques for multiple-station shared-use vehicle systems (2004)
2. Kek, A.G., et al.: A decision support system for vehicle relocation operations in carsharing systems. Transp. Res. Part E Logistics Transp. Rev. **45**(1), 149–158 (2009)
3. de Almeida Correia, G.H., Antunes, A.P.: Optimization approach to depot location and trip selection in one-way carsharing systems. Trans. Res. Part E Logistics Transp. Rev. **48**(1), 233–247 (2012)
4. Bruglieri, M., Colorni, A., Luè, A.: The relocation problem for the one-way electric vehicle sharing. Networks **64**(4), 292–305 (2014)
5. Weikl, S., Bogenberger, K.: Relocation strategies and algorithms for free-floating car sharing systems. Intell. Trans. Systems Mag. IEEE **5**(4), 100–111 (2013)
6. Uesugi, K., Mukai, N., Watanabe, T.: Optimization of vehicle assignment for car sharing system. In: Apolloni, B., Howlett, R.J., Jain, L. (eds.) KES 2007, Part II. LNCS (LNAI), vol. 4693, pp. 1105–1111. Springer, Heidelberg (2007)
7. Fan, W., Machemehl, R., Lownes, N.: Carsharing: dynamic decision-making problem for vehicle allocation. Transp. Res. Rec. J. Transp. Res. Board **2063**, 97–104 (2008)
8. Kortum, K., Machemehl, R.B.: Free-floating carsharing systems: innovations in membership prediction, mode share, and vehicle allocation optimization methodologies, Citeseer (2012)
9. Fangchun, Y., et al.: An overview of internet of vehicles. Commun. China **11**(10), 1–15 (2014)
10. Wang, S., Fan, C., Hsu, C.-H., Sun, Q., Yang, F.: A vertical handoff method via self-selection decision tree for internet of vehicles. IEEE Syst. J. **99**, 1–10 (2014)

A Method for Private Car Transportation Dispatching Based on a Passenger Demand Model

Wenbo Jiang[1(✉)], Tianyu Wo[1], Mingming Zhang[1], Renyu Yang[1], and Jie Xu[2]

[1] School of Computer Science and Engineering, Beihang University, Beijing 100191, China
{jiangwb,woty,zhangmm,yangry}@act.buaa.edu.cn
[2] School of Computing, University of Leeds, Leeds, UK
j.xu@leeds.ac.uk

Abstract. Although the demand for taxis is increasing rapidly with the soaring population in big cities, the number of taxis grows relatively slowly during these years. In this context, private transportation such as Uber is emerging as a flexible business model, supplementary to the regular form of taxis. At present, much work mainly focuses on the reduction or minimization of taxi cruising miles. However, these taxi-based approaches have some limitations in the case of private car transportation because they do not fully utilize the order information available from the new type of business model. In this paper we present a dispatching method that reduces further the cruising mileage of private car transportation, based on a passenger demand model. In particular, we partition an urban area into many separate regions by using a spatial clustering algorithm and divide a day into several time slots according to the statistics of historical orders. Locally Weighted Linear Regression is adopted to depict the passenger demand model for a given region over a time slot. Finally, a dispatching process is formalized as a weighted bipartite graph matching problem and we then leverage our dispatching approach to schedule private vehicles. We assess our approach through several experiments using real datasets derived from a private car hiring company in China. The experimental results show that up to 74 % accuracy could be achieved on passenger demand inference. Additionally, the conducted simulation tests demonstrate a 22.5 % reduction of cruising mileage.

Keywords: Spatial-temporal data · Data mining · Private car transportation · Vehicle scheduling

1 Introduction

The number of taxis in Beijing grows slowly during these years although demands for taking taxi has been sharply increasing with the soaring population. At present, it is extremely hard for citizens to hail a taxi especially on the occasion of congested traffic. However, at the same time, taxi occupancy is still very

C.-H. Hsu et al. (Eds.): IOV 2015, LNCS 9502, pp. 37–48, 2015.
DOI: 10.1007/978-3-319-27293-1_4

low even in busy time [9]. This contradictory phenomenon leads to great requirements for customers vehicle transportation.Consequently, there are companies such as Uber [1] which provides private car services to satisfy the above demands. In this manner, private cars become a supplementary to the current taxi service in which citizens could reserve a private car through mobile applications in any place at any time. However, taxicab is inefficient due to characteristics such as low capacity utilization, high fuel costs, heavily congested traffic, and a low ratio (*live index*) of *live miles* (miles with a fare) to *cruising miles*(miles without a fare) [2]. The same challenges exist in private car service as illustrated in Fig. 1. The rate represents the ratio of live miles to total miles. Apparently, the rate of approximately 80 % vehicles is no more than 0.6. The average rate is merely 0.4601, indicating a typical low live index among most private cars. Namely, idle cruising miles take a large portion of the total miles.

Despite some similar challenges within the taxi service and private car service, the differences and the root causes still need to be deeply investigated. As for the taxi service, drivers can pick up a passenger easily in rush hours. However, drivers might cruise several street blocks to pick up a passenger if they lack the experience of when and where the citizens usually expect to catch a taxi. Therefore, experienced taxi drivers have less cruising miles than the inexperienced [4]. To mitigate the empirical impacts, many solutions generate and leverage knowledge of experience by using historical GPS records, and recommend pick-up positions or cruising routes to drivers for the sake of reducing cruising miles [2,3,7]. However, unlike the taxi service, passengers reserve the cars by Apps in private car service. In fact, each passenger order includes sufficient information such as the current position, destination, appointment pickup time, car type and etc. Intuitively, the abundant information could provision sufficient space to dispatch and schedule private vehicles based on the knowledge mining, giving rise to significant optimization potentials. Without fully utilizing of these information, larger cruising miles and longer waiting time will be produced and will result in increasing operating costs and negative user experience. All these strongly motivate our proposal in this paper.

Additionally, determining a specific position or a region is extraordinarily important in scheduling private cars. Presently, a private car driver could only accept orders according to the instructions from the dispatch center passively. The problem will not protrude in rush time due to the large requirements of reservation orders. In other cases, however, drivers have great needs to be instantly notified a specific potential position in order to diminish the time and monetary waste before receiving a new order. Therefore, an effective scheduling should improve the cruising miles whilst reducing passenger waiting time. In order to accurately obtain the specifying designated places, the dispatch center chooses the places based on subjective cognition and the dynamic passengers' demand should be also taken into account when determining the specified places because it is variable along with time. To summarize, we have to cope with the following challenges appropriately: (1) predicting the time-varied and dynamic passenger demand accurately; (2) dispatching vehicles to satisfy the requirements of

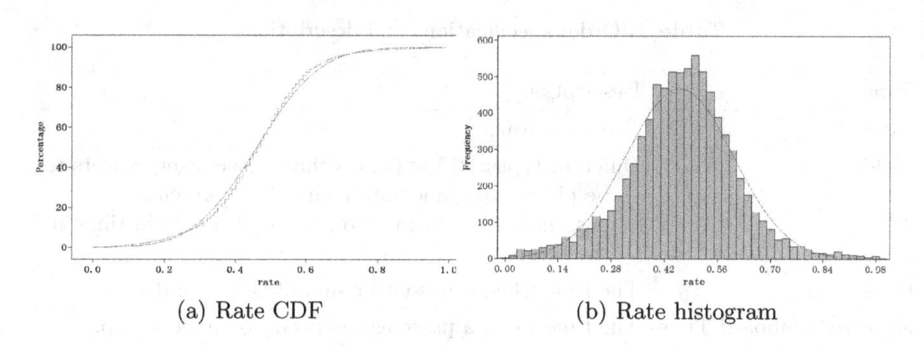

(a) Rate CDF (b) Rate histogram

Fig. 1. Rate of live miles compared to total miles

different passengers who are free to choose different type of cars; (3) scheduling in real-time manner and provisioning current orders as well as future orders. In particular, the major contributions of the work in this paper can be summarized as follows:

- We propose an innovative study to reduce private vehicle cruising miles and a scheduling method based on a passenger demand model.
- A demand specification model is advocated based on Locally Weighted Linear Regression, which could reach 74 % accuracy at most.
- The real-time scheduling is formalized as a weighted bipartite graph matching problem and 22.5 % cruising miles could be reduced by using the described scheduling method.

The rest of this paper is organized as follows. Section 2 describes the passenger demand model. Section 3 introduces algorithms to dispatch private cars. Furthermore, we evaluate the model and dispatching algorithms on a real dataset in Sect. 4. Section 5 concludes this paper and discusses the future works.

2 Demand Model

In this section, we elaborate the Passenger Demand Model in detail. Firstly, the training set, formulas and algorithms are respectively described to expose the model's details. We also demonstrate how to predict the passenger demand in real time manner. To begin with, we make some definitions in terms of the passenger demand in different contexts. In taxi service, passenger demand mainly indicates the citizen requirements to take a taxi; In comparison, we define passenger's demand in the private car service as the orders that preserve private cars in a region over a time slot. As Table 1 shows, an order mainly contains such information.

The region is a much more fine-grained space definition than district. We split Beijing urban area into R regions using a density based spatial clustering algorithm DBSCAN [8] in which $i(1 <= i <= R)$ represents the i-th region. A convex polygon is used to indicate a region space coverage. After statistical

Table 1. Order specifications and descriptions

Name	Description
Order ID	Order identifier
Order Type	Different types: "1" indicates that a passenger expects to be picked up immediately and "2" represents a passenger expects to be on board at a certain time in the future
Book Time	The time when a passenger submits a new order.
Expected Onboard Time	The time when a passenger expects to be picked up.
Start Time	The time when a passenger is picked up.
Start Longitude	The longitude of GPS coordinates where a private car picks up a passenger.
Start Latitude	The latitude of GPS coordinates where a private car picks up a passenger.
End Time	The time when a passenger is dropped off.
End Longitude	The longitude of GPS coordinates where a private car drops off a passenger.
End Latitude	The latitude of GPS coordinates where a private car drops off a passenger.
Car Type	Three types of cars which passengers can book with different charges.
Status	Order status: "0" indicates a new submited order, "1" indicates an already matched order to a private car, and "2" means a completed order.

analyzing of historical orders, one day could be divided into T time slots and $j(1 <= j <= T)$ represents the j-th time slot. In addition, each time slot's duration is regarded as an hour for the sake of understandability, and most works adopt this dividing method [5]. Therefore, the passenger demand in the i-th region over the j-th time slot could be expressed as $P_{i,j}(1 <= i <= R, 1 <= j <= T)$.

Furthermore, based on the mobility of cars and passengers, we assume that passenger demand in one region in the next time slot is relevant to orders in all regions over current time slots. In Sect. 4, we conduct some experiments on a real dataset to confirm these assumptions. In particular, the orders in each region over the current time slot is an argument while the passenger demand in one region in future time slot is independent variable. A set of coefficients can be used to estimatethe relation between the orders in current time slot and the predicted orders in the future time slot. To this end, the linear regression model targets our demand and the core philosophy is the coefficients estimation. With the input data stored in matrix X, and regression coefficients stored in vector w, predicted result can bydepicted as Eq. 1.

$$Y = X^T \cdot w \tag{1}$$

To achieve an ideal prediction, the error margin between the real demand and predicted one is supposed to be miminized as possible. The model is utilized to obtain the optimal set of regression coefficients w, given a training set X and Y. Typically, we aim to find the vector w with the minimum error. In order to improve the computation accuracy, we use the *squared error* instead of the *accumulated error* to represent the required difference. More specifically, the sum of the squared errors (SSE) can be described by Eq. 2. The utilized least squares approach find the optimal w with the least SSE.

$$SSE = \sum_{i=1}^{m} (y_i - x_i^T \cdot w)^2 \tag{2}$$

We also depict the SSE in matrix form, as shown in Eq. 3 shows.

$$SSE = (Y - X \cdot w)^T (Y - X \cdot w) \tag{3}$$

We solve SSE's derivative, and make it equal to zero (Eq. 4).

$$X^T (Y - X \cdot w) = 0 \tag{4}$$

The best regression coefficients w could be found by using Eq. 5.

$$\hat{w} = (X^T \cdot X)^{-1} \cdot X^T \cdot Y \tag{5}$$

It is the best estimation we could achieve from w. It is also worth noting that the above formula contains the requirement for the matrix inversion. Therefore, the equation only works on the occasion of the inversion of matrix $X^T \cdot X$ exists. Typically, the linear regression using the proposed least squares method could obtain the ideal result. However, it might be very likely to encounter with the under-fitting phenomenon. This is because it tries to get unbiased estimation of minimum mean square error and obviously we can not get the best infer result once the model is under-fitting.

To handle with this problem, we introduce the Locally Weighted Linear Regression(LWLR) to produce some bias in the estimation to reduce the mean square error of prediction. In LWLR, we assign certain weights to the sample points which are close to the predicted sample point. In this context, the closest sample point gets the biggest weight. If the weight of a sample point exceeds a certain threshold, it will be selected. Afterwards, the selected points will constitute a sample subset. We use least square method to learn the linear regression model on this subset.

We can use the following Equation to get the optimal regression coefficient:

$$\hat{w} = (X^T \cdot W \cdot X)^{-1} \cdot X^T \cdot W \cdot Y \tag{6}$$

W is a weight matrix with each element indicating every sample point's weight. In order to generate the weighted matrix W, LWLR uses the kernel function to ensure the fact that the sample point closest to the predicted sample point could get the maximum weight. Specifically, there are a number of types of kernel function and the most commonly-used Gaussian Kernel Function is chosen in our solution, as demonstrated in Eq. 7.

$$W(i,i) = exp(\frac{|x_i - x|}{-2k^2}) \tag{7}$$

In this way, we can build a weight matrix W, containing only diagonal element and the sample point x_i closer to x could finally get the greater weight. Additionally, k is a user-specified parameter, which determines the weight given to x_i.

Algorithm 1. predictPassengerDemandInDifferentRegions()

1: initialize training set X $Matrix$ $xMat$;
2: initialize training set Y $Matrix$ $yMat$;
3: initialize test set X $Matrix$ X;
4: **for all** $Region \in R$ **do**
5:    add p(i,j) to X ;
6: **end for**
7: create weights matrix W;
8: $xTx \leftarrow xMat.Transpose * (W * xMat)$;
9: **if** determinant of xTx equals to 0 **then**
10:    return $NULL$;
11: **else**
12:    $ws \leftarrow xTx.Inverse * (xMat.Transpose * (W * yMat))$
13:    return $X * ws$;
14: **end if**

Algorithm 1 is introduced to predict passenger demand in i region over $(j+1)$ time slot. Firstly, we build the training set and test set according to the above description. Secondly, we use the kernel method as described in Eq. 7 to generate weight matrix. Finally, after examining if xTx has an inverse matrix, we obtain the coefficients using Eq. 6 and predict the passenger demand in different regions. If xTx does not have an inversematrix, we cannot figure out the coefficients and the passenger demand. In such conditions, parameter estimation and optimization algorithms are used to facilitate the computation of the approximately optimal coefficients. Due to the fact that xTx's inverse matrix could be easily obtained in most cases, the parameter estimation is out of the scope of this paper and will not be discussed.

3 Dispatching Algorithm

In this section, we describe in detail the core idea of the private vehicle dispatching algorithm. In Sect. 2, the prediction model is advocated and the passenger demand count in different regions could be generated. Before introducing our solution, some fundamental definitions are given as follows:

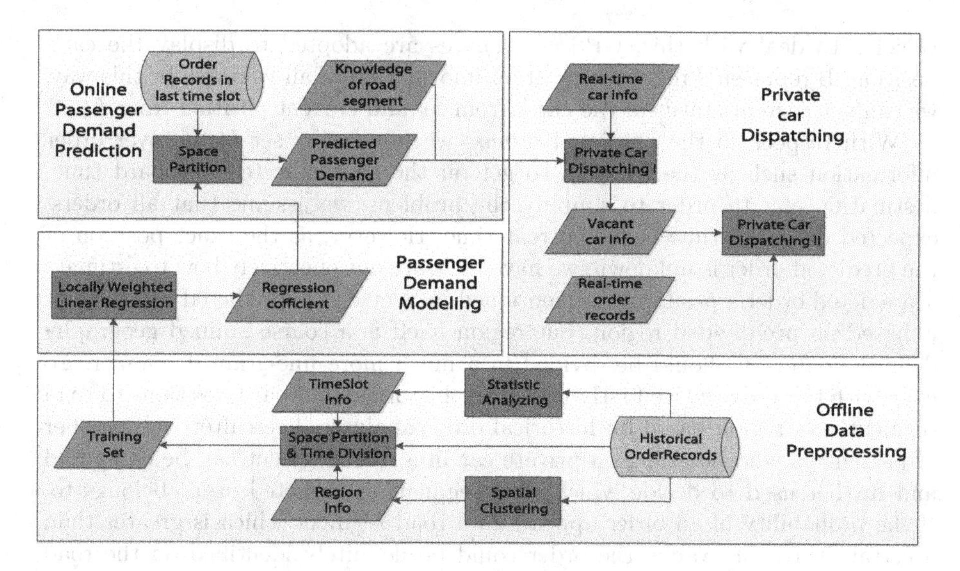

Fig. 2. Dispatching process overview

Definition 1 Bipartite Graph Matching. *Given a bipartite graph G = (C,O,E), C is a vertex set in which each vertex represents a vacant car and each vertex in set O represents a predicted order. M is a subgraph of G. Any two edges in M's edge set E are not attached to the same vertex and M is a match of vacant cars and predicted orders.*

Definition 2 Road Segment. *If a road meets a junction or the road offset exceeds a threshold, a new road segment is generated. A road segment could be uniquely identified by the combination of ID, start and end GPS coordinates.*

Assuming that we have estimated passenger demand $\hat{P}(i, j+1)$ in i region over $(j+1)$ time slot. What we need is to assign vacant cars to passengers with the minimum total cruising mileage. A bipartite graph model $G(C, O, E)$ is therefore used to formalize our problem. Intuitively, distance between a vacant car and a predicted order becomes the weight of an edge connecting those two kinds of vertices. Accordingly, a matching with the minimum weight could be formalized as weighted bipartite graph matching problem (Fig. 2).

Before performing the matching algorithm, how to represent the weight of the edges needs to be determined properly. Because the objectives are to find a scheduling method with the minimum cruising mileage, the distance between vacant cars and predicted orders could be used to represent the weights of the edges. In addition, C is the set of current vacant cars, with a vector S recording real-time status of all vehicles. Therefore, we can always get an updated C from S. In particular, "1" indicates that the current passenger vehicle is occupied, whereby "0" represents a vacant status of the current vehicle. Furthermore, in order to calculate the distance, we have to gain the real-time location of the vehicle. To deal with this, GPS coordinates are adopted to display the car's position. L represents real-time location information of all vehicles. In this way, we can get current status of the car k from S_k and current position from L_k.

With respect to the predicted orders, we use vertex set O to cover order information such as the location to get on the car, expected on-board time, destination etc. In order to simplify the problem, we assume that all orders' expected on-board time is the current time. However, as the exact position of the predicted order is unknown, we have to figure out effectively how to estimate a predicted order's position. As mentioned in Sect. 2, the predicted order count is based on pre-divided region, but region itself is a coarse-grained geography definition. Region should be divided in a much more fine-grained manner. To cope with these issues, we firstly match the historical on-board positions to road segment in a region based on historical order analysis. Thereafter, the number of passengers who boarding on private car in a road segment can be computed and further used to decide which road segment a predicted order belongs to. If the probability of an order appears in a road segment which is greater than a certain threshold value, the order could be definitely identified on the road segment. Eventually, the position of the predicted order can be determined by this road segment.

Moreover, we will discuss how to set the weight of edge. In bipartite graph G (C, O, E), E is the set of edges. Intuitively, the distance between the car and the order is used to represent the weight of edge between them. Despite other factors including the weather, real-time traffic etc., we use Baidu Web APIs to get the shortest distance between two GPS coordinates in our experiment for the clearness and simplicity of the model.

Given above definitions and descriptions, the scheduling problem could be formalize as a weighted bipartite graph matching problem. Kuhn-Munkras algorithm is used to find the optimal matching with minimum cost. Due to the special nature of KM algorithm, set C and O should have equal vertexes. In fact, we can add some vertexes to the minimum set on the occasion of two unequal set. Additionally, the weights of edges associated to those added vertexes are set to be zero. The detailed descriptions of the scheduling process could be found in Algorithm 2.

After the matching decision, vacant cars will be assigned to the road segments where predicted order exists.

In the real scenario, a real order might be placed while vacant cars are forwarding to a matched road segment. The real order contains information including the passenger's position, destination, etc. Meanwhile, vacant cars' information can be extracted from vector S and L. In case of the emerging real orders, the closest car will be scheduled and appointed to satisfy the real order demand. Particularly, the status and location of the vehicle need to be updated in real-time manner. Once a vehicle is assigned to a real order, the corresponding status related to real order will be occupied and updated. The status "1" will turn into "0" when the passenger gets off. Finally, the scheduling process proceeds until all private cars are out of service (finish working).

Algorithm 2. schedulePrivateCar()

```
 1: initialize each vertex in set C in j+1 time slot;
 2: predictPassengerDemandInDifferentRegions ();
 3: initialize each vertex in set O in j+1 time slot;
 4: find a minimum cruising mileage matching;
 5: while a real order occurs do
 6:    select vacant cars near the real order's position;
 7:    if a vacant car accepts the order then
 8:       update car's status;
 9:       update order's status;
10:       update L and S;
11:    end if
12: end while
```

4 Experimental Evaluation

In this section, we set up several experiments to evaluate the effectiveness of the proposed passenger demand model and dispatching algorithm.

4.1 Dataset and Experiment Environment

We use a real dataset from a private car hiring company (for commercial reasons we cannot identify the name), which contains 2720 vehicles and 356706 order records for a whole month in Beijing. The test machine is equipped with Debian 6.0 operation system, 8 GB Memory and Intel i5-4570 with a frequency of 3.20 GHZ is set up.

4.2 Data Processing

Because the real dataset can not be directly used, the data processing starts with data cleaning and pre-operations. Specifically, we filter the order with a time duration less than 10 min and larger than 60 min, resulting in the remaining 324512 records. Subsequently, spatial clustering is utilized to split Beijing urban area into multiple regions. To simplify our experiment, DBSCAN algorithm is used to divide urban areas in Beijing four-rings into 136 areas. Two parameters (Epsilon and MinPoints) are set to be 0.15 and 100 respectively according to the engineering experiences. Actually, the Epsilon indicates the density of a region, while MinPoints shows how many points a region need to possess at least. Besides, a day is partitioned into several time slots based on the order distribution in one day (Fig. 3).

Figure 4(b) illustrates a good example of regions partition near the Beijing Railway Station and the average count of orders in each time period. We select a time period in which the order count is larger than the average count, leading to a more convenient and reliable prediction. The time duration between 7:00 AM to 10:00 PM is selected as the experimental time interval.

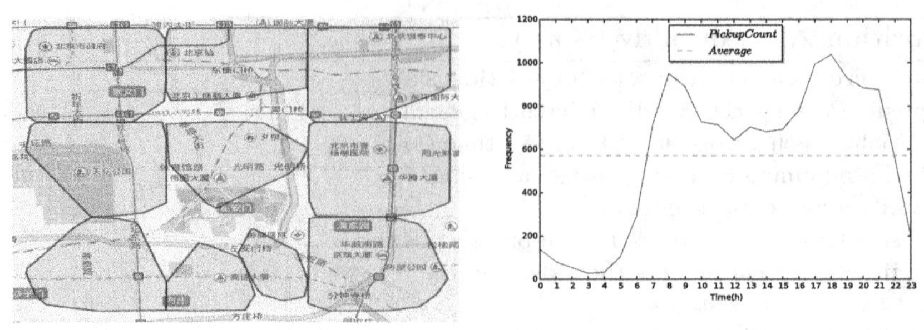

(a) Several regions near Beijing Railway Station

(b) Average order count in one day

Fig. 3. Regions and time slots

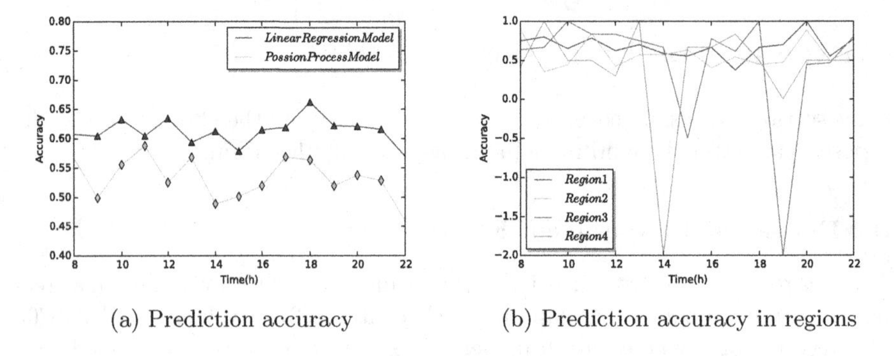

(a) Prediction accuracy

(b) Prediction accuracy in regions

Fig. 4. Passenger demand model evaluation

4.3 Model Evaluation

We firstly introduce Non-homogeneous Poisson Process (NHPP) Model. Poisson process is a stochastic process that is often used to study the occurrence of events. It assumes that the arriving rate of events λ is constantly stable, and has the Poisson distribution of counting and exponential distribution of inter-event time. In fact, NHPP [6] is a specific Poisson process with a time-dependent arriving rate function $\lambda(t)$. The model is more flexible and appropriate to depict the human-related activities because these activities often vary over time but have strong periodicity. We use ΔT to indicate time increment and $N(\Delta T)$ to represent order increment in ΔT. Specifically, $N(\Delta T + j) - N(j)$ follows the Poisson distribution with a parameter $\lambda \Delta T$.

$$P\{N(\lambda T + j) - N(j) = k\} = \frac{e^{-\lambda \Delta T}(\lambda \Delta T)^k}{k!} \tag{8}$$

Here we set ΔT to one hour and λ can be obtained by maximum likelihood estimation:

$$\hat{\lambda} = \frac{\overline{N}}{\Delta T} \tag{9}$$

where $\overline{N}$ can be denoted by average orders in one hour. Therefore, we could use the average order count to predict passenger demand. We use the following equation to calculate the prediction accuracy:

$$Accuracy = \frac{P_{i,j+1} - |P_{i,j+1} - \hat{P}_{i,j+1}|}{P_{i,j+1}} \tag{10}$$

It is observable from Fig. 4(a) that the prediction accuracy of our model outperforms NHPP, and the accuracy in different time periods varies in an acceptable range. Meanwhile, a peak accuracy could be achieved at 18:00. This is due to an obvious passenger demand trend around 18:00. Figure 4(b) also reveals the prediction accuracy of different regions. Apparently, the accuracy for Region 1 and 2 is steady all the time with little fluctuation. However, the inference accuracy in Region 3 and 4 suffers from some degrees of variations, even with some outlier results.This is because the number of orders in these Regions are so scarce that the proposed model could not capture a reasonable value. To solve this problem, we not only enlarge the size of required dataset, but increase the region's coverage area as well.

4.4 Dispatching Algorithm Evaluation

In order to replay the real behaviors of the moving vehicles and the user demands, we emulate their moving traces obtained from the historical data and apply them into our algorithms. As shown in Table 2, we choose the real orders and car traces from 08:00 to 13:00 on May 6th, 2015. At 8:00, there are 324 vacant cars and 296 predicted orders in this evaluation. The result illustrates that the proposed model could finish matching of vacant cars with predicted orders in 52 s. In addition, we leverage Baidu Web API to obtain the distance between a vacant car and a predicted order with average speed recorded. Once a real order emerges, the nearest vacant car could be selected to pick up the passenger. On average, 22.5 % cruising miles could be reduced in this simulated experiment.

Table 2. Dispatching results

Time	Vacant car count	Predicted order count	Reduced cruising miles
08:00	325	296	21.5 %
09:00	332	369	23.1 %
10:00	275	285	19.2 %
11:00	247	212	24.9 %
12:00	203	245	27.2 %
13:00	243	210	26.3 %

5 Conclusions and Future Work

At present, taxis in urban areas could not fully satisfy the booming citizens' demand for convenient transportation means. Fortunately, private car hiring is emerging as an alternative and offering a variety of ways for short trips. It is observed that serving vehicles have to spend a large portion of their service time on cruising around streets, resulting in large amountof monetary waste. To solve this problem, this paper presents a private car dispatching method based on a passenger demand model.We use Locally Weighted Linear Regression to depict passenger demands in detail, and formalize private car dispatching as a weighted bipartite graph matching problem. The experimental results demonstrate our model outperforms the non-homogeneous poisson process model with a more stable prediction accuracy. We also exploit a real dataset for simulating the private car dispatching process. By using our proposed approach, 22.5 % cruising mileage can be saved. In the future, we will further enlarge our dataset to optimize the proposed model and to take into account other factors affecting passenger demands, such as weather, holidays etc. We are also planning to parallelize the dispatching algorithm to reduce further the total execution time.

Acknowledgments. This work is supported in part by China 973 Program (2014CB3 40300), National Natural Science Foundation of China (91118008, 61170294), China 863 program (2015AA01A202), HGJ Program (2013ZX01039002-001), Fundamental Research Funds for the Central Universities and Beijing Higher Education Young Elite Teacher Project (YETP1092).

References

1. Uber. https://www.uber.com
2. Powell, J.W., Huang, Y., Bastani, F., Ji, M.: Towards reducing taxicab cruising time using spatio-temporal profitability maps. In: Pfoser, D., Tao, Y., Mouratidis, K., Nascimento, M.A., Mokbel, M., Shekhar, S., Huang, Y. (eds.) SSTD 2011. LNCS, vol. 6849, pp. 242–260. Springer, Heidelberg (2011)
3. Qu, M., Zhu, H., Liu, J., Liu, G., Xiong, H.: A cost-effective recommender system for taxi drivers. In: Proceedings of the 20th ACM SIGKDD International Conference on Knowledge Discovery and Data Mining. ACM (2014)
4. Yuan, J., et al.: Where to find my next passenger. In: Proceedings of the 13th International Conference on Ubiquitous Computing, pp. 109–118. ACM (2011)
5. Zhang, D., He, T., Lin, S., Munir, S., Stankovic, J.A.: Dmodel: online taxicab demand model from big sensor data in a roving sensor network. In: 2014 IEEE International Congress on Big Data (BigData Congress), pp. 152–159. IEEE (2014)
6. Zheng, X., Liang, X., Xu, K.: Where to wait for a taxi?. In: Proceedings of the ACM SIGKDD International Workshop on Urban Computing. ACM (2012)
7. Yuan, N.J., Zheng, Y., Zhang, L., Xie, X.: T-finder: a recommender system for finding passengers and vacant taxis. IEEE Trans. Knowl. Data Eng. **25**(10), 2390–2403 (2013)
8. Ester, M., et al.: A density-based algorithm for discovering clusters in large spatial databases with noise. In: Proceedings of International Conference on Knowledge Discovery & Data Mining, pp. 226–231 (1996)
9. Chang, H.W., Tai, Y.C., Hsu, Y.J.: Context-aware taxi demand hotspots prediction. Int. J. Bus. Intell. Data Min. **5**(1), 3–18 (2010)

Real-Time Location System and Applied Research Report

Yansong Cui and Jiayu Zhao(✉)

Laboratory of Intelligent Traffic Information and Network Technology,
Beijing University of Post and Telecommunications,
Beijing, China
zhaojiayu2013@163.com

Abstract. RTLS (Real-Time Location System, RTLS), is a real-time or nearly real-time positioning system by wireless communication technology in a designated space (office, venue, city, global). RTLS is usually based on ZigBee, Wi-Fi, UWB, Bluetooth, infrared, ultrasonic, etc. With the features of low cost, low power, low complexity, high positioning accuracy and ease of deployment, it is widely applied in logistics, medical and military fields. This paper presents an application of RTLS based on ZigBee, in order to achieve real-time positioning of the vehicle. The RTLS is built in indoor parking environment, and it is tested and optimized its positioning accuracy. Experimental results show that the vehicle positioning results in the present system can achieve high positioning accuracy, positioning accuracy in 13 meters within a range of positioning accuracy 91.7 % in the 7.8 m range positioning accuracy was 83.3 %, correct positioning accuracy was 79.2 %.

1 Introduction

Real Time Location System (RTLS) is composed of RTLS tag, RTLS reader, RTLS server and RTLS application system, the system structure shown in Fig. 1. RTLS system combines radio frequency identification and localization algorithm, which uses communication technology and positioning technology to collect data and upload, and then carry out real-time positioning according to the positioning algorithm. Each part of the function of RTLS work processes and systems are as follows: RTLS tag is installed in the location of a person or object, it according to certain time interval to send a certain frequency radio signal, RTLS to read and write device receives the signal transmitted pass RTLS server, by the RTLS server calculated the physical location of a target, available to RTLS application system.

The key technology of RTLS system is communication technology, positioning technology and positioning algorithm. Among them, the most important RTLS communication protocol is the communication protocol, which is the technology of RTLS tag and RTLS reader communication. It can use RFID, Wi-Fi, ZigBee, UWB, ultrasonic, infrared, Bluetooth and other technology, and the communication between RTLS server and RTLS application system using Service Web technology for data transmission. RTLS positioning technology can be classified according to various methods, according to whether based on ranging technology can be divided into [1], which is

C.-H. Hsu et al. (Eds.): IOV 2015, LNCS 9502, pp. 49–57, 2015.
DOI: 10.1007/978-3-319-27293-1_5

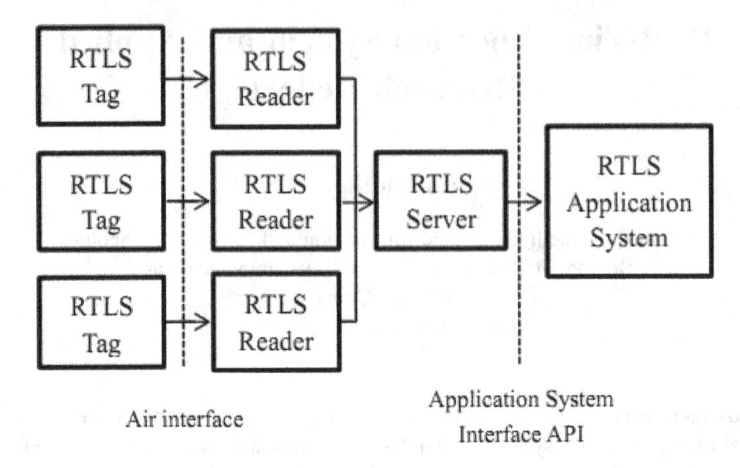

Fig. 1. RTLS Structure

based on [2], including RSSI, TDOA, AOA, NFER [3], TOA [4, 5], etc. RTLS positioning algorithm is mainly used in the circumferential positioning method, including the circular positioning method, including the three sides and the multilateral positioning method [6].

RTLS technology has a wide range of applications, not only can be used for the positioning of objects, but also can be used for personnel positioning, its main application areas in logistics, medical and military. Due to the cost of RTLS hardware and software to about 5 % to 20 % in a gradual decline, the market demand for the future RTLS will increase year by year.

ZigBee technology is a new emerging wireless network technology with short distance and low power consumption. It is a kind of technology between wireless RFID [6] (RFID) and Bluetooth [7] (Bluetooth). It is hoped to develop a kind of low cost, low power consumption and large scale wireless network. It fills the gaps in the market of low cost and low power wireless communication technology. ZigBee technology is a kind of RTLS communication technology, which has the characteristics of simple network nodes, easy to build, good robustness, scalability and so on.

In this paper, a real-time positioning technology is proposed, which can realize real-time positioning of the vehicle in the indoor vehicle networking system. RTLS system uses ZigBee as the communication technology, received signal strength RSSI as the ranging technology, the positioning algorithm using the proximity of the non-ranging technology and the combination of scene analysis method. The actual situation of the system is built, and the algorithm of the vehicle is modified and tested. The test sample is randomly generated, can cover the entire test environment, the test results are representative and feasible.

2 Application Environment

In this paper, the real-time positioning system (RTLS) is used to build the indoor parking lot in the downtown area of high buildings, and the use of GPS in such an environment has been greatly restricted. Because of the high buildings in the urban area, the single GPS receiver needs two minutes to calculate the first position, the positioning time is too long. In addition, the positioning and tracking of GPS terminal needs real-time, terminal power consumption, battery life is short [8, 9]. These factors have limited the application of GPS in the city. After the actual test, GPS cannot distinguish the specific floor of the vehicle parking, so the indoor parking is very suitable for the use of RTLS positioning.

In this paper, the RTLS test environment for indoor parking, a total of 3 layers. The number of spaces in which the number of three layers is: 78 parking spaces on floor 1, 66 parking spaces on floor 2, 54 parking spaces in floor 3.

3 RTLS Based on ZigBee

ZigBee has 3 kinds of network topologies, which are star shaped network, the tree net and the net. ZigBee network nodes can be divided into three types: ZigBee Coordinator (ZigBee coordinator, ZC), ZigBee routing node (ZigBee router, ZR) and ZigBee terminal node (ZigBee end device, zed) the ZC and ZR belongs to full function devices (full function device, FFD) can communicate with the FFD and RFD (reduced function device), zed belong to reduced function devices (RFD) only and FFD communication, each network can only have a ZC, is the initiator of a network and become the network control center and other nodes corresponding to join the network [10].

The real-time positioning system (RTLS) based on ZigBee communication technology is designed as follows: RTLS tags, RTLS reader by ZigBee network system node, where RTLS tags for ZigBee terminal node (ZED), RTLS reader for ZigBee coordinator or (ZC) ZigBee routing node (ZR). RTLS server and RTLS application system design commonly used Service Web technology. In this system, the RTLS server for data exchange with the ARM ZigBee system, RTLS application system for the background server.

In this paper, the real time positioning system (RTLS) system block diagram is shown in Fig. 2, which can be roughly divided into two parts, RTLS-ZigBee communication part and RTLS- positioning algorithm. RTLS-ZigBee communication part of the ZigBee communications to obtain RTLS tags (ZED) of the original positioning information (ZED emission signal intensity RSSI); RTLS- positioning algorithm based on the original positioning information to calculate the specific location of the RTLS tag and upload the background. RTLS work flow is as follows: RTLS tags by a certain time interval to around 1 RTLS reader (ZC) send request, the RTLS reader (ZR) back to the received ZigBee signal intensity (i.e. RSSI emission signal intensity ZED); around 1 jump distance RTLS reader (ZC) will measure the ZigBee signal intensity (i.e., ZED emission signal intensity RSSI), and send back to RTLS tags (ZED);

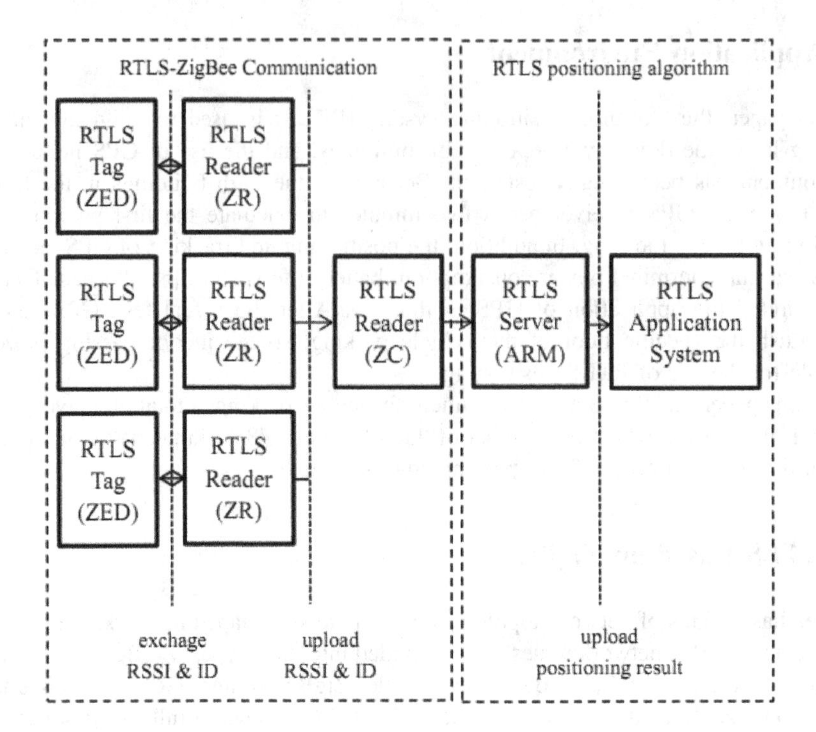

Fig. 2. RTLS system chart

RTLS tag collection received within a certain time interval RSSI information and its own ID RTLS (RTLS)., RTLS server according to the positioning algorithm to calculate the RTLS tag (ZED) of the specific location, submitted to the RTLS application system (background server).

RTLS system in the indoor parking lot is as follows: RTLS tags (ZED) installed in a taxi, a total of 400 taxis, test randomly selected 24 taxis as sample; RTLS reader in the indoor parking lot 9 devices, including 8 RTLS reader (ZR) and 1 RTLS reader (ZC).

4 Positioning Algorithm

The real-time positioning system (RTLS) is based on the RTLS (ZR) RTLS (ZED) signal intensity RSSI as the basis for positioning. The conversion relationship between the signal strength RSSI and the transmission node and the node spacing is derived as follows: [11]

The acceptance signal strength between RSSI and the transmission power of the node is in accordance with the formula,

$$\mathrm{RSSI} = 10 * \lg\left(P'_{rx} / P_{ref} \right) \tag{1}$$

Among them, the RSSI value is expressed as the received signal strength, the P_{rx} is received by the receiving node of the power, the reference value $P_{ref} = 1$ mW.

In ideal free space, $P'_{rx} = P_{rx}$, free space wave formula,

$$P_{rx} = P_{tx} * G_{tx} * G_{rx} * (\lambda / 4\pi d_{ij})^2. \quad (2)$$

P_{tx} means transmitting node power, P_{rx} means receiving the received power, G_{rx}, G_{tx} respectively means receive and transmit gain, d_{ij} is the distance between the transmission node and the receiving node.

By formulas (1) and (2) can be obtained,

$$\text{RSSI} \propto 1/\lambda^2. \quad (3)$$

In the practical application environment, due to multipath, diffraction obstacles and other factors, radio propagation path loss and theoretical value compared to some of the changes in the case [12]. And the test environment in the indoor parking lot, this change will become big, and the change in different time will keep changing. This will give a great difficulty to the design and application of the positioning algorithm based on ranging technology.

In the actual indoor parking lot, the RTLS location algorithm is based on the combination method of the location and scene analysis without the need of distance measurement. The actual situation of the system is built, and the algorithm of the vehicle is modified and tested.

Taking third layers in the indoor parking lot as an example, the method of the proximity and scene analysis of the test environment is described. This layer of the CPC has 54 parking spaces, the test environment of 8 RTLS reader (ZR) and 1 RTLS reader (ZC), the distribution of parking spaces and RTLS reader (ZR), as shown in Fig. 3. The original data of the localization algorithm is RTLS (ZR) receiver RTLS tag

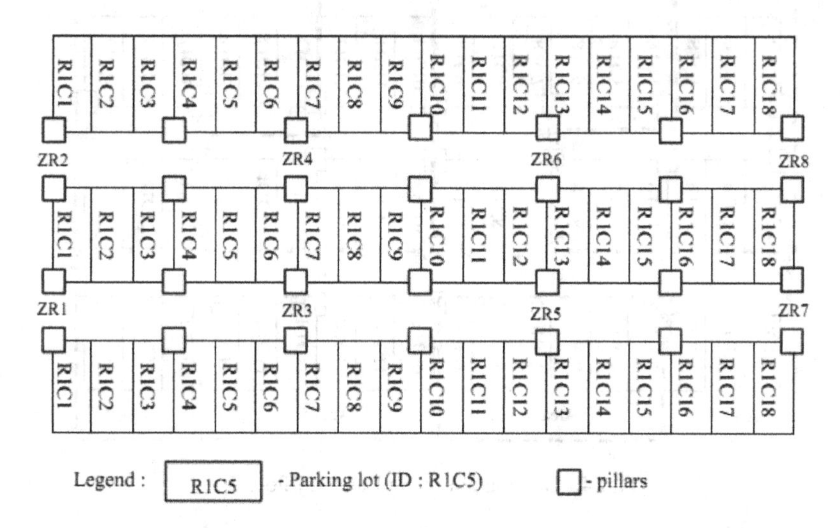

Fig. 3. The distribution of the third floor parking space and RTLS reader

signal strength RTLS, so the RSSI reader (ZR) distribution position and location algorithm, the installation position is shown in Fig. 3; the RTLS reader (ZC) is a ZigBee signal RSSI, so its distribution is independent of the location algorithm, the RTLS reader (ZC) is installed in ZR1 and ZR3.

The RTLS location algorithm is based on the combination method of the non-ranging technology and the method of scene analysis, which is based on the average value of 8 RTLS reader (ZR) in a certain period of time (1 min) to receive the signal intensity of RTLS tags RSSI average, square difference and upload data.

The positioning algorithm can be described as follows. First, the parking lot is divided into 8 areas, as shown in Fig. 4, according to the average value of the signal intensity of RTLS tag. At present, the RTLS expression can be accurately located in Area. After that, in each area, three spaces are found between the two posts in the same row, according to a large number of measurement data is obtained in Table 1. Table 1 indicates the judging conditions of different blocks within each block, and when the vehicle is stopped in accordance with Table 1. The RTLS tag can be positioned exactly to the block now. Finally, according to the probability density function of the stationary random process and the formula (4), the probability of the same block in the same block is calculated, and the maximum of the probability of the maximum of the three spaces is fixed.

$$\mathrm{F} = e^{(E_{x1}-E_{ZR1})^2/2V_{ZR1}^2} * e^{(E_{x2}-E_{ZR2})^2/2V_{ZR2}^2} * \ldots * e^{(E_{xS}-E_{ZRS})^2/2V_{ZRS}^2} \quad (4)$$

E_{x1}, E_{x2}, …, E_{x8} is the mean of $RSSI_{ZR1}$, $RSSI_{ZR2}$, … $RSSI_{ZR8}$. E_{ZR1}, E_{ZR2}, …, E_{ZR8} is the mean of $RSSI_{ZR1}$, $RSSI_{ZR2}$, …, $RSSI_{ZR8}$. V_{ZR1}, V_{ZR2},…, V_{ZR8} is the variance of $RSSI_{ZR1}$, $RSSI_{ZR2}$, …, $RSSI_{ZR8}$ (Fig. 5).

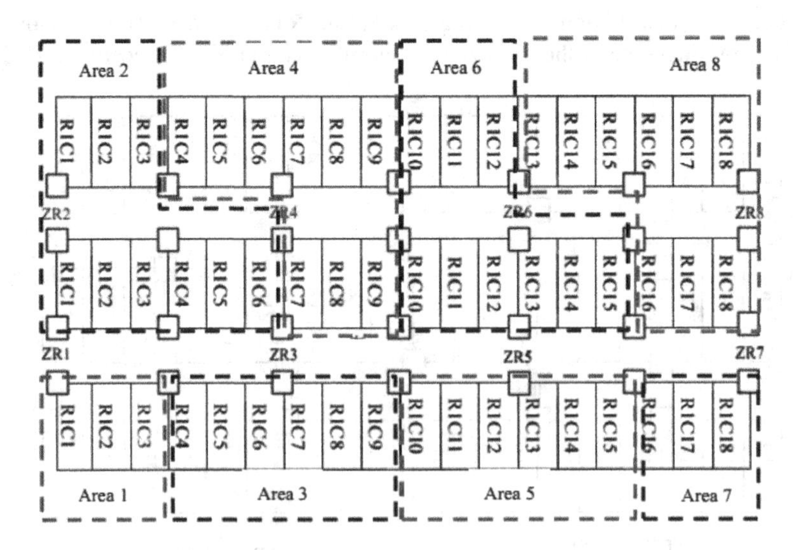

Fig. 4. Distribution map of area on floor 3

Table 1. Block Judgement on floor 3, $RSSI_{ZRX}$ is the reference value of RSSI(ZRX) which is between RTLS tag and RTLS reader, The range of $RSSI_{ZRX}$ values is 0 to 255

Area	Block	Judgement
Area 1	Block 1	Only one
Area 2	Block 1	$RSSI_{ZR1} > RSSI_{ZR3}$ & $RSSI_{ZR4} - RSSI_{ZR3} \leq 5$
	Block 2	$RSSI_{ZR1} \leq RSSI_{ZR3}$
	Block 3	$RSSI_{ZR1} > RSSI_{ZR3}$ & $RSSI_{ZR4} - RSSI_{ZR3} > 5$
Area 3	Block 1	$RSSI_{ZR1} > RSSI_{ZR5}$
	Block 2	$RSSI_{ZR1} \leq RSSI_{ZR5}$
Area 4	Block 1	$RSSI_{ZR2} \geq RSSI_{ZR6}$
	Block 2	$RSSI_{ZR2} < RSSI_{ZR6}$ & $RSSI_{ZR2} - RSSI_{ZR1} > 5$
	Block 3	$RSSI_{ZR2} < RSSI_{ZR6}$ & $RSSI_{ZR2} - RSSI_{ZR1} \leq 5$
Area 5	Block 1	$RSSI_{ZR3} > RSSI_{ZR7}$
	Block 2	$RSSI_{ZR3} \leq RSSI_{ZR7}$
Area 6	Block 1	$RSSI_{ZR4} > RSSI_{ZR8}$ & $RSSI_{ZR8} - RSSI_{ZR7} \leq 5$
	Block 2	$RSSI_{ZR4} \leq RSSI_{ZR8}$
	Block 3	$RSSI_{ZR4} > RSSI_{ZR8}$ & $RSSI_{ZR8} - RSSI_{ZR7} > 5$
Area 7	Block 1	Only one
Area 8	Block 1	$RSSI_{ZR6} \geq RSSI_{ZR8}$
	Block 2	$RSSI_{ZR6} < RSSI_{ZR8}$ & $RSSI_{ZR6} - RSSI_{ZR5} > 5$
	Block 3	$RSSI_{ZR6} < RSSI_{ZR8}$ & $RSSI_{ZR6} - RSSI_{ZR5} \leq 5$

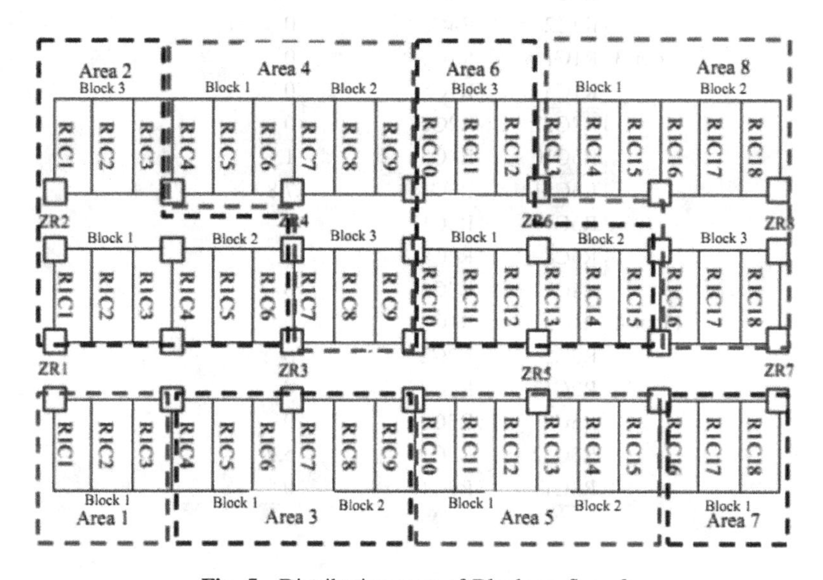

Fig. 5. Distribution map of Block on floor 3

5 Test Results

In the indoor parking lot after the layout of the RTLS, the installation of the RTLS tag taxi positioning function for the sample test. According to the number of parking spaces in each layer of the sampling test of at least 10 % of the parking spaces, the 2 layer of the 10 sampling inspection of 3 parking spaces, the 4 layer of 8 sampling inspection of 6 spaces, the test results are shown in Table 2. The test starts to get the results of the location for 1 min.

The experimental results show that the positioning accuracy of the vehicle can reach higher accuracy in the system, and the accuracy is 100 %. The accuracy is 83.3 % and the accuracy is 13 in the range of 7.8 m.

Table 2. Result of sampling test on floor 3

Floor	Position ID	Positioning result	deviation (m)
Floor 2	R1C2	R1C2	0
	R1C9	R1C9	0
	R1C11	R2C8	7.8
	R1C20	R1C20	0
	R2C2	R2C2	0
	R2C8	R2C8	0
	R2C20	R2C20	0
	R3C5	R3C5	0
	R3C21	R4C14	21.3
	R4C2	R4C2	0
Floor 3	R1C1	R1C1	0
	R1C5	R1C5	0
	R2C17	R2C17	0
	R3C10	R3C5	13
	R3C11	R3C14	7.8
	R3C17	R3C17	0
	R4C8	R4C8	0
	R4C11	R4C11	0
Floor 4	R1C3	R1C3	0
	R2C2	R2C2	0
	R2C17	R2C17	0
	R3C8	R3C8	0
	R3C12	R3C14	5.2
	R3C17	R3C17	0

6 Conclusion

In this paper, a real time positioning technology (RTLS) is proposed, which realizes the real-time positioning of the vehicle in the indoor vehicle networking system. RTLS system uses ZigBee as the communication technology, received signal strength RSSI as the ranging technology, the positioning algorithm using the proximity of the non-ranging technology and the combination of scene analysis method. The actual situation of the system is built, and the algorithm of the vehicle is modified and tested. The test sample is randomly generated, can cover the entire test environment, the test results are representative and feasible. The experimental results show that the positioning accuracy of the vehicle can reach higher accuracy in the system, and the accuracy is 100 %. The accuracy is 83.3 % and the accuracy is 13 in the range of 7.8 m. In this paper, the real-time positioning technology (RTLS) is used to solve the problem of real-time positioning of the RTLS tag in the test environment.

References

1. Wang, F.-B., Shi, L., Ren, F.-Y.: Self-localization systems and algorithms for wireless sensor networks. J. Softw. **16**(5), 857–866 (2006)
2. Saltzer, J., Reed, D., Clark, D.: End-to-End arguments in system design. ACM Trans. Comput. Syst. **2**(4), 195–206 (1984)
3. Ren, F.Y., Huang, H.N., Lin, C.: Wireless sensor networks. J. Softw. **14**(2), 1148–1157 (2003)
4. Harter, A., Hopper, A.: A distributed location system for the active office. IEEE Netw. **8**(1), 62–70 (1994)
5. Harter, A., Jones, A., Hopper, A.: A new location technique for the active office. IEEE Pers. Commun. **5**(4), 42–47 (1997)
6. Qian, Z.-H., Zhu, S., Wang, X.: An cluster-based zigbee routing algorithm for network energy optimization. Chin. J. Comput. **36**(3), 485–493 (2013)
7. Bluetooth[S/0L], http://ww.bluetooth.org
8. Xia, J., Ye, S., Liu, Y., Zhao, L.: Wi-Fi assisted GPS positioning with fixed geodetic height. Geomatics Inf. Sci. Wuhan Univ. **39**(1), 52–55 (2014)
9. Liu, Q., Yao, Y.-B., Zhang R., Wang, J.: A discussion of pulsar autonomous navigation positioning and timing for spacecraft. GNSS World of China **34**(2), (2009)
10. Cao, G.: Research on RTLS technology application and standard system. Inf. Technol. Stand. **39**(6), 38–42 (2012)
11. Zhan, J., Liu, H.-L., Liu, S.-G., Zhu, F.: The study of dynamic degree weighted centroid localization algorithm based on RSSI. Acta Electronica Sinica **39**(1), 82–88 (2011)
12. Chen, W., Li, W., Shou, H., Yuan, B.: Weighted centroid localization algorithm based on rssi for wireless sensor networks. J. Wuhan Univ. Technol. (Transportation Science & Engineering) **30**(2), 265–268 (2006)

Vehicular Cloud Serving Systems with Software-Defined Networking

Yao-Chung Chang[1(✉)], Jiann-Liang Chen[2], Yi-Wei Ma[3], and Po-Sheng Chiu[2]

[1] Department of Computer Science and Information Engineering, National Taitung University, Taitung, Taiwan
ycc@nttu.edu.tw

[2] Department of Electrical Engineering, National Taiwan University of Science and Technology, Taipei, Taiwan

[3] China Institute of FTZ Supply Chain, Shanghai Maritime University, Pudong, China

Abstract. This work proposes an SDN architecture that helps cloud customers receive highly efficient services in a vehicular cloud serving system. An application called SDNBroker is designed herein and superposed on the Software-Defined Networking (SDN) controller using a Northbound Application Programming Interface (Northbound API). When a customer requests a service, the developed SDNBroker schedules cloud serving resources using the linear programming algorithm and optimizes network routing using Dijkstra's algorithm. Three cloud-serving systems and three applications were utilized to validate SDNBroker. The proposed SDNBroker system is 5 % ~ 15 % more efficient than the specified cloud serving system which is a NonSDNBroker system with single routing path. Furthermore, the network link utilization of the proposed route procedure is increased from that of the specified cloud serving system (NonSDNBroker system with single routing path) by approximately 5.8 %.

Keywords: Software-Defined networking (SDN) · Broker · Vehicular cloud serving system · Linear programming algorithm · Dijkstra's algorithm

1 Introduction

Rapid development of cloud computing systems has facilitated service applications that no longer reside in a single domain, but are dispersed among many domains, and provided using virtualization and distributed computing technology [1]. Most cloud customers order cloud services from specific cloud providers and the service quality is fixed by the ordered cloud systems [2, 3]. However, an application can be implemented in many cloud systems. This work proposes a broker mechanism in vehicular cloud serving systems to help cloud customers obtain highly efficient services. A cloud broker application, called SDNBroker, is developed for scheduling cloud resources and planning network routing to provide highly efficient services. The SDNBroker is based on Software-Defined Networking (SDN) operations and superposed on the SDN controller device [4]. Highly efficient services are scheduled using the linear programming algorithm and Dijkstra's algorithm. The proposed SDNBroker system is more efficient than

C.-H. Hsu et al. (Eds.): IOV 2015, LNCS 9502, pp. 58–67, 2015.
DOI: 10.1007/978-3-319-27293-1_6

the specified cloud serving system with NonSDNBroker system. Furthermore, the network link utilization of the proposed route procedure is increased from that of the NonSDNBroker system.

2 Related Works

A heterogeneous vehicular cloud serving architecture can support a novel telematics application. Figure 1 shows the heterogeneous vehicular cloud serving system concept. Developers must design schemes that optimize the use of architectural and deployment paradigms to maximize the benefits that are provided by heterogeneous cloud computing. The proposed a Heterogeneous Vehicular Serving System with SDNBroker architecture comprises a "highly efficient" service scheme that incorporates vehicular cloud serving systems, an optimal resource allocation strategy and an SDN platform; these are related to SDN technology, cloud serving systems, SDN VANET, Dijkstra's routing algorithm and existing mechanisms, which are described below.

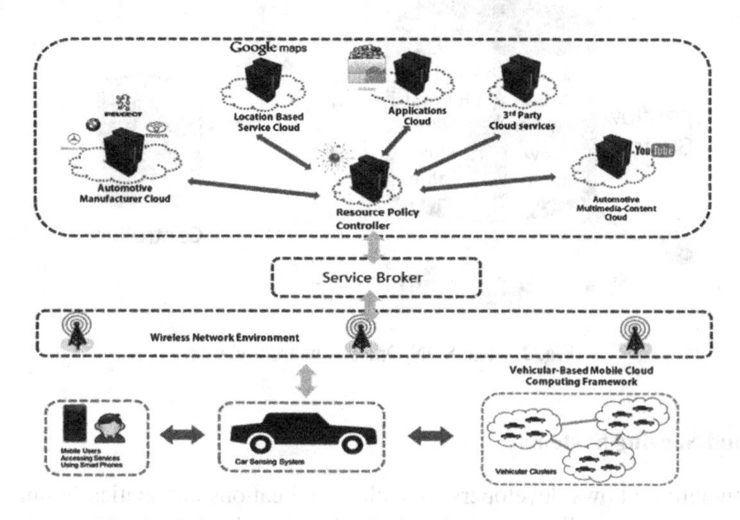

Fig. 1. A heterogeneous vehicular serving system concept

2.1 Software-Defined Networking

Conventional network architectures cannot meet the demands of firms, operators or end users. SDN increases the operational efficiency of cloud data centers, motivating Internet service providers, telecommunications operators and many research centers to develop related technologies to address more effectively current network problems by using SDN technology. SDN is an innovative networking method with tremendous potential, it separates data and control planes. In the SDN architecture, the network intelligence and state are logically centralized, and the underlying network infrastructure is abstracted from related applications. Consequently, characterized by its programmability, automation and network control, SDN enables operators to construct highly scalable, flexible networks that adapt readily to fluctuating network requirements. Based on SDN architecture, OpenFlow is a communication protocol for establishing forward tables in SDN-like switches, as presented in Fig. 2 [5]. Using the abstraction layer of the switch, OpenFlow defines the forward tables to achieve customization. A secure channel between a switch and a controller is used to manage and control SDN messages. The switch follows the rules in flow tables and the group table to look up or forward packets. These flow entries in flow tables can be added, deleted and updated by the controller. When arriving at the switch, a flow is matched against flow entries in the flow tables. Action is triggered when a flow is matched with flow entries in flow tables. If the packet does not match any entry in the flow tables, it may be sent to the controller over a secure channel.

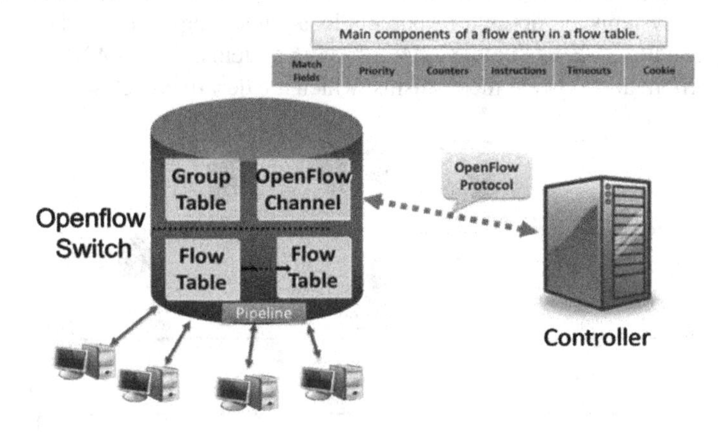

Fig. 2. The SDN/OpenFlow architecture

2.2 Cloud Serving Systems

Cloud computing allows developers to deploy applications automatically during task allocation and storage distribution by using distributed computing technologies on several servers [6, 7]. The cloud computing architecture is composed of three layers: The bottom layer, infrastructure as a service, is service-oriented. The middle layer, platform as a service, is a service platform that allows developers to deploy their own applications. The top layer, software as a service, allows users to access services in manner that meets their requirements. By using virtual devices, cloud computing reduces server power cost and minimizes hardware cost. Users can access multiple data storage databases from various cloud serving systems for the enormous number of data. Heterogeneous cloud computing supports the on-demand assignment of cloud resources to satisfy the SLA(Service Level Agreement) of cloud computing [8, 9]. Providers of a heterogeneous cloud computing service can combine an integrated cloud computing framework device to respond to "highly efficient" service requests.

2.3 SDN VANET

Vehicular Cloud Computing (VCC) is one of the solutions that are deployed to maintain and promote the Intelligent Transportation Systems (ITS) [10]. VCC emerges from the convergence of powerful vehicle resources, advances in network mobility, ubiquitous sensing and cloud computing [11]. Furthermore, the concept of SDN-based VANET architecture and operational mode adapt SDN to VANET environments are demonstrated in [12]. Software-Defined Networking (SDN) can be used to provide the flexibility and programmability to VANET. Besides, SDN VANET introduces new services and features to traditional VANET. Benefits of a SDN VANET are also presented in [12]. Moreover, a novel roadside unit (RSU) cloud for the Internet of Vehicles (IoV) is proposed in [13]. The proposed RSU cloud consists of traditional and specialized RSUs employing SDN to dynamically instantiate, replicate, and migrate services.

2.4 Dijkstra's Routing Algorithm

Dijkstra's algorithm is a link-state algorithm, in which a graphical map of the network provides the basic data that are used for each node. To generate its map, each node floods the entire network with information (such as throughput or delay) about the other nodes to which it can connect. Each node then independently assembles this information into a routing table. Using this table, each router independently determines the least-cost path from itself to every other node using a shortest paths algorithm. The result is a tree graph that is rooted at the current node, through which the path from the root to any other node is the least-cost path to that node. From this tree is constructed the routing table, which specifies the best next hop from the current node to any other node.

2.5 Existed Mechanisms

The feasibility of integrating network and computing resources has been widely examined, especially in grid computing. For most meta-computing systems, scheduling is not a single problem but a set of problems, owing to the various requirements and many characteristics of the various clouds of resources. This reference [14] designs utility-oriented federated cloud systems that can thoroughly elucidate application service behavior associated with intelligent down-scaling and up-scaling infrastructures. Unfortunately, users must address most federation-related issues before using their services, making the inter-cloud concept inapplicable to user scenarios. This reference [15] designed a scheme that supports the live mobility of virtual machines between clouds, while meeting the cloud insularity requirements of autonomy, privacy and security. Current methods fail to support policies for dynamically coordinating load distribution among cloud serving systems to optimize the cloud for hosting application services with high efficiency. In this work, an SDNBroker architecture that can satisfy the SLA of a cloud serving system and ensure the Quality-of-Service of service networking is designed.

3 SDNBroker Architecture

The developed SDNBroker is an SDN application, which communicates with an SDN controller using a Northbound API. The SDNBroker is initially activated to schedule the available cloud resources when a cloud customer requests a service. The scheduled plan is then sent to the Resource Allocation Policy module of the SDN controller. The Policy Module sends the new policy information to the Translation Unit and the Forwarding Unit, and then the SDN switches to update the Flow Tables of all of the SDN switches. The updated Flow Tables to the "highly efficient" cloud serving system. In this work, "highly efficient" refers to the minimal flow completion time (FCT) or the maximal relative FCT improvement given a limit on the cloud rental fee. Figure 3 the architecture of SDNBroker operation.

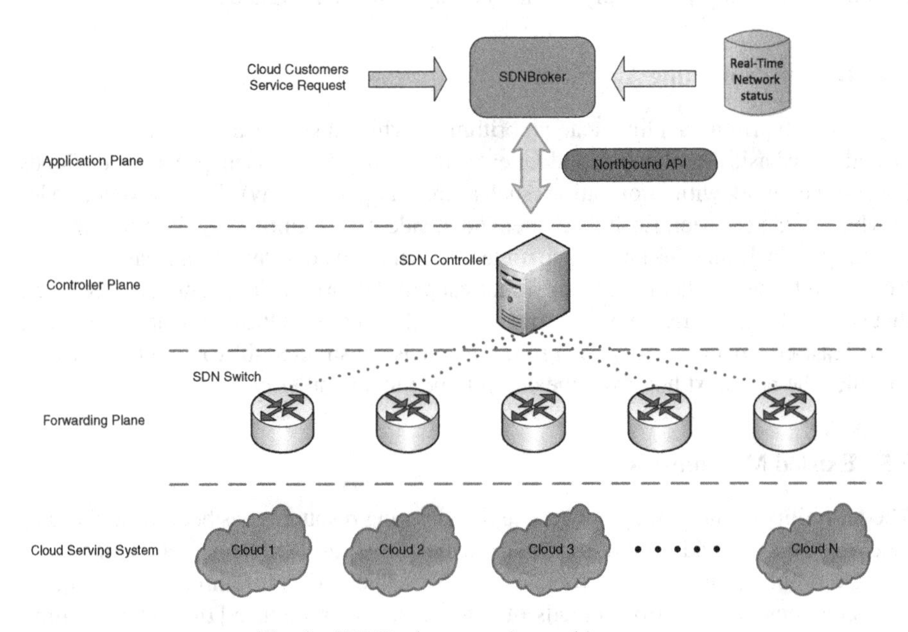

Fig. 3. SDNBroker operation architecture

The SDNBroker enables a scheduler that is based on the Linear Programming algorithm to generate a service plan for cloud customers. Figure 4 shows the algorithm and flow chart of the developed scheduler. The procedure generates the "highly efficient" service plan, which is then sent to the SDN controller to update the flow policy. Finally, the Flow Tables of all SDN switches are updated according to the flow policy that has been received from the SDN controller.

The scheduler is based on the linear programming algorithm for finding the local maxima and minima of a function subject to constraints. In this work, Eq. (1) defines the problem of optimizing the SLA. S_{i1} and S_{i2} are the relative SLAs per CPU unit and per storage unit of the cloud i. F_{i1} and F_{i2} are the improved flow completion times per CPU unit and per storage unit of the cloud i. X_{i1} and X_{i2} are the values that represent the resources allocated to the CPU and storage in cloud i. C_{i1} and C_{i2} are the rental fees for each CPU unit and each storage unit in cloud i. C is the total budget of the customer. F is the required minimal improved flow completion time. n is the number of cloud systems. If SLA is a maximum of a function, then there exists X_{i1} and X_{i2} that are stationary points in for the linear program.

$$\text{Maximum SLA} = \sum_{1}^{n} (S_{i1}X_{i1} + S_{i2}X_{i2}) \tag{1}$$

Subject to:

1. $\sum_{1}^{n} (C_{i1}X_{i1} + C_{i2}X_{i2}) \leq C$
2. $\sum_{1}^{n} (Fi1Xi1 + Fi2Xi2) \geq F$
3. $Xi1 \geqq 0$
4. $Xi2 \geqq 0$

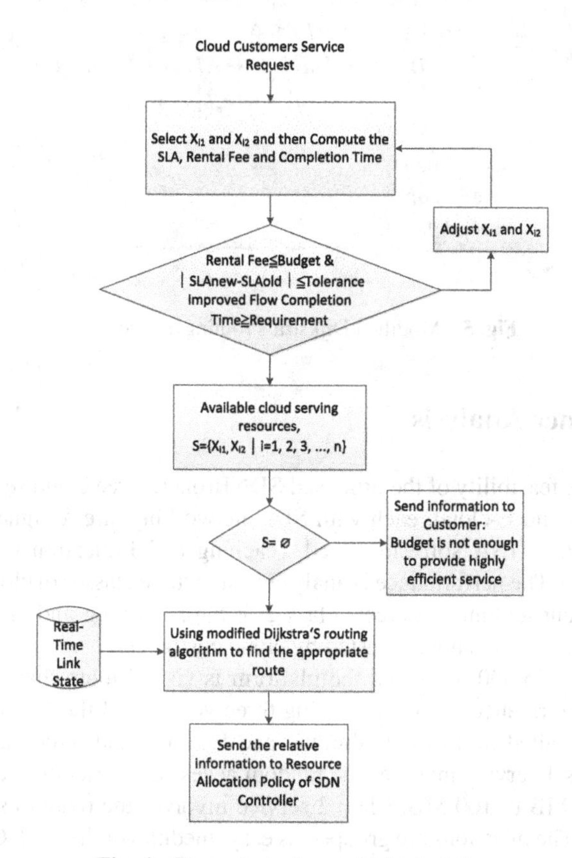

Fig. 4. Flow chart of proposed scheduler

Figure 5 shows the pseudo code of the modified Dijkstra's routing algorithm that is used in the SDNBroker. The algorithm is designed to determine the best route path that supports a highly efficient service. The modified Dijkstra's routing algorithm considers two SLA-related performance metrics throughput (T) and latency (L) in selecting the routing path.

Algorithm Modified Dijkstra

Require: $G = (V, E), s, T, w(e, t), \Delta$

$Q \Leftarrow V$

$L[s] \Leftarrow 0, L[v] \Leftarrow \infty \ \forall \ v \in V$ s.t. $v \neq s$

while $Q \neq \emptyset$ ***do***

  $u \Leftarrow x \in Q$ s.t. $L[x] = min_{y \in Q} L[y]$

  $u \Leftarrow Q/\{u\}$

  for all $e \in E$ s.t. $e = (u, v)$ ***do***

    if $w(e, L[u] + T) < \Delta$

      if $L[v] > (L[u] + w(e, L[u] + T))$ ***then***

        $L[v] \Leftarrow L[u] + w(e, L[u] + T)$

      end if

    end if

  end for

end while

Fig. 5. Modified Dijkstra's routing algorithm

4 Performance Analysis

To investigate the feasibility of the proposed SDNBroker, three cloud serving systems, T-cloud, V-cloud and E-cloud, each with SDN networking, are designed herein. The test platform is the ITSA software-based Teaching and Education Cloud Platform (www.itsa.org.tw). The performance is analyzed using three classes of cloud application (class 1: access curriculum resources, class 2: e-tutor learning and class 3: creative community project), which are requested by cloud customers.

ITSA has over 35,000 users and the platform is visited more than 100,000 times monthly. The performance is analyzed using three versions of the ITSA platform that are separately installed in T-cloud, V-cloud and E-cloud, and three classes of cloud application. Class 1 service involves the random access of curriculum resources and a flow size from 1 MB to 100 MB. Class 2 service involves the random selection of an e-tutor question. The questions are grouped as easy, medium or difficult. Class 3 service involves the random generation of a project and a discussion thereof within community. Table 1 presents information about the applications and issues related to cloud resources.

Table 1. Relationship between application and rental fee

Rental Fee \ Application	Curriculum Resources	e-tutor Learning	Creative Community Project
Storage	X	X	V
Computing	V	V	V
Networking	V	V	V

"V" denotes the kind of rental fee is require

This work compares the performance of SDNBroker with that of non-SDNBroker for three network routing mechanisms the single path routing algorithm, Dijkstra's routing algorithm and the modified Dijkstra's routing algorithm. Experimental results show the percentage improvement in the SLA that is achieved using various service classes. The performance metrics for each scheme are Maximum, Minimum and Average. As the result, the proposed mechanism of the class 2 service outperforms the non-SDNBroker mechanism by approximately 15 % (Average) because the latter requires more cloud computing resources. However, when the cloud service accesses only simple data, the performance each scheme is slightly improved by approximately 5 %.

Figure 6 shows the network route utilization. With the MDRA scheme, the network loading is more distributed than with the DRA or the SPRA scheme. Furthermore, with the MDRA scheme, the link utilization averages 75.2 % while with the DRA scheme, the link utilization averages 68.8 % and with the SPRA scheme, it averages 64.4 %.

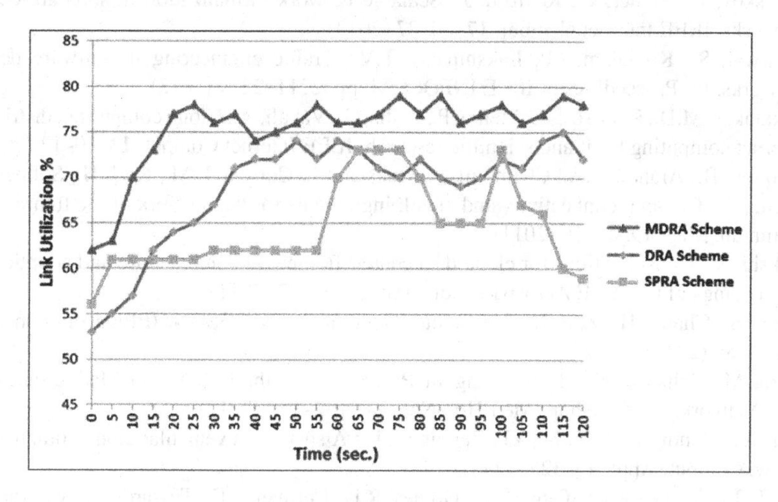

Fig. 6. Network route utilization

5 Conclusion

In this work, an SDNBroker system that is based on SDN networking, vehicular cloud serving systems, the linear programming algorithm and Dijkstra's routing algorithm, is designed. This system provides "highly efficient" service in a vehicular cloud serving system. In this version of SDNBroker, "highly efficient" refers to the maximization of the SLA. Only a cloud is used to provide a service to a customer in this stage. The next version of SDNBroker will consider economic dispatch suing several simultaneously serving systems.

Acknowledgment. The authors would like to thank the Ministry of Science and Technology Republic of China, Taiwan for financially supporting this research under Contract No. MOST 104-2221-E-143-004.

References

1. Papagianni, C., Leivadeas, A., Papavassiliou, S., Maglaris, V., Pastor, C.C., Monje, A.: On the optimal allocation of virtual resources in cloud computing networks. IEEE Trans. Comput. **62**, 1060–1071 (2013)
2. Li, A., Yang, X., Kandula, S., Zhang, M.: CloudCmp: comparing public cloud providers. In: Proceedings of the 10th ACM SIGCOMM Conference on Internet Measurement, pp. 1–14 (2010)
3. Ngan, L.D., Kanagasabai, R.: OWL-S Based semantic cloud service broker. In: Proceedings of IEEE 19th International Conference on Web Services, pp. 560–567 (2012)
4. Drutskoy, D., Keller, E., Rexford, J.: Scalable network virtualization in software-defined networks. IEEE Internet Comput. **17**, 20–27 (2013)
5. Agarwal, S., Kodialam, M., Lakshmanm T.V.: Traffic engineering in software defined networks. In: Proceedings of IEEE INFOCOM, pp. 2211–2219 (2013)
6. Dikaiakos, M.D., Katsaros, D., Mehra, P., Pallis, G., Vakali, A.: Cloud computing: distributed internet computing for it and scientific research. IEEE Internet Comput. **13**, 10–13 (2009)
7. Ahlgren, B., Aranda, P.A., Chemouil, P., Oueslati, S., Correia, L.M., Karl, H., Sollner, M., Welin, A.: Content, connectivity and cloud: ingredients for the network of the future. IEEE Commun. Mag. **49**, 62–70 (2011)
8. Bakshi, K.: Considerations for cloud data center: framework, architecture and adoption. In: Proceedings of the IEEE Aerospace Conference, pp. 1–7 (2011)
9. Chen, S., Chang, H., Zou, X.: Inter-cloud Operations via NGSON. IEEE Commun. Mag. **50**, 82–89 (2012)
10. Gerla, M.: Vehicular cloud computing. In: Proceedings of the 11th Annual Mediterranean Ad Hoc Networking Workshop (Med-Hoc-Net), pp. 152–155 (2012)
11. Md, W., Mehdi, S., Abdullah, G., Rajkumar, B.: A survey on vehicular cloud computing. J. Netw. Comput. Appl. **40**, 325–344 (2014)
12. Ku, I., Lu, Y., Gerla, M., Ongaro, F., Gomes, R.L., Cerqueira, E.: Towards software-defined vanet: architecture and services. In: Proceedings of 13th Annual Mediterranean Date of Conference Ad Hoc Networking Workshop (MED-HOC-NET), pp. 103–110 (2014)
13. Mohammad, A.S., Ala, A.F., Mohsen, G.: Software-defined networking for rsu clouds in support of the internet of vehicles. IEEE Internet Things J. **2**, 133–144 (2015)

14. Buyya, R., Ranjan, R., Calheiros, R.N.: InterCloud: utility-oriented federation of cloud computing environments for scaling of application services. In: Hsu, C.-H., Yang, L.T., Park, J.H., Yeo, S.-S. (eds.) ICA3PP 2010, Part I. LNCS, vol. 6081, pp. 13–31. Springer, Heidelberg (2010)
15. Sotiriadis, S., Bessis, N., Antonopoulos, N.: Towards Inter-cloud Schedulers: A Survey of Meta-scheduling Approaches. In: Proceedings of the 2011 International Conference on P2P, Parallel, Grid, Cloud and Internet Computing, pp. 59–66 (2011)

Temporal Centrality Prediction in Opportunistic Mobile Social Networks

Huan Zhou[1], Shouzhi Xu[1](✉), and Chungming Huang[2]

[1] College of Computer and Information Technology, China Three Gorges University, Yichang, China
zhouhuan117@gmail.com, xsz@ctgu.edu.cn

[2] Department of Computer Science and Information Engineering, National Cheng Kung University, Tainan, Taiwan
huangcm@locust.csie.ncku.edu.tw

Abstract. In this paper, we focus on predicting nodes' future importance under three important metrics, namely betweenness, and closeness centrality, using real mobility traces in Opportunistic Mobile Social Networks (OMSNs). Through real trace-driven simulations, we find that nodes' importance is highly predictable due to natural social behaviour of human. Then, based on the observations in the simulation, we design several reasonable prediction methods to predict nodes' future temporal centrality. Finally, extensive real trace-driven simulations are conducted to evaluate the performance of our proposed methods. The results show that the Recent Uniform Average method performs best when predicting the future Betweenness centrality, and the Periodical Average Method performs best when predicting the future Closeness centrality in the MIT Reality trace. Moreover, the Recent Uniform Average method performs best in the Infocom 06 trace.

1 Introduction

Recently, with the rapid proliferation of wireless portable devices (e.g., smartphones, ipad, PDAs) with bluetooth or wi-fi, Opportunistic Mobile Social Networks (OMSNs) begin to emerge [1–3]. Opportunistic Mobile Social Networks combine opportunistic networks and social networks together. Previous studies have shown that the performance of such networks depends highly on the user's social behavior as opportunistic networks and social networks share many common characteristics. Their common features motivate an increasing research interests in OMSNs, especially using the social network analysis technology to help the design of routing protocols [4].

Previous studies in OMSNs have proposed diverse metrics to measure the relative importance of a node in networks such as betweenness centrality, and closeness centrality [5]. However, when calculating such centrality metrics, the current studies focused on analyzing static networks that do not change over time or using aggregated contact information over a period of time. Actually, nodes in OMSNs are inherently dynamic, which is driven by natural social behaviour

C.-H. Hsu et al. (Eds.): IOV 2015, LNCS 9502, pp. 68–77, 2015.
DOI: 10.1007/978-3-319-27293-1_7

of human. Therefore, it is not prudent to assume stationary human behavior in the design of practical applications. In particular, many researchers have also observed that nodes in OMSNs have a regular pattern [6,9]. For example, a student will go to school with his neighbours at morning every day, and have classes with his classmates in the classroom, which brings a regular pattern of physical contacts, etc., which in turn provides the periodicity seen in the underlying communication processes. Here, we make a simple assumption: since a node's schedule is regular, if it is an important node in the network at current time, then it is highly possible that its importance in the future will be correlated with the importance at current time.

In this paper, we therefore focus on predicting nodes' future importance under two important metrics, namely betweenness, and closeness centrality, using different real mobility traces in OMSNs. Actually, some studies have attempted to capture the dynamic behaviour of human. Tang et al. [7] proposed temporal centrality metrics based on temporal paths in order to measure the importance of a node in a dynamic network. Kim et al. [8] proposed several methods to predict nodes' importance in the future. Our work is similar to the work in [8], but we ignore the lagged time, and only predict the centrality value of a single time window in the network model, which makes the problem more clear and the prediction results more reasonable. Our contributions in this paper are three-folds:

1. Through real trace-driven simulations, we show that nodes' centrality values are highly correlated with their past recent centrality values, and also have periodical behavior at 24 h difference.
2. Based on the observations, we design several intuitive prediction methods, to predict nodes' future temporal centrality.
3. We evaluate the performance of the proposed prediction methods using different real mobility traces, and show that the best-performing prediction functions are more accurate on average than just using the last centrality value.

The remainder of this paper is organized as follows. We present the preliminaries network model in Sect. 2. Section 3 analyzes the correlation between the past and future centrality value, and proposes several methods to predict the future temporal centrality in OMSNs. Extensive simulations are conducted to evaluate the performance of the proposed methods in Sect. 4. At last, we conclude the paper in Sect. 5.

2 Preliminaries

In this section, we first introduce the network model related to this paper, and then define the notation and terminology for centrality metrics. Finally, we introduce the generalized network centrality prediction problem which will be used in the rest of the paper.

2.1 Network Model

We assume that the time during which a network is observed is finite, from t_{start} until t_{end}; without loss of generality, we set $t_{start} = 0$ and $t_{end} = T$. A dynamic network contact graph $G_{0,T} = (V, E_{0,T})$ on a time interval $[0, T]$ consists of a set of vertices V and a set of temporal edges $E_{0,T}$, where stochastic contact process between a node pair u, $v \in V$ on a time interval $[t_s, t_e]$ $(0 \leq t_s \leq t_e \leq T)$ is modeled as an temporal edge $e^{uv}_{t_s,t_e} \in E_{0,T}$.

We divide the time interval $[0, T]$ into fixed discrete time windows $\{1, 2, ..., n\}$. We use $w = \frac{T}{n}$ to denote the size of each time window, expressed in some time unites (e.g., minutes or hours). In other words, a dynamic network can be represented as a series of static graphs at each time, $G_1, G_2, \ldots, G_n$. The notation G_t $(1 \leq t \leq n)$ represents the aggregate graph which consists of a set of vertices V and a set of edges E_t where an edge $e^{uv}_{t_s,t_e} \in E_t$ exists only if a temporal edge $e^{uv}_{t_s,t_e} \in E_{0,T}$ exists between vertices u and v on a time interval such that $t_e \leq wt$ and $t_s > w(t-1)$.

2.2 Network Centrality Measures

Centrality refers to a group of metrics that aim to quantify the "importance" or "influence" of a particular node (or group) within a network. There are several common methods to measure "centrality". In this paper, we only introduce two of them: betweenness, and closeness centrality. Formally, we use the standard definition of the betweenness, and closeness centrality, and the centrality value of a node i can be expressed as follows [5]:

Betweenness Centrality. Betweenness centrality measures the extent to which a node lies on the shortest paths linking other nodes in the network, which is calculated as the proportional number of shortest paths between all node pairs in the network, that pass through a certain node. Betweenness centrality of a certain node i can be expressed as:

$$Betweenness(i) = \sum_{u \neq i, v \neq i, i \in V}^{r} \frac{\delta_{u,v}(i)}{\delta_{u,v}} \tag{1}$$

where $\delta_{u,v}$ is the total number of shortest paths starting from the source node u and the destination node v, and $\delta_{u,v}(i)$ are the number of shortest paths starting from the source node u and the destination node v which actually pass through node i.

Closeness Centrality. Closeness centrality measures the distance a certain node to all other reachable nodes in the network, which is calculated as the average shortest path length between a certain node and all other reachable nodes. Closeness centrality can be regarded as a measure of how long it will take message to spread from a given node to other nodes in the network. Closeness centrality of a certain node i can be expressed as:

$$Closeness(i) = \frac{1}{|V|-1} \sum_{j \neq i, j \in V}^{r} \Delta_{i,j}(i) \qquad (2)$$

where $\Delta_{i,j}$ is the number of hops in the shortest path from node i to node j and V is the set of nodes in the network.

2.3 Centrality Prediction Problem

As shown in Fig. 1, we generalize the problem for centrality prediction in this paper as follows: Given a dynamic network $G_{1,r}$ observed during r past time intervals, we want to predict the average network centrality values of the nodes in the network in the $r+1$ time intervals. Therefore, the purpose of this paper is to propose several prediction methods to minimize the prediction error, and evaluate the impact of different parameters on the performance of the proposed prediction methods. In order to evaluate the effectiveness of the prediction methods, we use $Error(G_t)$ to denote the average error between the guessed centrality values and the true centrality values, which can be expressed as:

$$Error(G_t) = \frac{\sum_{i \in V} | C_t(i) - \hat{C}_t(i) |}{| V |} \qquad (3)$$

where $C_r(i)$ is node i's centrality value such as $Betweenness(i)$, or $Closeness(i)$, and $\hat{C}_t(i)$ to denote the node i's predicted centrality value in G_t.

3 Centrality Prediction

In this section, we first use two real mobility traces, *Infocom 06* [10] and *MIT Reality* [11] to test whether the centrality can be predicted. Then, based on the findings, we introduce several methods to predict the future temporal centrality.

3.1 Analysis of Correlation Between Past and Future Centrality

Note that human in reality always have regularity. Therefore, we hypothesize that the past centrality has high correlation with the future centrality. To test

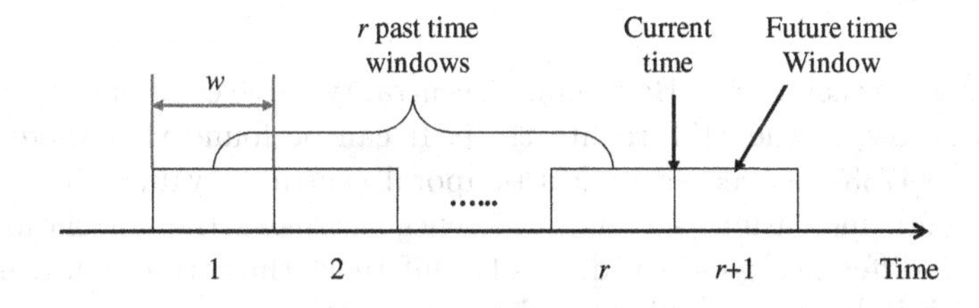

Fig. 1. Illustration of the past and future time windows.

this hypothesis, we use two real mobility traces, *Infocom 06* and *MIT Reality* collected from real environments. Users in these two traces are all carrying Bluetooth-enabled portable devices, which record contacts by periodically detecting their peers nearby. The traces cover various types of corporate environments and have various experiment periods. For simplicity, we only analyze the correlation of Betweenness Centrality between the past and future time intervals.

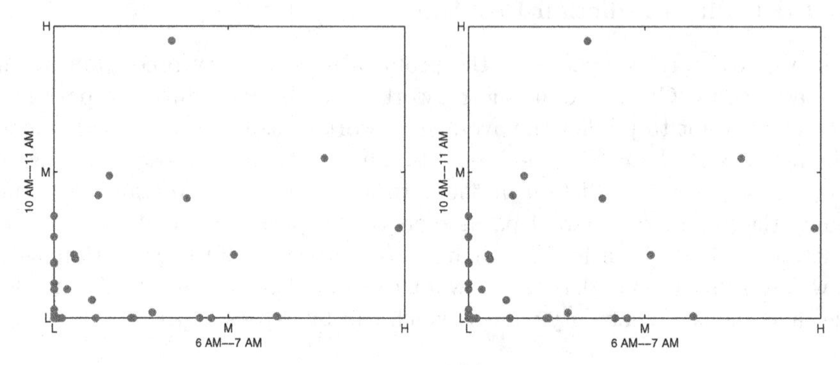

(a) Time difference: -4 hours, Average correlation: 0.4733 (b) Time difference: -8 hours, Average correlation: 0.3407

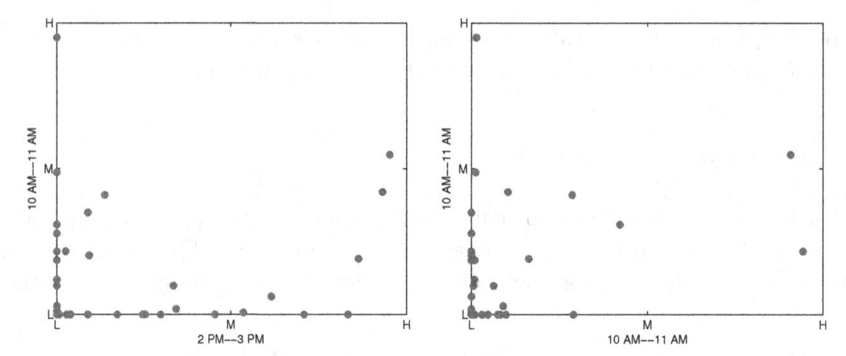

(c) Time difference: -20 hours, Average correlation: 0.2565 (d) Time difference: -24 hours, Average correlation: 0.4155

Fig. 2. Scatter plots depicting correlation between a fixed window (y-axis) and an increasingly distant window from the past (x-axis) every 4 h in the MIT trace

Figure 2 plots each nodes' Betweenness centrality value compared to its value in a past window, in the MIT reality trace. It can be found that there is high correlation (0.4733) between a node's temporal centrality value with its value 4 h ago; second, increasing the time difference decreases the correlation (e.g., 0.3407 at -8 h difference); and third, at 24 h difference the correlation rises again (0.432), which indicates possible periodic behaviour.

Figure 3 shows each nodes' Betweenness centrality value compared to its value in a past window, in the infocom 06 trace. It can be found that similar to the

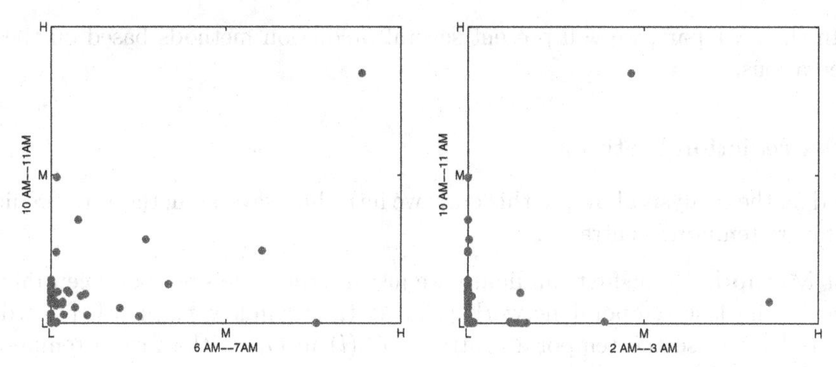

(a) Time difference: -4 hours, Average correlation: 0.5804

(b) Time difference: -8 hours, Average correlation: 0.2451

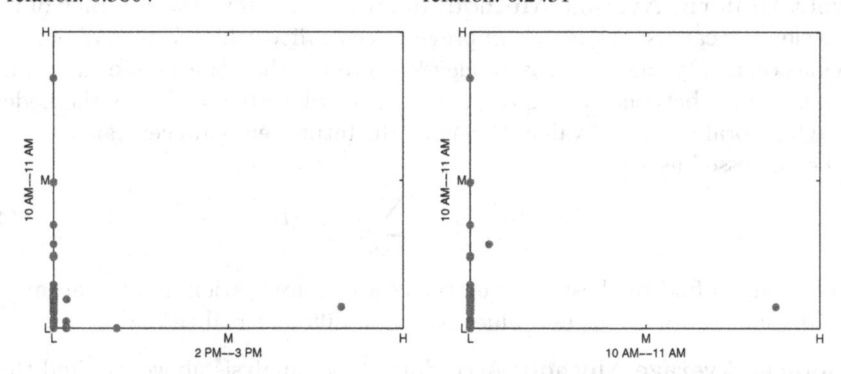

(c) Time difference: -20 hours, Average correlation: -0.0155

(d) Time difference: -24 hours, Average correlation: 0.0254

Fig. 3. Scatter plots depicting correlation between a fixed window (y-axis) and an increasingly distant window from the past (x-axis) every 4 h in the Infocom 06 trace

results in the MIT Realty trace, recent past temporal centrality values are highly correlated compared with more distant values(e.g., 0.5804 at -4 h difference). However, different from the results in MIT Realty trace, the pattern of repeated peaks with -24 h time difference seems rather weak in Infocom 06 trace. The main reason is that people attending the IEEE Infocom 2006 conference are more likely to seek out new colleagues to talk to at the breaks between sessions, rather than socialising with the same people, but students in the MIT campus are more likely to meet the same people when they are taking classes or walking in the campus.

Based on the results reported above, we have two key observations:

1. Recent past centrality values are highly correlated compared with more distant values in one day.
2. A node's temporal centrality value with its value at 24 h difference are highly correlated, which indicates possible periodic behaviour.

In the next part, we will present several prediction methods based on these observations.

3.2 Prediction Methods

Based on the analysis above, in this part, we introduce several methods to predict the future temporal centrality.

Last Method. As the first candidate, we just use the node's temporal centrality value in the last temporal network (G_r) at time window r. In other words, for $i \in V$, we use the temporal centrality $C_r(i)$ in G_r as the future temporal centrality $C_{fu}(i)$ in G_{r+1}.

Recent Uniform Average Method. In order to improve the accuracy of the prediction, we can use the node's m previous centrality values instead of one last previous centrality value. A reasonable idea is to use the node's uniform average centrality value between $G_{r-m+1}, \ldots, G_{r-1}, G_r$ where $0 < m \leq r$ as the node's future temporal centrality value. Formally, the future temporal centrality $C_{fu}(i)$ can be expressed as:

$$C_{fu}(i) = \frac{1}{m} \sum_{k=r-m+1}^{r} C_k(i) \tag{4}$$

We want to find the best m given the cost of computation and the accuracy of prediction, and will suggest values based on different real traces.

Periodical Average Method. According to the analysis above, we find that human activities are repeated periodically. Hence, an intuitive method is to use these periodical patterns to improve the accuracy of the prediction. For human contact network, reasonable periods are a day or week. Given the period p of a day or a week, we consider using the node's periodical average centrality value between $G_{r-m+1}, \ldots, G_{r-1}, G_r$ where $0 < m \leq r$ as the node's future temporal centrality value. Hence, we first have to find periodical time windows of the $(r+1)$-th time window. We define $a = min_{k=r-m+1}^{r}(r+1-k)w \; mod \; p$ as the time window which is the closest to the periodical time window of the $(r+1)$-th time window. Then, we define $f(k)$ as:

$$f(k) = \begin{cases} 1, & kw \; mod \; p \equiv a, \\ 0, & kw \; mod \; p \neq a. \end{cases} \tag{5}$$

Then, the future temporal centrality $C_{fu}(i)$ in G_{r+1} can be expressed as:

$$C_{fu}(i) = \frac{\sum_{k=r-m+1}^{r} f(k) C_k(i)}{\sum_{k=r-m+1}^{r} f(k)} \tag{6}$$

4 Performance Evaluation of Prediction Functions

In this section, we aim to evaluate the performance of the proposed methods above in the Infocom 06 and MIT Reality traces, and find the proper parameter values (e.g., m) of each prediction method at the same time. For each prediction method, as introduced above, we use $Error(G_t)$ to evaluate the prediction accuracy of the proposed methods.

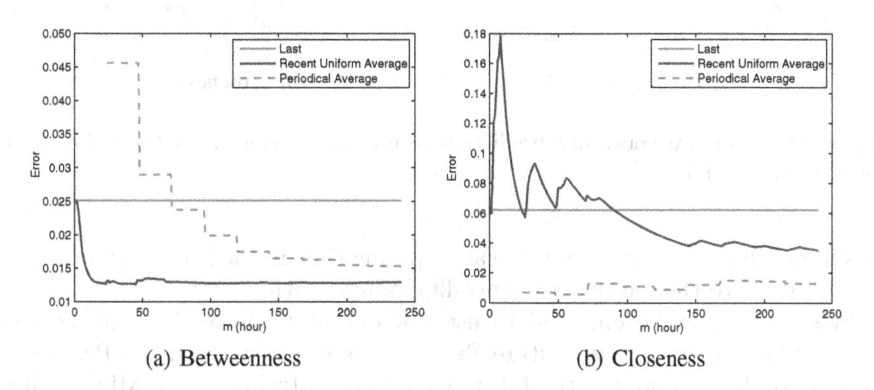

(a) Betweenness (b) Closeness

Fig. 4. The temporal centrality prediction results by varying m in the MIT Reality traces when w is 1 h

Figure 4 shows the temporal centrality prediction results of different prediction methods in the MIT Reality trace. It can be found that the Recent Uniform Average Method achieves the best results in Betweenness, and the Periodical Average Method achieves the best results in Closeness. Therefore, we recommend to use Recent Uniform Average method to predict the future Betweenness centrality, and use the Periodical Average Method to predict the future Closeness centrality in the MIT Reality trace. It is worth noticing that with the increasing of m, the Recent Uniform Average method achieves the minimum $Error(G_t)$ (0.01254) when m is around 45 h in Betweenness, and Periodical Average Method achieves the minimum $Error(G_t)$ (0.00569) when m is 49 h in Closeness. We would not recommend using Last method because its relative accuracy is not enough, although its computation cost is relatively cheap.

Figure 5 shows the temporal centrality prediction results of different prediction methods in the Infocom 06 trace. By contrast, it can be found that with the increasing of m, the Periodical Average method is not as good as that in the MIT Reality trace. Actually, we have already observed that there is no noticeable periodic patterns while the recent past centrality values are highly correlated in the Infocom 06 trace - Fig. 3 illustrates this. Therefore, we recommend to use Recent Uniform Average method to predict the future temporal centrality value in the Infocom 06 trace. Furthermore, the Recent Uniform Average method achieves the minimum $Error(G_t)$ (0.1095 in Betweenness, and 0.1415 in Closeness) when m is 3 h, and then the $Error(G_t)$ values will increase after this time interval.

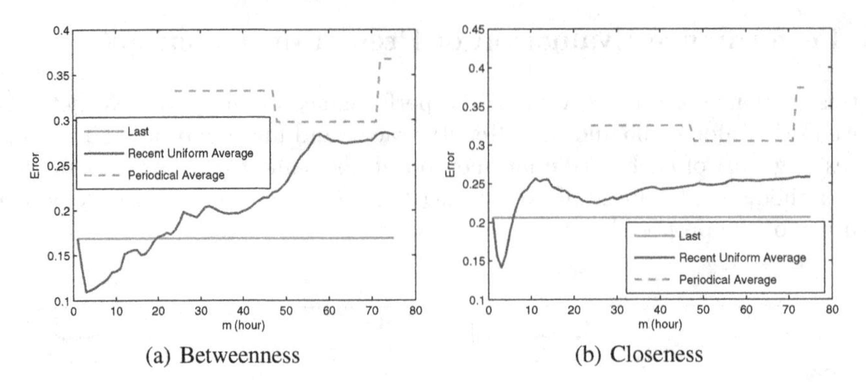

(a) Betweenness (b) Closeness

Fig. 5. The temporal centrality prediction results by varying m in the Infocom 06 traces when w is 1 h

This imply that the centrality value at a specific time in the Infocom 06 trace is highly related to the recent past centrality values in 3 h.

In summary, we recommend to use the Recent Uniform Average method performs best to predict the future Betweenness centrality, and the Periodical Average Method to predict the future Closeness centrality in the MIT Reality trace, while we recommend to use the Recent Uniform Average method to predict the future temporal centrality in the Infocom 06 trace.

5 Conclusions

In this paper, we have predicted nodes' future importance under two important metrics, namely betweenness, and closeness centrality, using real mobility traces in OMSNs. Through real trace-driven simulations, we find that nodes' centrality value are highly correlated with their past recent centrality values, and have periodical behavior at 24 h difference. Then, based on the observations in the simulation, we design several reasonable prediction methods to predict nodes' future temporal centrality. Extensive real trace-driven simulation results show that the Recent Uniform Average method performs best when predicting the future Betweenness centrality, and the Periodical Average Method performs best when predicting the future Closeness centrality in the MIT Reality trace. Moreover, the Recent Uniform Average method performs best in the Infocom 06 trace.

Acknowledgements. This research was supported in part by NSFC under grants 61174177, Hubei Key Laboratory of Intelligent Vision Based Monitoring for Hydroelectric Engineering under grant 2014KLA07, and Natural Science Foundation of Hubei Province of China under grant 2014CFB145. The corresponding author is Shouzhi Xu, Email: xsz@ctgu.edu.cn

References

1. Fan, J., Chen, J., Du, Y., Wang, P., Sun, Y.: DelQue: A socially-aware delegation query scheme in delay tolerant networks. IEEE Trans. Veh. Technol. **60**(5), 2181–2193 (2011)
2. Li, F., Wu, J.: MOPS: Providing content-based service in disruption-tolerant networks. In: IEEE ICDCS (2009)
3. Zhou, H., Zhao, H., Chen, J.: Energy saving and network connectivity tradeoff in opportunistic mobile networks. In: IEEE Globecom (2012)
4. Zhou, H., Chen, J., Zhao, H., Gao, W., Cheng, P.: On exploiting contact patterns for data forwarding in duty-cycle opportunistic mobile networks. IEEE Trans. Veh. Technol. **62**(9), 4629–4642 (2013)
5. Scott, J.: Social network analysis. Sociol. **22**(1), 109–127 (1988)
6. Zhou, H., Chen, J., Fan, J., Du, Y., Das, S.K.: ConSub: incentive-based content subscribing in selfish opportunistic mobile networks. IEEE J. Sel. Areas Commun. **31**(9), 669–679 (2013)
7. Tang, J., Kim, H., Anderson, R.: Temporal node centrality in complex networks. Phys. Rev. E **85**(2), 2181–2193 (2012)
8. Kim, H., Tang, J., Anderson, R., Mascolo, C.: Centrality prediction in dynamic human contact networks. Comput. Netw. **56**(3), 983–996 (2012)
9. Gao, W., Li, Q., Zhao, B., Cao, G.: Multicasting in delay tolerant networks: a social network perspective. In: Mobihoc, pp. 299–308. ACM (2009)
10. Scott, J., Gass, R., Crowcroft, J., Hui, P., Diot, C., Chaintreau, A.: CRAWDAD data set cambridge/haggle (v. 2009–05-29) (2009). http://crawdad.cs.dartmouth.edu/cambridge/haggle
11. Eagle, N., Pentland, A.S., Lazer, D.: Inferring friendship network structure by using mobile phone data. Proc. Nat. Acad. Sci. **106**(36), 15274–15278 (2009)

A Novel Data Sharing Mechanism via Cloud-Based Dynamic Audit for Social Internet of Vehicles

Zhiqiang Ruan[1], Wei Liang[2(✉)], Haibo Luo[1], and Hui Yan[1]

[1] Department of Computer Science, Minjiang University, Fuzhou, China
{rzq_911,rbhappy,yanhui0086}@163.com

[2] College of Software Engineering, Xiamen University of Technology, Xiamen, China
idlink@163.com

Abstract. In this paper, we leverage on the cloud-based data sharing paradigm for vehicles exchange information in Internet of Vehicles (IOV). Vehicles can join different interest groups to form a social network where they can facilitate to modify or share data in the cloud. A basic service for such application is ensuring data integrity and reliability in the presence of hardware/software failures or malicious activities. However, due to the highly dynamic of vehicles, group members may change extremely fast, which poses a great challenge on constructing security related data sharing system. To address this problem, we propose a novel online auditing method that can verify the integrity of shared data and consider signature refreshment by introducing a third party auditor in the cloud. Experiments results demonstrate that the proposed scheme can significantly improve efficiency of auditing tasks compared with the existing vehicle-based countermeasures.

1 Introduction

Nowadays, vehicles can be equipped with advanced technologies that make them easily interconnect and interoperate with surrounding vehicles, which is also referred to as Internet of Vehicles (IOV) [1]. There has been a growing interest in using IOV to build vehicular social network (VSN) where vehicles (including users) can participate in entertainment, public service, and emergency related data exchanges [2, 3].

For example, RoadSpeak [4] is a VSN designed for voice chat application where vehicle users can dynamically form time, location, route, and interest groups to interact messages with owner's group and friends through vehicular ad-hoc networks (VANETs). Kranz et al. [5] describe using vehicle embedded sensors to detect driving activities and later synchronize the status by human readable short texts shared by the owner's groups and mechanic via VANETs, where user has the authority to share the private information to the public or to a selected group.

Although IOV can process and storage large scale data of vehicles, humans, objects, and environments to meet various application services, the heterogeneous network and highly dynamic of vehicles make data sharing between smart things (i.e. vehicle-vehicle) very difficult, on the other hand, cloud computing as a new paradigm for the dynamic provisioning of computing and storage service that are available to vehicles in a high speed topology change.

C.-H. Hsu et al. (Eds.): IOV 2015, LNCS 9502, pp. 78–88, 2015.
DOI: 10.1007/978-3-319-27293-1_8

Specifically, with the storage and sharing services offered by the cloud, both vehicles and users can join an interest group by sharing data with each other. Moreover, group users can not only search and modify shared data, but also update the fresh data with group members. We envision that social IOV based on cloud-based mode would be an integral part of the future Internet of Things (IoT), that is, such application provide efficient and safe travel of vehicle passengers in near real-time while the online data ensure intelligent driving of the vehicles or big data analysis for the transport authorities.

Since shared data is no longer stored on local vehicles but outsourced to the cloud, for security purpose, integrity protection upon shared data is essential in open and public cloud. Plenty of works [6–12] have been proposed to protect the integrity of cloud data, a straightforward way is applied signature on each data segment (i.e. each chat record in RoadSpeak), and the integrity of shared data can be guaranteed unless all the signatures are correct. One common feature of these works is to introduce a third party auditor (TPA) who either could be an independent authority to offer audit services on data integrity for users or a client who want to use cloud data for specific goals (e.g. search, statistics, analysis, etc.). However, most of these works either focus on verifying the integrity of individual data [6–10] or how to protect identity privacy from auditors when verifying the integrity of shared data [11, 12]. Therefore, they are orthogonal to our paper.

In this paper, we consider a social IOV of highly dynamic with vehicles join or leave an interest group extremely fast, which means each time when a vehicle/user leaves the interest group or misbehaves, the signature created by him/her has to be regenerated in order to prevent this user obtain and modify the shared data any further. We argue that the vehicle-based self-verification approach based on literature [13, 15, 16] will incur a large amount of communication and computation cost by downloading, verifying, re-generation, and uploading entire data, especially when the group members join or leave dynamically and the limited resources of vehicle devices (e.g. sensing element). Therefore, such user-centered design (abbreviated by UCD) is unaffordable in the real data sharing services. Conversely, we exploit a variant proxy signature scheme [13] that achieves efficient re-signing on the same data segment signed by the left vehicle, so that he/she can not further involve in data exchange. More importantly, the vehicle can alleviate from heavy computation/communication and thus save energy consumption. In addition, the design properties guarantee that the cloud just on behave of valid users to regenerate a signature of revoked vehicle while keeping data segment nondisclosure to him.

2 System Model for IOV

2.1 Network Model and Security Requirements

The network model in our paper has four entities: the vehicles that consume or provide services/applications of IoV, the third party auditor (TPA), the gateway or access point (AP) inside of the vehicle, and the cloud. Users of cloud can be humans or intelligent sensing devices, where human include the people in vehicles such as drivers and passengers as well as the people in environment of IoV such as pedestrians, cyclists, and driver's family members. The generated data from sensing devices are first processed by the AP, which acts as a gateway between the intra-vehicle network and the road [1]. Furthermore, vehicles can further form different interest group (e.g. based on car model, common routes)) and request for verifying the integrity of shared data through the TPA, which then offers auditing services via a challenge-response protocol, as described in Fig. 1.

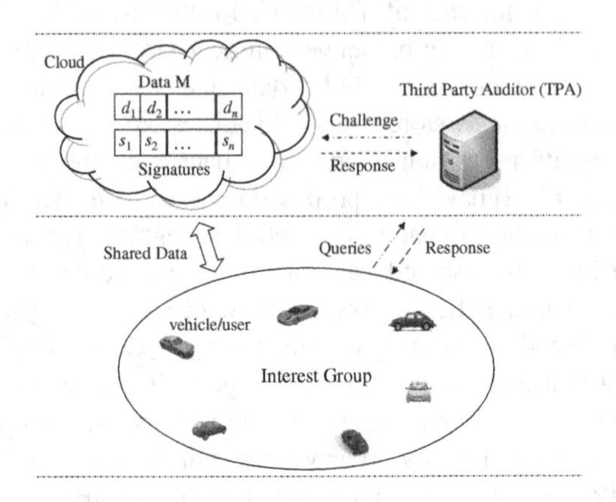

Fig. 1. The cloud-based data sharing model for IOV

In the interest group, each data has an original user, and the shared data is divided into several segments. Group members may modify a segment in shared data by executing basic (insert, update, and delete) operations. We assume the cloud is not a fully trust entity due to the hardware/software failures or malicious attacks, thus each segment is appended with a signature. Clearly, if a user changes a segment, a new signature on this segment must be regenerated with his/her private key. In addition, once a user leaves the interest group either by normal termination or misbehaves, all the signatures he/she created are no longer valid, and the previously produced signatures on data segments by that user have to be regenerated by using a valid user's private key, so that the data segments later can be checked by the public keys of the valid users.

2.2 Design Goals

The objective of this paper can be summarized as follows.

(1) Data Integrity and Correctness: The TPA can on behave of each participator to verify the integrity and correctness of shared data in the cloud.
(2) Efficiency. Group members may dynamically join or leave the interest group, when a user is removed from the interest group, the segments signed by him/her should be efficiently regenerated, so that the left user can not create valid signature on shared data any more.
(3) Online Audit: On the one hand, the TPA can verify the correctness of shared data without downloading the entire data, even though a few signatures have been re-calculated by the cloud. On the other hand, the cloud can only transform the signature of the left users to a valid user's signature on the same segment, but it cannot permit to sign other segments for them.

3 Design of the Proposed Scheme

In this section, we first introduce a new re-signature scheme with homomorphic properties, and then present the proposed secure audit mechanism for shared data in IOV. In this paper, we always assume that there is no collusion between the users and the cloud.

3.1 Proxy Re-signing Scheme with Homomorphic Properties

Since existing proxy re-signing approaches [13, 15] has to retrieve the whole dataset to examine their integrity, which inevitably increases the resource consumption of the system. We design PRSH, a Proxy RE-signing mechanism with Homomorphic properties (blind verification and unforgeability) to overcome this problem. PRSH has six operations:

Initialization: Let G_1 and G_2 denote two cycle groups of order p, g and q are the generator of G_1, a bilinear map e: $G_1 \times G_1 \rightarrow G_2$, and a one way hash function H:$\{0,1\}^* \rightarrow G_1^*$ which maps arbitrary binary strings into nonzero points in G_1. The system public parameters are (G_1, G_2, e, g, q, H).

GenKey: User A randomly selects a value $a \in Z_p^*$ as its private key sk_A, and calculates the public key $pk_A = g^a$ based on the system public parameters (G_1, G_2, e, g, q, H).

Sign: Given the data segment $d \in Z_p$, the segment identifier id, and user A's private key $sk_A = a$, the signature on segment d signed by user A is: $s = (H(id)q^d)^a \in G_1$.

Verify: Given the data segment d, the identifier id, public key of user A, and signature s, the TPA can verify the correctness of signature by computing $e(s, g) = e((H(id)q^d)^a, g) = e(H(id)q^d, pk_A)$, if the result outputs 1, the signature is valid, otherwise the signature is invalid.

RegenKey: Once user A leaves the interest group, a new key rk_{AB} should be generated by the cloud to resign segment d, this process has to involve another honest user B: (1) the cloud randomly chooses a value $\lambda \in Z_p^*$ and sends it to user A; (2) user A calculates λ/a with its private key and sends it to user B; (3) user B computes $\lambda b/a$ and sends it to the cloud, where $sk_B = b$; (4) the cloud obtains $rk_{AB} = b/a \in Z_p^*$.

ReSign: Given signature s, re-signing key rk_{AB}, segment $d \in Z_p$, public key pk_A, and segment identifier id, the cloud tests $Verify(pk_A, d, id, s) \stackrel{?}{=} 1$, if the outcome is 0, the cloud outputs $\perp$; otherwise, it outputs $s' = s^{rk_{AB}} = (H(id)q^d)^{a \cdot b/a} = (H(id)q^d)^b$.

As we can see, the cloud indirectly utilizes user B's private key to resign data segment d, however, the cloud does not know user B's private key b, and cannot use it to sign other segments on behave of user B. Now, we show that PRSH can support homomorphic property with blind verification and unforgeability. First, given user

A's public key pk_A, two signatures s_1, s_2, signed by user A, the TPA can check the integrity of $d' = a_1d_1 + a_2d_2$ without knowing d_1 and d_2 by performing

$$e(s_1^{a_1} \cdot s_2^{a_2}, g) \stackrel{?}{=} e(H(id_1)^{a_1} H(id_2)^{a_2} q^{d'}, pk_A) \tag{1}$$

where a_1 and a_2 are two random numbers, id_1 and id_2 are two segment identifiers. The correctness of Eq. (1) can be further verified according to bilinear maps:

$$\begin{aligned} e(s_1^{a_1} \cdot s_2^{a_2}, g) &= e(H(id_1)^{a_1} q^{a_1 d_1} H(id_2)^{a_2} q^{a_2 d_2}, g^a) \\ &= e(H(id_1)^{a_1} H(id_2)^{a_2} q^{d'}, pk_A). \end{aligned} \tag{2}$$

Clearly, PRSH can support blind verification. Second, if the cloud can generate a valid signature s' on segment $d' = a_1d_1 + a_2d_2$ without obtaining user A's private key sk_A, it has

$$\begin{cases} s' = s_1^{a_1} \cdot s_2^{a_2} \\ s_1^{a_1} \cdot s_2^{a_2} = (H(id_1)^{a_1} H(id_2)^{a_2} q^{d'})^a \\ s' = (H(id') q^{d'})^a \end{cases} \tag{3}$$

where $H(id') = H(id_1)^{a_1} H(id_2)^{a_2}$, this indicates that given the value of $\Gamma = H(id_1)^{a_1} H(id_2)^{a_2}$, one can easily find a segment identifier id' satisfy $H(id') = \Gamma$, which violates the truth that H is a one-way hash function. Therefore, the signatures created by the cloud are blind verification and unforgeability.

3.2 The Proposed Secure Auditing Scheme

Based on PRSH, we propose a Secure Mechanism for Auditing shared data in IOV with signature Refreshment (SMAR). In SMAR, each segment is appended with the signer's identity so that the proxy and the TPA can determine which key is needed during re-signature. SMAR includes seven steps:

Initialization: The initialization is the same as that in PRSH, with G_1 and G_2 are two cyclic groups of order p, g and q are the generator of G_1, bilinear map e: $G_1 \times G_1 \rightarrow G_2$, and a one way hash function H:$\{0,1\}^* \rightarrow G_1^*$. The public parameters are (G_1, G_2, e, g, q, H). Let $D = (d_1, \cdots, d_n)$ be the shared data, where n is the total number of segments. The size of each interest group is c.

GenKey: We suppose u_1 is the original user who created the shared data in the cloud, and each group member u_i generates his/her public and private key pairs. Specifically, u_i selects a random number $r_i \in Z_p^*$ as his/her private key, and generates the public $pk_i = g^{r_i}$.

Sign: When u_1 produces the shared data, he/she generates a signature on each segment, after that, whenever a user in the interest group makes some modifications, the signatures on those segments are also re-computed. Given segment $d_l \in Z_p$, the identifier id_l,

$l \in [1, n]$, user i's private key r_i, user i generates the signature on segment d_l as: $s_l = (H(id_l)q^{d_l})^{r_i} \in G_1$.

RegenKey: The cloud creates the re-signing key rk_{ij} as follows: (1) the cloud randomly selects a number $\lambda \in Z_p^*$ and sends it to user i; (2) user i calculates λ / r_i with its private key and sends it to user j; (3) user j computes $\lambda r_j / r_i$ and sends it to the cloud, where $sk_j = r_j$; (4) the cloud obtains $rk_{ij} = r_j / r_i \in Z_p^*$.

ReSign: When a user i leaves the group, a new signature has to be regenerated with the help of user j, the selection of user j is to ask the group manager (u_1) to build a priority list, which includes each user's id and sorted by priority. The cloud chooses the first user in priority list to convert the signature into his/her. Specifically, given signature s_l, re-signing key rk_{ij}, segment $d_l \in Z_p$, public key pk_i, and segment identifier id_l, the cloud tests $e(s_l, g) \stackrel{?}{=} e(H(id_l)q_{d_l}, pk_i)$, if the testing result outputs 0, the cloud returns $\perp$; otherwise, it outputs $s_l^{'} = s^{rk_{ij}} = (H(id_l)q^{d_l})^{r_i \cdot r_j / r_i} = (H(id_l)q^{d_l})^{r_j}$. At the end of resigning, u_1 removes user i's id from priority list and signs the new list with his/her private key.

GenProof: The data integrity is verified via a challenge-and-response protocol between the TPA and the cloud, that is, the cloud receives an inquiry from the TPA and creates a proof of shared data, and then the TPA can verify the correctness of proof sent by the cloud. More concretely, the TPA generates a challenge message as follows: (1) randomly selects a subset S of m elements from $[1, n]$, this subset contains the segments that will be audited by the TPA; (2) selects a random number $\rho_\tau \in Z_w^*$, for $\tau \in S$, where w is a small prime; (3) forms a auditing information (τ, ρ_τ).

The TPA sends the auditing information to the cloud, which then creates a proof on requested data as follows: (1) the cloud divides the set S into c subsets $S_1, \cdots, S_c$ according to the signer's identifier, where S_i represents the segments that signed by user i. Obviously, $S = S_1 \cup \cdots \cup S_c$ and $S_i \cap S_j = \emptyset$, for $i \neq j$; (2) the cloud generates $\Omega_i = \sum_{\tau \in S_i} \rho_\tau d_\tau \in Z_p$ and $\Psi_i = \prod_{\tau \in S_i} s_\tau{}^{\rho_\tau} \in G_1$ for each S_i; (3) the cloud forms the final proof as $< \Omega, \Psi, (id_\tau, v_\tau)_{\tau \in S} >$, where $\Omega = (\Omega_1, \cdots, \Omega_c)$ and $\Psi = (\Psi_1, \cdots, \Psi_c)$, and sends to the TPA.

Verify: After receiving the auditing information (τ, ρ_τ) and the proof $< \Omega, \Psi, (id_\tau, v_\tau)_{\tau \in S} >$, where v_τ is the signer's identifier. The TPA retrieves all group user's public keys $(pk_1 \cdots, pk_c)$ from dedicated key server (we assume that all public keys are stored separately from data server), then the TPA verify the correctness of the proof by performing

$$e(\prod_{i=1}^{c} \Psi_i, g) \stackrel{?}{=} \prod_{i=1}^{c} e(\prod_{\tau \in S_i} H(id_\tau)^{\rho_\tau} \cdot q^{\Omega_i}, pk_i). \tag{4}$$

If this outputs 1, the TPA can trust that all segments in D are reliable. Or else, outputs 0, and the TPA will report a failure message to the group manager. Equation (4) can be explained as

$$
\begin{aligned}
e(\prod_{i=1}^{c}\Psi_i, g) &= \prod_{i=1}^{c} e(\prod_{\tau\in S_i} s_\tau{}^{\rho_\tau}, g) \\
&= \prod_{i=1}^{c} e(\prod_{\tau\in S_i}(H(id_\tau)q^{d_\tau})^{r_i\rho_\tau}, g) \\
&= \prod_{i=1}^{c} e(\prod_{\tau\in S_i}(H(id_\tau)^{\rho_\tau}\cdot\prod_{\tau\in S_i} q^{d_\tau\rho_\tau}, g^{r_i}) \\
&= \prod_{i=1}^{c} e(\prod_{\tau\in S_i} H(id_\tau)^{\rho_\tau}\cdot q^{\Omega_i}, pk_i)
\end{aligned} \tag{5}
$$

Theorem 1. SMAR can resist the cloud forge the audit proof and pass the verification as long as the Discrete Logarithm problem holds.

Proof. Assume that the cloud received the auditing information (τ, ρ_τ) and forged a proof $< \Omega', \Psi, (id_\tau, v_\tau)_{\tau\in S} >$ on incorrect shared data D' that can pass the verification in Eq. (3), where $\Omega = (\Omega_1, \cdots, \Omega_c)$ and $\Omega_i' = \sum_{\tau\in S_i}\rho_\tau d_\tau'$. Let $\Delta\Omega_i = \Omega_i' - \Omega_i$, $i \in [1, c]$, there must exist a nonzero element in $\Delta\Omega_i$, if the forge proof passed the TPA verification, then we have

$$
e(\prod_{i=1}^{c}\Psi_i, g) = \prod_{i=1}^{c} e(\prod_{\tau\in S_i} H(id_\tau)^{\rho_\tau}\cdot q^{\Omega_i'}, pk_i). \tag{6}
$$

Since the correct proof is

$$
e(\prod_{i=1}^{c}\Psi_i, g) = \prod_{i=1}^{c} e(\prod_{\tau\in S_i} H(id_\tau)^{\rho_\tau}\cdot q^{\Omega_i}, pk_i), \tag{7}
$$

we have

$$
\prod_{i=1}^{c} q^{\Omega_i r_i} = \prod_{i=1}^{c} q^{\Omega_i' r_i}, \quad \prod_{i=1}^{c} q^{r_i\Delta\Omega_i} = 1. \tag{8}
$$

Let $q^{r_i} = \alpha^{\chi_i}\beta^{\delta_i}$ and $\beta = \alpha^x$, where $\alpha, \beta \in G_1$, $x \in Z_p$, χ_i and δ_i are random number of Z_p, then we have

$$
\prod_{i=1}^{c}(\alpha^{\chi_i}\beta^{\delta_i})^{\Delta\Omega_i} = \alpha^{\sum_{i=1}^{\eta}\chi_i\Delta\Omega_i}\cdot\beta^{\sum_{i=1}^{\eta}\delta_i\Delta\Omega_i} = 1, \tag{9}
$$

we can find a solution for Diffie Hellman problme as:

$$
\beta = \alpha^{-\frac{\sum_{i=1}^{\eta}\chi_i\Delta\Omega_i}{\sum_{i=1}^{\eta}\delta_i\Delta\Omega_i}}, \quad x = -\frac{\sum_{i=1}^{\eta}\chi_i\Delta\Omega_i}{\sum_{i=1}^{\eta}\delta_i\Delta\Omega_i}. \tag{10}
$$

Since p is a large prime, the probability that the denominator is zero with $1/p$, thus, we can find an algorithm to figure out Discrete Logarithm problem with $1-1/p$, which is non-negligible. □

The communication cost of SMAR mainly incurred by the challenge and response protocol, the challenge message (τ, ρ_τ) is $m \cdot (|n| + |w|)$ bits, where m is the number of segments selected for auditing, $|n|$ is the size of shared data, and $|w|$ is the element size of Z_q. The receipted proof $< \Omega, \Psi, (id_\tau, v_\tau)_{\tau \in S} >$ has length of $2c \cdot |p| + m \cdot |id|$ bits, where c is the group size, $|p|$ is the element size of G_1, $|id|$ is the length of segment identifier. Hence, the overall communication cost of SMAR is $2c \cdot |p| + m \cdot |(id| + |n| + |w|)$. The computation cost of SMAR mainly cost by signature replacement during ReSign process, which introduces $2pair + 2\exp_{G_1} + hash_{G_1} + mul_{G_1}$, where $pair$ represents bilinear maps operation on e: $G_1 \times G_1 \rightarrow G_2$, $\exp_{G_1}$, $hash_{G_1}$, and mul_{G_1} represent exponentiation operation, hash operation and multiplication operation in G_1, respectively. According to Eq. (5), the computation cost of SMAR is $(c+1)pair + (m+c)\exp_{G_1} + mhash_{G_1} + (m+2c)mul_{G_1}$.

4 Performance Evaluation

We conduct the performance evaluation of SMAR by using VS2010 with the following parameters: the element size of G_1 or Z_p is $|p| = 160$ bits, $|w| = 80$ bits, the size of segment identifier is $|id| = 80$ bits, the number of segments in shared data is $|n| = 10000$, each segment has equal length of 2 KB. We assume the bandwidth between each vehicle is 500 Kbps, and the communication range of each vehicle is 300 m. We also assume that there are no more than 5 % of group members dynamic drop out of the group in each fixed interval of time. We compare SMAR with UCD, which is also an online auditing scheme based on bilinear maps. Three parameters that impact the evaluation results are measured: (1) the number of re-signing segments (l), (2) the number of vehicles in an interest group (c), and the number of segments audited by the TPA (m). We evaluate the performance of SMAR in terms of re-signing time, auditing time, and communication overhead.

Figure 2 shows the re-signing time of SMAR and UCD as the function of re-signed segments (l), as we can see, both SMAR and UCD are increases with the increases of re-signed segments l, and SMAR outperforms UCD as the number of segments increases. For example, when the number of re-signed segments is 100 (1 % of all shared data), the re-signing time of SMAR and UCD are 6 s and 9 s, respectively. This gap is larger when l reaches 1000 (10 % of shared data), with 13 s and 33 s for the two schemes. This is because the AP in UCD has to perform a series operation of downloading, verification, re-signing, and uploading the signatures on behaving of the intra-vehicle sensing devices, which incurs tremendous of resources. Based on this figure, it can be concluded that SMAR can reduce user's verification time.

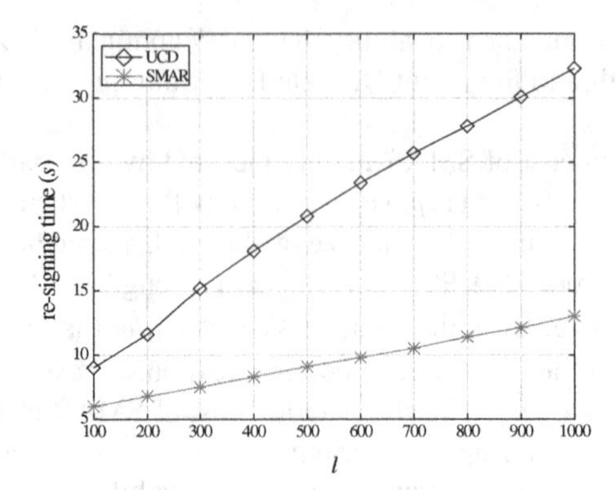

Fig. 2. Impact of l on re-signing time

Figure 3 shows how auditing time varies when group member (c) grows, where we fix the number of resigning segments (l = 500). As we can see, the auditing time of SMAR is slightly increases as the number of group members increases for SMAR. In addition, the auditing time of UCD is much more than that in SMAR, even for the high case (i.e., c = 50), SMAR only has 1/3 audit time in that of UCD under the same l. We can conclude that SMAR is efficient for supporting large groups.

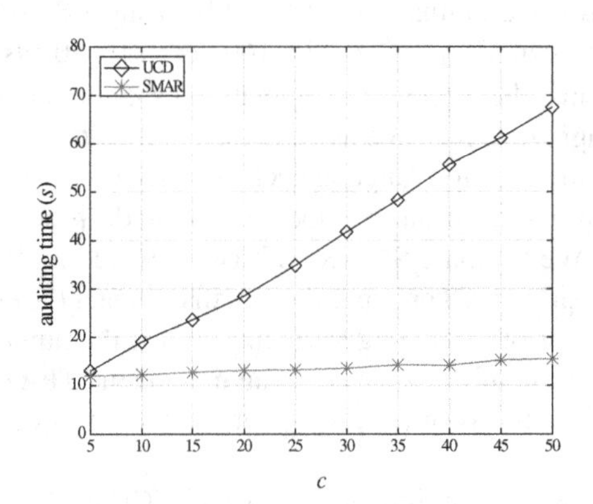

Fig. 3. Impact of c on auditing time

Figures 4 and 5 show the impact of parameter m on the auditing time and communication cost for SMAR and UCD, respectively, where we fix l = 100 and c = 20. Note that the TPA can choose a number of random segments instead of selecting all segments in shared data to enhance the efficiency of auditing. Previous work [14] has proved that a verifier can detect forgery attacks with a probability of 95 % when the number of selected segments m approximately reaches 300. To further improve the detection probability (i.e., 99 %), we can either increase the number of random segments in one auditing task or carry out several sampling audit on the same shared data. However, these two methods require more communication overhead during auditing, as can be observed from Figs. 4 and 5.

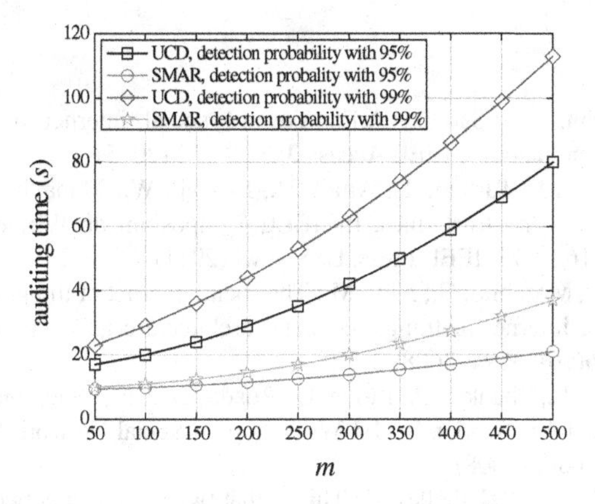

Fig. 4. Impact of m on the auditing time

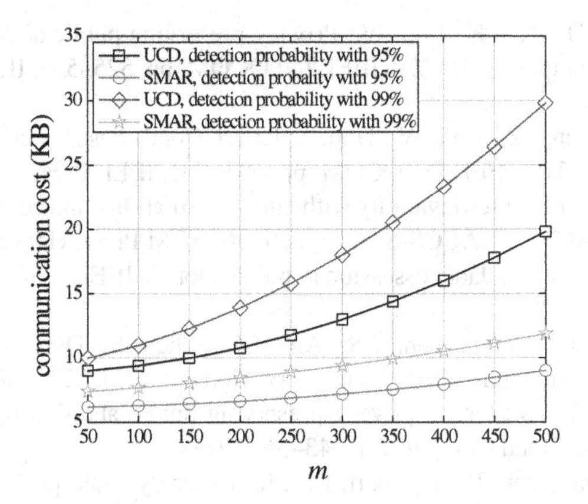

Fig. 5. Impact of m on communication cost

5 Conclusions

In this paper, we present a new auditing scheme to ensure that the users of IOV share their data in the cloud with integrity protection in mind. We allow the cloud to re-sign segments that were previously signed by the left users, and by carefully design the public verifier can dynamic audit the correctness of shared data without consuming extra resources of existing users. Experiment results show that the cloud can promote the efficiency of data sharing and utilization. In the future, we will design a re-signing mechanism that can resist collusion attack between the proxy and the users without sacrificing the desirable features of the proposed mechanism.

Acknowledgements. This work is supported by the China Postdoctoral Science Foundation funded project (Grant no. 140778), the Natural Science Foundation of Fujian Province (Grant no. 2014J05079), the special funding from Fujian Provincial Education Department (Grant no. JK2013043).

References

1. Alam, K.M., Saini, M., Saddik, A.E.: Toward social internet of vehicles: concept, architecture, and applications. IEEE Access **3**, 343–357 (2015)
2. Abbani, N., Jomaa, M., Tarhini, T., Artail, H., El-Hajj, W.: Managing social networks in vehicular networks using trust rules. In: IEEE Symposium on Wireless Technology and Applications, pp. 168–173. IEEE Press, Langkawi (2011)
3. Atzori, L., Iera, A., Morabito, G., Nitti, M.: The social internet of things (SIoT)-When social networks meet the internet of things: concept, architecture and network characterization. Comput. Netw. **56**(16), 3594–3608 (2012)
4. Smaldone, S., Han, L., Shankar, P., Iftode, L.: RoadSpeak: enabling voice chat on roadways using vehicular social networks. In: 1st Workshop on Social Network Systems, pp. 43–48. IEEE Press, New York (2008)
5. Kranz, M., Roalter, L., Michahelles, F.: Things that twitter: social networks and the internet of things. In: 8th International Conference on Pervasive Computing, pp. 1–10. IEEE Press, Mannheim (2010)
6. Wang, C., Wang, Q., Ren, K., Lou, W.: Privacy-preserving public auditing for data storage security in cloud computing. In: 29th IEEE INFOCOM, pp. 525–533. IEEE Press, San Diego (2010)
7. Cao, N., Yu, S., Yang, Z., Lou, W., Hou, Y.T.: LT Codes-based secure and reliable cloud storage service. In: 31th IEEE INFOCOM, pp. 693–701. IEEE Press, Orlando (2012)
8. Yuan, J., Yu, S.: Proofs of retrievability with public verifiability and constant communication cost in cloud. In: ACM ASIACCS-SCC, pp. 19–26. ACM Press, Hangzhou (2013)
9. Wang, H.: Proxy provable data possession in public clouds. IEEE Trans. Serv. Comput. **6**(4), 551–559 (2013)
10. Zhu, Y., Ahn, G.-J., Hu, H., Yau, S.S., An, H.G., Chen, S.: Dynamic audit services for outsourced storage in clouds. IEEE Trans. Serv. Comput. **6**(2), 227–238 (2013)
11. Li, H., Wang, B., Li, B.: Oruta: privacy-preserving public auditing for shared data in the cloud. IEEE Trans. Cloud Comput. **2**(1), 43–56 (2014)
12. Tate, S.R., Vishwanathan, R., Everhart, L.: Multi-user dynamic proofs of data possession using trusted hardware. In: 3rd ACM CODASPY, pp. 353–364. ACM Press, San Antonio (2013)
13. Boldyreva, A., Palacio, A., Warinschi, B.: Secure proxy signature schemes for delegation of signing rights. J. Cryptology **25**(1), 57–115 (2012)
14. Ateniese, G., Burns, R., Curtmola, R., Herring, J., Kissner, L., Peterson, Z., Song, D.: Provable data possession at untrusted stores. In: 14th ACM CCS, pp. 598–610, ACM Press, Alexandria (2007)
15. Kumar, R., Verma, H.K., Dhir, R.: Analysis and design of protocol for enhanced threshold proxy signature scheme based on RSA for known signers. Wirel. Pers. Commun. **80**(3), 1281–1345 (2015)
16. Liu, X., Zhang, Y., Wang, B., Yan, J.: Mona: secure multi-owner data sharing for dynamic groups in the cloud. IEEE Trans. Parallel Distrib. Syst. **24**(6), 1182–1191 (2013)

A Statistical-Based Anomaly Detection Method for Connected Cars in Internet of Things Environment

Mee Lan Han, Jin Lee, Ah Reum Kang, Sungwook Kang, Jung Kyu Park, and Huy Kang Kim(✉)

Graduate School of Information Security, Korea University, 5-Ga Anam-Dong, Seongbuk-Gu, Seoul 136-701, Republic of Korea
{blosst,lee-jin,armk,kangsungwook,pjk0825,cenda}@korea.ac.kr

Abstract. A connected car is the most successful thing in the era of Internet of Things (IoT). The connections between vehicles and networks grow and provide more convenience to users. However, vehicles become exposed to malicious attacks from outside. Therefore, a connected car now needs strong safeguard to protect malicious attacks that can cause security and safety problems at the same time. In this paper, we proposed a method to detect the anomalous status of vehicles. We extracted the in-vehicle traffic data from the well-known commercial car and performed the one-way ANOVA test. As a result, our statistical-based detection method can distinguish the abnormal status of the connected cars in IoT environment.

Keywords: Connected car · Anomaly detection · ANOVA · Internet of things

1 Introduction

The connectivity of the vehicles among the network is on the rise, as the traditional vehicle evolved into the smart vehicle with the Information and Communication Technology (ICT). The wireless technology such as Bluetooth, Wi-Fi and 4G apply to the vehicle and then support various smart services. In-Vehicle Infotainment (IVI) system and telematics service are working with Electronic Control Units (ECUs) and their connected network. For the In-Vehicle communication, there are well-known communication methods such as Controller Area Network (CAN), Local Interconnect Network (LIN), Media Oriented Systems Transport (MOST) to deliver control messages efficiently [1,2].

The electro-computerized devices and their network become a cause of malicious attacks that attackers can control the vehicles remotely. The connected cars over the networks can be easily affected by attacks from outside. At the same time, the security problems occurred in the car can affect other connected cars and networks in IoT environment. The current security threats against the connected cars can be categorized into the direct physical attack and indirect

C.-H. Hsu et al. (Eds.): IOV 2015, LNCS 9502, pp. 89–97, 2015.
DOI: 10.1007/978-3-319-27293-1_9

physical attack according to the attack vector. The direct physical attack is commonly monitoring the car network or sending malicious packets via On-Board Diagnostics-II (OBD-II) port. The indirect physical attack aims at vulnerable hardware and software (e.g. telematics device, media player) via Bluetooth or Internet connectivity for the remote attacks [3,4]. However, security safeguards for the connected cars vehicle are not prepared enough. Moreover, various IoT devices will be connected with the cars in the very near future. Many IoT devices generate a large amount of network traffic; a connected car does not have enough computing power to analyze these traffic. Especially, it will be more difficult to distinguish attack traffic from normal traffic if a connected car has more network interface to other IoT services. For this reason, we propose the method to classify the attack patterns by analyzing vehicle's movement patterns [5].

2 Literature Review

Traditionally, a car's internal network is a closed network, and therefore, all systems are safe from the external threats. But nowadays, external internet connections with a car bring threats to a car that only have occurred in ICT environments such as DoS attack, remote buffer overflow attack, privacy infringement, and so forth.

Among threats in a car, most threats occur because of the insecure control of the OBD-II port (which is designed to access CAN) or insecure design for the update module of firmware or using telemetric services without well-designed access control mechanism. These threats are fatal when they occur in the connected car environments. It can do harm both of the car's security and human driver's safety. Moreover, a compromised connected car can propagate malicious attacks to the other connected devices in IoT environment. Many studies have been conducted for enhancing a connected car's security. So far, existing car security studies mainly focus on the vulnerability analysis by exploiting the OBD-II port. ETRI progresses developing car communication technologies using multi-hop methods based on WAVE [4]. It did not aim to develop security technologies, but it focused on establishing a communication environment. The EVITA project made HSM for in-car network security in Europe [6]. These studies emphasize the importance of cipher algorithms for protecting ECU while there were few algorithms or security protocols to detect anomaly behaviors in the CAN and ECUs.

Table 1 shows the security trends of cars from 2010 to 2014. In 2010, Several studies analyzed external threats of ECU using CAN network and experimented with wireless network security for vehicles [4,12,18]. From 2011 to 2014, most studies were focused on threats of telemetric services and smartphones connected to vehicles. We also analyzed the trend of car security threats and categorized into three access types in Table 2. First, attackers access vehicles directly via connecting the OBD-II port. Second, attackers access vehicles indirectly via exploiting car subsystems. Third, attackers access vehicles remotely via hacking telemetrics devices through Bluetooth, Wi-Fi, and 3G.

Table 1. Security trends for cars from 2010 to 2014

Years	Security trends of cars
2014	Various remote attacks against cars are introduced [7,10]
2013	A brake system was hacked in a moving car [16]
2012	Hitag-2 immobilizer encryption key was decrypted [15,17]
2011	Android phones can control the engine by sending malformed text messages [13] Telemetric service can be hacked by SMS messages sent by a smart phone [14]
2010	A case study was made on wireless network security and threats for vehicles: TPMS [4,18] American hackers remote controlled car horns and engines on 100 cars in Texas [12]

Direct access attacks can affect all of the systems in the vehicles. It can control steering systems, braking systems, display systems, and ECUs. A typical ECU communicates with other devices via CAN BUS. So far, there is no strong authentication method to protect CAN packets. Therefore attackers can do sniffing and spoofing CAN packets easily [8,9]. Direct access attacks are critical but rarely happened because direct access attacks require physical access to OBD-II ports of the target car. Because of the low scalability of the direct access attack, this threat usually underestimated. Indirect access attacks took an advantage of a weak point of internal threats in the vehicles. For the first, indirect access attacks exploit vehicle's subsystem by doing a buffer overflow attack. For example, attackers can gain the high privilege of the audio system by using malware embedded music files [4]. After this process, attackers can affect the other critical systems connected to the vulnerable subsystem.

Remote access attacks focus on hacking a communication channel between telemetric service or telemetric modules and smartphones. The latest vehicles come with built-in Bluetooth and wireless module. Security issues can occur when car vendors do not take care of the threats and risks of Bluetooth and wireless technologies. When telemetric service has a security vulnerability, then attackers can make scalable attacks [10]. Recent years, Charlie Miller and Chris Valasek continuously released hacking tools and attack methods from 2010 to 2014 [9,11]. They showed that attackers can control the steering wheel, braking, acceleration, and display system of the major car vendors. They also suggested detection methods regarding the attacks.

3 Methodology

Computers and networks in usual ICT environment adopt multi-layered security model to protect themselves from various security threats. They implement security safeguards in the physical layer, network layer, OS layer, application layer and data layer from the viewpoint of defense-in-depth. However, it is difficult to implement the security safeguards in a connected car because of the limitation of computation power of embedded devices in the car. The complex security system can degrade the overall performance of car subsystems and do harm the

Table 2. Security threat based on access types for vehicles

Access type	Method	Effect
Direct	OBD-II	Controlling braking and wipers Manipulating ECU
Indirect	Media system	Damaging connected systems with CAN network
Remote	Smartphone Telemetric service Bluetooth service	Starting the car engine Manipulating car services of on load services

availability of CAN [19,20]. To avoid this problem, we suggest the light-weight method based on the statistical approach.

3.1 Data Collection and Dimension Reduction

We extracted the driving data of the real vehicle produced by the A-company (anonymized to protect the car manufacturer) in South Korea. We used the OBD-II scanner named CarbigsP[1] [21]. We can observe forty-one types of the vehicle data for the engine, fuel, gear and a steering wheel. A total of six participants drove on the same driving section. They each conducted for parking, driving in downtown and driving in the route of the lower range of fluctuation as an expressway. We used the factor analysis method for understanding the relevance of the extracted amounts of data. Forty-one types of the extracted data were reduced to several factors. It is the method that used to summarize information and to reduce a complexity of the data. It combines with the several similar factors into one factor by correlation. We reduced the forty-one variables to four variables condensed as the engine, fuel, gear and a steering wheel. It shows that new variables[2] contain some of the extracted variables from OBD-II scanner in Table 3.

3.2 One-way ANOVA for the Driving Pattern Test

The driving pattern varies across drivers and drivers intents and conditions. Therefore, it is difficult to explain the normal and abnormal pattern of the vehicle data just via the maximum speed or fuel efficiency opened by product. For this reason, we conducted the homogeneity test and independence test through the one-way ANOVA for the extracted vehicle data. The one-way ANOVA is the method used to determine the homogeneity of variance and to ascertain the difference of average. It means that the value of the average clustered to the same

[1] http://www.carbigs.com.

[2] Engine-related = Intake air pressure (kPa) + Calculated load value (%) + Engine torque + Accelerator position (%) + Flywheel torque (Nm) + Fuel consumption (mcc).

Table 3. The formation of new variables via factor analysis

New Variables	Variable extracted from CarbigsP (OBD-II scanner)
Engine-related	Intake air pressure (kPa), Calculated load value (%), Engine torque, Accelerator position (%), Flywheel torque (Nm), Fuel consumption (mcc)
Fuel-related	Wheel speed, Engine idle target speed (rpm), Vehicle speed (km/h), Torque convertor speed (rpm), Torque convertor turbine speed (before filtering)
Gear-related	Current gear position
Wheel-related	Steering wheel angle (%), Accelerator speed-lateral

set is statistically not a difference, in contrast, the value of the average clustered to the different set is a difference. We verified the abnormal vehicle pattern such as a burst of acceleration and braking on the authority of the adopted variables through the homogeneity test and independence test. It shows results of the same group analysis of the one-way ANOVA in Table 4. We found out about a significance level and the adopted variables based on the result from P-value (0.05). The result of the homogeneity test indicates that driving pattern of the six participants is similar just in case of the adopted variable in Table 4. Furthermore, it shows that the distribution of driving pattern is different in the engine-related variable as the adopted variable via the independence test. Figure 1 shows the distribution of driving data correspond to an adopted variable, based on the findings of homogeneity and independence test in Table 4. Figure 1(A) shows the result of homogeneity verification. Figure 1(B) shows the result of independence verification. Figure 1(A-1) is the data distribution of the six participants under parking; the variable 'Wheel-related' has the best homogeneity. Figure 1(A-2) and (A-3) are the data distribution of the six participants under constant speed

Table 4. The analysis of the one-way ANOVA for homogeneity and independence test

Homogeneity Test	Adopted variable	Significance level
Parking (A-1)	Wheel-related	0.082
Constant speed (A-2)	Engine-related	0.328
Driving in downtown (A-3)	Engine-related	0.147

Independence Test	Adopted variable
Parking (A-1)	Engine-related
Constant speed (A-2)	Engine-related
Driving in downtown (A-3)	Engine-related

and driving in downtown; the variable 'Engine-related' has the best homogeneity. Figure 1(B) is the data distribution of the three driving pattern; the variable 'Engine-related' has the best independence.

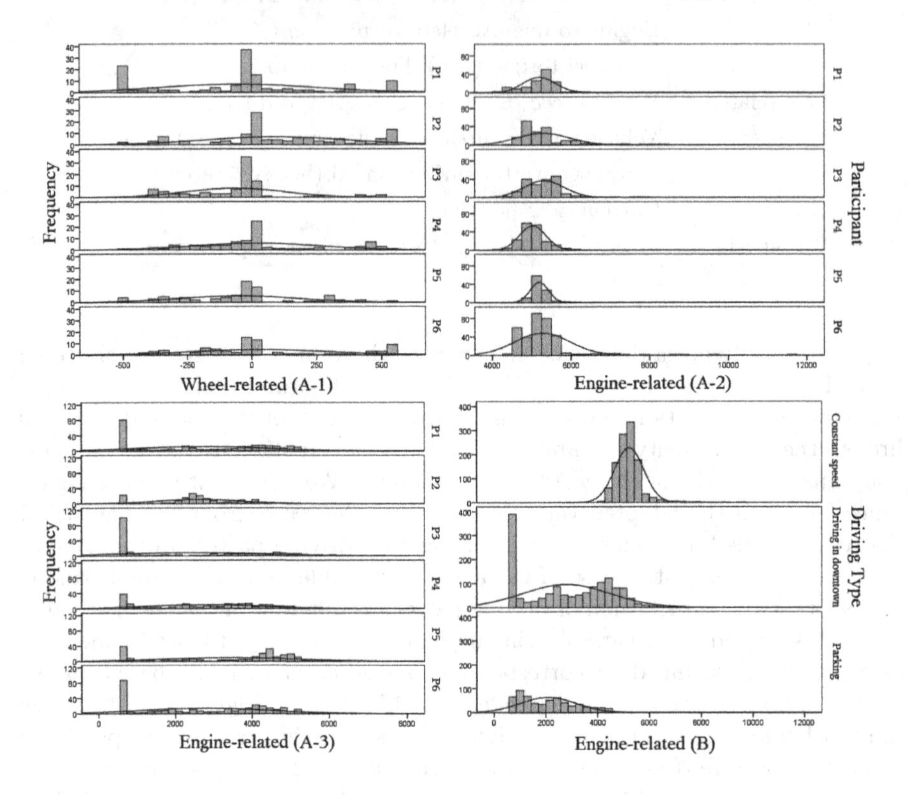

Fig. 1. Data distribution by driving patterns

4 Evaluation

The Burst of Acceleration and Braking of the Car. Our research has been conducted to detect the abnormal events of the vehicle such as the burst of acceleration and braking. The burst of acceleration event in a short time can occur in the case of an accident. Also, the burst of the braking event in a short time can occur in the case of a possible accident. To show the difference in normal driving status and abnormal status, we summarized the in-vehicle event data in Fig. 2. The rise of engine-related event per second is not observed in normal status (blue line). We found out that the rise of engine-related event per second is bigger in the burst of acceleration and braking cases (red line).

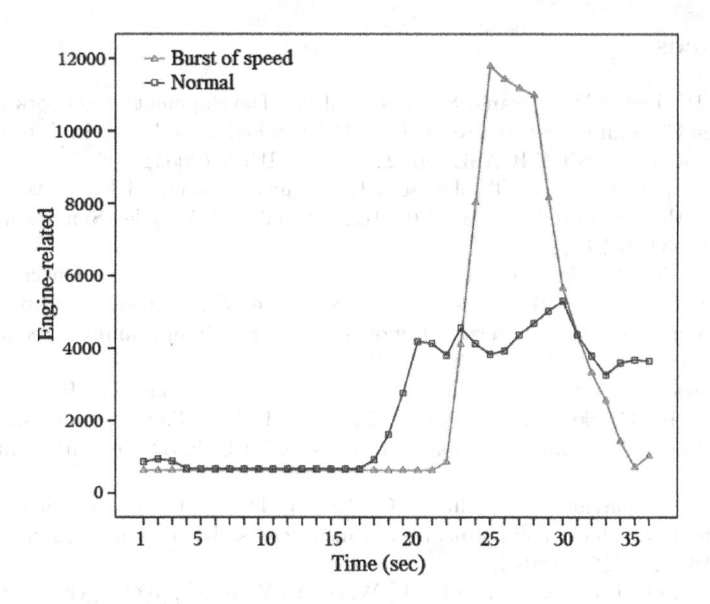

Fig. 2. The comparison of the normal and the abnormal vehicle data about engine-related variable (Color figure online)

5 Remarks

One of the difficulties for the car security research is obtaining a dataset. Because of the lack of open driving dataset, we extracted driving dataset from the real commercial car. It is possible to generate both of normal and abnormal driving pattern data. But it is hard to generate extreme driving pattern (e.g. car crash accident or vehicle rollover).

We tried to obtain the abnormal driving pattern data such as car accidents as best as we could although it was quite dangerous. To lower all possible risk, all test drivers wore safety equipment like a helmet and a seat belt and did drive test on the decommissioned roadway.

6 Conclusion

In this paper, we extracted the real in-vehicle traffic data from the well-known commercial car and performed the one-way ANOVA test. As a result, our statistical-based detection method can distinguish the abnormal status of the connected cars. This analysis can help to develop a light-weight algorithm based on the statistical method to detect anomalous attacks in CAN traffic. As a future work, we plan to extract more in-vehicle data from the various car vendors to generalize our method.

Acknowledgments. This work was supported by Samsung Research Funding Center of Samsung Electronics under Project Number SRFC-TB1403-00.

References

1. Seo, S.H., Lee, S.W., Hwang, S.H., Jeon, J.W.: Development of network gateway between CAN and FlexRay protocols for ECU embedded systems. In: International Joint Conference SICE-ICASE, pp. 2256–2261. IEEE (2006)
2. Kleberger, P., Olovsson, T., Jonsson, E.: Security aspects of the in-vehicle network in the connected car. In: 2011 IEEE Intelligent Vehicles Symposium (IV), pp. 528–533. IEEE (2011)
3. Koscher, K., Czeskis, A., Roesner, F., Patel, S., Kohno, T., Checkoway, S., McCoy, D., Kantor, B., Anderson, D., Shacham, H., Savage, S.: Experimental security analysis of a modern automobile. In: IEEE Symposium on Security and Privacy (SP), pp. 447–462. IEEE (2010)
4. Checkoway, S., McCoy, D., Kantor, B., Anderson, D., Shacham, H., Savage, S., Koscher, K., Czeskis, A., Roesner, F., Patel, S., Kohno, T.: Comprehensive experimental analyses of automotive attack surfaces. In: USENIX Security Symposium (2011)
5. Kolios, P., Panayiotou, C., Ellinas, G.: ExTraCT: expediting offloading transfers through intervehicle communication transmissions. IEEE Trans. Intell. Transp. Syst. **16**, 1238–1248 (2014)
6. Henniger, O., Ruddle, A., Seudi, H., Weyl, B., Wolf, M., Wollinger, T.: Securing vehicular on-board IT systems: The EVITA project. In: VDI/VW Automotive Security Conference (2009)
7. Car-hacking: Remote access and other security issues - Computerworld (2012)
8. Hack the diagnostics connector, steal yourself a BMW in 3 minutes - Extreme Tech (2012)
9. Miller, C., Valasek, C.: Adventures in automotive networks and control units. In: DEF CON 21 Hacking Conference. Las Vegas, NV: DEFCON (2013)
10. Oka, D.K., Furue, T., Langenhop, L., Nishimura, T.: Survey of vehicle IoT bluetooth devices. In: IEEE 7th International Conference on Service-Oriented Computing and Applications (SOCA), pp. 260–264. IEEE (2014)
11. Miller, C., Valasek, C.: A survey of remote automotive attack surfaces. Black Hat USA (2014)
12. Poulsen, K.: Hacker disables more than 100 cars remotely. Internet (2010). http://www.wired.com/threatlevel/2010/03/hacker-bricks-cars
13. Hackers steal Subaru Outback with smartphone, USA Today (2011)
14. Hack of telematics device lets attackers, Car Driver (2011)
15. Strobel, D., Driessen, B., Kasper, T., Leander, G., Oswald, D., Schellenberg, F., Paar, C.: Fuming acid and cryptanalysis: handy tools for overcoming a digital locking and access control system. In: Canetti, R., Garay, J.A. (eds.) CRYPTO 2013, Part I. LNCS, vol. 8042, pp. 147–164. Springer, Heidelberg (2013)
16. 7 On Your Side: How cars can get hacked, ABC7news (2013)
17. Verdult, R., Garcia, F.D., Balasch, J.: Gone in 360 seconds: Hijacking with Hitag2. In: Proceedings of the 21st USENIX Conference on Security Symposium, pp. 37–37. USENIX Association (2012)
18. Roufa, I., Miller, R., Mustafaa, H., Taylora, T., Oh, S., Xua, W., Gruteserb, M., Trappeb, W., Seskarb, I.: Security and privacy vulnerabilities of in-car wireless networks: A tire pressure monitoring system case study. In: 19th USENIX Security Symposium, Washington DC, pp. 11–13. (2010)
19. Larson, U.E., Nilsson, D.K., Jonsson, E.: An approach to specification-based attack detection for in-vehicle networks. In: 2008 IEEE Intelligent Vehicles Symposium, pp. 220–225. IEEE (2008)

20. Hoppe, T., Kiltz, S., Dittmann, J.: Applying intrusion detection to automotive IT-early insights and remaining challenges. J. Inf. Assur. Secur. (JIAS) **4**, 226–235 (2009)
21. OBD-II scan program, The Carbigs program supports the safety diagnosis and the driving record for all of the type of car of Hyundai-Kia Motor Group produced since (2004)

Intelligent Mobility

Target Localization and Navigation with Directed Radio Sensing in Wireless Sensor Networks

Ju Wang[1](✉), Hongzhe Liu[2], Hong Bao[2], Brian Bennett[1], and Cesar Flores-Montoya[1]

[1] Engineering and Computer Science, Virginia State University, 1 Hayden Rd, Petersburg, VA, USA
[2] Beijing State Key Labratory of Information Service Engineering, Beijing United University, Beijing 100101, China
jwang@vsu.edu

Abstract. Autonomous mobile robots guided by a set of interconnected RF sensors present a interesting scenario of vehicle localization and navigation. We investigate the feasibility of a novel RF sensing based method to address the target and robot localization for situations when conventional localization means fail to deliver. Our method allows the robot to discover the location of a target sensor and execute a navigate plan using the networked RF beacon nodes. The location of the target sensor is estimated by matching the beacon observations against an RF map. A particle filtering algorithm is used to track the location of target sensor node. The algorithm demonstrates a beyond-the-grid accuracy even only a coarse RF map is used.

Keywords: RF mapping · Robot localization · Navigation · Particle filtering

1 Introduction

Wireless Sensor Network (WSN) has become a valueble tool in numerous applications to provide realtime field data. One interesting yet challenging application of WSNs is to provide localization and navigational aid to autonomous intelligent vehicles. Many scenarios would benefit immensely by using a WSN-aided navigation solution. For instance, search and rescue of a missing person in heavy vegetation has always been difficult with conventional locating methods. GPS signal might be unusable due to lack of cellular coverage, or blocked satellite signal. Aerial camera and other vision equipments are limited by resolution and

J. Wang—This work was supported by NSF under Award Award No. 1040254, The National Natural Science Fundation of China No. 61271370, and the Project of Construction of Innovative Teams and Teacher Career Development for Universities and Colleges under Beijing Municipality IDHT20140508.

C.-H. Hsu et al. (Eds.): IOV 2015, LNCS 9502, pp. 101–113, 2015.
DOI: 10.1007/978-3-319-27293-1_10

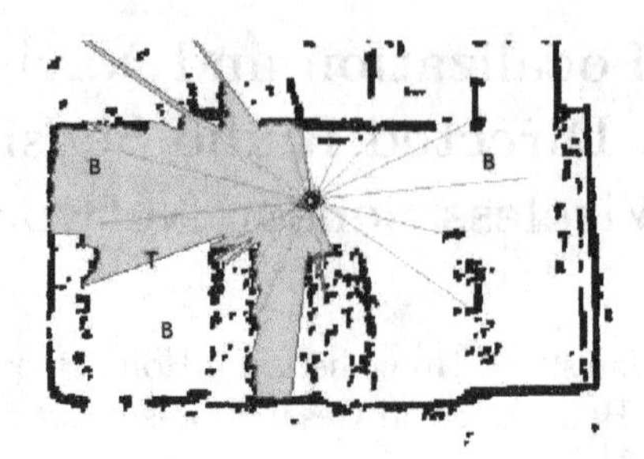

Fig. 1. Sensor localization in an indoor WSN with three beacon nodes (B) and one target node (T).

obstacled field of view. On the other hand, if the missing person carry a small RF sensor and there is a WSN in the area to relay the RF signal, the searching and localization could be greatly improved by use of intelligent vehicles (robots) to following the RF signal. Figure 1 provides an example scenario.

We investigate a localization and navigation solution based on distributed RF sensing. The goal is to accurately locate and navigate to a target WSN sensor of unknown physical location using only RF measurements provided by a network of WSN sensor nodes in its surrounding area. Radio sensing based localization involves measuring radio signal strength (RSS) of a target transmitter to estimate the distance between the transmitter and the receiver. The method has gained much attention recently [4,6–8] for its potential applications in GPS-denied scenarios and search and rescue missions. The main research challenge however is the accuracy of the technique: could RF sensing technology provide sufficient accuracy to navigate a robot towards a wireless sensor device with only coarse RF profiling?

There are several recent works to improve the accuracy of RF localization [5], such as using a comprehensive RF profiling. However RF sensing is widely regarded as a low accuracy localization method, especially for a moving robot.

Our approach is distinct in that we use RF measurement from a dynamic set of observing sensors and which are relayed through the network for more precise location estimate. We adopt a modified particle filtering [11] for robot self-localization. The method represents the location estimate by a set of evolving particles, driven by new RSS observations and robot movement. To estimate the posterior of the robot location, we calculate the likelihood of the RF observation conditioned on the robot's trajectory. The location of the target sensor is estimated by matching the RSS signature of the target sensor with an RF map.

Our second contribution is a novel path planning method, referred to as RF mode, based on RF displacement matching, as compared to standard mode which finds a global path based on the shortest distance. The RF mode utilizes a scanning directional antenna to minimize RF matching displacement. Our experimental data shows that the RF mode results in more robust navigation behaviors in the presence of large localization error, which might occur if the RF map is too coarse.

In the rest of the paper, Sect. 3 discusses the RF mapping process and sensor localization. Section 4 presents the modified particle filtering algorithm for robot localization. Section 5 discusses the RF matching based navigation mode and its performance.

2 Background and Literature Review

Localization problem represents a major research challenge in both wireless communication and robotics research communities. In most case, an estimation of the robot (or mobile phone) location will be obtained in a coordination frame. In an infrastructure environment such as a mobile cellular network [10], the time-of-flight (TOF), angle-of-arrival (AoA), or signal strength of beacon radio signals could be used for location estimation. In robotic applications the scenarios are slightly different, mainly because the cellular infrastructure is not available. The location knowledge must be obtained by analyzing local measurements from an odometer sensor, a range sensor, vision cues (landmarks, or special tags [8]), sound echoes (as in sonar), or a radar sensor.

Radio signal strength (RSS) localization is deemed as a less and inaccurate localization technique due to the complexity in radio propagation pattern caused by issues such as multi-path propagation. Despite this drawback, using RF sensing as localization and navigation technology in WSN has several advantages. For one instance, RF sensing is less limited by terrain and weather conditions than GPS or vision based technologies. The RF sensing solution also is more scalable in a large WSN network since the RF measurements can be relayed to robot through the network [6].

The feasibility of radio sensing as a localization technique for WSN is discussed in [2,4,7]. Bahl et al. [4] reports that in a complex indoor environment, RSS based empirical methods have higher accuracy than model-fitting based methods. There are several improvements on the RSS localization technique. In [5], a distributed sensing algorithm is used to cancel RF measurement errors for an indoor environment.

Localization with sequence of observations from a noisy sensor can be improved by Extended Kalman Filter (EKF) [9]. Due to its low computational complexity, Kalman filter is widely implemented as a de-noising tool to correct low level sensor reading (such as odometer sensor). EKF based localization methods usually operate a Process (Motion) Model which obtains odometer reading, and then execute an Observation Model which incorporates readings from ranger sensor or GPS sensor to perform an posterior update. More sophisticated methods typically use a population based method such as Particle Filter [1] or significance sampling technique [3,9,11]. A similar process and update model would be used, however upon all particles in the pool.

3 System Components and RF Mapping

Most localization and navigation systems rely on an accurate map and the amount of information embedded in the map. In our investigation, the localization and navigation will use an RF map created by a mobile robot equipped

Fig. 2. (a) P3DX robot, (b) Equiped with directional RF sensor, (c) WSN nodes and base station.

with a directional antenna. This allows us to create a more detailed RF map and offer some unique advantages that are not considered in previous WSN location studies. Of particular importance, the mobile robot can perform RSS measurements of multiple beacon sensors from different angle.

Figure 2 shows the main components in our system setup. The WSN service robot is a modified P3DX robot with a customized physical structure on top, which is a multi-level supporting deck constructed by light-weighted metal wires, welded together to achieve desired strength. The robot is controlled by a netbook running Ubuntu 12.04 and ROS Hydro. The robot is equipped with build-in odometer sensor and laser range sensor for obstacle avoidance, but the main equipment for localization and navigation is a scanning directional RF unit. The wireless sensor and beacon nodes are Memsic Iris motes running TinyOS firmware. A Memsic base station unit is connected to the robot control computer for both interfacing with the WSN and RSS measurement from the beacon nodes.

3.1 Network Layer Support

The WSN network layer is modified to support localization and navigation function. The new network layer support transmitting, detection, and relaying of the beacon measurement at various network entities. The new network layer includes four accessing primitives and services:

1. *Beacon Trigger Command (BTC):* This command (packet) is initiated from the mobile robot or base station. The packet directs one beacon node to start broadcasting beacon packets. The command is relayed through the network via data/command dissemination protocol (STP).
2. *Beacon Stop Command (BSC):* Upon receiving this command, a beacon node will stop broadcasting to save battery.
3. *Beacon Measurement Collect (BMC):* This command directs a specific wireless sensor node to measure the beacon packet.
4. *Beacon Measurement Report (BMR):* A wireless sensor node reports it current RSS measurements to the base station.

When receiving a BTC packet, the selected wireless beacon nodes will broadcast beacon packets in channel $0 \times 0B$ (2405 MHz) of ZigBee frequency band.

Other channels could be used to reduce the interference from local WIFI networks. Each beacon packet contains a local sequence number and a sensor node ID to distinguish it from other beacon packets (there are multiple beacon nodes in). The beacon nodes are positioned about 2 ft above the ground. It is notable that we assume the beacon nodes locations to be the same during the mapping and the navigation period. The coordinates of the beacon nodes however is not required. The respective locations of the beacon/sensors in our experiment setup are shown in Fig. 1. The beacon nodes transmit BEACON packets every 200 ms by default.

The receiver at the mobile robot is a Memsic MIB510 wireless base station. Instead of a usual dipole antenna, the receiver is connected to a panel antenna for directional sensing. The MIB510 network layer is also modified to a 'Monitor mode' to capture all packets in the wireless channel. The physical layer packet from MIB510 base station includes a received signal strength indicator (RSSI) and the link-quality-indicator (LQI) fields. RSSI is reported in scale from 0 (−91 dBm) to a maximum observation of 84 (−7 dBm).

3.2 RF Map

The RF mapping process creates an RF profile (or map) of the WSN under study. The WSN is assumed to form a typical ad-hoc meshed topology with a few data sinks. During the RF mapping process, the mobile robot will be programmed to drive through a set of predetermined grid points and record the RF signature from nearby beacon nodes. Depending on the actual set of beacon nodes observable at the robot's location, the RF map might contain different number of useful measurements while surveying the area. The resultant RF map is a high dimension matrix defined on a 2D grid. The RF map in our experiment setup is shown below in Fig. 5. The measurement of a particular beacon reaches the strongest point when the robot is close to or at the corresponding sensor node. The signal strength decreases as the robot moves away from the beacon node. The strongest signal strength is in the range of 40 (−51 dBm). The lowest RSSI observed is 12 (−79 dbm) before packet becomes completely undecipherable. The resultant RF map also shows strong multipath effect in the indoor office setting. We find that the standard deviation of slow fading is normally small (within 2 dBm).

3.3 Target Sensor Localization Using RF Map

The estimation of the target sensor location is relative straigh forward. The following steps illustrate this process:

1. The robot issues a BTC command to relevant beacon nodes to start beacon generating.
2. After acks from all beacon nodes, the robot send a BMC command to the target sensor node to start RSS measurement.

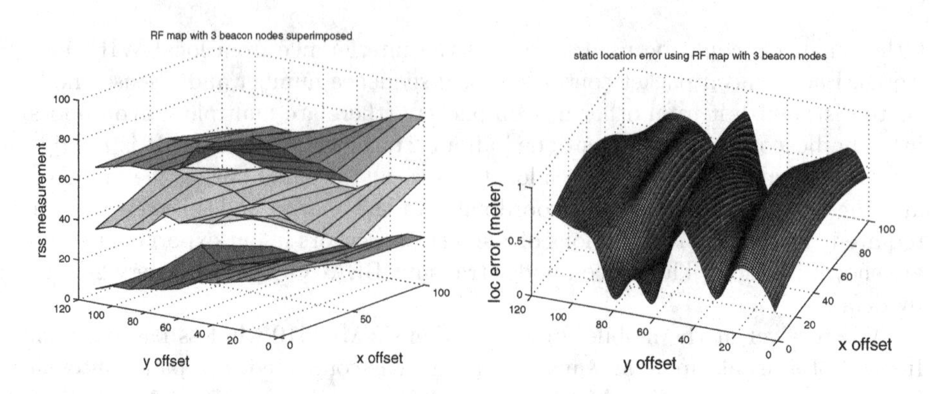

Fig. 3. (a) RF map consists of RSSI Measurement with three beacons at each grid point, (b) Static Sensor location error based on RF Map

3. The target sensor node will perform RF measurement for a period of time, 20 s in a typical scenario, and send the average RSS reading to the robot via BMR packet.
4. The robot compares the received RSS vector against previously established RF map to produce a location estimate.

Let (x, y) represent the sensor location and SS represent the measured RSS of the beacon set. The estimation of the robot location is one such that:

$$(\hat{x, y}) = argmax_{d \in (x,y)} \{ \Pi_{i \in R} Pr_i(\hat{SS_i} | (x, y) \} \quad (1)$$

The process above produce a reasonablely estimation so far as at least three beacon nodes are used and placed apart from each other. Figure 3b shows the average location error in our testing envirionment. The average location error is less than 1 m.

4 Robot Location with Particle Filtering

The location of the robot can be estimated in a similar way as the sensor location, using the method discussed above. However the accuracy of the robot location can be further improved by taking advantage of robot mobility, which allows more diverse RSS measurements.

The computation tool for this purpose is the particle filtering algorithm, which has been successfully used in robot localization with laser range sensor (Lidar). The similar principle however could be used here even our sensor is far less accurate than typical Lidar sensor. Particle filter is a Bayesian based filter that samples the whole robot work space by a weight function derived from the belief distribution. The particle filtering framework works as following: the algorithm maintain a set of particles (robot location candidates); at the arrival of a new RSS measurement, the posterior probability of each location candidate are re-evaluated; a new population is re-sampled based on posterior probability [7,9].

One notable difference of our implementation is that the observation data feed to the particle filter is a sparse vector value instead of the conventional scalar value. In the following discussion, we represent a location estimate by a particle $<s, w>$, where $s = (x_r, y_r, \theta)$ describe the x-y location in a 2D map and θ the orientation of the robot. Each particle also contains a weight value w which is proportional to the probability of the location estimate.

4.1 Motion Model

We assume a simple Motion (or Process) Model where the robot position is update by odometry difference between consective time

$$s(k+1) = s(k) + \Delta(x, y, \theta) + w_0 \tag{2}$$

where k is the discretized time and w_0 is the odometry noise,

To estimate $\hat{s(k)}$ we use sensor observations $SS_i(s(k))$, which is the RSS measurement of the i-th beacon node at the current robot location. SS_i follows observation model:

$$\hat{SS_i}(s(k)) = SS_i(s(k)) + n_i \tag{3}$$

4.2 Particle Weight Estimation

The set of N particles $P = \{< s^i, w^i >\}_{i=1:N}$ approximates the distribution of the robot location and will be updated as new sensory observations arrive. The conditional probability of the robot location is calculated based on current RSS observations and its previous location. As in typical particle filtering implementation, P is re-sampled from the prior location estimation using the motion model and odometer record. The posterior distribution of the robot location s_t at time current t is obtained by the joint probability of previous location probability $Pr(s_{t-1})$ and the conditional probability $Pr(SS_t|s_t)$, which represent the probability of the new beacon measurement under the current estimated robot location. Here s_t is calculated from the mobile model in Eq. (2). We have the following equation:

$$Pr(s_t|\bar{SS}_{1:t}, u_{0:t}) = \tag{4}$$

$$Pr(s_{t-1}|\ \bar{SS}_{1:t-1}, u_{0:t-1}) Pr(\bar{SS}_t|s_t) \tag{5}$$

$$Pr(\bar{SS}_t|s_t) = \ \Pi_{j \in B} Pr(SS_t(j)|s_t, M) \tag{6}$$

The computation of $Pr(\bar{SS}_t|s_t)$ requires the RF map M obtained earlier. The beacon set B consists of all beacon nodes whose measurement exceeds a predefine threshold, which we choose 6 dBm in this study. For simplicity, a Gaussian distribution is used to calculate $Pr(SS_t(j)|s_t, M)$. The selection of the active beacon set B could have significant impact on the results, which is an ongoing investigation. Other possible strategies are:

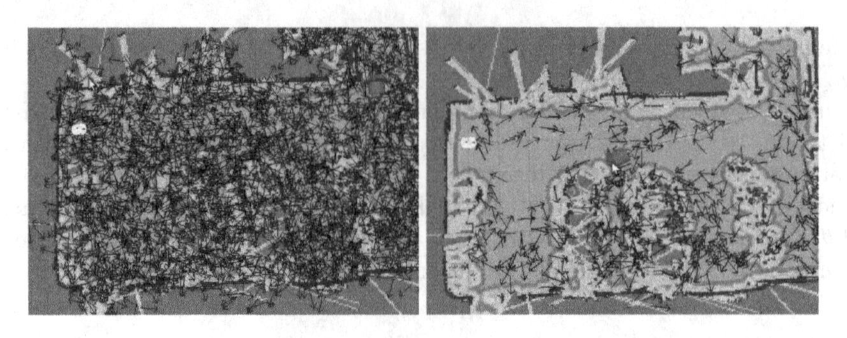

Fig. 4. Particle updating (a) Initial particle set, (b) After 5 iterations. Pink arrow is the mean location among all particles (Color figure online).

- **Strongest RSS:** B only contain the beacon nodes with the highest RSS reading. The rational of this strategy is that high RSS reading means less noise interference.
- **Closest to Robot:** here B will contain the beacon nodes that is physically close to the target sensor.
- **Highest gradient first:** here B will contain the beacon nodes whose RSS at the estimated robot position is the steepest.

The probability $Pr_(s_t)$ of all particles are then normalized and used as the weight w for each particle. During the re-sampling stage, the weight value determines how many new particles will be produced from each of the old particles. It is notable that the new particle is not the exact copy of the existing previous particle. A random location displacement will be introduced on top of the parent particle.

4.3 Particle Updating Behavior

Figure 5 shows two snapshots of the particle filtering process in one experiment run. Each brown arrow represents a particle. The direction of the arrow indicates the estimated orientation of the robot. We re-randomize the particle set every 1 foot to emulate a robot kidnapping event. The result validated the efficiency of particle filtering process in terms of converging speed. The particle population is always maintained to 1000. There is significant reduction of the particle cluster after 5 iterations, indicating many initial location estimates are removed due to new observations (Fig. 4).

5 Navigation with RF Matching

Navigating to a wireless target sensor node could be done in one of the two modes:

1. standard mode: The navigation path is calculated based on the estimated sensor location and current robot location. This usually results in a shortest path in Euclidean space.
2. RF Matching mode: the navigation path is one that minimizes the RSS distance between the robot and the target sensor. This mode is aided by a scanning directional antenna to provide an accurate motion direction of the target node.

The standard mode usually requires high fidelity in robot odometer and rotational tracking. This sometimes could be a problem if the wheel-ground traction is poor. The RF matching mode derives the motion plan differently by examining the shortest path in the *RSS space*. The RF matching mode would be particularly effective when the robot is in the vicinity of the target node. With RF matching, we compare local beacon measurements at the robot and that at the target sensor node. These measurements are then matched against the RSS profile to determine θ.

The directional antenna plays a critical rôle in the navigation algorithm. Depending on whether the robot can hear the target sensor node or not, two different modes of navigation are possible: (1) Mode 1 (direct matching): If the RF of the target node is functional and the robot is close enough to hear beacon packets from the target node, the robot will scan the directional antenna and record beacon packets from the target node. The antenna-bodyframe angle with the strongest RSS reading is selected as θ. The robot will rescan after it moves forward by 1 foot. (2) Mode 2 (indirect matching): If the target node can't be heard, the robot is guided by matching the RSS reading of the beacon nodes between the robot and the target node. We refer to this navigation mode as indirect matching since we are comparing the RSS of beacon packets instead of packets directly from the target node.

The rationale behind the *indirect matching* is because of the 2-D nature of outdoor field and (rotational) odometer drifting in some terrain, which makes navigation more complicated than the indoor case. The use of a directional antenna alleviates both problems and is the main source of improved locating accuracy.

5.1 Navigation Base on RF Matching

The indirect matching method allows the target sensor node to be several hops away from the robot, assuming that the beacon measurement $\boldsymbol{SS}_t$ from the target node is successfully relayed to the robot via WSN. If the target node fails to provide current beacon measurement, the multi-hop navigation algorithm can still utilize WSN network topology to determine a closest neighbor node as an alternate navigation goal. This topic however is beyond the scope of this paper.

With N beacon nodes, we assume a partial RF profile field function

$$\boldsymbol{SS}(\mathbf{x},\mathbf{y}) =< \mathbf{ss_1}(\mathbf{x},\mathbf{y}), \mathbf{ss_2}(\mathbf{x},\mathbf{y}), ...\mathbf{ss_N}(\mathbf{x},\mathbf{y}) > \quad (7)$$

where $ss_i(x, y)$ is the RSS profile of the $i - th$ beacon signal at (x, y). If point (x, y) is one of the preselected measurement points, referred to as grid points, the actual measurement value is used. For a non-grid point, $ss()$ is interpreted linearly from its neighboring grid points.

The navigation loop can be described as follow:

1. Denote $\boldsymbol{SS}^t = (ss_1^t, ss_2^t, ...ss_n^t)$ as the beacon measurement at the target node, the location of the target wireless sensor node $(\hat{x_t}, \hat{y_t})$ is estimated by

$$(\hat{x_t}, \hat{y_t}) = argmax_{(x,y)}\{\sum Pr_i(ss_i^t|(x, y)\} \tag{8}$$

   This is obtained with the Maximum-Likelyhook estimation algorithm discussed in Sect. 3.
2. The RSS measurement at the robot is denoted by $\boldsymbol{SS}^r(\theta) = (ss_1^r(\theta), ss_2^r(\theta), ...ss_n^r(\theta)$. The angle of the directional antenna is taken from a range $\theta \in [0^o\ 270^o]$.
3. Resample the particle set for the robot location P, the location of the robot itself $(\hat{x_r}, \hat{y_r})$ is estimated using a processed version $\boldsymbol{SS}^r = (max(ss_1^r(\theta)), max(ss_2^r(\theta)), ...max(ss_n^r(\theta)))$.

$$(\hat{x_r}, \hat{y_r}) = argmax_{(x,y)\in P}\{\sum Pr_i(ss_i^t|(x, y)\} \tag{9}$$

4. calculate $\delta_{SS} = |\boldsymbol{SS}_t - \boldsymbol{SS}_r|$, if $\delta_{SS} < \zeta$, stop.
5. Else, an optimum movement direction θ^* is calculated such that the RSS discrepency δ_{SS} will be reduced the most.
6. The robot will travel along θ^* for D meters.

The optimun θ^* is selected by a local greedy procedure:

- Descretize θ_i in the scanning range $[0\ 270^o]$ at a step size of 0.06^o.
- calculate a potential location

$$P_i = (\hat{x_r} + D. \sin\theta_i, \hat{y_r} + D. \cos\theta_i) \tag{10}$$

- calculate the projected RSS at location P_i based on RSS profile:

$$\boldsymbol{SS}(P_i)$$

- calculate the potential gain: $\delta_{SS_i} = |\boldsymbol{SS}_t - \boldsymbol{SS}(P_i)|$

$$\theta^* = argmin_{\theta_i}\delta_{SS_i} \tag{11}$$

5.2 Scanning Directional Measurement

A planner directional antenna assembly is mounted on a FLIR Pan-Tilt Unit (PTU) to provide directional measurement. The PTU offer continuous pan rotation and fine control of pointing angle through its serial command interface. The RSSI reading from the directional antenna is strongest if the radio source is in

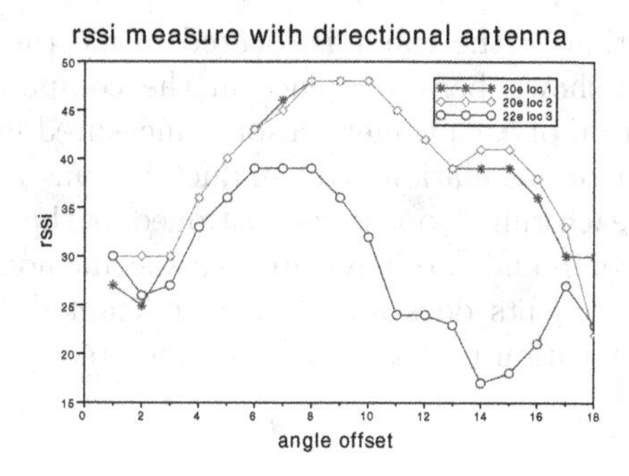

Fig. 5. RSS vs Angular offset measured by directional antenna in room HM22E, target-robot distance: 3 ft.

the middle of the antenna's field of view. The readings fall off as the target move to the side of the field view. To obtain good angular sensitivity for the directonal antenna, a minimum transmitter-receiver distance is required. Configuring the transmitter at default power, our tests show that the received signal strength becomes saturated and the pointing angle indistinguishable when the TR distance is less then 1 ft. The figure below shows the measured RSS of a source node three feet away from the robot.

5.3 Navigation Experimental Results

Our experiment is conducted in the basement section of HuntMC building at VSU, including a hallway and several inter-connected computer labs. The testing WSN consists of 6 data nodes and three beacon nodes, each programmed with a unique ID. The navigation algorithm will stop if it believes itself within 1 foot from the target. The actual distance to the target is measured manually. The test area is about 100 ft in length. The Beacon packets can still be received outside the testing area minus noticeable packet drops. Figure below shows the RF matching based navigation trials and the measured errors for different target nodes (Fig. 6).

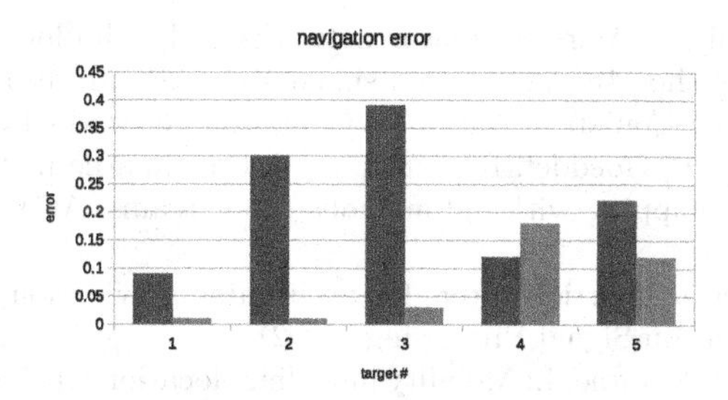

Fig. 6. Navigation error of four target location in room HM22E

The starting position of the robot is located inside the robotic lab, while the two target nodes shown here are place in the computer lab next to the robotic lab. The location of each target sensor is measured manually and serves as ground truth for error calculation. We conduct 10 runs for each target node independently. After each run, robot is re-positioned at the same starting point (home location) and commanded to navigate to a specific node ID. We note that the result of the experiments does not show significant difference at different starting location. All location measurements are meters.

6 Conclusions

We present a location and navigation method for intelligent vehicle based on collaberative RF measurement in a WSN-assisted applications. The method utilizes a vector of RSS measurements to estimate the location of the target sensor and robot itself. We further discussed a novel navigation method using the RSS measurement together with a directional antenna. Our result shows that RSS based navigation can achieve reasonable accuracy in a coarsely profiled field.

References

1. Grisetti, G., et al.: Improving grid-based slam with rao-blackwellized particle filters by adaptive proposals and selective resampling. In: Proceedings of the 2005 IEEE International Conference on Robotics and Automation (2005)
2. Thrun, S., Fox, D., Burgard, W., Dellaert, F.: Robust monte carlo localization for mobile robots. Artif. Intell. **128**, 99–141 (2001)
3. de Freitas, N., et al.: Diagnosis by a waiter and a mars explorer. In: Proceedings of the IEEE, Special Issue on Sequential State Estimation, April 2003
4. Bahl, P., Padmanabhan, V.: Radar: An in-building RF-based user location and tracking system. In: INFOCOMM (2000)
5. Moore, D., Leonard, J., Rus, D., Teller, S.: Robust distributed network localization with noisy range measurements. In: Proceedings of ACM Sensys 2004, November 2004
6. Whitehouse, K., Karlof, C., Culler, D.: A practical evaluation of radio signal strength for ranging-based localization. SIGMOBILE Mob. Comput. Commun. Rev. **11**(1), 41–52 (2007)
7. Alippi, C., Vanini, G.: Wireless sensor networks and radio localization: a metrological analysis of the MICA2 received signal strength indicator. In: 29th Annual IEEE International Conference on Local Computer Networks LCN 2004 (2004)
8. Olson, E., Strom, J., Goeddel, R., Morton, R., Ranganathan, P., Richardson, A.: Exploration and mapping with autonomous robot teams. ACM Commun. **56**(3), 62–70 (2013)
9. Gustafsson, F., et al.: Particle filters for positioning, navigation, and tracking. In: IEEE Transactions on Signal Processing (2002)
10. Liu, T., Bahl, P., Chlamtac, I.: Mobility modeling, location tracking, and trajectory prediction in wireless ATM networks. IEEE J. Sel. Areas Commun. **16**(6), 922–936 (1998)

11. Kotecha, J.H., et al.: Gaussian particle filtering. IEEE Trans. Sig. Process. **51**(10), 2592–2601 (2003)
12. Batalin, M.A., Sukhatme, G., Hattig, M.: Mobile robot navigation using a sensor network. Robotics and Automation, 2004. In: 2004 IEEE International Conference on Proceedings of ICRA 2004, vol. 1. IEEE (2004)

Predicting Pedestrian Injury Metrics Based on Vehicle Front-End Design

Benjamin Lobo[1](✉), Ruosi Lin[1], Donald Brown[1], Taewung Kim[2], and Matthew Panzer[2]

[1] Department of Systems and Information Engineering, University of Virginia, Charlottesville, VA 22904, USA
bjl2n@Virginia.edu
[2] Center for Applied Biomechanics, 4040 Lewis and Clark Dr, Charlottesville, VA 22911, USA

Abstract. Simulations provide vehicle designers with the capability to evaluate the safety of their designs in a wide variety of scenarios. However, the high-fidelity simulations required for safety assessment carry significant computational costs. As such, the engineering team must carefully select automotive designs to simulate, and use the results obtained to accurately predict the performance of new designs over a wide range of metrics. This paper describes the modeling of automotive simulation outputs to accurately predict a large number of widely used pedestrian injury metrics given the vehicle front-end design. The models in this paper allow the vehicle designer to identify and focus on the variables that most affect the different injury metrics, and determine which variables are most important to the overall safety performance of the vehicle.

1 Introduction

Designing the front-end of a vehicle to minimize the risk of injury to a pedestrian in a collision is a complex task faced by vehicle designers. The large number of design variables that can be adjusted and the fact that each design variable typically has a continuous range of values that it can take on results in a combinatorial number of potential designs. Vehicle designers have turned to modeling and simulation as a way of effectively and efficiently testing and evaluating font-end designs. Even though modeling and simulation does not require physically building the proposed design before testing it, the models can be computationally costly, sometimes taking up to a week to run a single simulation. As such, it is preferable for the designer to fully understand the effect that different design variables have on a pedestrian when the vehicle is involved in a front-end collision with a pedestrian, so that only those designs that are potentially viable are investigated.

This work looks at the task of accurately predicting commonly used pedestrian injury metrics given a specific vehicle front-end design, and determining which front-end design variables play a significant role influencing the injury metrics. In particular, there are 24 different design variables that can be adjusted,

C.-H. Hsu et al. (Eds.): IOV 2015, LNCS 9502, pp. 114–126, 2015.
DOI: 10.1007/978-3-319-27293-1_11

and 44 different injury metrics that can be recorded for a given design. Thus the problem is to accurately predict the 44 different injury metrics given a set of 24 specific design variable values. Building accurate predictive models for this purpose will allow the search space to be effectively and efficiently searched for a design that satisfies multiple competing criteria (in addition to considering the pedestrian impact, vehicle designers must also consider aerodynamics, aesthetics, etc.). This paper begins with a literature review, which is followed by a discussion of the data and the methods used to build the predictive models. Results are then presented which show the success of these methods on a specific data set.

2 Literature Review

Much research has been conducted to understand the relationship between vehicle front-end design and pedestrian injuries. Niederer and Schlumpf (1984) analyze how four hood models affect pedestrian kinematics at different impact speeds and claim that both vehicle front geometry and stiffness influence pedestrian head impacts. In particular, they find that front shape dominates the gross motion, while the deformability affects the acceleration level during the direct contact. Similarly, Han et al. (2012) study the effects of vehicle impact speed and front design on pedestrian injury risks. They notice that effects on pedestrian body regions are mostly influenced by variation in vehicle front designs. Moreover, Han et al. (2012) conclude that the minicar has the best front-end geometry in terms of lowering the overall injury risk. Besides the primary impact with the car, pedestrians may also suffer from the secondary impact with the ground. Crocetta et al. (2015) investigate the role of vehicle front-end design when considering pedestrian-ground contact. Their results show that once the vehicle impact speed exceeds 40 km/h, low front vehicles like sedans are no longer advantageous in reducing the severity of head-ground impact (Crocetta et al. 2015, p. 68). These studies all confirm that vehicle front geometry is a vital factor to consider in pedestrian safety.

The idea of optimization has been applied in studies that aim to mitigate the negative effects on pedestrians involved in a collision. Using the nonparametric Radial Basis Function (RBF) to implement response surfaces, Zhao et al. (2010) developed an optimal vehicle front-end geometry to protect pedestrians' heads. By validating values obtained from the RBF model through simulations, Zhao et al. (2010) conclude that the RBF-based response surface model handles the non-linearity in head injury criterion (HIC) scores well, with strong predictive capability (p. 149). Likewise, Kausalyah et al. (2014) looked at lowering the risk of head injuries to both child and adult pedestrians. The authors first propose separated front designs for adults and children, as children are more likely to be run over by cars. By using genetic algorithms (GAs) to find the optimal solution and multinomial logistic regression (MLR) to deal with run-overs, Kausalyah et al. (2014) produce an integrated front end design that effectively reduces the HIC values for both types of pedestrian. The two studies represent common research interests on reducing pedestrian head impacts, as head injuries are mostly fatal (Crandall et al. 2002). Lower extremities are also well studied, since they are also commonly injured regions (Crandall et al. 2002). Lv et al. (2015) examine the reliability design optimization for designing an optimal vehicle front-end structure so as to minimize pedestrian lower extremity injury risks. By using the multi-objective particle swarm optimization (MOPSO) algorithm to incorporate probabilistic bounds into the optimization problem, the authors are able to capture uncertainty in design variables.

3 Simulation Setup

A multi-body human model and a vehicle model were used to simulate the vehicle impacting the pedestrian. The pedestrian model was developed by combining the upper body parts from MADYMO's scalable 50th percentile human male occupant model (ver 4.10, TNO) and the lower extremity part from Kerrigan (2008) and Hall (1998). The bio-fidelity of the pedestrian model was improved using cadaveric blunt impact test data (Rawska et al. 2015). A parametric vehicle model was developed using variables to characterize its geometry and structural stiffness. Four vehicle regions (bumper, grill, hood, and windshield, see Fig. 1) were defined using five landmarks. The definition of the eight variables representing the geometry of the vehicle followed those of Mizuno et al. (2005), and the ranges of the variables were selected to represent sedans, hatchbacks, and sport utility vehicles. The contact stiffness of each of the four regions was characterized using four parameters, namely p1, p12, F1, and F12, for a total of 16 variables (see Fig. 1). The ranges for each of the 24 design variables can be found in Table 1.

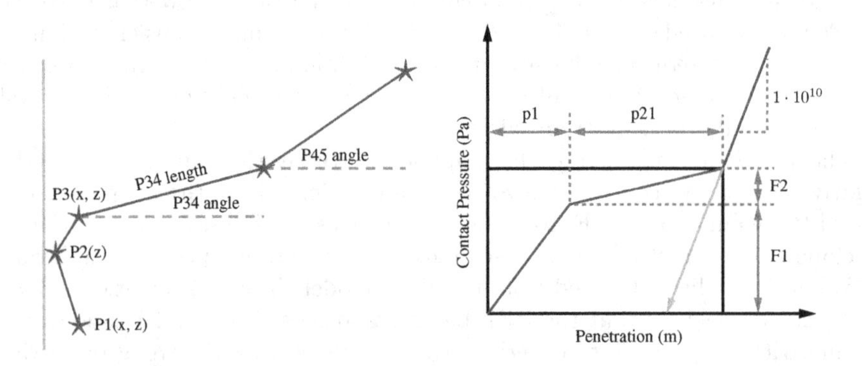

Fig. 1. Vehicle geometry and stiffness design variables

Each simulation required that each of the 24 different design variables (8 geometry related variables and 16 stiffness related variables) be assigned a specific value. Due to the size of the search space and the computational cost of each simulation, a Latin Hypercube sample of the design space was obtained

Table 1. Ranges for the 24 different design variables

Variable	Design Variable	Parameter	Units	Lower Bound	Upper Bound
v_1	P1(x)	Bottom depth	mm	0.0	95.3
v_2	P1(z)	Bottom height	mm	197.6	520.0
v_3	P2(z)	UBRL height	mm	521.2	667.6
v_4	P3(x)	Bonnet leading	mm	20.5	210.4
v_5	P3(z)	BLERL height	mm	703.9	1111.0
v_6	P34L	Bonnet length	mm	814.6	1474.1
v_7	P34 angle	Bonnet angle	°	7.5	19.0
v_8	P45 angle	Windscreen angle	°	21.0	42.1
v_9	F1	Bumper	kPa	0.5	6.0
v_{10}	p1	Bumper	mm	10.0	20.0
v_{11}	F21	Bumper	kPa	0.0	6.0
v_{12}	p21	Bumper	mm	20.0	100.0
v_{13}	F1	Grill	kPa	6.0	10.0
v_{14}	p1	Grill	mm	20.0	60.0
v_{15}	F21	Grill	kPa	0.0	2.0
v_{16}	p21	Grill	mm	0.0	40.0
v_{17}	F1	Hood	kPa	0.5	4.0
v_{18}	p1	Hood	mm	10.0	40.0
v_{19}	F21	Hood	kPa	0.0	5.0
v_{20}	p21	Hood	mm	20.0	100.0
v_{21}	F1	Windshield	kPa	0.05	2.0
v_{22}	p1	Windshield	mm	10.0	40.0
v_{23}	F21	Windshield	kPa	0.0	2.0
v_{24}	p21	Windshield	mm	20.0	200.0

using the maximinLHS() function in R. The resulting sample is a set of 1000 24-by-1 unique vectors, where each vector contains the specific value for each one of the 24 design variables. The simulation described above was run 1000 times, where on each run a different one of the input vectors was used.

Given an input vector, the simulation produced peak values of 44 raw physical quantities which are widely used as injury metrics. These responses are categorized by the 10 body regions presented in Table 2. For each one of these 44 outputs, there are 1000 values corresponding to the 1000 unique input vectors. These 44 data sets containing 1000 data points each are what the analysis is performed on.

Table 2. Physical quantities used as injury metrics, grouped by body region

Location	Metrics
Head	HIC 15 (Eppinger et al. 1999)
	BrIC (Takhounts et al. 2013)
	Linear & Angular Accleration
	NIJ (Tension only) (Eppinger et al. 1999)
Face	Contact force (Cormier et al. 2011)
Neck	NIJ (Lund 2003)
Thorax	Half Cmax at 4 levels (Viano et al. 1989)
	VCmax at 2 levels
Abdomen	VCmax
	VCmaxCmax
Pelvis	Lateral Force (Viano et al. 1989)
Thigh	Bending Moment at proximal, mid, and distal shafts (Kerrigan et al. 2004)
Knee	Angle
Leg	Bending Moment at proximal, mid, and distal shafts (Kerrigan et al. 2004)
Ankle	Angle
	Force (Funk et al. 2002)

4 Methodology

4.1 Data

There was minimal correlation between the 24 predictor (design) variables. The 44 response variable data sets could be characterized by 4 different types of distributions, namely skewed left, skewed right, normally distributed, and 'single' value. The 'single' value distribution is one in which the vast majority of data points are clustered around a single value, while the remaining minority are outliers. Examples of the four different types of response variable data set distributions can be seen in Fig. 2 (skewed left: thorax_L_lowest_VCMax, skewed right: femur_R_dist, normally distributed: ankle_R_ang, 'single' value: abdomen_F_low).

4.2 Modeling Approach

The modeling approach randomly divides each response variable data set into a training set and a test set. This process is repeated multiple times, so that for a given response variable, the data set composed of 1000 values generates multiple (random) training and test sets. This approach gives the mean and standard deviations of both model goodness of fit metrics and model performance metrics.

Model Goodness of Fit Metric. The model goodness of fit metric is the traditional R^2 value in the case of the linear regression models, and the 'pseudo'-R^2 value in the case of random forest regression models. The 'pseudo'-R^2 value is calculated as

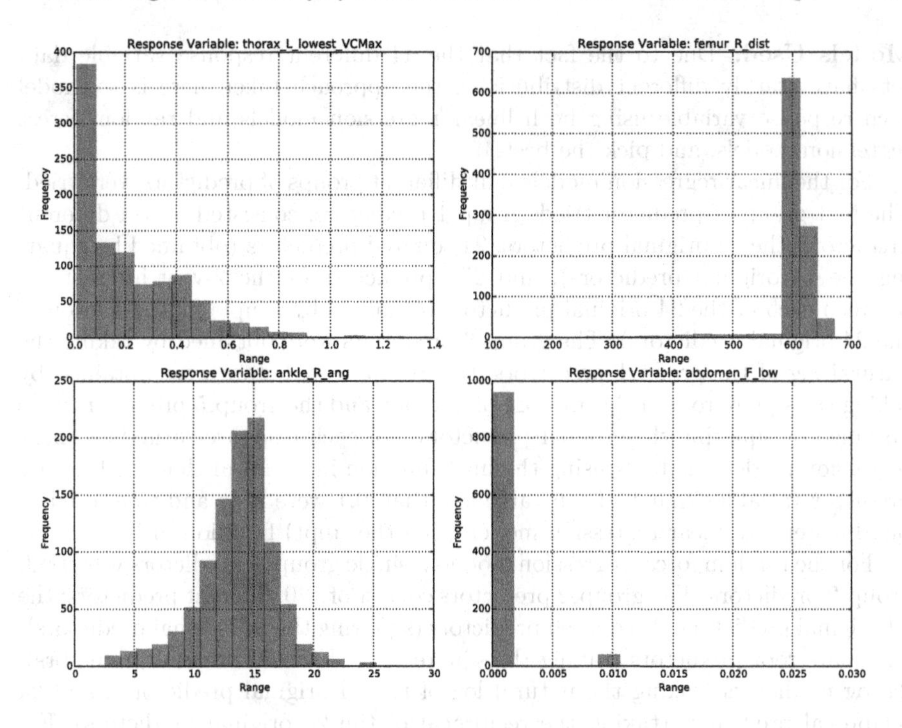

Fig. 2. Examples of the four different types of response variable distributions

$$1 - \frac{\text{MSE}}{\text{Var}(y_{\text{train}})} \tag{1}$$

where MSE is the mean squared error and $\text{Var}(y_{\text{train}})$ is the variance of the response variable training set.

Model Performance Metric. The metric to judge the predictive performance of a model is the normalized root mean squared error (NRMSE). NRMSE is calculated as

$$\sqrt{\frac{\text{MSE(Model)}}{\text{MSE(Null Model)}}} \tag{2}$$

where the Null Model predicts the average response variable value of the training set ($\bar{y}_{\text{train}}$) regardless of the setting of the predictor variables. The Null Model is a minimum information model and forms a benchmark against which the performance of other (hopefully better) models can be evaluated. Given the definition of NRMSE, the closer the value of NRMSE is to zero for a given model, the better the model being evaluated is. For example, a Model with a NRMSE value of 0.89 corresponds to a MSE ratio value of 0.80 and thus has a MSE that is 20 % less than the Null Model MSE. For the results presented later, a model was considered to be "good" if it had a NRMSE value of 0.90 or less.

Models Used. Due to the fact that the 44 different response variable data sets have starkly different distributions, the approach taken here is to model each response variable using both linear regression models and random forest regression models, and pick the best.

For the linear regression models four different groups of predictors were tried. The first group of predictor tried, group_1 predictors, consisted of 324 different predictors: the 24 original predictors, 24 squared predictors (obtained by squaring the 24 original predictors), and 276 predictors of the 2-way interactions between each of the 24 original predictors (obtained by simple multiplication of the 24 original predictors). The group_3 predictors were obtained by taking the natural log of the group_1 predictors, the group_4 predictors were obtained by taking the square root of the group_1 predictors, and the group_5 predictors were obtained by squaring the group_1 predictors. For each response variable a linear regression model was built using the lm() function in R. From that model, only predictor variables that had a p-value less than 0.1 were kept and were used to build a stepped linear regression model using the step() function in R.

For the random forest regression models a single group of predictors was tried, group_2 predictors. The group_2 predictors consist of 120 different predictors: the 24 original predictors, 24 squared predictors (squaring the 24 original predictors), 24 square root predictors (taking the square root of the 24 original predictors), 24 log predictors (taking the natural log of the 24 original predictors), and 24 reciprocal predictors (taking the reciprocal of the 24 original predictors). For each response variable a random forest regression model was built using the randomForest() function in R.

For the response variable data, four different transformations were considered: none (the data was left alone), the square root transformation, the natural log transformation, and the squaring transformation. Thus for each response variable there were 16 different linear regression models built (4 groups of predictors and 4 types of response variable transformations), and 4 different random forest models built (1 group of predictors, and 4 types of response variable transformations).

4.3 Generating the Training and Testing Sets

In general a single training and testing set pair for a particular response variable was generated as follows. A random sample of size 700 data points, along with their associated predictor variable vectors, was taken from the data set of 1000. The testing set was then composed of the remaining 300 data points and their associated predictor variable vectors. However, for those response variables whose data sets fell under the 'single' value distribution description, the training and testing set generation process was slightly different.

Training and Testing Sets for Highly Skewed Data. Taking the response variable abdomen_F_low as an example, the histogram in the lower right hand corner of Fig. 2 illustrates that the vast majority (959 data points) are zero and there are a few (41) outlier/extreme points that are non-zero.

For response variables like this, the training set was constructed by taking the n extreme data points and their associated predictor variable vectors. A set of 3n extreme jittered data points and jittered predictor variable vectors were created using the jitter() function in R. A random sample of 2*3n non-extreme data points and their accompanying predictor variable vectors together with the 3n jittered data points formed the training set so that the training set composition was one-third of jittered extreme data points and two-thirds non-extreme data points. The testing set was constructed by taking the n extreme data points and their associated predictor variable vectors together with the 1000-n-2*3n non-extreme data points and associated predictor variable vectors that were NOT used in the training set. The testing set has 1000-2*3n data points. For the response variable abdomen_F_low which has 41 extreme data points, the training set had 369 data points (123 extreme jittered and 246 non-extreme), while the testing set had 754 data points (41 extreme and 713 non-extreme).

5 Results

5.1 Predictive Modeling

Table 3 contains the overall results of the experimentation.

Each row contains:

1. The name of the response variable,
2. The type of model: (a) 'initial' for an initial linear regression model, (b)'stepped' for a stepped linear regression model, (c) 'original_rf' for a random forest regression model that used the regular training and testing sets, or (d) 'skewed_rf' for a random forest regression model that used the training and testing sets for highly skewed data
3. The type of predictor used (one of "group_1", "group_2", "group_3", "group_4", or "group_5"),
4. The type of transformation applied to the response variable data: (a) "none", (b) "sqrt" for the square root transformation, (c) "sqrd" for the squared transformation, or (d) "log" for the natural log transformation,
5. The mean and standard deviation (in parenthesis) of the R^2 (for linear regression models) or pseudo-R^2 (for random forest regression models) goodness of fit metric, and
6. The mean and standard deviation (in parenthesis) of the NRMSE model predictive performance metric.

The table shows the best model for each response variable if the average NRMSE value is less than 1, where the "best" model is defined as the one that has the smallest average NRMSE value. The table is sorted in ascending order based on the mean NRMSE column.

Table 3. Best model overall for each response variable

Response variable	Model type	Predictor group	Resp. trans.	R^2	NRMSE
thorax_L_low	original_rf	group_2	none	0.9661 (0.0017)	0.1915 (0.0134)
thorax_L_lowest	stepped	group_4	none	0.9729 (0.0020)	0.2014 (0.0116)
abdomen_R_highest	stepped	group_4	sqrt	0.9685 (0.0020)	0.2222 (0.0111)
thorax_L_lowest_VCMax	original_rf	group_2	sqrt	0.9496 (0.0024)	0.2343 (0.0152)
thorax_L_high	original_rf	group_2	none	0.9270 (0.0038)	0.2874 (0.0202)
neck_L	skewed_rf	group_2	sqrd	0.8944 (0.0133)	0.3369 (0.0682)
thorax_R_lowest_VCMax	skewed_rf	group_2	sqrt	0.9457 (0.0058)	0.3369 (0.0220)
head_5	skewed_rf	group_2	sqrd	0.8951 (0.0263)	0.3390 (0.1205)
pelvis_L	skewed_rf	group_2	sqrt	0.8459 (0.0255)	0.3911 (0.1220)
thorax_L_highest	original_rf	group_2	none	0.8672 (0.0101)	0.3987 (0.0311)
thorax_R_low_VCMax	skewed_rf	group_2	sqrt	0.9208 (0.0058)	0.3999 (0.0247)
thorax_R_low	stepped	group_1	none	0.9023 (0.0080)	0.4036 (0.0343)
thorax_R_lowest	original_rf	group_2	none	0.8705 (0.0085)	0.4044 (0.0360)
abdomen_L_highest	original_rf	group_2	sqrt	0.8503 (0.0066)	0.4346 (0.0245)
pelvis_R	original_rf	group_2	sqrt	0.8493 (0.0072)	0.4369 (0.0247)
head_2	skewed_rf	group_2	sqrd	0.8723 (0.0353)	0.4551 (0.1455)
thorax_L_low_VCMax	original_rf	group_2	sqrt	0.8002 (0.0142)	0.4966 (0.0590)
thorax_R_high	stepped	group_4	none	0.8195 (0.0102)	0.5761 (0.0404)
face_R	original_rf	group_2	sqrt	0.7744 (0.0096)	0.5788 (0.0326)
face_L	original_rf	group_2	sqrt	0.7743 (0.0099)	0.5791 (0.0319)
tibia_L_prox	stepped	group_1	log	0.7642 (0.0177)	0.6605 (0.0426)
knee_R	skewed_rf	group_2	sqrt	0.6335 (0.0828)	0.6632 (0.1760)
thorax_R_highest	original_rf	group_2	sqrt	0.6915 (0.0137)	0.7005 (0.0329)
abdomen_F_low	skewed_rf	group_2	log	0.8554 (0.0095)	0.7020 (0.0400)
femur_L_dist	stepped	group_1	sqrd	0.7223 (0.0221)	0.7073 (0.0622)
abdomen_F_high	original_rf	group_2	sqrt	0.6960 (0.0129)	0.7310 (0.0358)
ankle_R_force	skewed_rf	group_2	sqrd	0.7966 (0.0320)	0.7396 (0.0794)
tibia_L_mid	stepped	group_1	log	0.6930 (0.0281)	0.8052 (0.0618)
ankle_L_force	skewed_rf	group_2	sqrd	0.7043 (0.0884)	0.8740 (0.0503)
ankle_L_ang	original_rf	group_2	none	0.5625 (0.0241)	0.9489 (0.0600)

Of the 44 different response variables, the table shows that there are 29 response variables that have "good" models, i.e., models where the average NRMSE is less than 0.9. What the table does not illustrate is that for many of the response variables, those that had a "good" linear regression model tended to also have a good random forest model. Thus if insights for a particular response variable need to be gained from a linear regression model, but the "best" model is deemed to be a random forest model, a linear regression model can be constructed that will also likely be "good".

5.2 Variable Importance

In addition to predicting new response values, the "best" random forest regression models can also be used to estimate the importance of the 24 original predictor (design) variables. This is important to designing the front end of the car as it allows the car designers to focus on specific design variables if they are interested in minimizing the effect of specific types of injuries.

The randomForest() function in R provides a measure of variable importance, %IncMSE. For a given predictor variable the metric indicates how much MSE increases when that variable is randomly permuted. The more important a variable is to the model, the more the model will degrade (i.e., the more MSE will increase) when that variable is randomly permuted. Thus the larger the value of %IncMSE for a variable is in a model, the more important that variable is to the model.

Because group_2 predictor variables are used in the random forest regression models, there are five different predictor variables to each one of the 24 design v_i, namely (a) original variable, $v_{i,\text{orig}}$, (b) the squared version of the variable, $v_{i,\text{sqrd}}$, (c) the square root version of the variable, $v_{i,\text{sqrt}}$, (d) the natural log version of the variable, $v_{i,\text{log}}$, and (e) the reciprocal version of the variable, $v_{i,\text{recip}}$. For a particular training and testing set t, the %IncMSE for design variable v_i is

$$mse\_inc(v_{i,t}) = \sum_{j \in \{\text{orig, sqrd, sqrt, log, recip}\}} \frac{mse\_inc(v_{i,t,j})}{5} \tag{3}$$

The average %IncMSE for design variable v_i over all training and testing sets $t \in T$ is

$$mse\_inc(v_{i,t}) = \sum_{t \in T} \frac{mse\_inc(v_{i,t})}{|T|} \tag{4}$$

Figure 3 is a bar plot that shows the average %IncMSE calculated for each design variable v_i in the random forest models for response variable thorax_L_low. The average %IncMSE is calculated as outlined above, with $|T| = 100$. The 8 geometry related design variables are colored yellow, while the 16 stiffness related design variables as colored gray. This plot allows the reader to quickly identify (a) which type of design variables (geometry or stiffness) are most important to predicting the response variable, and (b) the relative importance of each design variable. For instance Fig. 3 shows that while geometry design variables are more important than stiffness variables, it also illustrates that by far design variables v4 and v5 are the most important variables to the model.

Although individual variable importance plots provide a sense of which design variables are important to predicting specific response variable values, they do not show which design variables are important overall (i.e., over all response variables under consideration). This can be achieved by averaging the average %IncMSE value for each design variable v_i over all response variables. Doing so yields Fig. 4, which shows the overall design variable importance plot. This figure illustrates that in general the designers should focus primarily on the geometry

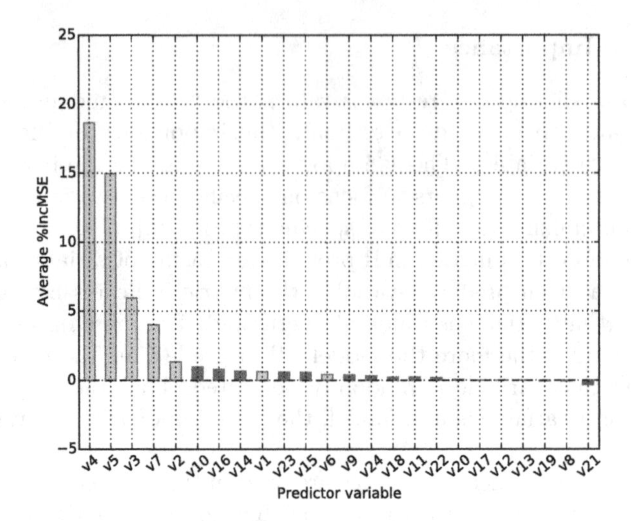

Fig. 3. Variable importance for response variable thorax_L_low

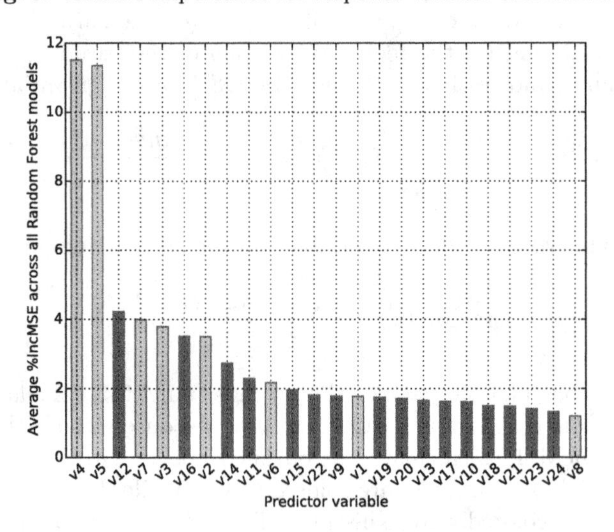

Fig. 4. Overall variable importance

related variables (specifically v4 and v5 and to a lesser extent v7 and v3), and also some specific stiffness related variables (perhaps v12, and v16).

6 Conclusion

The task of designing the front-end of a vehicle is a complex and time-consuming process. Modeling and simulation have been adopted in order to reduce the time and effort required to test and evaluate a single proposed design. This research used modeling and simulation to build a data set which was used to build predictive models. These models predict various pedestrian injury metrics given a specific car design without the need for physically building and testing the design, or even simulating the design (which itself can be computationally intensive).

Our work does have some limitations. There are still response variables for which the current methods do not yield "good" models (models where NRMSE < 1). Further research is required to determine how best to address these response variables. Given the good predicitve models, the next step is optimization – using the predictive models to search the design space for promising front end designs.

This work is the first step in a process where the output from an initial run of simulations is used to build good predictive models. In turn, these predictive models will guide the next run of simulations, and help the vehicle designer focus only on those designs that have potential to be the final design. Overall the results from this research will help reduce the ever increasing cost of designing a car to meet, among other criteria, the required safety standards.

References

Cormier, J., Manoogian, S., Bisplinghoff, J., Rowson, S., Santago, A., McNally, C., Duma, S., Bolte, J.: The tolerance of the maxilla to blunt impact. J. Biomech. Eng. **133**(6), 064501 (2011). 4th

Crandall, J.R., Bhalla, K.S., Madeley, N.J.: Designing road vehicles for pedestrian protection. BMJ Br. Med. J. (Int. Ed.) **324**(7346), 1145–1148 (2002)

Crocetta, G., Piantini, S., Pierini, M., Simms, C.: The influence of vehicle front-end design on pedestrian ground impact. Accid. Anal. Prev. **79**, 56–69 (2015)

Eppinger, R., Sun, E., Bandak, F., Haffner, M., Khaewpong, N., Maltese, M., Saul, R.: Development of improved injury criteria for the assessment of advanced automotive restraint systems II. In: National Highway Traffic Safety Administration, pp. 1–70 (1999)

Funk, J.R., Srinivasan, S.C., Crandall, J.R., Khaewpong, N., Eppinger, R.H., Jaffredo, A.S., Potier, P., Petit, P.Y.: The effects of axial preload and dorsiflexion on the tolerance of the ankle. Stapp Car Crash J. **46**, 245–265 (2002)

Hall, G.W.: Biomechanical Characterization and Multibody Modeling of the Human Lower Extremity (Doctoral Dissertation). University of Virginia, Charlottesville, VA (1998)

Han, Y., Yang, J., Mizuno, K., Matsui, Y.: Effects of vehicle impact velocity, vehicle front-end shapes on pedestrian injury risk. Traffic Inj. Prev. **13**(5), 507–518 (2012)

Kausalyah, V., Shasthri, S., Abdullah, K.A., Idres, M.M., Shah, Q.H., Wong, S.V.: Optimisation of vehicle front-end geometry for adult and pediatric pedestrian protection. Int. J. Crashworthiness **19**(2), 153–160 (2014)

Kerrigan, J.R., Drinkwater, D.C., Kam, C.Y., Murphy, D.B., Ivarsson, B.J., Crandall, J.R., Patrie, J.: Tolerance of the human leg and thigh in dynamic latero-medial bending. Int. J. Crashworthiness **9**(6), 607–623 (2004)

Kerrigan, J.R.: A computationally Efficient Mathematical Model of the Pedestrian Lower extremity (Doctoral Dissertation). University of Virginia, Charlottesville, VA (2008)

Lund, A.: Recommended procedures for evaluating occupant injury risk from deploying side airbags. In: The Side Airbag Out-of-Position Injury Technical Working Group (A Joint Project of Alliance, AIAM, AORC and IIHS). Lund (IIHS), Chairman, First Revision (2003)

Lv, X., Gu, X., He, L., Zhou, D., Liu, W.: Reliability design optimization of vehicle front-end structure for pedestrian lower extremity protection under multiple impact cases. Thin-Walled Struct. **94**, 500–511 (2015)

Mizuno, Y.: Summary of IHRA Pedestrian Safety WG Activities (2005)-Proposed test methods to evaluate pedestrian protection afforded by passenger cars. In: Proceedings International Technical Conference on the Enhanced Safety of Vehicles. vol. 2005. National Highway Traffic Safety Administration (2005)

Niederer, P.F., Schlumpf, M.R.: Influence of vehicle front geometry on impacted pedestrian kinematics. Paper Presented at the SAE Technical Papers (1984)

Rawska, K., Kim, T., Bollapragada V., Nie B., Crandall, J., Daniel, T.: Evaluation of the biofidelity of multibody pediatric human models under component-level, blunt impact and belt loading conditions. In: IRCOBI Conference Proceedings (2015)

Takhounts, E.G., Craig, M.J., Moorhouse, K., McFadden, J., Hasija, V.: Development of brain injury criteria (Br IC). Stapp Car Crash J. **57**, 243 (2013)

TNO: MADYMO Human Body Models Manual, Madymo Facet occupant model version 4.0 Release 7.5, s.l. (2013)

Viano, D.C., Lau, I.V., Asbury, C., King, A.I., Begeman, P.: Biomechanics of the human chest, abdomen, and pelvis in lateral impact. Accid. Anal. Prev. **21**(6), 553–574 (1989)

Zhao, Y., Rosala, G.F., Campean, I.F., Day, A.J.: A response surface approach to front-car optimisation for minimising pedestrian head injury levels. Intl. J. Crashworthiness **15**(2), 143–150 (2010)

SANA: Safety-Aware Navigation Application for Pedestrian Protection in Vehicular Networks

Taehwan Hwang[1] and Jaehoon (Paul) Jeong[2](✉)

[1] Department of Digital Media and Communications Engineering, Sungkyunkwan University, Suwon, Gyeonggi-Do 440-746, Republic of Korea
taehwan@skku.edu

[2] Department of Interaction Science, Sungkyunkwan University, Suwon, Gyeonggi-Do 440-746, Republic of Korea
pauljeong@skku.edu

Abstract. This paper proposes a Safety-Aware Navigation Application (SANA) for pedestrian protection in vehicular networks. Because the distracted walking by the smartphone usage in a street or crossroad usually causes road accidents and casualties, it is necessary to design an energy-efficient safety service for a smartphone to warn a pedestrian of possible danger. SANA provides smartphone users with such a safety service. This service calculates the collision possibility that is modeled from the travel delay (i.e., moving time from a position to another position) of both a vehicle and a pedestrian. It also generates an alarm to warn both the vehicle and pedestrian that are relevant to a possible collision. It considers the encounter time of the vehicle and pedestrian for maximum sleeping time to save energy. This paper proposes a scheduling algorithm for optimizing such a sleeping time, considering the filtering of irrelevant smartphones to minimize false positive alarms. The results of the simulation prove that our SANA outperforms legacy schemes in terms of energy consumption and alarm delay (i.e., time difference between the expected alarm time and the actual alarm time).

Keywords: Smartphone · Alarming · Safety app · Collision prediction · Energy saving · Scheduling

1 Introduction

Recently, smartphones have been popularly used by people for various applications (e.g., text messaging, voice calling, email, and web surfing) in road networks. This can cause the collision between pedestrians and vehicles by distracted walking and driving. In 2012, 4,743 pedestrians were killed in traffic crashes in the United States, and another 76,000 pedestrians were injured [1]. This statistics indicates one crash-related pedestrian death every 2 h, and a pedestrian injury every 7 min.

To the best of our knowledge, this paper is the first smartphone-based alarming system called Smartphone-Assisted Navigation Application (SANA) for

C.-H. Hsu et al. (Eds.): IOV 2015, LNCS 9502, pp. 127–138, 2015.
DOI: 10.1007/978-3-319-27293-1_12

pedestrian protection. The proposed alarming system is designed to ensure the safety of pedestrians with minimal disturbance, while minimizing smartphone energy consumption. In particular, we choose a way of providing effective visual or audio alarm for texting users or music-listening users while walking in streets. The proposed alarming system consists of two-level alarms, such as pre-warning and warning. A pre-warning is an additional warning before a warning is generated in order that a pedestrian can react to a warning promptly to avoid an accident. After the delivery of the pre-warning, a warning is delivered to the pedestrian for accident avoidance. In this paper, we design this two-level alarming system and evaluate our design in terms of energy consumption and alarm delay.

In this paper, we focus on the reduction of energy consumption and alarm delay in our SANA alarming system. The frequent updates of GPS location improve the synchronization, but also cause worse energy efficiency and shorten battery life [2]. To achieve the high synchronization and energy-efficiency simultaneously, this paper provides collision prediction procedure and energy-efficiency scheduling through filtering for collision prediction. SANA can prevent the pedestrians from colliding with vehicles when they cross a street. To prevent this collision, SANA gives a warning to both a vehicle and a pedestrian as soon as they arrive in the range of the safety distance. To give a successful warning, SANA needs frequent updates of GPS location for high synchronization of both vehicles and pedestrians. Note that this paper is the enhanced version of our early paper with simulation-based evaluation [3]. The contributions of this paper are as follows:

- An architecture for smartphone-based alarming system. A pedestrian's smartphone communicates with a driver's smartphone via road-side unit (RSU) in Dedicated Short Range Communications (DSRC) [4] for alarming service.
- A probability model for travel delay and collision. The travel delays of a pedestrian and a vehicle are modeled with Gamma distribution. The collision probability of a pedestrian and a vehicle is computed based on Gamma distribution.
- A scheduling algorithm for DSRC communications for energy efficiency. The scheduling algorithm performs the filtering for irrelevant smartphones from possible collisions and determines working and sleeping time for communications between the smartphones of a pedestrian and a driver.

The remaining of the paper is constructed as follows. Section 2 describes our design of SANA for pedestrian safety. Section 3 presents energy consumption modeling for operations in wireless communication and GPS devices. Section 4 shows safety performance modeling of alarm delay. Section 5 evaluates the performance of SANA in terms of energy consumption and alarm delay. Finally, Sect. 6 concludes the paper along with future work.

2 The Design of SANA

In this section, we explain our design of SANA for pedestrian protection. The pedestrian protection is very important to reduce the fatality around school zones

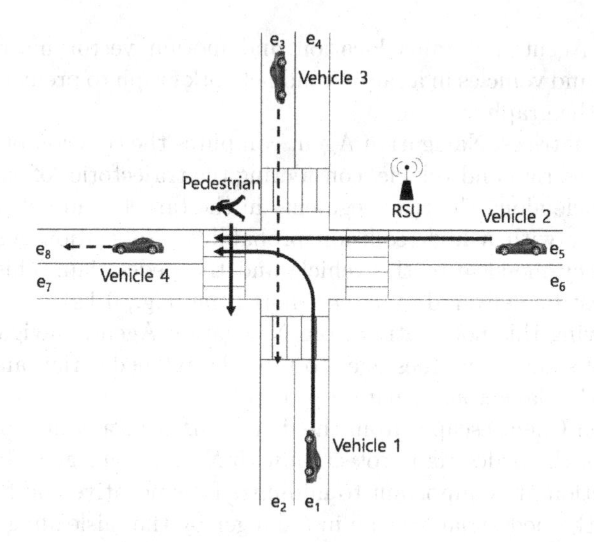

Fig. 1. Efficient collision prediction by vehicle filtering

and downtown streets. Nowadays most of people are carrying a smartphone as either a pedestrian or driver every day.

The pedestrian protection can be performed through the communication between the smartphones of a pedestrian and a driver when the vehicle approaches the pedestrian. If two smartphones share their trajectories and motion vectors, we can expect the possibility that the pedestrian and the vehicle will collide with each other. When the vehicle may hit the pedestrian just in a couple of seconds, SANA can give alarm to both the pedestrian and the driver so that they can avoid a collision. Figure 1 shows the situation that the pedestrian can be hit by Vehicle 1 or Vehicle 2. Pedestrian protection can be achieved by the collision prediction of a vehicle and a pedestrian. We can predict collision by modeling the travel delay of the vehicle and the pedestrian, as discussed in Sect. 2.2.

2.1 SANA Collision Prediction Procedure

For the collision prediction between a pedestrian and a vehicle, we assume the reliable interaction between the pedestrian's smartphone and the driver's smartphone through communication infrastructure. It is assumed that one Navigation Clients are running on the pedestrian's smartphone and the driver's smartphone. Also, Navigation Agent is running on an RSU as a middle cloud [5] near the pedestrian in order to reduce the interaction delay between Navigation Clients. Navigation Client and Navigation Agent can communicate with each other by the cellular link through 4G-LTE [6] or vehicular communication link based on DSRC [4]. The procedure for pedestrian protection is as follows:

1. As Navigation Client, a vehicle or pedestrian with navigator periodically (0.1 s) reports to a nearby Navigation Agent its location, direction, and speed during its travel from its source to its destination.

2. Navigation Agent maintains location and motion vector matrices for the pedestrians and vehicles in a target road network graph to predict the possible collision in the graph.
3. With these matrices, Navigation Agent computes the collision probability for a pair of pedestrian and vehicle, considering the trajectories of the pedestrian and the vehicle along the road segments in the target road network.
4. For each pair with a high collision probability, Navigation Agent delivers the emergency message to the vehicle and the pedestrian. This emergency message must be delivered in a very short time (e.g., 0.1 s).
5. When receiving this notification from Navigation Agent, Navigation Clients immediately show alarm messages to the relevant pedestrian and the driver to react to the dangerous situation promptly.
6. If Navigation Client escapes from the dangerous situation, it repeats Steps 1 through 5 for the pedestrian protection with Navigation Agent. In this pedestrian protection, it is important to minimize false negative and false positive. Otherwise, the pedestrian can be in a danger by the misleading guidance.

2.2 Travel Delay and Collision Probability

In this section, we model the travel delay and collision probability of vehicle and pedestrian on a road segment. We assume that road statistics is available, such as the average travel delay μ and the travel delay standard deviation σ of vehicles and the pedestrians. We also assume that the travel delays of vehicles and pedestrians follow the Gamma distributions such that $P \sim \Gamma(\kappa_p, \theta_p)$ and $V \sim \Gamma(\kappa_v, \theta_v)$ [7–9]. Figure 2 shows the distributions of pedestrian delay P and vehicle delay V. Every node notifies RSU of its current position, and RSU gets the μ and σ value of a road segment from Trafic Control Center (TCC) [10] periodically. To predict the collision on a crosswalk, the travel delay from current position to a crosswalk is required. Let δ_{ped_i} be the travel delay that the pedestrian i will move from its current position to the entrance of the next crosswalk and δ_{veh_j} be the travel delay that the vehicle j will move from its current position to the border of the next crosswalk. The travel delay δ to next crosswalk of the node can be calculated with the ratio of its current position (i.e., offset from the entrance of the road segment) x [m] to the length l [m] of the road segment. Because we know the average travel delay μ [sec] of the road segment, we can calculate the travel delay δ [sec] as follows:

$$\delta = \frac{x \cdot \mu}{l}. \tag{1}$$

If a pedestrian and a vehicle are expected to arrive at the same time, the collision probability must be high. Let $T_{collision}$ be the duration that the collision may happen. We can compute the collision probability of the pedestrian i and the vehicle j as $P_{ped_i,veh_j} = P[\delta_{veh_j} - T_{collision} \leq \delta_{ped_i} \leq \delta_{veh_j}]$. Assuming that the pedestrian travel delay distribution and the vehicle travel delay distribution are independent of each other, the collision probability P_{ped_i,veh_j} is computed as follows:

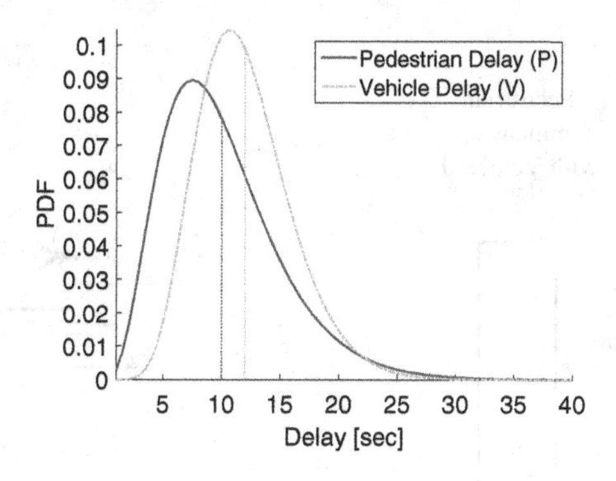

Fig. 2. Distibutions of pedestrian delay and vehicle delay

$$P_{ped_i,veh_j} = P[\delta_{veh_j} - T_{collision} \leq \delta_{ped_i} \leq \delta_{veh_j}] = \int_0^{\infty} \int_{v-T_{collision}}^{v} f(p)g(v)dpdv, \quad (2)$$

where $f(p)$ is the probability density function (PDF) of the pedestrian travel delay, and $g(v)$ is the PDF of vehicle travel delay. We will consider that the pedestrian and the vehicle will be safe when the collision probability $P_{ped,veh}$ is less than 80 %. So far, we have explained the collision probability based on the road statistics and the trajectory information. In the next section, we will design an energy-efficient scheduling using the collision probability discussed in this section.

2.3 Energy-Efficient Scheduling through Filtering

A key idea for energy-efficient scheduling is to filter out irrelevant vehicles for a specific pedestrian. This filtering lets RSU compute an optimal sleeping schedule of a pedestrian's smartphone for the message exchange via vehicular communications, such as V2I. From Fig. 1, Vehicles 1 and 2 are relevant to Pedestrian in that they can collide with Pedestrian. However, Vehicles 3 and 4 are irrelevant to Pedestrian, so they are filtered out in the computation of the sleeping periods of Pedestrian.

In Fig. 3, since Vehicles 1 is relevant to Pedestrian, the work and sleep schedule of Pedestrian's smartphone is computed considering the encounters with RSU and vehicles. At time t_1, Pedestrian communicates with RSU for the duration δ to get work and sleep schedule for collision prevention and smartphone energy saving. At time t_2, Pedestrian communicates with Vehicle 1 to exchange the location and direction information to prevent possible collision. Therefore, the work and sleep scheduling for Pedestrian's smartphone can be performed with the trajectories and mobility characteristics (e.g., speed and direction) of the pedestrians and vehicles. Also, for the work and sleep schedule for vehicles, the same procedure can be applied using the algorithm discussed in this section.

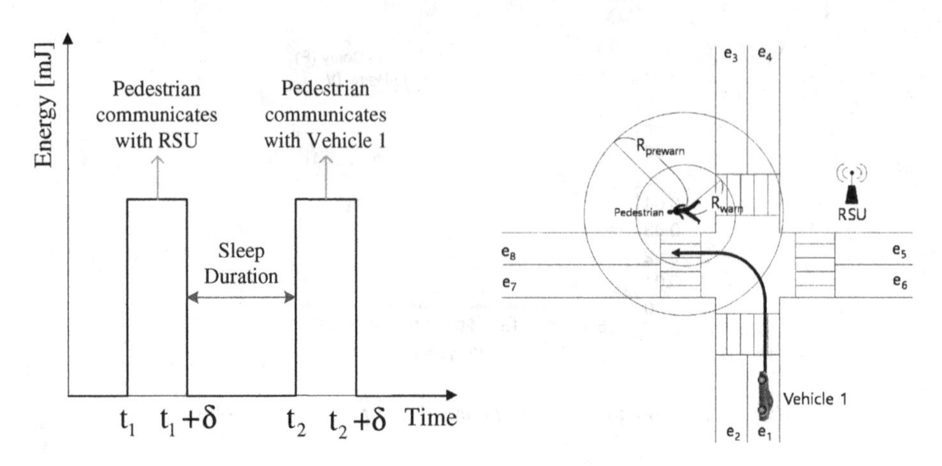

Fig. 3. Communication scheduling

Fig. 4. Pedestrian protection area

2.4 Definition of Pedestrian Protection Area

In SANA, a pedestrians smartphone will get warning messages from an RSU when it encounters a vehicle soon. These warning messages are generated when the vehicle enters an area including the pedestrian. In this paper, this area for warning message generation is defined as pedestrian protection area. This pedestrian protection area consists of the following two types of areas: (i) Warning area and (ii) Pre-warning area.

Definition 1. (Warning Area). *Let* ***Warning area*** *be the area around a pedestrian through which a vehicle can reach the pedestrian in safety time. That is, the safety time is the time taken for the vehicle to hit the pedestrian from the perimeter of the warning area. The safety time depends on the speed of the vehicle and the safety distance for pedestrian protection. Most nations in Europe commonly say 2 s rule as the safety time [11]. SANA also uses 2 s as the safety time. If a vehicle enters the* ***Warning area****, SANA will give a warning message to both the pedestrian and the vehicle for pedestrian protection.*

Definition 2. (Pre-warning Area). *Let* ***Pre-warning area*** *be the area around a pedestrian which a vehicle toward the pedestrian can reach the perimeter of the warning area for the pedestrian in safety time. A pre-warning message is a message to warn beforehand the generation of a message for the warning area. It must be delivered at the twice of the safety time before the vehicle hits the pedestrian. That is vehicles and pedestrian should be alarmed before 4 s until estimated collision time. If SANA has predicted a vehicle to enter the* ***Pre-warning area****, SANA will wake the pedestrian's smartphone to exchange the location with the vehicle and give a pre-warning to both the pedestrian and the vehicle. This pre-warning can let the pedestrian and the driver recognize the actual warning corresponding to the warning area in a prompt and reliable way [12]. Figure 4*

Table 1. Average additional energy cost of various DSRC and GPS activities

Variable	Description	Test Value
E_e (J/sec)	Additional energy cost of establishing a connection	10.00
E_i (J/sec)	Additional energy cost of maintaining the IDLE mode	8.83×10^{-1}
E_s (J/sec)	Additional energy cost of maintaining the SLEEP mode	4.20×10^{-2}
E_d (J/byte)	Additional energy cost of downloading data in IDLE mode	7.50×10^{-6}
E_u (J/byte)	Additional energy cost of uploading data in IDLE mode	8.70×10^{-6}
E_g (J/sec)	Additional energy cost of maintaining GPS	1.43×10^{-1}
E_a (J/sec)	Additional energy cost of changing from SLEEP mode to IDLE mode	8.00×10^{-3}

shows the protection areas whose outer circle is ***Pre-warning area*** *and whose inner circle is warning area.*

2.5 Collision Prediction Algorithm

In this section, we present the collision prediction algorithm for pedestrians and vehicles. The collision prediction algorithm is based on the travel delay model. The RSU gathers the mobility information from nodes in the road network. Based on the collision probability suggested in Sect. 2.2, the collision prediction algorithm works as follows:

1. RSUs are deployed near crosswalks. The TCC [10] gives each RSU its location information, adjacent road segments, and vehicular traffic statistics (e.g., the average travel delay μ and travel delay deviation σ of its adjacent road segments).
2. A node i (e.g., vehicle or pedestrian) sends its current location to RSU to get the safety information including collision probability at a crosswalk.
3. The travel delay (δ_{ped_i} or δ_{veh_i}) will be calculated by RSU. The method to compute the travel delay is presented in Sect. 2.2. RSU tells the node its travel delay to the next crosswalk.
4. The collision probability will be calculated by RSU. RSU uses two different $T_{collision}$'s that are 2 s for the warning area and are 4 s for the pre-warning area, respectively. The method to compute the collision probability is presented in Sect. 2.2. RSU will calculate all collision probabilities between the node i and the other nodes, and then select the maximum collision probability.
5. RSU gives the travel delay and the collision probability to the node i.
6. The node i is regarded as safe, if the collision probability is less than 80 %.
7. The node i is in the pre-warning area, if the probability collision is equal or greater than 80 % and $2 < \delta_{ped_i}$ or $\delta_{veh_i} \leq 4$.
8. The node i is in the warning area, if the probability collision is equal or greater than 80 % and $0 \leq \delta_{ped_i}$ or $\delta_{veh_i} \leq 2$.

3 Energy Consumption Modeling

In this section, we present the energy cost for the activities of wireless communication and GPS. We refer to these energy cost values from several measured results of other analyses [13,14]. We also present a simple model for total energy consumption in simulation time. For energy consumption model, our vehicular networks need the energy parameters of DSRC. In the real world, the graph of DSRC wireless energy consumption has complex values. Since this work focus on energy efficient scheduling, we assume that the set of modes consists of Establishment, IDLE, and SLEEP. Furthermore, the energy consumption value has its average value for each mode. In each mode, the additional energy consumption is required for download, upload, GPS operation, and mode change. We assume that the wireless module of smartphone needs additional energy to transit the current mode to SLEEP or IDLE. We refer to the measured results of power consumption from the reports of related analyses [13,14]. The average value of each state is presented in Table 1. We have built simple energy models for wireless communication, assuming three modes, Establishment, IDLE, and SLEEP. Additionally, we need to consider the energy for each mode such as, download, upload, GPS operation, and mode change, as shown in Table 2. The total power consumption in the simulation time is the sum of power consumptions to maintain modes and additional processing as follows:

$$P_{total} = P_e + P_i + P_s + P_d + P_u + P_g + P_a. \tag{3}$$

The energy consumption is one of performance metrics to evaluate the scheduling method. SANA is compared with two baselines, such as always-awake and duty-

Table 2. Parameter values for energy consumption of mobile node

Variable	Description
T_e (sec)	Time of establishing a connection
T_i (sec)	Time of maintaining IDLE mode
T_s (sec)	Time of maintaining mode
N_d (byte)	Data size of downloading data
N_u (byte)	Data size of uploading data
T_g (sec)	Time of maintaining GPS
T_a (sec)	Time of changing SLEEP mode to IDLE mode
Pe ($= E_e \cdot T_e$) (J)	Total energy consumption for establishing a connection
Pi ($= E_i \cdot T_i$) (J)	Total energy consumption for maintaining the IDLE mode
Ps ($= E_s \cdot T_s$) (J)	Total energy consumption for maintaining the SLEEP mode
Pd ($= E_d \cdot N_d$) (J)	Total energy consumption for downloading data in IDLE mode
Pu ($= E_u \cdot N_u$) (J)	Total energy consumption for uploading data in IDLE mode
Pg ($= E_g \cdot T_g$) (J)	Total energy consumption for maintaining GPS
Pa ($= E_a \cdot E_a$) (J)	Total energy consumption for changing to IDLE mode in SLEEP mode
P_{total} (J)	Total energy consumption of simulation time

cycle. In the next section, we will present the method to measure the safety performance of scheduling in the vehicular networks.

4 Safety Performance Modeling

The safety performance is a key indicator to evaluate the scheduling method in our vehicular networks. To avoid the collision between a pedestrian and vehicle, RSU gives the pre-warning and warning to both pedestrian and vehicle. We can guarantee the safety when the alarm message is delivered to smartphones in time. Thus, we use the delay of alarm message as a safety performance metric. In the ideal case, a mobile node i receives a pre-warning at $T_{prewarn_i}$ and a warning at T_{warn_i}, if there is no delay of alarm. In the real world, the pre-warning and warning are received with delay. The delayed pre-warning is received at $t_{prewarn_i}$ and the delayed warning is received at t_{warn_i}. The delay time of pre-warning $\delta_{prewarn_i}$ and warning δ_{warn_i} can be calculated as follows:

$$\begin{aligned} \delta_{prewarn_i} &= t_{prewarn_i} - T_{prewarn_i}, \\ \delta_{warn_i} &= t_{warn_i} - T_{warn_i}. \end{aligned} \tag{4}$$

The total alarm delay of a node is δ_{alarm_i}. The total delay of all nodes is δ_{total} in the simulation time. If the number of nodes is n, δ_{alarm_i} and δ_{total} can be calculated as follows:

$$\begin{aligned} \delta_{alarm_i} &= \delta_{prewarn_i} + \delta_{warn_i}, \\ \delta_{total} &= \sum_{i=1}^{n} \delta_{alarm_i}. \end{aligned} \tag{5}$$

We will use the total delay of all nodes δ_{total} as a safety performance metric in Sect. 5.3.

5 Performance Evaluation

In this section, we evaluate the performance of SANA in terms of the energy consumption and the average alarm delay. We compare SANA with two baselines, such as Always-On and Duty-Cycle. Always-On is a scheduling scheme which send its location information at every 0.1 s. Always-On can achieve the best safety performance, but it consumes the battery most quickly. Duty-Cycle is a scheduling scheme to allow a mobile node to work according to a fixed period. The refresh period of the duty-cycle in our simulation is 0.3 s which means that a mobile node wakes up every 0.3 s and sleeps after finishing its work. However, if warning or pre-warning message is delivered from RSU, the mobile node does not sleep until it gets a safe message from RSU. Our simulation map has 200 m-by-200 m area. We check the impact of vehicle speed as we set the maximum vehicle speed limit to 40, 50 and 60 km/h. The maximum pedestrian speed is

2 km/h. 4 cross walks exist on an intersection. We assume taht pedestrian and vehicle start to move at different road segments. Our simulation increases the vehicle inter-arrival time from 10 to 60 s by 5 s. The pedestrian inter-arrival time is 10 s constantly. The simulation time is 1,000 s. We assume the communication range of RSU can cover whole area of the target road network.

5.1 Simulation Design

In this section, we present how the simulation is implemented for SANA. The simulation of SANA is implemented in Veins [15] which is an open source framework for Inter-Vehicular Communication (IVC) simulation in a data network simulator called OMNeT++ [16], cooperating with a road network simulator called SUMO [17] via Traffic Control Interface (TraCI). This allows for the bi-directionally coupled simulation of road traffic and network traffic. The movement of vehicles in SUMO is reflected in the movement of nodes in OMNeT++ via Veins.

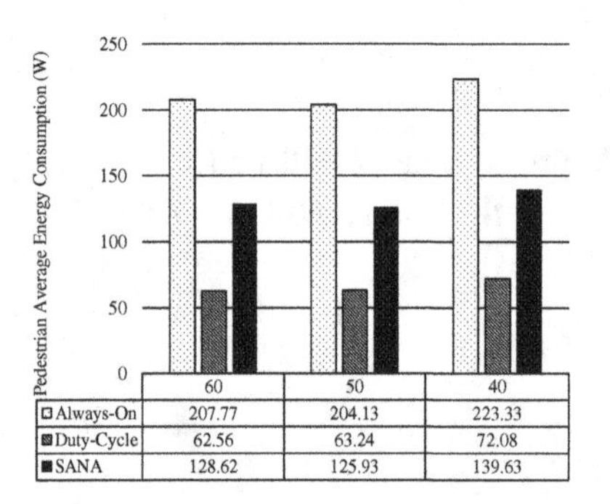

Fig. 5. Pedestrian average energy consumption per vehicle inter-arrival time

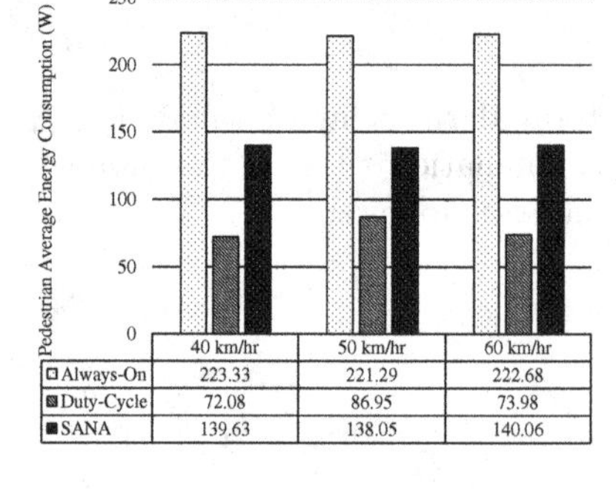

Fig. 6. Pedestrian average energy consumption per vehicle maximum speed

5.2 The Impact of Energy Consumption

In this section, we investigate the impact of vehicle inter-arrival time and vehicle maximum speed on energy consumption. Figures 5 and 6 show the energy consumption of three scheduling schemes according to vehicle inter-arrival time and vehicle maximum speed, respectively. Always-On consumes battery most quickly. In Fig. 5, the longer inter-arrival time vehicles have, the less energy the pedestrian node consumes. This is because the longer inter-arrival time means less traffic, so the number of vehicles which should communicate with a pedestrian decreases. For example, as shown in Fig. 5, for the vehicle inter-arrival time of 40 s, Always-On consumes 223.33 W, Duty-Cycle consumes 72.08 W, and SANA consumes 139.63 W. SANA reduces only 63 % of the energy consumption to the Always-On. In Fig. 6, we can see that the energy consumptions of both Always-On and SANA are not affected by vehicle maximum speed, but that of Duty-Cycle is a little affected. Thus, Duty-Cycle is the most energy-efficient scheme from the simulation result.

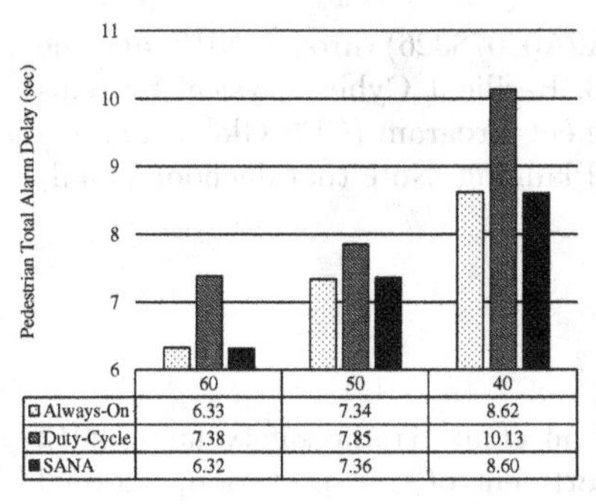

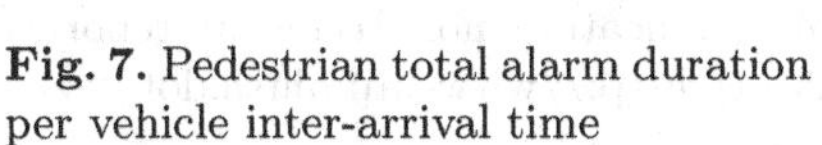

	60	50	40
Always-On	6.33	7.34	8.62
Duty-Cycle	7.38	7.85	10.13
SANA	6.32	7.36	8.60

Fig. 7. Pedestrian total alarm duration per vehicle inter-arrival time

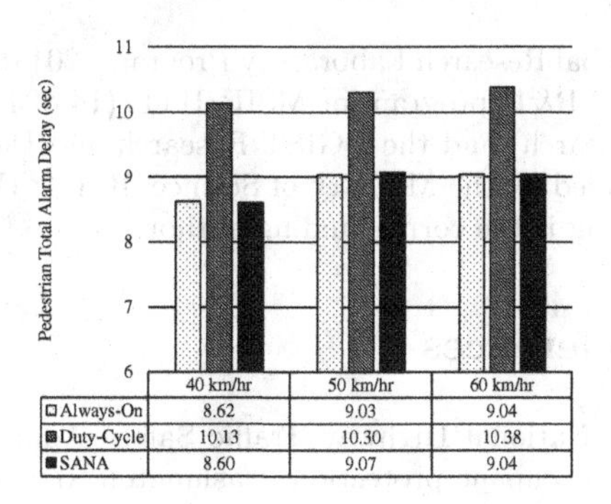

	40 km/hr	50 km/hr	60 km/hr
Always-On	8.62	9.03	9.04
Duty-Cycle	10.13	10.30	10.38
SANA	8.60	9.07	9.04

Fig. 8. Pedestrian total alarm duration per vehicle maximum speed

5.3 The Impact of Alarm Delay

In this section, we investigate the impact of vehicle inter-arrival time and vehicle speed on alarm delay. Figures 7 and 8 show how the alarm delay of pedestrian is related to each scheduling scheme. Always-On is the most safe scheme from the simulation result. For example, as shown in Fig. 7, for the vehicle inter-arrival time of 40 s, Always-On has 8.62 s delay, Duty-Cycle has 10.13 s delay, and SANA has 8.60 s delay. SANA reduces 15 % of the alarm delay of Duty-Cycle. Duty-Cycle is the most dangerous scheme from the simulation result, even though it is most evergy-efficient. Therefore, from the simulation results, it can be concluded that SANA is a promising pedestrian protection App in the optimization of both energy consumption and alarm delay.

6 Conclusion

In this paper, we proposed our design of Safety-Aware Navigation Appplication (called SANA) for the road safety. For road safety, DSRC will be common communication technology in vehicular networks. In this paper, we investigate a pedestrian protection App running in smartphones with DSRC or 4G-LTE device. To design our SANA service framework, we first described our delay modeling for a mobile nodes travel delay. We then explained our energy-aware SANA service for pedestrian protection. SANA makes RSU gather the mobility information of nearby mobile nodes. RSU can predict the collision among nearby mobile nodes. With the prediction, RSU can schedule the working time of mobile nodes to reduce energy consumption. The scheduling can be optimized to minimize the safety performance of alarm delay. As future work, we will enhance our collision prediction algorithm in order to reduce false positive and false negative alarms for a more reliable alarming service.

Acknowledgments. This research was supported by Basic Science Research Program through the National Research Foundation of Korea (NRF) funded by the Ministry of Science, ICT & Future Planning (2014006438). This research was supported in part by

Global Research Laboratory Program (2013K1A1A2A02078326) through NRF, and the ICT R&D program of MSIP/IITP (14-824-09-013, Resilient Cyber-Physical Systems Research) and the DGIST Research and Development Program (CPS Global Center) funded by the Ministry of Science, ICT & Future Planning. Note that Jaehoon (Paul) Jeong is the corresponding author.

References

1. National Highway Traffic Safety Administration et al. Traffic safety facts: 2012 occupant protection. washington, dc: Us department of transportation, national highway traffic safety administration: 2014. publication no. Technical report, DOT-HS-811-892.[cited 2014 September 8] (2014). http://www-nrd.nhtsa.dot.gov/Pubs/811892.pdf
2. Mane, P.S., Khairnar, V.: Power efficient location based services on smart phones. Int. J. Emerg. Technol. Adv. Eng. **3**(10), 350–354 (2013)
3. Hwang, T., Jeong, J.P., Lee, E.: SANA: Safety-aware navigation app for pedestrian protection in vehicular networks. In: 2014 International Conference on Information and Communication Technology Convergence (ICTC), pp. 947–953, October 2014
4. Morgan, Y.L.: Notes on dsrc & wave standards suite: its architecture, design, and characteristics. Commun. Surv. Tutor. IEEE **12**(4), 504–518 (2010)
5. Huang, D., Xing, T., Huijun, W.: Mobile cloud computing service models: a user-centric approach. Netw. IEEE **27**(5), 6–11 (2013)
6. Monfreid, C.: The lte network architecture-a comprehensive tutorial. In: Some content may change prior to final publication. Alcatel-Lucent White Paper (2009)
7. Polus, A.: A study of travel time and reliability on arterial routes. Transp. **8**(2), 141–151 (1979)
8. Berry, D.S., Belmont, D.M., et al.: Distribution of vehicle speeds and travel times. In: Proceedings of the Second Berkeley Symposium on Mathematical Statistics and Probability, pp. 589–602. The Regents of the University of California (1951)
9. DeGroot, M.H., Schervish, M.J.: Probability and Statistics. Prentice Hall, Upper Saddle River (2002)
10. Philadelphia Department of Transportation. Traffic Control Center. http://philadelphia.pahighways.com/philadelphiatcc.html
11. CEDR. Safe Distance Between Vehicles (2010). http://www.cedr.fr/home/fileadmin/user_upload/Publications/2010/e_Distance_between_vehicles.pdf
12. Fan, J., McCandliss, B.D., Sommer, T., Raz, A., Posner, M.I.: Testing the efficiency and independence of attentional networks. J. Cogn. Neurosci. **14**(3), 340–347 (2002)
13. Perrucci, G.P., Fitzek, F.H.P., Widmer, J.: Survey on energy consumption entities on the smartphone platform. In: 2011 IEEE 73rd Vehicular Technology Conference (VTC Spring), pp. 1–6. IEEE (2011)
14. Rahmati, A., Zhong, L.: Context-for-wireless: context-sensitive energy-efficient wireless data transfer. In: Proceedings of the 5th International Conference on Mobile systems, Applications and Services, pp. 165–178. ACM (2007)
15. Veins. open source Inter-Vehicular Communication (IVC) simulation. http://veins.car2x.org/documentation/
16. OMNeT++. Network Simulator. http://www.omnetpp.org/
17. SUMO. Simulation of Urban Mobility. http://sumo-sim.org/userdoc/Downloads.html

A Cooperative Collision Warning System Based on Digital Map and Dynamic Vehicle Model Fusion

Lei Cai[1(✉)], Ping Wang[1], Chao Wang[1], Xinhong Wang[1], and Qingquan Zou[2]

[1] Department of Electronics and Information Engineering, Tongji University, Shanghai, China
leicai1225@163.com,
{pingwang,chaowang,wangxinhong}@tongji.edu.cn
[2] Shanghai Automotive Industry Corporation, Shanghai, China
zouqingquan@saicmotor.com

Abstract. The vehicle collision warning system (CWS), enabled by recent advances in positioning systems and wireless communication technologies, plays a significant role in reducing traffic accidents. Since vehicle position is the most frequently exchanged information in vehicular networks and the key information in collision prediction, it becomes crucial to establish a strong level of trust in the positions, especially in active safety applications. To this end, in this paper we proposed a cooperative collision warning system (CCWS) based on digital map and dynamic vehicle model fusion. In CCWS, vehicles broadcast self-relevant information to neighbor vehicles periodically. In contrast to conventional methods that use only vehicle's dynamic models, the proposed system takes the information provided by a digital map into consideration to improve the accuracy of vehicle position estimation. Each vehicle calculates whether a potential collision may occur by fusing the dynamic vehicle model, digital map and the information received. If such a dangerous situation is predicted, an alarm will be triggered to inform the driver. Our experiment results show that the CCWS has a better performance in vehicle position estimation and collision prediction compared with the CWS without considering cooperation.

Keywords: Collision warning system (CWS) · Vehicular networks · Cooperative CWS (CCWS) · Digital map · Position estimation

1 Introduction

Nowadays, there are more and more vehicles on the road. Consequently, road traffic accidents occur frequently and cause millions of casualties each year worldwide. Improving driving safety and reducing the amount of traffic accidents on the road are still the first priority in general and particularly in the Intelligent Transportation System

This work is supported by the 2013 joint SAIC-Shanghai Committee of Science and Technology project (No. 13DZ1108000).

C.-H. Hsu et al. (Eds.): IOV 2015, LNCS 9502, pp. 139–151, 2015.
DOI: 10.1007/978-3-319-27293-1_13

(ITS) domain. A collision warning system (CWS) mounted on a vehicle monitors the surrounding environment, including the state and motion of the neighboring vehicles, and predicts the possibility of collisions in order to help addressing the above issues.

So far many kinds of CWSs have been developed. Most of them use environment sensors such as radar [1], lidar, sonar and video camera to obtain information, usually including relative distance, position and speed, regarding neighbor vehicles [2]. Then such information is processed to calculate the time to collision (TTC) to determine whether potential collision may happen. Albeit widely adopted, the reliability and performance of these techniques rely heavily on the capability and performance of sensors. Since these sensors are easy to be affected by bad weather, complex road condition and other uncontrollable factors, a satisfactory performance is usually hard to guarantee. Recently, with the rapid development of vehicular communication technologies, such as the dedicated short-range communication (DSRC) technique [3], a cooperative-driving concept has emerged as a solution to improve the performance of CWS [4]. Typically, the CCWS consists of two modules [5]. A self-sensing module, which includes global positioning system (GPS), sensors, on-board diagnostic and a communication module that exchanges the self-relevant information attained via the self-sensing module between surrounding vehicles. Through processing the information obtained from other vehicles and from its own module, a CCWS estimates the threat of potential collisions. Compared with the conventional CWSs without considering cooperation, the CCWS has a wider coverage range and better performance in terms of collision prediction.

Since vehicle positions play a key role in collision prediction, errors in exchanging such information between vehicles may disrupt the functionality of CCWS and thus lead to false alarm or even severe accidents. Hence the level of transmission accuracy is usually required to be sufficiently high [6]. And it is also necessary to develop better vehicle position estimation schemes in order to provide an efficient and accurate way of implementing active safety applications based on CCWS.

To this end, we propose a new CCWS in this paper. Based on location history and expected movement trajectory, a vehicle's position is first estimated using dynamic information including velocity, GPS position, acceleration, and course angle. Then the estimated position is enhanced by the integration of the roadmap topology, and is used to define a plausible geographic area within which a vehicle should reside. A fusion model is used in the integration of information of roadmap and vehicle's dynamic state. Finally, according to the processed vehicle position, CCWS adopts the CPA (Closet Point of Approach) algorithm to predict the possibility of collision [7]. The experiment results show that the CCWS has a high accuracy in position estimation and collision prediction.

The rest of this paper is organized as follows: Sect. 2 presents the proposed CCWS architecture. Section 3 describes the key algorithm of the CCWS. Section 4 presents the system performance. Finally, the conclusions are given in Sect. 5.

2 Cooperative Collision Warning System Architecture

The CCWS consists of four core parts: Data Acquisition, Communication Module, Position Estimation, and Collision Warning Algorithm. Figure 1 shows the architecture of the CCWS.

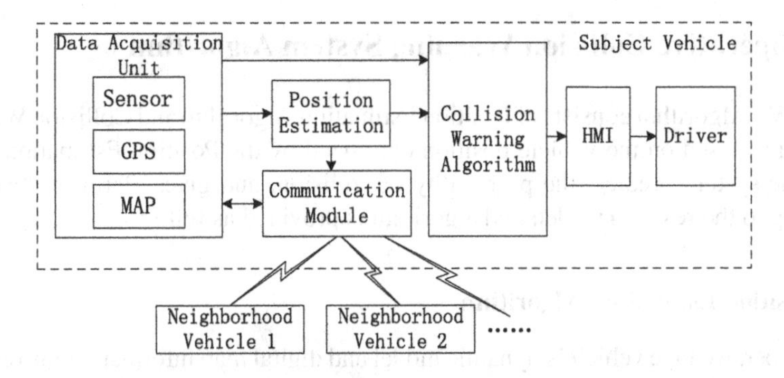

Fig. 1. System architecture

Data Acquisition Unit collects the information of sensor, GPS and digital map. All this information is then used to estimate position and broadcast through Communication Module which connects the subject vehicle and surrounding vehicles. Then the Collision Warning Algorithm will calculate the possibility of collision. If such a dangerous situation is predicted, an alarm will be triggered to inform the driver on HMI (Human Machine Interface).

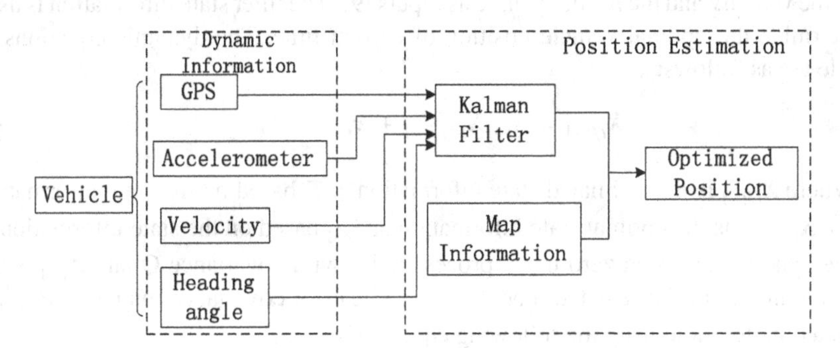

Fig. 2. Position estimation framework

Each vehicle is equipped with a GPS device, an accelerometer sensor, and a DSRC device. Generally, vehicle's state information is the foundation of the system, which includes GPS position, acceleration, heading angle, and road topology. Such information is delivered to the DSRC device at any time and broadcast periodically. The DSRC device connects the subject vehicle and neighboring vehicles in real time. According to the vehicle's dynamic state information, vehicle position is firstly estimated through a Kalman Filter (KF) method. Then the data of digital map is used to estimate the vehicle position based on the roadmap topology. Finally, we can get optimized vehicle position based on the fusion of those two estimation results. Figure 2 shows the part of Position Estimation in detail.

With the roadmap topology information and optimized position of subject vehicle and neighboring vehicles, the subject vehicle then determines the possibility of collision.

3 Cooperative Collision Warning System Algorithm

The CCWS Algorithm consists of Position Estimation Algorithm and Collision Warning Algorithm. Based on the vehicle position optimized by the Position Estimation Algorithm, the system predicts the probability of collision, and gives alarm to the driver according to the result. The detailed algorithm is provided as follows.

3.1 Position Estimation Algorithm

In this paper, we take vehicle's dynamic model and digital map information into consideration in vehicle position estimation. Firstly, the position estimation procedure uses the vehicle's previously received position updates to estimate vehicle position with the dynamic state of the vehicle. Then roadmap topology is used to estimate vehicle position. Finally, a fusion model is used to integrate the vehicle position estimated by roadmap topology and vehicle's dynamic state information [8].

(1) *Vehicle Dynamic Model*

In this approach, we integrate the model using a Kalman linear filter with the positions, the velocity and the heading angle as inputs [9]. The filter state information is used to determine the optimal vehicle position of current time. The dynamic equations of vehicle are as follows:

$$X_{t/t-1} = A_{t-1}X_{t-1/t-1} + w_{t-1} \tag{1}$$

Where $X_{t/t-1}$ is the estimated state information at T_t based on the state information at T_{t-1}. $X_{t-1/t-1}$ is the optimal state information at T_{t-1} based on the state information at T_{t-1}. w_{t-1} is the Gaussian zero-mean process noise with covariance Q and A_{t-1} is the transition matrix of the selected model at T_{t-1}. The error covariance matrix of $P_{t/t-1}$ is estimated in iteration using the following equation:

$$P_{t/t-1} = A_{t-1}P_{t-1/t-1}A_{t-1}^T + Q \tag{2}$$

In our case, X is expressed as follow:

$$X = (x, y, v, \varphi)^T \tag{3}$$

Where (x, y) denotes the vehicle position, v means the velocity of the vehicle, and ϕ denotes the heading angle of the vehicle. And the observation vector Z received form GPS is equal to:

$$Z = (\hat{x}, \hat{y}, \hat{v}, \hat{\varphi})^T \tag{4}$$

In our case, the dynamic equations of the vehicle are as follows:

$$\begin{cases} x_t = x_{t-1} + v * \Delta t * \cos \varphi \\ y_t = y_{t-1} + v * \Delta t * \sin \varphi \end{cases} \tag{5}$$

In the equations, $\Delta t = T_t - T_{t-1}$ denotes the prediction time. With the constant-input assumption, the dynamic equation of the vehicle is as follows:

$$X_{t/t-1} = \begin{pmatrix} 1 & 0 & \Delta t * \cos \phi_{t-1} & 0 \\ 0 & 1 & \Delta t * \sin \phi_{t-1} & 0 \\ 0 & 0 & 1 & 0 \\ 0 & 0 & 0 & 1 \end{pmatrix} \begin{pmatrix} x_{t-1} \\ y_{t-1} \\ v_{t-1} \\ \phi_{t-1} \end{pmatrix} + w_{t-1} \tag{6}$$

And the optimal estimation of vehicle state information is then calculated using this equation:

$$X_{t/t} = X_{t/t-1} + Kg_t(Z_t - X_{t/t-1}) \tag{7}$$

Where Kg_t is the optimal Kalman gain:

$$Kg_t = P_{t/t-1} \left(P_{t/t-1} + R\right)^{-1} \tag{8}$$

Where R is the covariance of Gaussian zero-mean observation noise. And regarding the calculated gain, we update the error covariance matrix P:

$$P_{t/t} = \left(I - Kg_t\right) P_{t/t-1} \tag{9}$$

I is the identity matrix [10, 11].

(2) *Digital Map-Guided Position Verification*

It is realistic to assume that the vehicle is going to follow the road geometry and certain road rules as the time lapses and cannot jump from one place to another one with no connection. Map-Guided position estimation uses inputs generated from GPS and supplements this with data from a high-resolution digital map to provide an enhanced position output. The static and deterministic road topology information helps to identify the correct road segment on which the vehicle is traveling and to determine the vehicle position on that segment. By doing so, we can get more accurate and reasonable position of the vehicle.

The map and road topology information will be processed as follows. Each road R_i is divided into several segments $(S_1, S_2, \ldots \ldots, S_n)$, and each S_n has four basic parameters: the road width W_{S_n}, the road length L_{S_n}, the road's heading angle H_{S_n}, and D_{S_n}, the distance between the current segment and the next intersection. The coordinates of each point of S_n are recorded by vehicles. So each vehicle has all information of all the digital map information around the vehicle $R_i(coordinates, S_n, W_{S_n}, L_{S_n}, H_{S_n}, D_{S_n})$ [12]. Coordinates mean all the points in R_i. The Fig. 3 shows the information of the digital map. And the Position Estimation's processing flow is shown as Fig. 4.

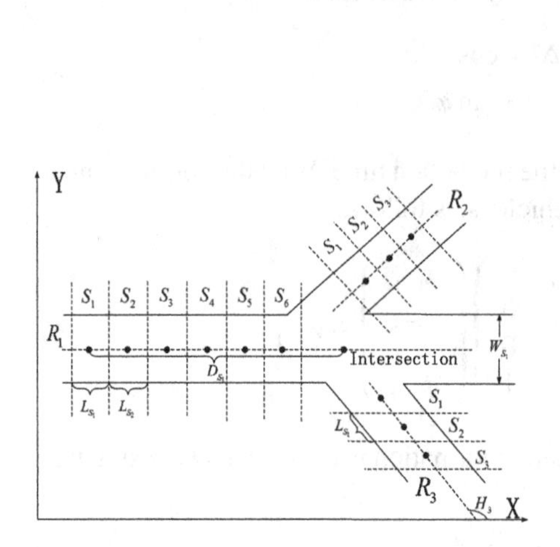

Fig. 3. Information of digital map **Fig. 4.** Flow of position estimation

In this algorithm, vehicles are assumed to store and update the periodic position information sent by other vehicles and all the digital map information around it. Once receiving a position information $(\hat{x}_t, \hat{y}_t)$ from GPS, the system will check whether the position $(\hat{x}_t, \hat{y}_t)$ matches the map. If not, the position information received from GPS is considered to be an error information. The Fig. 5 shows the error position of the GPS. Then the system begins to estimate the vehicle position with vehicle's optimized state information estimated by the system at T_{t-1} and digital map.

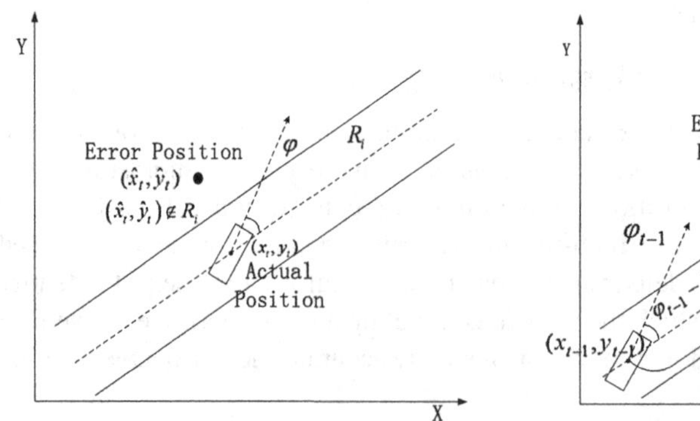

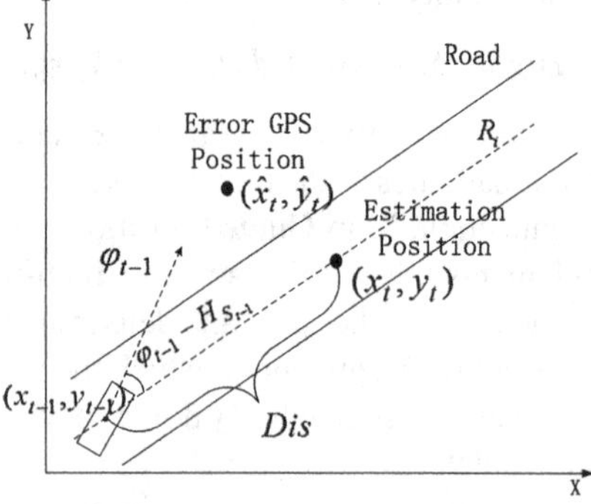

Fig. 5. Error position of the GPS **Fig. 6.** Estimated position

The system will search a road segment which the vehicle should reside using the last reliable vehicle state information and GPS position [13]. And the *Dis* is calculated to estimate the distance traveled by the vehicle during the $\Delta t = T_t - T_{t-1}$. Figure 6 shows how to calculate the distance and estimate vehicle position. Because the vehicle drives on the middle of the road, so according to the Triangle Law, *Dis* is calculated by the following equation:

$$Dis = (v_{t-1}\Delta t + \frac{1}{2}a_{t-1}\Delta t^2)\cos(\phi_{t-1} - H_{S_{t-1}}) \tag{10}$$

If $Dis \leq D_{S_i}$, we can determine that the vehicle is still on the road R_{t-1} at T_t. So the vehicle position (x_t, y_t) can be estimated with the previous state information of the vehicle and the information of the digital map. And if the $Dis > D_{S_i}$, it means that vehicle is not on the road R_{t-1} at T_t. To determine in which road segment the vehicle should most probably be found if it traveled with the previous state information, the system will search for a road whose $H_{S_i} \approx \hat{\phi}_{t-1}$. Because the update frequency of the GPS is 10 HZ and the interval of T_t and T_{t-1} is small, so we can determine the road segment according to the $H_{S_i} \approx \hat{\phi}_{t-1}$ approximately. Then we can calculate the Dis', which means the distance between the vehicle and the intersection that vehicle passed just now, by the following equation:

$$Dis' = Dis' - D_{t-1} \tag{11}$$

Then the vehicle position will be estimated as the Fig. 6 shows.

If the vehicle position $(\hat{x}_t, \hat{y}_t)$ is on the road, the system will check whether the position is reasonable by comparing the $Dis_{changed}$ and *Dis*. $Dis_{changed}$ is the distance between the optimal position (x_{t-1}, y_{t-1}) at T_{t-1} and GPS position $(\hat{x}_t, \hat{y}_t)$ received at T_t. And the $Dis_{changed}$ is expressed as follows:

$$Dis_{changed} = \sqrt{(\hat{x}_t - x_{t-1})^2 + (\hat{y}_t - y_{t-1})^2} \tag{12}$$

If $Dis_{changed} > 2Dis$, the $(\hat{x}_t, \hat{y}_t)$ will be considered as an error position because generally the distance traveled by the vehicle won't be greater than 2*Dis* during the interval ($\leq$1s). Also 2*Dis* is the optimal threshold obtained from the test. Or else, the GPS position $(\hat{x}_t, \hat{y}_t)$ will be considered as the optimal estimation position.

(3) *Information Fusion*

The combination of information given by the map-guided position and the dynamic vehicle model position is needed for estimating a more realistic and optimal vehicle position. The following equations show how to fuse those two information:

$$P = P_1^{-1} + P_2^{-1} \tag{13}$$

Where P is the variance matrix of estimated error of the fusion of the two information's estimation error variance.

$$\begin{cases} x_f = P^{-1}(P_1^{-1}x_1 + P_2^{-1}x_2) \\ y_f = P^{-1}(P_1^{-1}y_1 + P_2^{-1}y_2) \end{cases} \tag{14}$$

(x_f, y_f) is the fusion of the two vehicle position estimated by the dynamic model and the map-guided estimation.

3.2 Collision Warning Algorithm

Vehicles transmit and receive information through DSRC device periodically. CWA is calculated at host vehicle when relevant information is received from other vehicles. And the CWA computes whether the collision between target and host vehicles will occur or not based on the concept of Closet Point of Approach (CPA) [14]. If the collision will occur, an alarm will be given to the driver, otherwise the system returns to the waiting state for the next oncoming information. However, there are some situations that calculation is not needed in order to save computing resource. We can filter those situations with the assist of the digital map information.

i. Two vehicles are in the two adjacent lanes, but there is a median barrier in the central of those two adjacent lanes.
ii. Two vehicles are in the two different roads, but those two roads have no intersection.
iii. Two vehicles are heading for different directions and their distance increases with time.

After filtering, received information will be sent to calculation in order to determine whether two vehicles may collide or not. The main idea of the CPA refers to the position at which two vehicles reach their closest possible distance. Two vehicles move along two trajectories in space, however their closest distance is not the same as the closest distance between the trajectories since the distance between the vehicles must be computed at the same moment in time. So, even two vehicles' future trajectories that intersect, vehicles move along these trajectories may remain far apart. The Fig. 7 shows the CPA.

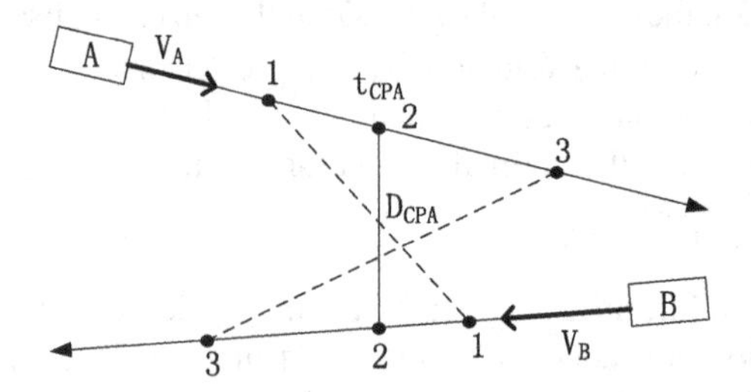

Fig. 7. Closest point of approach

The coordinates of the two vehicles are expressed as follows:

$$\begin{cases} x_A(t) = x_{A_0} + (v_A \Delta t + 0.5 a_A \Delta t^2) \cos \varphi_A \\ y_A(t) = y_{A_0} + (v_A \Delta t + 0.5 a_A \Delta t^2) \sin \varphi_A \end{cases} \tag{15}$$

$$\begin{cases} x_B(t) = x_{B_0} + (v_B\Delta t + 0.5a_B\Delta t^2)\cos\varphi_B \\ y_B(t) = y_{B_0} + (v_B\Delta t + 0.5a_B\Delta t^2)\sin\varphi_B \end{cases} \tag{16}$$

Where $x_A(t)$ means the current coordinate of the vehicle A and x_{A_0} means the original coordinate of the vehicle A. The $(v_A\Delta t + 0.5a_A\Delta t)\cos\varphi$ means the component of the distance traveled by the vehicle A on X axis. And the distance between two vehicles is illustrated as follows:

$$D(t) = \sqrt{(x_A - x_B)^2 + (y_A - y_B)^2} \tag{17}$$

$$D_{CPA} = \text{minimum}\,\{D(t)\} \tag{18}$$

Because D(t) is a complex quartic equation so that we choose to use Matlab to calculate the D_{CPA}. Then we can calculate the T_{CPA} with D_{CPA}. After calculation, we can say that vehicle A and vehicle B will collide after T_{CPA} time units if D_{CPA} is less than the size of the vehicle [15].

Based on the calculation of T_{CPA}, D_{CPA} and threshold set according to the different environment, we can know whether collision between two vehicles may occur or not, and when it will happen.

4 Performance Analysis

In order to evaluate the performance of the CCWS, three electric vehicles Roewe E50 were adopted to test this system in practice. The MK4 (Developed by Cohda Wireless and NXP) was adopted as the DSRC device, and we used an IPC (Industrial Personal Computer) as the upper computer and the CPU of the CCWS. Figure 8 shows one of the three testing vehicles.

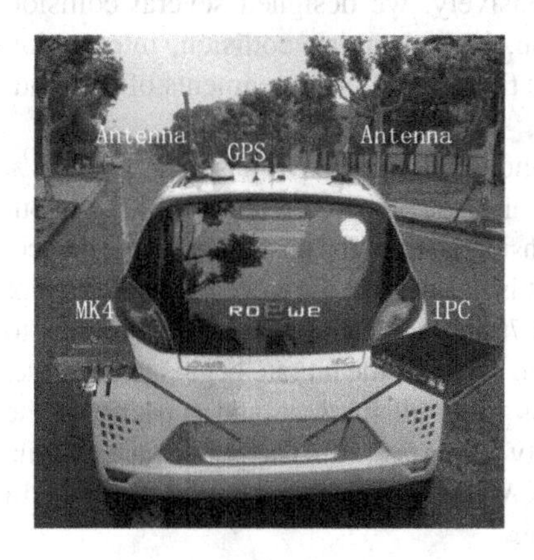

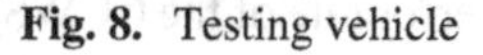
Fig. 8. Testing vehicle

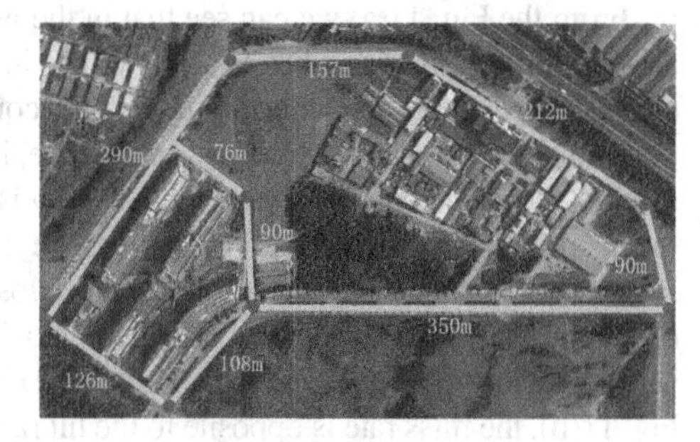

Fig. 9. Testing environment

And the Fig. 9 shows the testing environment in Tongji University, Shanghai, China. The average width of the road is 5 m, and length of each road segment is marked in the Fig. 9.

More than 300 collision tests have been conducted under various conditions including different speed, drivers, and routes. Each vehicle broadcasted the self-state information and received the state information of other vehicles periodically. In the test of Position Estimation, vehicle moved on the road center line mostly. The Fig. 10 shows the GPS trajectory and vehicle trajectory optimized by the Position Estimation Algorithm approximately.

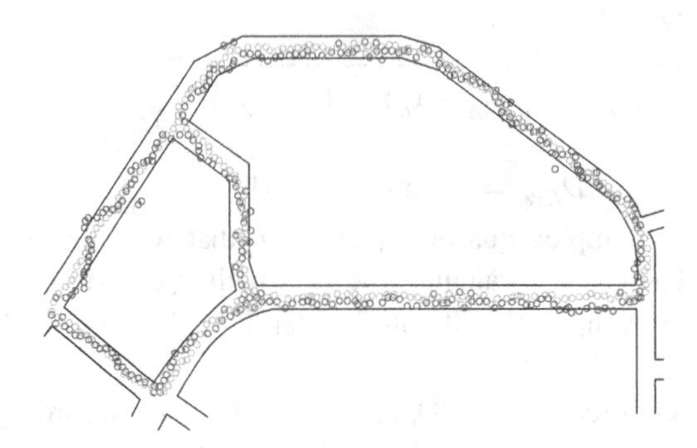

Fig. 10. The trajectory of GPS (blue) and optimized trajectory of vehicle (red) (Color figure online)

From the Fig. 10 we can see that the trajectory of GPS (blue) deviates from road center line while the optimized trajectory of vehicle (red) is more smooth and close to the real trajectory of vehicle. Consequently, GPS causes large error in accurate positioning which cannot satisfy the requirement of collision warning. But the trajectory optimized by the Position Estimation Algorithm has higher precision and accords with the vehicle's trajectory in practice much more than GPS trajectory.

In order to analyze the CCWS comprehensively, we designed several collision scenes in the test, including rear-end collision, lane-changing collision, intersection collision, etc. Each scene was tested at least 50 times at different segments of the road. And Fig. 11 shows the results of the test.

From the Fig. 11(a) we can see that in the whole test, the CCWS has more than 92 % alerting hit rate and has a better performance in hit rate compared with the CWS without considering cooperation of the digital map. In consideration of the unstability of the device, transmission delay, and packet loss in practice, it is normal that the alerting hit rate cannot reach 100 %. And when the vehicle's velocity is 70 km/h, the alerting hit rate increases to the maximum 97 %. Then with the increase of the velocity, the hit rate begins to decrease. Because of the Doppler Effect, the packet loss rate will increase and capacity of the channel will decrease when the vehicle's velocity increases [16]. The performance of the MK4 will influence the performance of the CCWS and vehicular network directly. In Fig. 11(b), the miss rate is opposite to the hit rate.

The alerting error rate is depicted in Fig. 11(c). The error rate of the CCWS is lower than 2.2 % and it increases when the velocity increases. One of the reason is that we set a low threshold when the vehicle drives at high speed in consideration of the safety. And the lower the threshold is, the easier the system alerts. So it increases the error rate

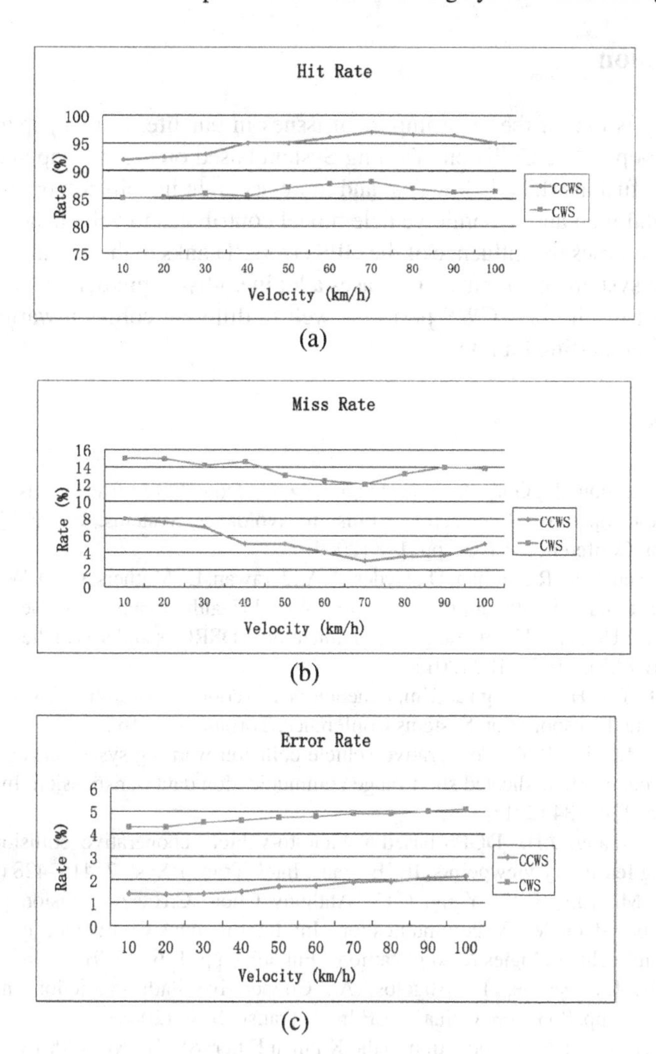

(a)

(b)

(c)

Fig. 11. (a) The alerting hit rate, (b) The alerting miss rate, (c) The alerting error rate

accordingly. Another reason is that the communication performance of the vehicular network decreases when the velocity increases because of the influences caused by Doppler Effect. The error rate of the CWS is more than 4 % while the CCWS is lower than 2.2 %. Obviously, the CCWS has a better performance in error rate.

Above all, we can see that the CCWS based on the digital map and vehicle dynamic model has a better performance in vehicle position estimation and collision prediction compared with the CWS without considering cooperation.

5 Conclusion

Driving safety is one of the most important issues in our life. In this paper, we have proposed a Cooperative Collision Warning System based on digital map and dynamic vehicle model fusion which is low cost and more accurate in collision prediction. The fusion of digital map and dynamic vehicle model contributes to vehicle accurate positioning and decreases the influence of the GPS errors. Thanks to the position estimation algorithm, the system performance increases a lot in collision prediction. According to the experiment result, the CCWS performs well in different collision warning scenes and has a higher alerting hit rate.

References

1. Fischer, J., Menon, A., Gorjestani, A., Shankwitz, C., Donath, M.: Range sensor evaluation for use in Cooperative Intersection Collision Avoidance Systems. In: IEEE Vehicular Networking Conference (VNC), pp. 1–8 (2009)
2. Peden, M., Scurfield, R., Mohan, D., Hyder, A.A., Jarawan, E., Mathers, C., O. World Health: World report on road traffic injury prevention. World Health Organization, Geneva (2004)
3. Kenney, J.B.: Dedicated Short-range communications (DSRC) standards in the united states. Proc. IEEE **99**(7), 1162–1182 (2011)
4. Huang, J.H., Tan, H.S.: Design and implementation of a cooperative collision warning system. In: Intelligent Transportation Systems Conference, Toronto, pp. 1017–1022 (2006)
5. Huang, C.-M., Lin, S.-Y.: Cooperative vehicle collision warning system using the vector-based approach with dedicated short range communication data transmission. Intel. Transp. Syst, IET **8**, 124–134 (2014)
6. Tan, H.S., Huang, J.H.: DGPS-based vehicle-to-vehicle cooperative collision warning: engineering feasibility viewpoints. IEEE Trans. Intel. Transp. Syst. **7**, 415–428 (2006)
7. Huang, C.-M., Lin, S.-Y., Yang, C.C., Anthony Chou, C.H.: A collision pre-warning algorithm based on V2 V communication. In: International Conference on Ubiqutious Information Technogologies & Applicaitions, Fukuoka, pp. 1–6 (2009)
8. Pannagiotis, L., George, T., Angelos, A.: Cooperative Path Prediction in Vehicular Environments, pp. 803–808. China, IEEE Intel. Transp. Syst. (2008)
9. Greg, W., Gary, B.: An Introduction to the Kalman Filter. SIGGRAPH 2001 Course 8, Los Angeles, CA, August 12–17, 2001
10. Miller, C., Allik, B., Piovoso, M., Zurakowski, R.: Estimation of mobile vehicle range & position using the tobit Kalman filter. In: IEEE 53rd Annual Conference on Decision and Control (CDC), Los Angles, pp. 5001–5007 (2014)
11. Quddus, M.A., Ochieng, W.Y., Noland, R.B.: Current map-matching algorithms for transport applications: State-of-the-art and future research directions. Transp. Res. Part C: Emerging Technol. **15**(5), 312–328 (2007)
12. Abu-Elkheir, M., Hassanein, H.S., Elhenawy, I.M., Elmougy, S.: Map-guided trajectory-based position verification for vehicular networks. In: IEEE Wireless Communications and Networking Conference (WCNC), China, pp. 2538–2542 (2012)
13. Zhao, D.C., Yang, Y., Huang, J., Liu, Y.H.: Vehicle position estimation using geometric constants in traffic scene. In: IEEE International Conference on Service Operations and Logistics, and Informatics (SOLI), China, pp. 90–95 (2014)

14. Huang, C.-M., Lin, S.-Y.: An advanced vehicle collision warning algorithm over the DSRC communication environment. In: IEEE Advanced Information Networking and Appliciations, Barcelona, pp. 696–702 (2013)
15. Wang, Y.P., W.J. E, Tian, D., Lu, G.Q., Yu, G.Z., Wang, Y.F.: Vehicle collision warning system and collision detection algorithm based on vehicle infrastructure integration. In: The 7th Advanced Forum on Transportation of China, China, pp. 216–220 (2009)
16. Ho, I.W.–H., Soung, C.L., Lu, L.: Feasibility study of physical-layer network coding in 802.11p VANETs. In: IEEE International Symposium on Information Theory (ISIT), Honolulu Hi, pp. 646–650 (2014)

Sensor Data Management for Driver Monitoring System

Chee Een Yap[1] and Myung Ho Kim[2](✉)

[1] School of Computing, Soongsil University, Seoul, Korea
ceyap@ssu.ac.kr
[2] School of Software, Soongsil University, Seoul, Korea
kmh@ssu.ac.kr

Abstract. Road accident becomes a threat to all drivers around the world. According to the study, fatigue or drowsiness is one of the causes to road accident. As the rapid development of the mobile devices and sensor networks, mobile based driver monitoring system has been widely proposed and discussed as an effort to reduce road accident rate around the world. Sensors such as EEG, temperature or respiration sensor are used to collect the signal from the driver to alarm the driver if drowsiness is likely to happen. However, the sensor data management of the collected data(signals) is not being paid enough attention. In this paper, we propose a sensor data management mechanism for the mobile based driver monitoring system to handle the data in a more efficient manner.

Keywords: Sensor data management · Driver monitoring · Invertible bloom filter

1 Introduction

The concept of sensor networks (SNs) was introduced in 2002 [1] following advances in telecommunication technologies. A sensor network can be defined as a composition of number of sensor nodes that spatially distributed to cooperatively monitoring physical, environmental, or human condition. These group of sensor nodes are normally small in size, low power, low cost with limited computational and wireless communication capabilities [2]. Sensor network has been widely implemented in various application such as military, habitat monitoring, health monitoring, traffic monitoring and also logistic management [3–5].

Sensor network applications can be ranged from wide geographical area application such as Great Duck Island (GDI) system [6] from UCB/Intel Research Laboratory where over 100 sensor nodes (Mica) are placed on the Great Duck Island to monitor micro-climat of nesting burrows of Storm Petrel. On the other hand, sensor network application could be also applied in a small area such as human body to monitor the condition of health. For example, Smart Sensors and Integrated Microsystems (SSIM) project [7], a sensor network application on human retina where 100 micro-sensors are built and implanted within human's eye to assist patients who suffer from vision problems.

C.-H. Hsu et al. (Eds.): IOV 2015, LNCS 9502, pp. 152–163, 2015.
DOI: 10.1007/978-3-319-27293-1_14

Sensor networks generate a very large amount of data which make data analysis a challenging task. Huge data volume from sensor network also attract the attention from researchers to discuss the issues in Big Data [8,9]. According to study, the amount of data on earth exceeded one zettabyte in 2010 [8]. By end of 2011, the data volume grew up to 1.8ZB. And it is expected to reach 35ZB in year 2020. According to Gartner [9] in 2012 "Big data is high volume, high velocity, and/or high variety information assets that require new forms of processing to enable enhanced decision making, insight discovery and process optimization." The term of big data sometimes is associated with 5V's characteristics: Volume, Variety, Velocity, Veracity and Value.

Sensor data management has been a research focus by researchers to handle the issues during managing the huge amount of data generated by sensors [10]. Database management system has been used to manage a large set of data, and it also been proposed as a tool to handle the sensor data. Database management system provides a high level of abstraction and also provides querying capability for the user to access the data. It also equipped with the security facilities which protect the data privacy of data stored. However, database management system suffers from several shortcomings for the application on sensor data. First, the database management system is too heavy to install for the sensor network especially on the driver monitoring system. The resources and computing power needed by database management system is far from the supported capability by a mobile phone. Second, database management system might not able to handle the noisy data from sensor as the sensors usually distributed and noise might be included in during the data acquisition stage. Database management system usually handle the data which well prepared and designed.

One of the major challenges in sensor data management is data ingest. The data from sensors are from various types of sensors. Each of the data has its own characteristic and representation format, it is important for the data management tool enable the storing of data in a common storage and preserving the properties of each data. As the data collected from various sensors, it is important to combine and compare the collected data for analysis purposes. The sensor data management should also support the query capability where information should be generated from the collected data through the querying process.

In this paper, we propose a sensor data management mechanism using an invertible bloom filter to manage the data from sensors. Our contributions can be summarized as below:

- Show the implementation of Invertible Bloom Filter as storage rather than a membership query algorithm.
- Storing data from different sensors in a common storage and further support query operation. The stored data can then be reveals for persistency and learning process.
- Introduction of redundancy detection to solve the data redundancy issue in a sensor network.

This paper is organized as follows. Section 2 is about the background study and motivation of this paper. In Sect. 3, we describe the traditional Bloom Filter and also Invertible Bloom Filter. In Sect. 4, we discuss the proposed sensor data management to manage the data from a mobile based driver monitoring system. Section 5 presents the experiment result and performance analysis. In Sect. 6, we conclude the paper and outline the future works.

2 Background Study and Motivation

Driving safety became one of the main concerns by society and also government around the world. According to World Health Organization report in 2013 [11], the total number of road traffic deaths hits 1.24 million a year which has become a threat to the society. According to study, one of the causes to the high number of accident case is the drowsiness or fatigue during driving. As then, many research and study focused in this domain [12–15]. As the rapid development of sensor network and mobile devices, many apps for driver safety monitoring system were proposed [16,17].

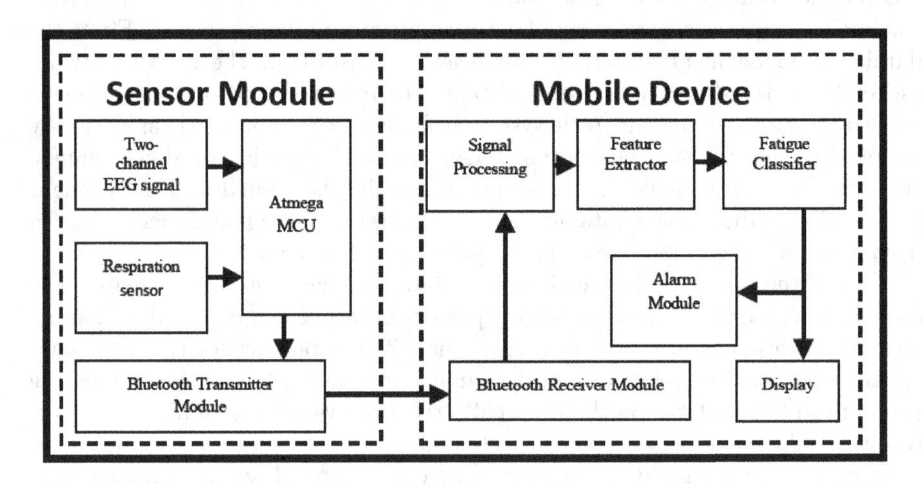

Fig. 1. Block diagram of system architecture for mobile-device-based fatigue indicator application

One of the recent publications from Boon-Giin [18] presents an application for in-vehicle non-intrusive mobile-device-based automatic detection of driver sleep-onset in real time as the detection of the drowsiness during driving. In this system, data from EEG sensor and respiration sensor are collected and analyzed to detect the drowsiness. In this paper, a sensor data management mechanism is proposed to support this driver monitoring system.

This system (Fig. 1) consists of two major parts: Sensor module and Mobile device. In the Sensor module, two channel EEG signal and respiration signal from respiration sensor are sent to Atmega MCU and further transmitted to Mobile device through bluetooth for further processing. The collected signals are pre-processed and sent to the feature extractor to extract the needed feature and served as input to classifier. The alarm shall be triggered if the computed vigilance reaches threshold set.

However, in the driver monitoring system from [18], the sensor data management of the detected signals after the decision from classifier is not being discussed. The data collected is important for the purposes of behavior pattern learning for a driver in the future. And also as an important input to further improve the performance of the classifier in the future. The data management in the driver monitoring system will face the similar challenges in big data as they do share some similarities in terms of volume, variety, velocity, veracity and value.

In order to handle the challenges in sensor network as discussed above, we propose a sensor data management mechanism based on Invertible Bloom Filter. The proposed mechanism can further solve the redundancy issue.

3 The Bloom Filters

3.1 Traditional Bloom Filter

Bloom filter proposed by Burton H. Bloom [19] as a simple space-efficient randomized data structure which enables a fast membership queries operation in 1970. Bloom filter has been widely applied in various field of applications such as network application [20,21], database application [22]. Bloom Filter consists of a vector of m bits and k hash functions $(h_1, h_2, \ldots, h_k)$ with the range of [0, m-1]. Each of the hash functions is independent from each other. Initially, each element of m is set to 0 to indicate an empty element in the vector. During the insertion process, each element $x \in S$ is used to generate hash value $h_1(x), \ldots, h_k(x)$ using the hash functions mentioned earlier. Each of the bits at positions $B[h_1(x)], \ldots, B[h_k(x)]$ in the vector of m is set to 1. To check either y is a member in the bloom filter, hash value $h_1(y), \ldots, h_k(y)$ shall be generated. If $B[h_1(y)] = 1, \ldots, B[h_k(y)] = 1$, we can conclude that y should be in the bloom filter.

Bloom filter has an advantage in term of storage where amount of memory used in storage is maintain even some number of elements are added into the set. Because of this characteristic, storage capacity of Bloom Filter usually is greater than other types of data structure such as self-balancing binary search tree and also hashes tables when the same number of memory located.

Bloom filter does not have the false negative problem (will never get a FALSE from query if the element is in the bloom filter), but suffers from the false positive problem (might get a TRUE even if the element is not in the bloom filter). The false positive rate in bloom filter can be reduced to a number that is as small as possible. Assume that each of the hash function used is independent and randomly distributed; the false positive rate of bloom filter can be calculated using the equation as below:

$$[f \approx (1 - (1 - \frac{1}{m})^{kn})^k \approx (1 - e^{-kn/m})^k] \quad (1)$$

m is bits in vector
n is number of elements inserted
k is number of hashed functions

The false positive rate of Bloom filter can be minimized when $k = ln2$ x m/n, and $f_{min} = (\frac{1}{2})^k \approx ((0.6185)^{m/n})$. As m increases in proportion to n, f will decrease. Deletion process is not supported in traditional bloom filter as deletion of any item in the list will reset $h_i(x)$ of the data to 0. If the related $h_i(x)$ also part of the result of hashing function from some other data, the bloom filter will no longer be consistent, as the deletion process will affect the other stored data in the bloom filter. This cause the false negative happens in the bloom filter. In the application of traditional bloom filter, deletion is not used as the traditional bloom filter is main to serve as membership query storage which to determine either an element exist in the storage or not.

3.2 Invertible Bloom Filter

As extension to counting Bloom Filter [23,24], Invertible Bloom Filter (IBF) has been introduced by David Eppstein and Michael T.Goodrich as a data structure which supports addition, deletion and even inversion process to regenerate the set inserted efficiently [25]. IBF is based on the peeling argument and been used for the set reconciliation.

In order to support the deletion process for bloom filter, counting bloom filter was introduced by Fan [26]. In the counting bloom filter, m bits in the traditional bloom filter replaced with a count. Any insertion of data will increase the count in the vector. And the count also set to shrink when the removal of data happened in the counting bloom filter. By having the count, the deletion process which reset the $h_i(x)$ to 0 will no longer affecting the other data who shared the same slot in the vector. However, the deletion process will not be carried out will the count is 0 which means no item exist in the particular slot. In order to determine either the data is exist in the counting bloom filter, all the count in $h_i(x)$ must not be 0. The counting bloom filter used the same configuration of hash functions as the traditional bloom filter discussed earlier (Fig. 2).

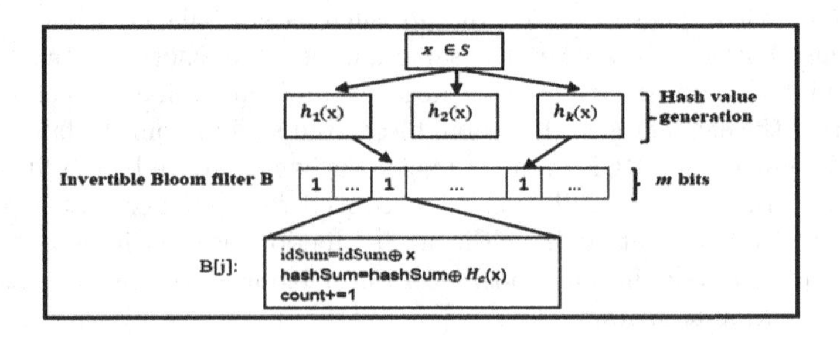

Fig. 2. Overview of Invertible Bloom Filter

In the Invertible Bloom Filter, three important fields are introduced to support the manipulation operation. The "count" field is introduced in Invertible Bloom Filter, and this field is incremented in each of hashed location when there is an element inserted to the particular index. "count" is used in counting Bloom Filter to support the addition and deletion of the elements in counting Bloom Filter. The value of "count" shall be decremented when an element is removed from the Bloom Filter. A query for whether y is in the Bloom Filter shall return only if all locations that y hashes to have a non-zero-count values. Besides the "count", Invertible Bloom Filter has "idsum" field in each cell. When an element is added into a particular cell, the IDs of the elements shall be added to "idSum" using XOR operation. Invertible Bloom Filter also uses "hashSum" as the third field. "hashSum" is used to store all the hash value that maps to the cell.

For the IBF encoding process, assume that there are input x, and k independent hash function used to generate hash value and the hash value further mod to cell using n. For IBF decoding, "pure" cell is used. "pure" cell is a cell where the count value is either 1 or -1 and the hashSum field must be equal to $H_c(idSum)$. Decoding process will return a list which contains all the elements inserted into the Invertible Bloom Filter after the process is completed. The "pure" list should be updated when any removal or addition process happen in IBF.

Table 1. Comparison between traditional bloom filter and Invertible bloom filter

Features	Traditional bloom filter	Invertible bloom filter
Support data insertion	Yes	Yes
Support data deletion	No	Yes
Support data regeneration	No	Yes

Data management mechanism must support addition and also deletion of the data. In the proposed data management, it should also support the data regeneration which is important to reveal the stored data and then further persist in the permanent storage. In the proposed sensor management mechanism, Invertible Bloom Filter is chosen over traditional bloom filter due to several features of Invertible Bloom Filter. Table 1 above shows the comparison between traditional bloom filter and Invertible Bloom Filter.

4 Proposed Solution

In order to have an effective sensor data management on the mobile based driver monitoring system with variety number of sensors, a sensor data management mechanism based on Invertible Bloom Filter is proposed. The target data are the data after the decision making process from the driver monitoring system. Similar to Invertible Bloom Filter XOR function is used in the insertion and also deletion of elements to and from the storage. We start the discussion of the proposed mechanism with the fields involved. Compare to the original IBF, one additional field name category is used. Four fields are used which are:

- "Count" – This is a counter field to determine the number of elements in a specific cell of storage. During the insertion process and deletion process, the counter shall be increased and decreased by 1.
- "idSum" – This field is used to store the sum of all the reading from the sensors which the reading should be hashed into some index in the storage.
- "hashSum" – This field is used to store the sum of all the hash values which hashed into a specific cell in the storage.
- "category" – This is a unique field used to identify the belonging of the signal. We used XOR operation to XOR all the hash value generated from k hash function for one data together with the type of the sensor.

The proposed sensor data management mechanism consists of data encoding for storing and also data decoding for retrieving purposes. The data from sensors shall be encoded using k number of hash function. Each type of the sensor shall be labeled with a number for the classification purposes.

Algorithm 1. Encoding Algorithm

for $x \in S$ **do**
    category = category $\oplus$ type
    redundancy_check(x)
    for j in HashToDistinctIndices(x, k, n) **do**
        B[j].idSum = B[j].idSum $\oplus$x
        B[j].hashSum = B[j].hashSum $\oplus$Hc(X)
        B[j].count = B[j].count + 1
        category = category $\oplus$ Hc(X)
    for i in HashToDistinctIndices(x, k, n) **do**
        B[i].category = category

Besides the encoding, in the proposed mechanism also provides a function to detect the duplication of data as data duplication is one of the major challenges in the sensor network scenario.

As in the said driver monitoring system, more than one sensor is used, it is important to differentiate the types of data in order to perform the data redundancy removal. If the data stored is redundant, it will likely increase the chances of collision happen in Bloom Filter.

For the retrieval process, "pure" cell is used which is similar to the original Invertible Bloom Filter. The "pure" list shall be updated when any removal process happened.

For the subtraction process, we use XOR operation to remove the elements inserted into the filter. As "pure" cell indicating a cell with counter = 1, we can confirm that there is only one element in the cell and the idSum of that cell is the data inserted.

From the idSum, we are able to trace the hash value generated from the data and further continue with the subtraction process to remove the element from the filter. Algorithm for the subtraction process is as below:

Algorithm 2. Redundancy Check

```
for i in HashToDistinctIndices(x, k, n) do
    if B[j].count > 0 then
        if B[j].idSum == x then
            inner_counter ++;
             outer_counter++;
        if (outer_counter == 3)&&(inner_counter > 0 then
            return true
        else
            return false
```

Algorithm 3. Decode Process

```
for i = 0 to i < n (IBF.number of cell) do
    if B[j] is pure then
        Add i to pureList
while pureList !=0 do
    i = pureList.dequeue()
    add (B[i].idSum, B[i].category) to output list
    perform IBF subtraction(B[i].idSum, B[i].hashSum, B[i].category)
for i = 0 to i < n(IBF.number of cell) do
    if B[i].idSum != 0 || B[i].hashSum != 0 || B[i].count1=0 then
        return FAIL
    else
        return SUCCESS
```

Algorithm 4. IBF Subtraction

```
Input : x
for j in HashToDictinctIndices(x, k, n) do
    B[j].idSum = B[j].idSum ⊕ x
    B[j].hashSum = B[j].hashSum ⊕ Hc(X)
    B[j].count = B[j].count - 1
    category = category ⊕ Hc(X)
for i in HashToDistinctIndices(x, k, n) do
    B[i].category = B[i].category ⊕ category
```

5 Experiment and Discussion

The dataset used in this evaluation are the actual data collected from the Mobile based driver monitoring system as stated in the background study. Driver monitoring system will analyses the data every 3 s, which consists of 603 data (300 data from EEG1(100 Hz) sensor, 300 data from EEG2(100 Hz) sensor and 3 data from Respiration (1 Hz)sensor). Based on the existing system, we prepared dataset as below:

Table 2. Dataset used in experiments

DataSet No	Data Collection Time (Sec)	Data Size
1	1	201
2	2	402
3	3	603
4	4	804
5	5	1005

5.1 Execution Time

Mobile based driver monitoring application is a real-time apps where the respond time is very important to the user. Users are expecting the result immediately and the storing process should be also efficient. In the driver monitoring system discussed in background study, the driver monitoring system will process the data from sensors every 3 s.

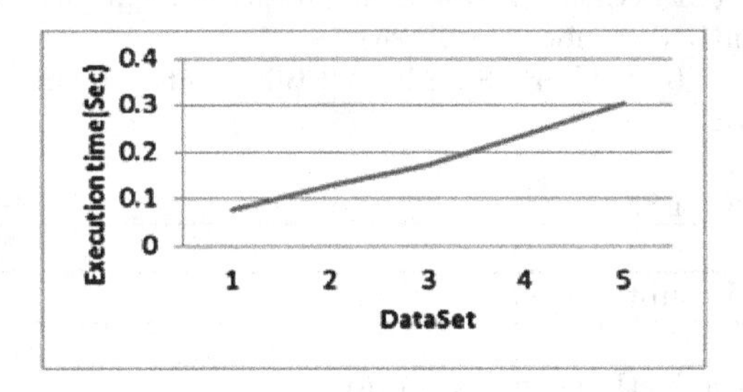

Fig. 3. Execution time in proposed sensor data management

The collection of data from the sensors should then be stored for record keeping and learning in the future. In Fig. 3, we plot the execution time used to keep the data into the memory level in the system. 0.1743 s is needed to check the redundancy and store the data for dataset 3 which representing data collected from sensors every 3 s. We also test the proposed management mechanism using different dataset which representing data from sensors every 1 to 5 s. We not consider anything more than 5 s as the driver monitoring system is a real time system, anything more than that is not reasonable. From the experiments, it shows the processing time is not more than 0.35 s even with dataset of 5 s.

5.2 Number of Redundancy

From the experiments, we realized that there are high probability of getting redundant data from sensors if the data collection environment remains unchanged for certain period of time. By referring to Fig. 4, we observed that the frequency of redundant data increases with the increment of the data size. For this, the redundancy mechanism is important to handle the big volume of data from sensors.

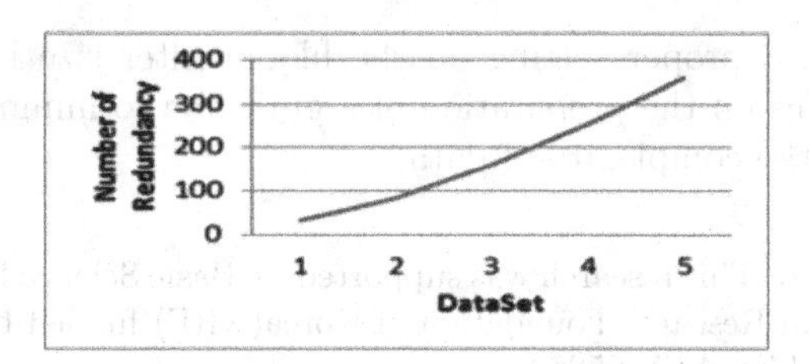

Fig. 4. Number of redundancy data

5.3 Collision Rate

Collision referring to the data that cannot be removed from the bloom filter once inserted. As discussed in many papers utilizing bloom filters, one of the ways to reduce the collision rate is through the increment of filter size. In the experiments, we show the relationship between the filter size and also the collision rate. The reading of collision drops with the increment of the filter size. From the experiment, we found that filter size with 50 times larger than the data size will have a less than 0.1 % collision rate.

5.4 Query Support Capability

We can further improve the driver monitoring system through the query support from the proposed mechanism. With the proposed mechanism, simple query such as highest reading for a specific time frame and the simple statistical analysis can be performed directly. The query result can further use to support the decision making algorithm in the driver monitoring system. As such, we can provide more accurate and precise decision for the driver.

6 Conclusion

Sensor data management should be paid attention to in order to increase the efficiency of a system. A proper sensor data management mechanism can increase the performance of data storing and retrieval. As the rapid development of sensors network, sensor data management on different sensors become more challenging especially the volume of the data increases rapidly. An efficient method is crucial to ensure the successful of the sensor network application. In this paper, we proposed a sensor data management mechanism based on Invertible Bloom Filter to manage different types of signal from variety sensors. Through the proposed mechanism, we can reduce the frequency to store the data to the permanent storage which also can save the power and energy consumption as the target application is a mobile based application. We also show how the redundant data could be removed easily with the proposed idea. In the experiment, we show the execution time of processing. However, the proposed idea suffers from collision problem. Some of the stored data could not be reveals if collision happens. For this, a proper setting on the bloom filter size is needed. In the future, we will focus on the preparation of a guideline to minimize the collision rate according to the complexity of data.

Acknowledgments. This research was supported by Basic Science Research Program through the National Research Foundation of Korea(NRF) funded by the Ministry of Education (NRF-2014R1A1A2058695).

References

1. Akyildiz, I.F., et al.: A survey on sensor networks. Commun. Mag. IEEE **40**(8), 102–114 (2002)
2. Lorincz, K., et al.: Sensor networks for emergency response: challenges and opportunities. IEEE Pervasive Comput. **3**(4), 16–23 (2004)
3. Zhang, R., et al.: Logistics transportation vehicle system for information acquisition based on wireless sensor network. Procedia Eng. **29**, 3954–3958 (2012)
4. Basu, D., et al. : Wireless sensor network based smart home: sensor selection, deployment and monitoring. In: Sensors Applications Symposium (SAS). IEEE (2013)
5. Alemdar, H., Ersoy, C.: Wireless sensor networks for healthcare: a survey. Comput. Netw. **54**(15), 2688–2710 (2010)
6. Mainwaring, A., et al.: Wireless sensor networks for habitat monitoring. In: Proceedings of the 1st ACM International Workshop on Wireless Sensor Networks and Applications, pp. 88–97. ACM, Atlanta, Georgia, USA (2002)
7. Maloberti, F., Malcovati, P.: Microsystems and smart sensor interfaces: a review. Analog Integr. Circ. Sig. Process. **15**(1), 9–26 (1998)
8. IBM: What is big data? (2012). http://www-01.ibm.com/software/in/data/bigdata/
9. Laney, D.: The Importance of Big Data: A Definition (2012)
10. Balazinska, M., et al.: Data management in the worldwide sensor web. Pervasive Comput. IEEE **6**(2), 30–40 (2007)
11. Organization, W.H.: Global status report on road safety 2013 (2013)
12. Sigari, M.-H., Fathy, M., Soryani, M.: A driver face monitoring system for fatigue and distraction detection. Int. J. Veh. Technol., pp. 11 (2013)
13. Rogado, E., et al.: Driver fatigue detection system. In: IEEE International Conference on Robotics and Biomimetics, ROBIO 2008 (2008)
14. Wen-Chang, C., et al.: A fatigue detection system with eyeglasses removal.In: 15th International Conference on Advanced Communication Technology, ICACT 2013 (2013)
15. Horn, W.-B., Chen, C.-Y.: A real-time driver fatigue detection system based on eye tracking and dynamic template matching. Tamkang J. Sci. Eng. **11**(1), 65–72 (2008)
16. Jin, Z., D. Jun, and Y. Honglue.: Driving Status' Monitoring and Alarming System Based on Information Fusion Technology. in Intelligent Control and Automation, WCICA, The Sixth World Congress on. 2006 (2006)
17. Aadi, M.F.K.a.F.: Efficient Car Alarming System for Fatigue Detection during Driving. International Journal of Innovation, Management and Technology, 3(4), 6 pages (2012)

18. Lee, B.-G., Lee, B.-L., Chung, W.-Y.: Mobile healthcare for automatic driving sleep-onset detection using wavelet-based EEG and respiration signals. Sensors **14**(10), 17915–17936 (2014)
19. Bloom, B.H.: Space/time trade-offs in hash coding with allowable errors. Commun. ACM **13**(7), 422–426 (1970)
20. Osano, T., Y. Uchida, and N. Ishikawa.: Routing Protocol Using Bloom Filters for Mobile Ad Hoc Networks. in Mobile Ad-hoc and Sensor Networks, MSN 2008. The 4th International Conference on. 2008. (2008)
21. Mitzenmacher, A.B.a.M.M.a.A.B.I.M.: Network Applications of Bloom Filters: A Survey. Internet Mathematics, 10 pages (2002)
22. Ross, M.C.a.C.A.L.a.G.A.M.a.K.A.: Buffered Bloom filters on solid state storage. in In First Intl. Workshop on Accelerating Data Management Systems Using Modern Processor and Storage Architectures (ADMS*10). (2010)
23. Li, W., et al.: Accurate Counting Bloom Filters for Large-Scale Data Processing. Mathematical Problems in Engineering, 2013, 11 pages (2013)
24. Yongsheng Hao, Z.G.: Redundancy Removal Approach for Integrated RFID Readers with Counting Bloom Filter. Journal of Computational Information Systems, 9(5),8 pages(2013)
25. Eppstein, D. and M.T. Goodrich.: Straggler Identification in Round-Trip Data Streams via Newton's Identities and Invertible Bloom Filters IEEE Trans. on Knowl. and Data Eng., 23(2)297–306 (2011)
26. Fan, L., et al.: Summary cache: a scalable wide-area web cache sharing protocol. IEEE/ACM Trans. Netw. **8**(3), 281–293 (2000)

Nighttime Vehicle Detection for Heavy Trucks

Zhijuan Zhang[1], Hui Chen[1](✉), Zhiguang Xiao[1], Linlin Chen[2], and Reinhard Klette[3]

[1] School of Information Science and Engineering, Shandong University, Jinan, China
huichen@sdu.edu.cn
[2] China National Heavy Duty Truck Group Co., Ltd., Jinan, China
[3] School of Engineering, Auckland University of Technology, Auckland, New Zealand

Abstract. This paper presents a method for detecting vehicles at nighttime, particularly for an application in heavy trucks. Researchers suggested detecting vehicles at nighttime based on symmetry of taillights, or by training a classifier. The headlights of heavy trucks are very bright which causes that taillights of a vehicle in front appear as being asymmetrical. The bright headlight also defines disturbing information that affects the training of a classifier. We propose an improved threshold algorithm that can effectively remove most non-taillight regions and strengthen the shapes of taillight pairs. Positive and negative samples are extracted from these thresholded (pre-processed) images to train a vehicle classifier by using Haar-like features and AdaBoost. We then detect vehicles in these pre-processed images. Experiments are performed using two alternative methods. Results show that our method, i.e. combining threshold pre-processing and training a classifier, is more accurate and robust than the other two methods.

1 Introduction

A *driver-assistance system* (DAS) can improve traffic safety by providing useful information about scenes around the *ego-vehicle* (i.e. the vehicle where the system is operating in), or improves driving comfort. Common input data for a DAS are digital images (e.g. CCD or CMOS), infrared images, Lidar, radar, ultra-sound, or GPS data. A *vision-based* DAS (VB-DAS) uses computer vision as its defining component. It may aim at detecting vehicles, pedestrians, obstacles, traffic signs, or other traffic-related patterns around the ego-vehicle [1,4]. The detection of vehicles in front of the ego-vehicle is essential for reducing the possibility of accidents. Because of insufficient lighting, incorrect detections are more likely at nighttime. The detection of vehicles in front of an ego-vehicle, under challenging lighting conditions, is a current research topic in computer vision [10]. In particular, a heavy truck on a highway, often driving at high-speed and with very bright headlights, has an increased chance of a collision with another vehicle at nighttime. This paper considers vehicle detection at nighttime for heavy trucks, taking the particularities of such an ego-vehicle into account.

C.-H. Hsu et al. (Eds.): IOV 2015, LNCS 9502, pp. 164–175, 2015.
DOI: 10.1007/978-3-319-27293-1_15

Because of insufficient light, features applied for vehicle detection at daytime are typically not suitable for detection at nighttime. At nighttime, a vehicle in front shows a pair of bright taillights which is basically the most available information for vehicle detection at nighttime [5]. Bo Liu et al. use color and motion information of taillights to locate vehicles [6]. They analyze the R, G and B channel information in taillight areas and extract those pixels with more red intensity. Then they locate vehicles by spatial relationships between two taillights. Junfu Guo et al. consider symmetry for a pair of taillights, the luminance character in taillight areas, and the red-channel information to detect vehicles at nighttime [2]. These authors set a reasonable threshold in HSV color space to extract taillights and to get rid of non-objects by evaluating the histograms of 300 taillight images. Then, they locate a taillight pair by areas of highlights and their spatial relationship. Umesh Kumar [5] aims at detecting vehicles in front of the ego-vehicle by following the standard Viola-Jones approach (see, e.g., [3]), i.e. by training a classifier with Haar-like features and using subsequently an AdaBoost algorithm. Kumar also applied an accurate threshold algorithm to remove noise from the images and to simplify the detection process.

In many nighttime scenarios, two taillights of a front vehicle are not symmetrical, and also not at the same height, for example, when the vehicle is driving around a corner. In these cases, Junfu Guo's method [2] cannot detect vehicles correctly. Besides, in recorded CCD images, the taillights usually do not appear red but close to white, especially when the taillights are further away from the ego-vehicle (a heavy truck in our case). When this occurs, Bo Liu's method is not very effective. Training a classifier for headlights and taillights is an alternative

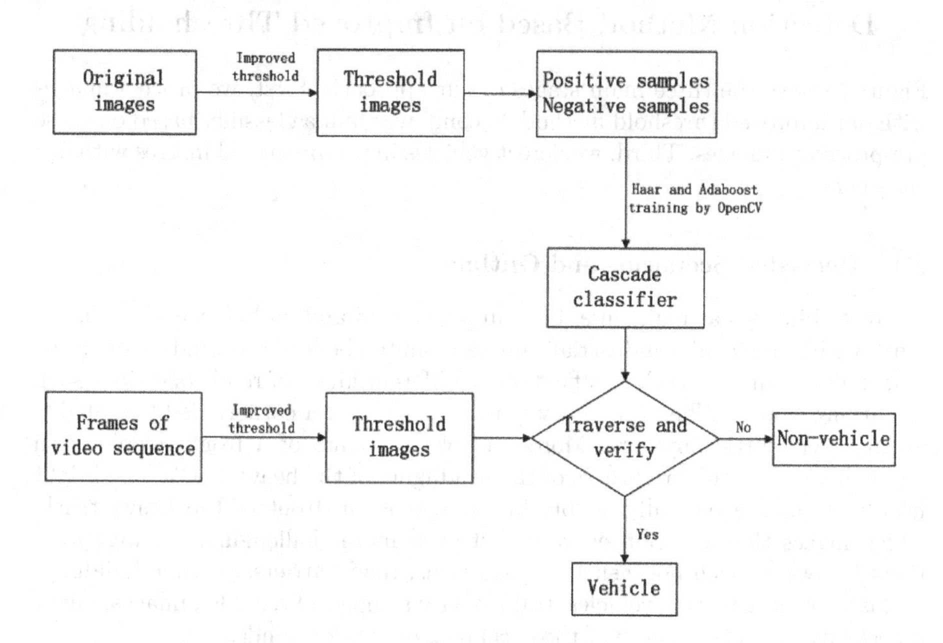

Fig. 1. Sketch of our proposed method

practical solution. Considering that our goal is to detect vehicles in front of heavy trucks, we need to be aware that headlights of the ego-vehicle are so bright that there is a large intensively illuminated area in front. This causes a higher false-detection rate when detecting vehicles directly in unprocessed images.

In this paper, we propose an improved threshold algorithm based on Kumar's method. Our algorithm performs pre-processing for removing most of the noise while pairs of taillights of vehicles front remain in the images. This helps to obtain a simplified background and to make the outlines of a taillight pair more obvious. Furthermore, this method also weakens Haar-like features of no-vehicles, such as for road signs, road barriers, buildings, and so on. We also train a classifier and detect vehicles based on thresholded pre-processed images. The threshold pre-process simplifies the training samples as well as the detection environment, and makes the classifier to pay more attention to the taillight area, all of which improves the *robustness* (i.e. at times when the environment changes suddenly) and accuracy of the detection. We demonstrate the effectiveness of our method by means of experiments.

The remaining sections of this paper are organized as follows. Section 2 provides a description of our detection method and shows our improved threshold algorithm. It firstly gives a detection flow chart, describes the overall detection ideas, and then proposes our improved threshold algorithm based on Kumar's threshold method. Section 3 briefly recalls Haar-like features, their time-efficient calculation, and the AdaBoost meta-algorithm. Section 4 contains experimental results and their discussion. Section 5 concludes.

2 Detection Method Based on Improved Thresholding

Figure 1 shows the three main stages of our approach. First, we process images with our improved threshold method. Second, we train a classifier based on these pre-processed images. Third, we detect vehicles in pre-processed images with our classifier.

2.1 Recorded Scenarios and Outline

We record images at nighttime, both in a test field and on highways in a heavy truck with a monocular industrial camera mounted behind the windscreen of the truck. The available truck test field offers different kinds of road conditions such as varying slopes. When the heavy truck goes around a corner, visible taillights are not symmetric anymore. Moreover, two taillights of a front vehicle often merge because of the brightness of the headlights of the heavy truck. The bright headlights create basically a "big bright region" in front of the heavy truck, which makes the detection environment even more challenging. On highways, there is always much noise such as road signs, road barriers, or some buildings. A classifier for detecting vehicles, trained with samples of recorded images, might detect false objects because of those challenging backgrounds.

For videos or images recorded in this driving environment, we propose a threshold algorithm which can remove most of the background but which keeps different kinds of shapes of front-vehicle taillights. We segment many of these different shapes as our positive samples, and segment other regions as our negative samples in the threshold-processed images. We use the samples to train a strong cascade classifier based on Haar-like features and AdaBoost; this classifier consists of many weak classifiers.

2.2 Our Improved Threshold Algorithm

For simplifying recorded images and intended detection, we aim at finding a reasonable threshold value for removing noise (as much as possible).

We tested Otsu's algorithm ([8]; see also description in [3]). Otsu's algorithm calculates a threshold value by minimizing the intra-class variance, which corresponds to maximising the inter-class variance.

We also tested Kumar's method [5]. He uses the maximum difference of bright values to the standard deviation as the threshold value for eliminating non-candidate objects. Let T_t be a threshold value for the image (or *frame*) I_t recorded at time t. It is defined by

$$T_t = \max_{(x,y)\in\Omega} \{I_t(x,y) - \sigma_t\} = \left\{ \max_{(x,y)\in\Omega} I_t(x,y) \right\} - \sigma_t = M_t - \sigma_t \tag{1}$$

with

$$\sigma_t^2 = \frac{1}{|\Omega|} \sum_{(x,y)\in\Omega} [I_t(x,y) - \mu_t]^2 \quad \text{and} \quad \mu_t = \frac{1}{|\Omega|} \sum_{(x,y)\in\Omega} I_t(x,y) \tag{2}$$

where $I_t(x,y)$ is the value of the pixel at (x,y) in the current frame I_t, and Ω is the image carrier (i.e. the set of all pixel locations, following [3] in notation). The maximum M_t in Eq. (1) simply equals $G_{\max}$ in general, the general maximum intensity value for the considered images, thus we could also simply use $T_t \approx G_{\max} - \sigma_t$.

Kumar's method is suitable for images whose intensity distribution of pixels is shown as in the histogram in Fig. 2, with peaks at two gray-levels. One of those

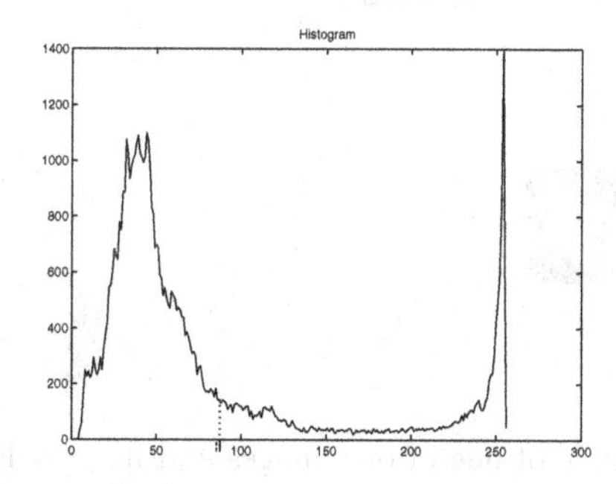

Fig. 2. Kumar's method corresponds to the shown histogram

is at M_t, the maximum intensity in image I_t, defined by vehicle lights and other bright interference such as road lamps. The second peak is basically defined by the mean intensity (or a minor shift towards 0) which corresponds in the night to the nearly uniform "gray-ness" in the image at non-illuminated regions; see [5]. Otsu's method follows from a theoretical derivation. Kumar's method follows from experimental evidence. Kumar's method requires less calculations than Otsu's algorithm. However, as illustrated by Fig. 3, upper left, our images are dominantly defined by a situation where there are no additional lights (such as, e.g., road lamps) besides vehicle lights. For such a case, the shape of a gray-level histogram is the one shown in Fig. 3, upper right, with minor "bumps" at the upper end of the scale rather than one dominant peak.

The intensities of most pixels in Fig. 3, upper left, are lower gray-levels, about between 10 and 20. The histogram does not repeat the one of Fig. 2 because of very few bright pixels. The histogram of the whole image does not show just two dominant peaks, and Kumar's method is not suitable for this kind of images.

Consider a window just showing one taillight, as shown in Fig. 3, lower left. For its histogram, see Fig. 3, lower right. This histogram shows basically two peaks or clusters, a peak for bright pixels and a cluster for lower gray-levels (about between 30 and 80). In such a local case, we can again derive a local threshold following the Kumar's method. Besides, the size of a car light at a distance is smaller than if near, and a smaller car light is basically always in the upper half of the image. Thus, we evenly divide an image into several parts from the first row to the last. We scan every part with a small window of a suitable size, such as 11×11 or smaller. The size of the window is different for every part,

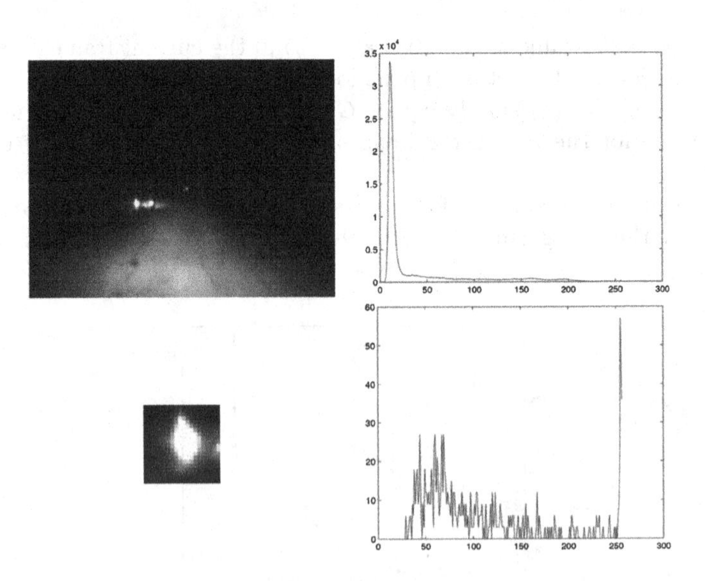

Fig. 3. *Upper row*: Example of one of our images and its gray-level histogram. *Lower row*: Window showing just one taillight, and its gray-level histogram

and it grows from the top towards the bottom. Because the maximum gray-level in a taillight region is always greater than an easily adjustable threshold M (e.g. $M = 250$), we just calculate thresholds at pixel locations where the maximum gray-level in the window is greater than M, and the mean intensity in the window is greater than the mean gray-level of the lower half of the whole image. Assume that, for the *ith* part, we calculated thresholds, in this way, $T_1, T_2, \ldots, T_j, j > 0$, in different j window positions of the *ith* part. We choose the maximum as the threshold of the *ith* part, named as T_{ti}. In this way, we can find i thresholds for i parts. More formally: For gray-level image I_t of size $N_{\mathrm{cols}} \times N_{\mathrm{rows}}$, obtain c thresholds $T_{\mathrm{t1}}, T_{\mathrm{t2}}, \ldots, T_{\mathrm{c}}$ as follows:

1. Calculate the mean μ_l of the lower half of I_t. Let $i = 1$, and the value of i is always less than or equal to c.
2. Set rectangle $(0, N_{\mathrm{rows}} \times (i-1)/c, N_{\mathrm{cols}}, N_{\mathrm{rows}}/c)$ as current processing area. Let $j = 1$.
3. In the current region, select a small window of size $(m \times i) \times (n \times i)$, with $m \ll N_{\mathrm{cols}}$, and $n \ll N_{\mathrm{rows}}$ for scanning the current part. If the mean gray-level μ_j of pixels in the window is greater than μ_l and the maximum value M_j in the window is greater than M, then let $T_j = M_j - \delta_i$, let $j = j+1$, and go to the next position of the window until the end of the current rectangular region is reached.
4. Let $T_{\mathrm{ti}} = \mathrm{mean}\{T_1, \ldots, T_j\}$. If we do not obtain a threshold in the current part, let $T_{\mathrm{ti}} = 255$.
5. Let $i = i+1$ and reset $j = 1$. Continue to calculate a threshold for the next rectangular region with a window of size $(m \times i) \times (n \times i)$.

The threshold operation (i.e., let $I_t(x, y) = 255$ if it was $I_t(x, y) \geq T_{ti}$ before, and let $I_t(x, y) = 0$ otherwise) produces an 8-bit gray-level image rather than a binary image, still allowing us to take gray-level samples when later detecting vehicles in these pre-processed images.

3 Haar-like Features and AdaBoost

A taillight pair is the most obvious characteristic of a vehicle in front at night that can be represented by Haar-like features. AdaBoost provides a tool for training a strong classifier, which selects the most suitable combinations of Haar-like features as weak classifiers at different stages of the generated strong classifier.

A classifier based on Haar-like features is in general much faster than one based on the feature defined by key points and descriptors [12]. There are several kinds of Haar-like templates, and basic templates are divided into classes for edge features, central features, linear features, and directional features. Every template includes white and black rectangles. Intensity values of pixels in the white and black areas are accumulated separately. The feature value is defined as a weighted combination of these two sums [9].

Haar-like features can be computed very fast based on an intermediate representation of the image, called the *integral image* [11]. The integral image at location (x, y) equals the sum of the pixel values above and to the left of (x, y):

$$I_{\mathrm{int}}(x,y) = \sum_{i \le x, j \le y} I(i,j) \tag{3}$$

where $I_{\mathrm{int}}(x,y)$ is the integral image and $I(x,y)$ the original image.

The AdaBoost meta-algorithm was improved based on a boosting algorithm by Freund and Schapire [13]. The AdaBoost algorithm cascades several weak classifier to form a strong classifier. A weak classifier [13] is defined as follows:

$$h(x,f,\Box,\theta) = \begin{cases} 1 & \text{if } f(x) < \Box\theta \\ 0 & \text{otherwise} \end{cases} \tag{4}$$

where x is the sample, $f(x)$ is the feature of a training sample (it is a Haar-like feature in our case), θ is a threshold value, and $\Box \in \{+,-\}$ specifies the direction of the inequality.

The AdaBoost algorithm finds a most suitable weak classifier for the current training stage to minimize the error, and assigns a coefficient for this weak classifier. Then it changes the weights of samples in the training set based on results of the current classification. Finally it generates a weighted summation to form a strong classifier [13].

4 Experiments and Analysis

We used a heavy truck and a Firefly MV camera to shoot video data (8-bit gray-level images). About half of the data was recorded in an HOWO T7H truck on the national Vehicle Testing Ground at DingYuan, Anhui, China; It has a highway forming an elliptic loop and standard longitudinal slope (20 % ∽ 60 %). The remaining data were recorded in an HOWO ZZ3257M3847C truck on the Jiqing highway and Jinghu highway in times between dusk and night.

We implemented our method in C/C++, OpenCV 2.4.7, and Visual Studio 2010. Our experiments included three tasks: (a) Compare results using Otsu's algorithm or Kumar's method with our improved threshold method. (b) Train a classifier with samples of the threshold pre-processed images from the Vehicle Testing Ground and detect vehicles in threshold pre-processed images. Train a classifier with samples from the original images and detect vehicles in the original images. Define two test image sets, one of which consists of images from the Vehicle Testing Ground, and the other one consists of images from highways. Detect vehicles in both image sets and compare the results. (c) Detect vehicles only based on symmetry of taillight pairs, and compare.

4.1 Comparing Thresholding Methods

We process the images by four threshold values to obtain an 8-bit gray-level image with two values 0 and 255, for convenience for the next step of our experiments. Figure 4 shows results obtained by Otsu's algorithm, Kumar's method, and our improved threshold method. Here, we evenly divide the image into four rectangular regions from top to bottom. The sizes of the template windows are

Fig. 4. Comparison of the three threshold methods. *Left*: Otsu's algorithm. *Center*: Kumar's method. *Right*: Our improved thresholding method

4×4, 8×8, 12×12, and 16×16, respectively. The threshold values for the four parts are 255, 255, 224, and 255.

By comparing the three results, it shows that our method generates better results than both Otsu's algorithm or Kumar's method. By the proposed threshold algorithm, most of the background of the image can be removed, while different kinds of shapes of taillight pairs remain which will be cut to make positive samples later in the trained classifier. We also detect vehicles in the pre-processed images.

4.2 Training Two Classifiers in Traditional or Our Method

Training a classifier requires to collect positive and negative samples, to generate a description file of positive samples, and to train a classifier from these samples. We adopt application programs provided by OpenCV to acquire a classifier, i.e. `opencv_createsamples.exe` for making a description file of positive samples, and `opencv_haartraining.exe` for training the classifier [7]. We use images from the Vehicle Testing Ground to train our classifier.

First we train a classifier in the traditional method based on Haar-like features and the AdaBoost algorithm. We prepare samples of original images shot at the testing ground. In these images, we cut vehicle regions as positive samples and other non-vehicle regions as negative samples. The size of positive samples is 20×20. We made 945 positive samples and 2400 negative samples. We run `opencv_createsamples.exe` to make our positive sample description file. Then we run `opencv_haartraining.exe` to train our Haar-feature-based classifier. Thus we obtain a classifier based on a traditional method, named `cascade1`. Finally we apply `cascade1` for detecting vehicles in the original images.

Second we train a classifier following our method. We obtain different shapes of vehicle taillights which are cut from the threshold pre-processed images to establish our positive samples. Other candidate parts of the pre-processed images are selected for our negative samples. We also set the size of positive samples as 20×20. We cut 1205 taillight pairs as our positive samples for constructing a positive sample set. We cut non-taillight regions as negative samples to obtain a set of negative samples. We made 2540 negative samples. By training on these samples, we obtain our classifier, named as `cascade2`. We apply `cascade2` on image which is processed in our improved threshold method and then mark the detection results in the original images.

We detect vehicles at different distances by changing the size of the detection template. We use our classifier to detect vehicles in data collected on the testing ground or on the highway. A detection result of a pre-processed image is shown in Fig. 5, left, and it is also marked in the original image as shown in Fig. 5, right. The image in the first row was taken at the testing ground, and that in the second row on the highway. Figure 5 shows that our method is suitable to data both from the testing ground and the highway. Figure 6 shows an image from the testing ground, the detection results of the traditional method, and of our method. This example shows that our method is more accurate because we removed most of the noise like the road sign.

Moreover, we prepared two test image sets to compare and analyze the performance of the classifiers above. One set consists of 200 images from Vehicle Testing Ground, which included 222 vehicles at different distances. Another consists of 200 images from the highway including 303 vehicles.

In a comparative experiment made with images from the testing ground, we change scales by a factor of 1.1, which means that the size of the detection template reduces 10 percent between two detection scans. We compared the detection rate and the false detection numbers of the two classifiers. Results are shown in Table 1. In the experiment with the highway image data, we set the

Fig. 5. Detection results. *Top row*: In pre-processed image from testing ground, and result marked in input image. *Bottom row*: In pre-processed image from highway, and result marked in input image

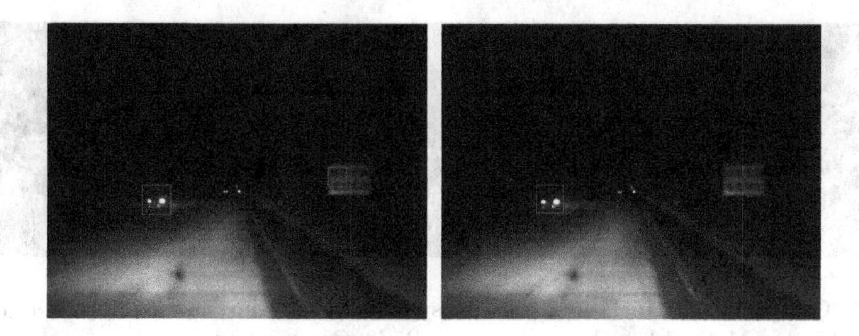

Fig. 6. Detection results. *Left*: Traditional method. *Right*: Our method

Table 1. Detection results of two methods with images from testing ground

Training and detection	Detection rate	# Miss detections	# False detections
No pre-processing	89.2 %	24	89
Improved method	94.1 %	13	28

Table 2. Detection results of two methods with images from highway

Training and detection	Detection rate	# Miss detections	# False detections
No pre-processing	16.2 %	205	98
Improved method	89.4 %	32	12

scales at 1.2 to detect vehicles. Results are shown in Table 1. By adjusting all the training parameters and the number of positive and negative samples, we obtained several classifiers following the traditional method, and `cascade1` is the best of those. When we make the training samples equivalent to the number of those used for training `cascade2`, unexpectedly, the performance of the final classifier is worse (detection rate of just 69.8 %) than that of `cascade1`.

The results of our comparisons demonstrate that our `cascade2` classifier has typically a higher detection rate and a reduced false-detection rate compared to the `cascade1` classifier, trained on samples cut from the unprocessed images. Table 2 also illustrates that our classifier is also suitable for the highway environment, and that it is more robust than when trained in the traditional way. We conclude that our improved threshold method removes most of the noise and simplifies the objects. Both results demonstrate that we obtain a better classifier by combining our improved threshold algorithm with the training of a classifier. Our method of training with samples from the pre-processed images, and detecting in pre-processed images, is more robust than other variants.

4.3 Detection Based on Taillight Pair Symmetry

We wrote another C/C++ program to detect taillight pairs by the symmetry feature. This program segments the gray-level image processed with our threshold method. For every segmented objective block, we calculate its area size and

Fig. 7. Comparison of method based on symmetry and our method. *left*: Original image. *Center*: Method based on symmetry. *Right*: Our method

centroid. By comparing their area sizes and centroid coordinates, based on symmetry, we determine a pair of taillights which belong to a vehicle. We mark the detected pairs in the threshold pre-processed images so as to compare the results. We calculate two types of results, either with the method of symmetry, or with our method combing threshold pre-processing and training of a classifier. Detection results are illustrated in Fig. 7.

By comparing the two groups of detection results, we can conclude that taillight pair detection fails often when following the symmetry feature, while our trained classifier can locate the vehicles well even in challenging cases, such as when driving on a sloped road, taillights are merging, and so forth.

4.4 Discussion

Experiment 4.1 shows that our threshold method is more suitable for the recorded images. It removes most of the noise and keeps the shape of vehicle's taillight pairs, which simplifies samples as well as the detection task. The comparison of results in 4.2 shows that a higher detection rate and a reduced number of false detections is possible when detecting vehicles in threshold pre-processed images by the classifier, which is trained with samples from the pre-processed images. This is more robust because the improved threshold method simplifies the samples and the detection scene and strengthens the outline of taillights. In some cases, especially when the ego-vehicle drives on a steep slope, or the taillights merge together, taillights of a vehicle in front are not at the same height and are also not symmetric. The method based on symmetry cannot detect vehicles correctly in such cases while our method can work well. Experiments reported in 4.3 verify this point.

All these experimental results show that our method can detect affront vehicles at night more accurately and robustly. Experimental results also verify the effectiveness of the proposed algorithm.

5 Conclusions

Considering the characteristics of images taken in a heavy truck on a test field or on highway, we proposed an improved threshold algorithm based on Kumar's method which simplifies the background and keeps the shapes of taillight pairs of front vehicles. Our image pre-processing procedure contributes to enhance the feature of taillight pairs and to weaken the feature of non-vehicle objects. Thus, theoretically, it has advantages over the training method with samples cut from the original images. Experiments also show that our method is more accurate and robust compared to others, and that our method has better adaptability.

The proposed vehicle detection system at nighttime for trucks can assist truck drivers to avoid collisions with other vehicles in front, and improves road safety.

Acknowledgments. This work is supported by the Natural Science Found of Shandong NSFSD under No. ZR2013FM032.

References

1. Yen-Lin, C., Chuan-Yen, C.: Embedded vision-based nighttime driver assistance system. In: Proceedings of IEEE International Conference on Transactions of Computer Communication Control and Automation, pp. 199–200 (2010)
2. Guo, J.B., Wang, J.Q.: Vehicle detection in nighttime based on monocular vision. Automot. Eng. **36**, 574–585 (2014). (in Chinese)
3. Klette, R.: Concise Computer Vision. Springer, London (2014)
4. Klette, R.: Vision-based driver assistance systems. Webster, J. (ed.), Wiley Encyclopedia Electrical Electronics Engineering (2015). doi:10.1002/047134608X. W8272
5. Kumar, U.: Vehicle detection in monocular night-time grey-level videos. In: Proceedings of IEEE International Conference on Image Vision Computing New Zealand, pp. 214–219 (2013)
6. Liu, B., Zhou, H.Q., Wei, M.X.: Vehicle detection in nighttime based on color and motion information. J. Image Graph. **10**, 187–191 (2005). (in Chinese)
7. OpenCV. www.opencv.org.cn
8. Otsu, N.: A threshold selection method from gray-level histograms. IEEE Trans. Syst. Man Cybern. **9**, 62–66 (1979)
9. Rakate, G.R., Borhade, S.R.: Advanced pedestrian detection system using combination of Haar-like feature, AdaBoost algorithm and edgelet-shapelet. In: Proceedings of IEEE International Conference on Computational Intelligence Computing Research, pp. 1–5 (2012)
10. Rezaei, M., Terauchi, M., Klette, R.: Robust vehicle detection and distance estimation under challenging lighting conditions. IEEE Trans. Intell. Transp. Syst. **16**(99), 1–21 (2015)
11. Paul, V., Michael, J.: Rapid object detection using a boosted cascade of simple features. In: Proceedings of IEEE Conference on Computer Vision Pattern Recognition, pp. 511–518 (2001)
12. Xue, P., Tong, G., Jing, Q.: Face detection in video based on AdaBoost algorithm and eye location. Video Eng. **35**, 114–117 (2011). (in Chinese)
13. Zhu, C.: Face detection and tracking based on OpenCV. Comput. Eng. Appl. **48**, 157–158 (2012). (in Chinese)

A Quality Analysis Method for the Fuel-level Data of IOV

Daxin Tian[1,4,5], Yukai Zhu[1], Haiying Xia[2(✉)], Jian Wang[1], and He Liu[3]

[1] Beijing Key Laboratory for Cooperative Vehicle Infrastructure Systems and Safety Control, School of Transportation Science and Engineering, Beihang University, Beijing 100191, China

[2] Research Institute of Highway Ministry Transport, Beijing 100088, China
hy.xia@rioh.cn

[3] The Quartermaster Equipment Institute of the General Logistics Department of CPLA, Beijing 100010, China

[4] Key Laboratory of Urban ITS Technology Optimization and Integration, The Ministry of Public Security of China, Hefei, China

[5] Jiangsu Province Collaborative Innovation Center of Modern Urban Traffic Technologies, Beijing 100191, China

Abstract. With the development of Internet of Vehicles (IOV), data mining on vehicle running status data has been a hot field of research, and the data quality has an important effect to the result of data mining. In this paper, we have investigated the problem of multidimensional analysis of vehicle running state data, especially the abnormal fuel-level data. In order to screen out the vehicles with abnormal sensors or equipments and evaluate the credibility of the data, we propose a bayesian classification algorithm to efficiently assess the data quality and screen out the abnormal vehicles in our database, with coefficient of variance (COV) and dispensation (COD) as feature attributes. Moreover, the accuracy indicators F-score and PPV of the classifier are used to determine the optimal threshold of the classifier. Our experiments on large real datasets show the feasibility and practical utility of proposed methods.

Keywords: IOV · Fuel level data · Naive bayes · Data quality control

1 Introduction

With the development of Internet of Vehicles(IOV), people pay more and more attentions to utilization of the vehicle running status data. Particularly, since the data mining techniques has been well-developed, the vehicle running data has revealed high worthiness in commercial spheres and city planning and designment [5]. On the other hand, it is well know that the data quality has strong relation to the result of data mining. In recent years, with the wide use of wireless communication and IOV technology, the data from vehicles which is low-cost and abundant, has become a very important data source for data mining in traffic

C.-H. Hsu et al. (Eds.): IOV 2015, LNCS 9502, pp. 176–185, 2015.
DOI: 10.1007/978-3-319-27293-1_16

engineering. But due to the uncertain causes, some sensor of vehicle is not accurate to some extent. Wireless communication may also cause loss of data. Hence, the vehicle data may be incorrect or incomplete and thus affect the accuracy of vehicle data mining. Therefore, quality control and analysis on these data is indispensable. In this paper, Vehicle running data consists of a large number of sensors and collects huge amount of multidimensional data, e.g. locations, time, velocity and fuel level. In some cases, the vehicle sensors occasionally or consistently report abnormal readings and atypical data (i.e., outliers), such data may imply the abnormalities of sensors or the unreliability of wireless communication environment.

Analysis of vehicle running data quality belongs to data quality control. Generally, data quality is defined by correctness, completeness, consistency. Data cleaning aims to solve the error, reduplication and conflict among data. In the field of data mining, data cleaning is a regular procedure in preliminary. Machine learning method is now widely applied in large amounts of fields, including knowledge discovery, sensors network organization and other fields of engineering. L.A. Tang and J.W. Han use data mining techniques in multi-dimensional sensor network data analysis and integration [12]. Support Vector Machine (SVM) and neural network has been widely used in fault diagnosis of sensors [4,8]. Supervised learning algorithms, as a category of machine learning methods, is suitable for digging regular patterns about outliers through the massive data in our case.

In this paper, we explain the data cleaning in vehicle running data quality respectively in preliminary. Vehicle running data need to be of high quality to reflect the operation pattern and running status of the vehicle, including vehicle fuel consumptions and travelling route [1]. As a result, the vehicle running status data need to be collected which means that there are bare error in the generation, transmission, collection of these data. In order to evaluate the quality level of the data or to diagnose the sensor fault, the study object's vehicle running status data is output in database and used as Bayesian filtering model's input files to screen out the vehicle with inferior data quality. The database and outliers situation are introduced in Sect. 2, the related formulas and indicators are introduced in Sect. 3. Then we preprocess the data in Sect. 4 first, and compute two indicators to evaluate the classification accuracy and finally determine the optimal threshold.

2 Backgrounds and Preliminaries

2.1 Database Introduction

The database used in this paper contain the vehicle running state data of more than 5,000 vehicles, which are mainly trucks, van trucks and cargo-buses distributed throughout China. The vehicle running state data consists of GPS coordinates, speeds, driving direction angle, fuel level value, abd uploading time of each record.

2.2 Typical Abnormal Situation

It's well know that there should be positive correlation between the correct fuel consumption and the speed data, which is the fuel level should decline when the vehicle is running. Contrary to the common sense, the fuel level of some vehicles remain unchanged while the vehicle is moving over an extended period. In some other conditions same values appear frequently, while the uploaded fuel-level data are float numbers which should always fluctuate in a lesser range. We suppose that the possible causes of these abnormal situation are in the following items.

- The sensors of vehicles may have too much noise or error of them are too large.
- The terminal equipments may upload or read wrong values or error codes in communication with the sensors.
- The information processing system on the server side may read and update wrong or null values into the database.

2.3 Naive Bayes

Naive Bayes methods are a set of supervised learning algorithms based on Bayes theorem by assuming that the attributes are conditional independent [10]. Given a class label y and a dependent feature vector x_1 through x_n, Bayes theorem can be stated as the following relationship:

$$P(y|x_1, ..., x_n) = \frac{P(y)P(x_1, ..., x_n|y)}{P(x_1, ...x_n)} \tag{1}$$

Using the conditional independence assumption that:

$$P(x_i|y, x_1, ..., x_{i-1}, x_{i+1}, ..., x_n) = P(x_i|y) \tag{2}$$

For all i, this relationship is simplified to :

$$P(y|x_1, ..., x_n) = \frac{P(y)\prod_{i=1}^{n} P(x_i|y)}{P(x_1, ...x_n)} \tag{3}$$

Since $P(x_1, ..., x_n)$ is constant given the input, we can use the following classification rule:

$$\hat{y} = \arg\max_{y} P(y)\prod_{i=1}^{n} P(x_i|y) \tag{4}$$

and we can use Maximum A Posteriori estimation to estimate $P(y)$ and $P(x_i \mid y)$; the former is then the relative frequency of class y in the training set [15]. Since Naive Bayes is a supervised learning method, the labels of the samples is needed previously.

2.4 Gaussian Model

Bayesian classifier is widely used in spam filtering and medical diagnosis. In these fields, the values of the attributes are usually binary or discrete, so the likelihood of the features can be computed simply by Bernoully Naive Bayes(BernoullyNB) or MultinomialNB model [13]. But in this paper, the vehicle running data consist of continuous numeric attributes (e.g. speed, fuel level, GPS-coordinates). Hence we use GaussianNB model to compute likelihood of the features, which implements the GaussianNB algorithm for classification. The likelihood of the features is assumed to accord with Gaussian probability-density function:

$$P(x_i \mid y) = \frac{1}{\sqrt{2\pi\sigma_y^2}} \exp\left(-\frac{(x_i - \mu_y)^2}{2\sigma_y^2}\right) \tag{5}$$

The parameters σ_y and μ_y are estimated by Maximum Likelihood Estimation (MLE). Then the threshold criterion for classifying a vehicle's running data as credible becomes:

$$\frac{P(y)\prod_{i=1}^{n} P(x_i|y)}{\sum P(y)P(x_1, ...x_n)} > T \tag{6}$$

3 Experiments

3.1 Data Preprocessing

Bayesian classifier requires a number of vectors of feature parameters, where the class label is drawn from some finite set. In other word, we need some feature parameters associated with two class of vehicles as input items. It's well known that feature selection can lower computing cost and improve the classifier's performance in the process of constructing the classifier, which can reduce the amount of processed data, save the processing time, reduce the effect of noise in the data, and improve the performance of information processing system [11]. From the above, data preprocessing is a pivotal procedure in our work, and we need to solve three problems in this procedure:

- Except for the fuel level, how to assess the data quality of the other attributes of the dataset.
- What features of the dataset should be reference value in evaluating the accuracy of the fuel level data.
- How to organize these classification features.

Evaluating the Accuracy of the Speed Data. In order to assess the data quality of the speed data, we compute the travel speed according to the GPS coordinates, and compare it with the speed which is uploaded by the terminal equipments. The travel speeds is compute by the following formula:

$$V_t i = \frac{111.199[(\varphi_i - \varphi_{i+1})^2 + (\lambda_i - \lambda_{i+1})^2 \cos(\frac{\varphi_i + \varphi_{i+1}}{2})]}{t_i - t_{i+1}} \tag{7}$$

φ denote the latitudes and λ denote the longitudes,and we compute the relative error ratio of the uploaded speeds v_u with the calculated travel speeds v_t:

$$\varepsilon = \frac{\sum_{i=1}^{n} v_{ui} - \sum_{i=1}^{n} v_{ti}}{\sum_{i=1}^{n} v_{oi}} \tag{8}$$

Then we found the minimum ε of all vehicles in the database is higher than 95 % which means that the two speeds are approximate extremely, thus we assume that the velocity data and the GPS coordinates are mainly correct as reference values in evaluating the accuracy of the fuel level data.

Features Selection. Since the dataset contains multidimensional feature vectors, feature subset selection is indispensable to reject redundant attributes. Redundant attributes are provide no more information than the attributes which have been selected, and irrelevant features give no useful information in the classification [6]. How to select and define the appropriate features for the classifier is an important issue. On one hand, under the previous assumption in the above subsection, there should be strong positive correlation between the correct fuel consumption and the speed data. Hence we use the Pearson correlation coefficient [9] between fuel consumption q_i and speed v_i of each vehicle as an important feature:

$$\hat{r_{q,v}} = \frac{n \sum q_i v_i - \sum q_i \sum v_i}{\sqrt{n \sum {q_i}^2 - (\sum q_i)^2} \sqrt{n \sum {v_i}^2 - (\sum v_i)^2}} \tag{9}$$

On the other hand, because the fuel consumption should vary in a certain range, the discrete degree of the fuel level data should also be in a certain range. In some abnormal conditions of the dataset, the fuel level of a sample vehicle remain unchanged while the vehicle is moving over an extended period which make the degree of dispersion diminish. While in some other conditions, the frequent error indication may boost the dispersion. Similar to the previous indicator, we assume that the coefficient of variation of the fuel level values of one vehicle ought to vary in an appropriate range. The coefficient of variation can be estimated by using the ratio of the sample standard deviation s, to the mean fuel level value $\bar{q}$ of the sample vehicle:

$$\hat{c_v} = \frac{s}{\bar{q}} = \frac{\sqrt{\frac{1}{n} \sum_{i=1}^{n} (q_i - \bar{q})}}{\bar{q}} \tag{10}$$

The distribution of these feature values is shown in next subsection as Fig. 2.

3.2 Experimental Methodology

Constructing Training Sets. As a supervised learning in machine learning field, bayesian classifier require that each example is a pair consisting of an input object such as a number of vectors and a desired output value which is also called supervisory signal [7]. The algorithm analyzes the training data and produces an inferred function, which is used in mapping new datasets. Since Naive Bayes is a supervised learning method, the labels of the samples is needed previously, which means we should stick labels to each sample manually before we construct the classifier. In this study, we tried to find the vehicles with large amounts of outliers which are defined as typical samples with abnormal data through observing the fuel level-time curve.

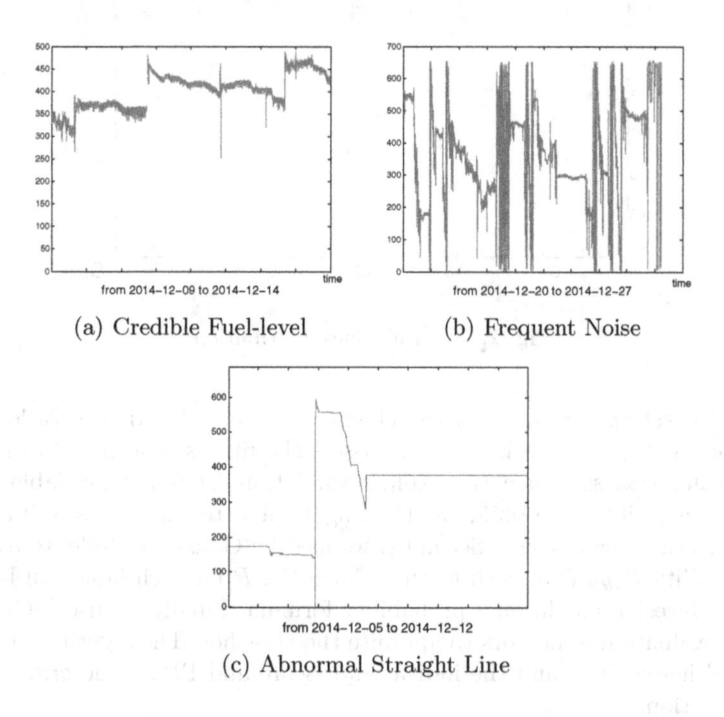

(a) Credible Fuel-level

(b) Frequent Noise

(c) Abnormal Straight Line

Fig. 1. Typical fuel level curves

The typical fuel level-time curves of two sample sets are as Fig. 1. From the comparison of these figures, Fig. 1(a) shows a typical fuel level curve of the sample with relatively credible data, while Fig. 1(b) and (c) shows the opposite situation. As the curves above show, the fuel-level values of credible vehicle should be in an appropriate range of dispersion, and fuel consumption should be fit for speed data to some extent. According to these factors, finally we find two typical sample sets consist of fifty vehicles with relatively credible data as training set y_1 and other fifty vehicles in opposite situation as training set y_2. The distribution of the feature values of two set is as Fig. 2, which the red points denote the positive instances set y_1 and the blue points denote the negative one.

Classifying Algorithm. We define the Pearson correlation coefficient between speed v and q as x_1, the coefficient of dispersion of fuel level data q as x_2.

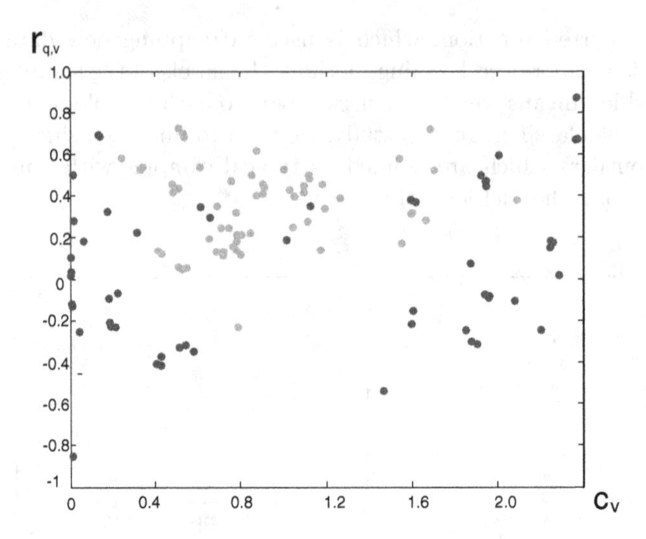

Fig. 2. Feature values distribution

In order to test the accuracy of the classifier, we use the Monte Carlo Cross-Validation(MCCV), which is a way to assess the fitness of a model to a hypothetical validation set when the explicit validation set is not available [2]. At first of the algorithm, we build the training set and testing set,as well as compute the feature vectors $x_{k,i}$. Secondly, we use the Gaussian Model to estimate the probability $P(y|x_i)$ of each feature, Then the P for each lable can be computed by Bayesian conditional probability formula. Finally we use MCCV and compute evaluation indicators to appraise the classifier. The algorithmic flow is illustrated hereinafter, and the indicators F-score and PPV is described in the next subsection.

3.3 Identification of the Classifier

F-Score. F-score (also called F-measure or F1-score) is a measure to evaluate a test's accuracy, which considers both the precision and the recall of the test to compute the score. The F-score can be interpreted as a weighted average of the precision and recall, where an F-score reaches its best value at 1 and worst at 0. As an indicator which is usually used in the field of information retrieval for document classification and query classification performance, F-score is also used in machine learning, especially in classification problem [3]. The balanced F-score F used in this paper(F1 score) is the harmonic mean of precision P and recall R, which is defined as:

$$F = 2 \cdot \frac{precision \cdot recall}{precision + recall} \tag{11}$$

Where *precision* is the number of correct positive results divided by the number of all positive results, and *recall* is the number of correct positive results

Algorithm 1. Bayesian Classifier

1: Let $x_{1k} = r(q_k, v_k), x_{2k} = cod(q_k), k = 1, ..., n$ be the feature vector of the kth instance.
2: Create training set y_k with label y_1, y_2 and testing set $u_j, j = 1, ..., m$.
3: Training the Classifier.
4: **for** k=1 **to** n **do**
5: Construct Gaussian Model $G(x_1), G(x_2)$ for each feature
6: Estimate σ_y, μ_y for each feature attribution using maximum likelihood estimation.
7: **end for**
8: Testing the Classifier using cross-validation.
9: **for** each T (thresholds which are set manually) **do**
10: **for** j=1 **to** m **do**
11: Compute probability for each feature $P(u_j|x_i) = G(x_i)$.
12: Compute $P(u_j) = \prod_{i=1}^{2} P(u_j|x_i)$.
13: **if** $[P(u_j) > T$ **then**
14: **return** the lable $l_j = y_1$.
15: Verify the label$l$ and update the result list.
16: **end if**
17: Compute the F-score and PPV.
18: **end for**
19: **end for**
20: Compare the indicators with each T and find the optimal one.

divided by the number of positive results that should have been returned, they are both common indicators in the information retrieval and machine learning field. The larger the F-score is, the more effective the classifier is likely [14]. In this paper, the F-score is used to judge whether the bayesian classifier is effective. After this preliminary assessment process, we found the classifier is actually effective. Then we came to a new problem that whether this indicator is the most appropriate evaluation criteria to this classification issue.

Optimal Threshold. Since F-score is a synthetical index in common classification problems, it can just indicate that the classifier is effective to some extent. For purpose of finding the optimal threshold T of the classifier, we define the second evaluation indicator as Positive Predictive Value (PPV) which is ratio of the number of true positives to the number of positive calls:

$$PPV = \frac{\sum TP}{\sum TP + \sum FP} \tag{12}$$

TP (True Positives) which means the number of the positive instances are measured accurately, while FP (False Positives) denotes the number of the negatives which are classified as positive wrongly. In general classification scenes, the ratio of TP to total positives which is called precision is more widely used. But in this paper, the purpose of the classification is to find the vehicles with credible data which is suitable for data mining and screen out the sensors and uploading equipments which are working in abnormal operational statement. As a result, we aim to find vehicles with high quality data more precisely and screen out more negative ones, the PPV indicator is appropriate to some extent. To find the optimal threshold T, we take an amount of experiments with differentiated T by interval number of 0.05, the experimental instances are selected by Monte Carlo cross-validation method (MCCV). The experimental result is shown in Table 1, total P and N is fifty each.

Table 1. Experimental results

Threshold	TP	TN	FP	FN	F-SCORE	PPV
0.40	44	37	13	6	0.813	0.772
0.45	43	38	12	7	0.818	0.782
0.50	41	42	8	9	0.825	0.835
0.55	41	44	6	9	0.877	0.872
0.60	38	44	6	12	0.807	0.863

From the experimental results, we found that both indicators of F-score and PPV come to maximum when the threshold $T = 0.55$. F-score and PPV both indicate that this bayesian classifier is feasible and efficient to our dataset. Hence we determine the optimal threshold and apply this classifier to our whole database. Then we found data of around 2,200 vehicles is of lower credibility.

4 Conclusion

In this paper, we have investigated the problem of multidimensional analysis of vehicle running state data. Some statistical indicators are introduced and applied in data quality evaluation, and a bayesian classification algorithm is proposed to efficiently assess the data quality and screen out the abnormal vehicles in database. Moreover, a proposed accuracy indicator of the classifier is used to determine the optimal threshold of the classification algorithm. Our experiments on large real datasets show the feasibility and practical utility of proposed methods. This paper is our first step in the vehicle running state data analysis. In the further research, the classification algorithm used in data quality assessment will be extended to a semi-supervised learning method. We will also develop the work of data analysis to support more complex applications, such as the atypical event detection and trustworthiness analysis based on our database. Otherwise, we are interested in applying more data mining methods in further study, such as vehicle mobility patterns identification in road nets.

Acknowledgments. This research is supported by the National Natural Science Foundation of China under Grant No. 91118008, and the Foundation of Key Laboratory of Road and Traffic Engineering of the Ministry of Education in Tongji University.

References

1. Ahn, K., Rakha, H., Trani, A., Van Aerde, M.: Estimating vehicle fuel consumption and emissions based on instantaneous speed and acceleration levels. J. Transp. Eng. **128**(2), 182–190 (2002)
2. Arlot, S., Celisse, A.: A survey of cross-validation procedures for model selection. Stat. Surv. **4**, 40–79 (2009)
3. Beitzel, S.M.: On understanding and classifying web queries. Ph.D. thesis, Citeseer (2006)
4. Bishop, C.: Pattern recognition and machine learning. IEEE Trans. Pattern Anal. Mach. Intell. (PAMI) **16**(4), 049901 (2006). Springer
5. Biswas, S., Tatchikou, R., Dion, F.: Vehicle-to-vehicle wireless communication protocols for enhancing highway traffic safety. IEEE Commun. Mag. **44**(1), 74–82 (2006)
6. Christopher, M.B.: Pattern Recognition and Machine Learning. Company New York 16(4), 049901 (2006)
7. Mohri, M., Rostamizadeh, A., Talwalkar, A.: Foundations of Machine Learning (2012). http://Kioloa08.mlss.cc
8. Ni, J., Zhang, C., Yang, S.X.: An adaptive approach based on KPCA and SVM for real-time fault diagnosis of HVCBS. IEEE Trans. Power Delivery **26**(3), 1960–1971 (2011)
9. Rodgers, J.L., Nicewander, W.A.: Thirteen ways to look at the correlation coefficient. Am. Stat. **42**(1), 59–66 (2012)
10. Tan, P.N., Steinbach, M., Kumar, V.: Introduction to Data Mining, Pearson education/Addison. Server (Part 26(25)), 236–238 (2006)
11. Provost, F., Fawcett, T.: Robust classification for imprecise environments. Mach. Learn. **42**(3), 203–231 (2001)
12. Tang, L.A., Yu, X., Kim, S., Han, J., Peng, W.C., Sun, Y., Gonzalez, H., Seith, S.: Multidimensional analysis of a typical events in cyber-physical data. In: 2012 IEEE 28th International Conference on Data Engineering, pp. 1025–1036 (2012)
13. Metsis, V.: Spam filtering with naive bayes - which naive bayes?. In: Third Conference on Email and Anti-Spam (CEAS) (2006). Telecommunications
14. Zemmoudj, S., Kemmouche, A., Chibani, Y.: Feature selection and classification for urban data using improved F-score with support vector machine. In: 2014 6th International Conference of Soft Computing and Pattern Recognition (SoCPaR), pp. 371–375. IEEE (2014)
15. Zhang, H.: The optimality of naive bayes. Int. J. Pattern Recogn. Artif. Intell. **19**(2), 183–198 (2005)

A Safety Guard for Driving Fatigue Detection Based on Left Prefrontal EEG and Mobile Ubiquitous Computing

Jian He[1,2(✉)], Mingwo Zhou[1], Chen Hu[1], and Xiaoyi Wang[1,2]

[1] School of Software Engineering, Beijing University of Technology, Beijing 100124, China
jianhee@hotmail.com
[2] Beijing Engineering Research Center for IoT Software and Systems, Beijing 100124, China

Abstract. According to the portable and real-time problems on the driving fatigue prevention based on electroencephalogram (EEG), a headband integrated with Thinkgear EEG chip, tri-axial accelerometer, gyroscope and Bluetooth is developed to collect the subject's left prefrontal Attention, Meditation EEG and head movement data. The relation between Attention and Meditation EEG when the subject is in the state of concentration, relaxation, fatigue and sleep is analyzed firstly. As a result, a new method for driving fatigue detection based on the correlation coefficient between subject's Attention and Meditation EEG is proposed. Meanwhile, the slide windows and k-Nearest Neighbors (k-NN) algorithm are introduced to classify the correlation coefficient between the subject's Attention and Meditation EEG, so as to detect driving fatigue and alert. Lastly, a software running on an Android smart device is developed based on the above technologies, and the experiment proves that it has noninvasive and real-time advantages, while its sensitivity and specificity are 80.98 % and 90.43 % respectively.

Keywords: Driving fatigue detection · Prefrontal lobe EEG · Ubiquitous computing · K-NN

1 Introduction

With the rapid expansion of cars in China, transportation safety has been of increasing concern. How to avoid or reduce transportation accidents has become a hot research field recently. Driving fatigue is a large factor in transportation accidents because of the marked decline in the driver's perception, recognition, and vehicle control while fatigued. For example, driver's fatigue is believed to account for 35 %–45 % of all vehicle accidents [1]. According to statistics, if the driver's response time could be half a second faster, 60 % transportation accidents could be avoided [2]. Hence, developing an accurate and noninvasive real-time driving fatigue detection system would be highly desirable, particularly if this system can be further integrated into an automatic warning system [3].

Different research has been reported on driving fatigue detection including methods identifying physiological associations between driver's fatigue and the corresponding

C.-H. Hsu et al. (Eds.): IOV 2015, LNCS 9502, pp. 186–197, 2015.
DOI: 10.1007/978-3-319-27293-1_17

patterns of the electroocculogram (EOG) (eye movement), electroencephalogram signals (brain activity), and electrocardiogram (ECG) signals (heart rate) [4]. Schmidt found that the eye blink duration and the blink rate typically increases while blink amplitude decreases as function of the cumulative time, and the saccade frequencies and velocities of EOG decline when people get fatigued [5]. And many research suggested that the change in the cognitive state can be associated with significant changes in the EEG frequency bands, such as delta (δ: 0–4 Hz), theta (θ: 4–8 Hz), alpha (α: 8–13 Hz), and beta (β: 13–20 Hz) [6, 7], or their combinations [8]. Besides, heart rate variability was found to be applicable for the detection of fatigue using the ECG power spectrum [9]. Although studies based on EOG signals showed that eye-activity variations were highly correlated with human fatigue and can accurately and quantitatively estimate alertness levels, the step size (temporal resolution) of those eye-activity based methods is relatively long (about 10 s) to track slow changes in vigilance [10]. On the contrary, the step size of the EEG-based methods can be reduced to about 2 s to track second-to-second fluctuations in the subject's performance [11]. However, the existing majority of EEG-based driving fatigue detection needs many wire sensors, which often makes the subject uncomfortable [12]. So the practical application on EEG-based driving fatigue detection technology faces great challenges, such as portability and real-time requirements.

According to those challenges, this paper is proposing a wearable driving fatigue detection system by using a mobile smart device, and a headband integrated with ThinkGear EEG chip, tri-axial accelerometer, gyroscope and Bluetooth to detect driving fatigue and issue alarm. The system incorporates an array of features, such as issuing alarm, automatic playing music as soon as it detects driving fatigue and subject's nod with agreement.

The rest of this paper is organized as follows. Section 2, the relation between the left prefrontal Attention and Meditation EEG is analyzed, and then a new driving fatigue detection method is proposed. Section 3 introduces the methodology to deploy a system based on the correlation coefficient of subject's left prefrontal Attention and Meditation EEG. Section 4, an implement of the driving fatigue system based on the above technologies is discussed. Section 5 shows the experiment and the result analysis. Section 6 summarizes the study and provides future research ideas.

2 Foundation

Prefrontal lobe plays a key role for the attentive regulation, thinking and reasoning. It accepts and processes information from sensory, motor and other brain regions, and then sends back the processed message so as to control activities of the related brain regions [13]. Hence, the state of fatigue while driving could be detected as long as the EEG data from the driver's frontal lobe could be real-time monitored. Since MindWave EEG headset integrated ThinkGear EEG chip, can collect the subject's left prefrontal EEG from FP1 [14] with 512 Hz frequency, and produce the Attention and Meditation EEG with 1 Hz frequency. The value of attention ranges from 0 to 100, the higher the value indicates that one's attention is more concentrated. The value of meditation also ranges from 0 to 100, the higher the value indicates that the brain activity of the user is lower. Therefore, MindWave EEG headset is used to study the relation between the Attention and Meditation EEG from the subject's prefrontal lobe.

Four kinds of 20 min scenarios (namely Concentration, Relaxation, Fatigue and Sleep) are designed to analyze the relationship between the subject's left prefrontal Attention and Meditation EEG. Concentration means the brain is in a state of concentration, like the subject is reading a book, or solving a math problem. Relaxation means the brain is in a state of relaxation, like walking or meditating. Fatigue is when the subject just finishes lunch, and wants to sleep but could not sleep. Sleep is when the subject falls asleep. Four graduate volunteers, two males and two females, aged from 22 to 25 are selected to put on MindWave EEG headset and carry out the four scenarios test. Meanwhile, each volunteer tests 3 times for each scenario, and the Attention and Meditation EEG from the subject's left frontal lobe are collected during the experiment. Since volunteers need time to transfer into a specific state after putting on their MindWave, only ten minutes of data, which is in the middle of each experimental data, is extracted. The relationship between the Attention and Meditation EEG in the state of concentration, relaxation, fatigue and sleep are shown respectively in Fig. 1

$$r = \frac{\sum_{i=1}^{n}(Xi - \overline{X})(Yi - \overline{Y})}{\sqrt{\sum_{i=1}^{n}(Xi - \overline{X})^2}\ \sqrt{\sum_{i=1}^{n}(Yi - \overline{Y})^2}} \tag{1}$$

It can be seen from Fig. 1 that there is a symmetrical relationship between Attention and Meditation EEG. Therefore, the correlation coefficient is introduced to analyze the relationship between Attention and Meditation in these four scenarios. The correlation coefficient r can be calculated according to the formula (1). X_i and Y_i present the value of Attention and Meditation EEG from the subject's left prefrontal lobe respectively at i moment, $\overline{X}$ and $\overline{Y}$ present the average value of the Attention and Meditation EEG respectively.

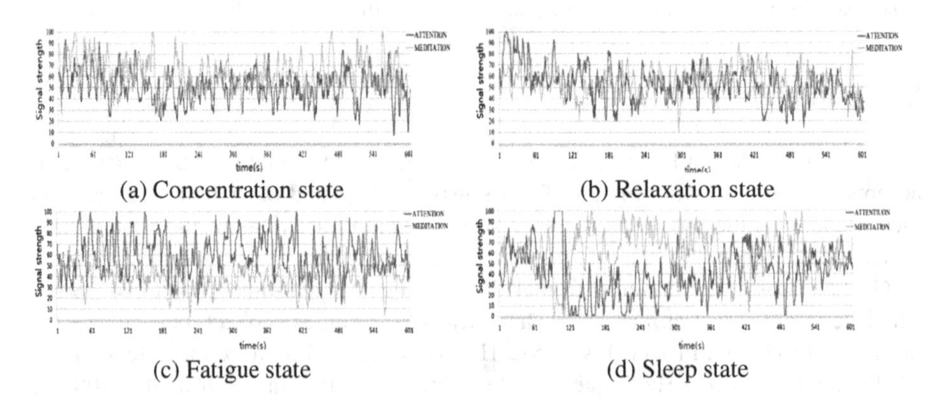

(a) Concentration state (b) Relaxation state

(c) Fatigue state (d) Sleep state

Fig. 1. The relationship between Attention and Meditation EEG when the subject is in the state of concentration, relaxation, fatigue and sleep shown respectively.

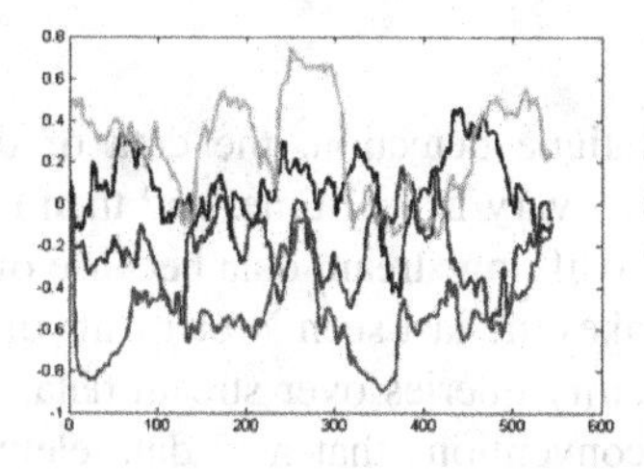

Fig. 2. Four correlation coefficients between Attention and Meditation EEG

Figure 2 shows the four correlation coefficients between Attention and Meditation EEG which are calculated by formula (1). Among them, the green curve, the black curve, the blue curve, and the red curve correspond correlation coefficients between Attention and Meditation EEG using concentration, relaxation, fatigue and sleep scenarios respectively. Figure 2 illustrates that the four correlation coefficient curves could be identified respectively as long as an appropriate classifying method is chosen.

Figure 3 shows the trends of Attention and Meditation EEG when the subject is in states of normal driving, shifting gears, turning, and fatigue. Since the duration for shifting gears and turning is very short, the sample duration is 10 s in Fig. 3. It can be seen from Fig. 3 that the symmetric relationship of fatigue between Attention and Meditation EEG in each scenario is quite different from each other. It also proves that it is feasible to detect the driving fatigue according to the correlation coefficient of driver's left prefrontal Attention and Meditation EEG.

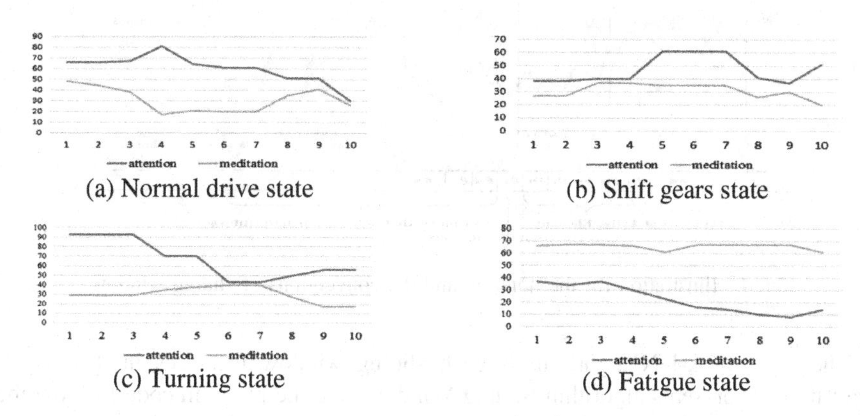

Fig. 3. The trends of Attention and Meditation EEG when the subject is in states of normal drive, shift gears, turning, and fatigue.

3 Methodology

Being a lazy learning algorithm, which uses the similarity function to answer queries, k-NN algorithm has been widely used as an effective classification model. Hence, it is introduced to classify the correlation coefficient between the subject's left prefrontal Attention and Meditation EEG, and detect driving fatigue.

3.1 Feature Extraction

In the process of driving fatigue detection, the data of the driver's left prefrontal Attention and Meditation EEG vary in real-time, and then makes up stream data. This causes great challenges in classifying stream data because of its infinite length. Hence, sliding window which just takes the last seen N elements of the stream into account is introduced to maintain similarity queries over stream data.

Figure 4 illustrates the conventions that new data elements are coming from the right and the elements at the left are ones already seen. The sliding window covers a time period of $T_S \times n$, which T_S is the sampling period. Each element of sensor data stream has an arrival time, which increments by one at each arrival, with the leftmost element considered to have arrived at time 1. Since the ThinkGear chip produces the Attention and Meditation EEG with 1 Hz frequency. The sample period equals to 1 s. Besides, most duration of driving operation is less than 10 s, so n equals to 10. So the length of the sliding window is 10.

For an illustration of this notation, consider the situation presented in Fig. 4. The start time of the sliding window is 4, the current time instant is 13, and the last seen element of the stream data is e_{13}. Each element e_i consists of the Meditation and Attention EEG collected by sensors at time i.

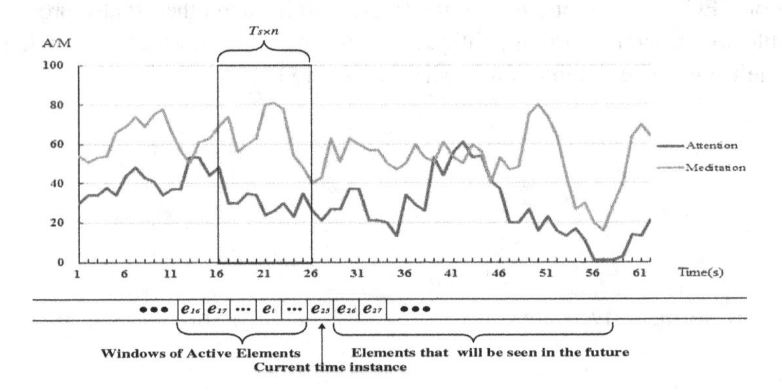

Fig. 4. Illustration for the notation and the conventions of sliding window

The input of k-NN algorithm is each sliding window instances in this paper. Algorithm 1 represents algorithm Sliding Window, and the program code for how the sliding window slides through the data stream see Fig. 5.

3.2 k-NN Algorithm for Driving Fatigue Detection

k-NN measures the difference or similarity between instances according to a distance function. Given a test instance x, its k closest neighbors $y_1,\ldots, y_k$, are calculated, and a vote is conducted to assign the most common class to x. That is, the class of x, denoted by $c(x)$, is determined by the formula (2) [15].

```
Input:Sensor data stream
Output:Type(label) of a slide instance
01  label=
02  Swidth=10, // width of sliding window
03  for (Sref=0; size(Sref+Swidth)>=Swidth; t++)
04        label = kNN(Dtrain, Sref+Swidth,k)
05  end for
06  return label;
```

Fig. 5. Algorithm 1: sliding window algorithm pseudo-code.

$$c(x) = argmax \sum\nolimits_{n=1}^{k} \delta(\mathrm{c}, \mathrm{c}(y_i)) \tag{2}$$

Where $c(y_i)$ is the class of y_i, and δ is a function that δ (u,v) =1 if u = v.

Since there is only one feature (namely the correlation coefficient between Attention and Meditation EEG) used for classifying, Manhattan distance defined in formula (3) is selected as the distance function. Among formula (3), D is the Manhattan distance, t and x are n-dimensional real vector.

$$D(t,x) = \sum\nolimits_{i=0}^{n} |t_i - x_i| \tag{3}$$

The k-NN algorithm for driving fatigue detection consists of the training and classifying phase. The training phase is as follows:

1. Samples the subject's Attention and Meditation EEG in 1 Hz frequency for ten seconds when the subject is normally driving, and calculates the correlation coefficient between Attention and Meditation EEG, and labels it as normal drive.
2. Samples the subject's Attention and Meditation EEG in 1 Hz frequency for ten seconds when the subject is in fatigue drive, calculates the correlation coefficient between Attention and Meditation EEG, and labels it as fatigue.
3. Repeats steps (1) and (2) until enough training samples are produced.

The classifying phase is as follows.

1. Constructs sliding window W_A [10] and W_M [10], of which are used to cache the last ten seconds Attention and Meditation EEG respectively.
2. Samples the subject's Attention and Meditation EEG in 1 Hz frequency, and appends them into the tail of W_A and W_M respectively.
3. Judges whether both W_A and W_M are full or not. If they are not full, then goes to (2); otherwise goes to (4).
4. Calculates the current correlation coefficient between Attention and Meditation EEG according to formulas (1).
5. Judges whether the current correlation coefficient between Attention and Meditation EEG is fatigue or not according to the formula (2). If it is normal driving, then it goes to (2); If it is fatigued driving, then it will issue an alert.

4 Implementation

Referencing the Mindwave, a headband integrated with ThinkGear EEG chip, tri-axial accelerometer and gyroscope is developed to collect subject's left prefrontal EEG and head movement data. Besides, the headband is integrated with Bluetooth module, of which makes it able to wireless send data. Since most smart devices (such as Android smart phone and ipad) are also integrated with the Bluetooth module as well, and they have strong calculation capability. Android smart device integrated with Bluetooth is introduced to receive the left prefrontal EEG and head movement data from the headband, and software running on it is developed based on above technologies to detect driving fatigue and issue alert.

4.1 System Architecture

Figure 6 shows the system architecture for driving fatigue detection. Among them, the headband samples the subject's left prefrontal Attention and Meditation EEG, the resultant acceleration and angular velocity, and transfers the data to an Android smart device via Bluetooth. The software running on the smart device will calculate the correlation coefficient between the Attention and Meditation EEG and judge whether the subject is fatigue or not by calling the k-NN algorithm, as soon as it receives the subject's left prefrontal Attention and Meditation EEG. The smart device will immediately issue an alert, and call the bus monitor or send a short message to the bus schedule center as long as it detects driving fatigue. Besides, the smart device will automatic play music as soon as it detects the subject's nod with agreement by a threshold algorithm [16].

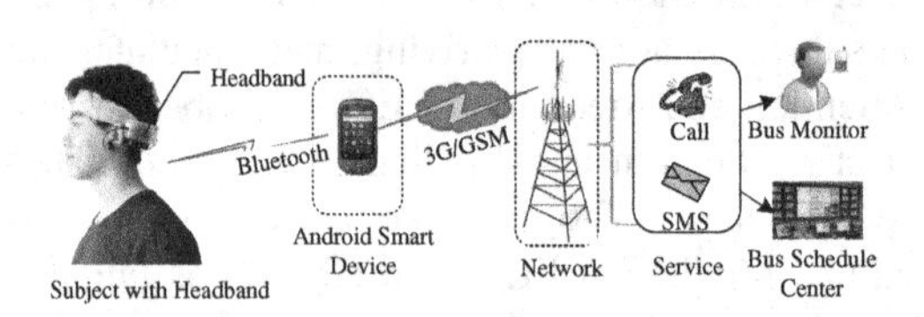

Fig. 6. The system architecture for detecting driving fatigue

4.2 Hardware Design

Figure 7 shows the sensor board which measures 25 mm × 55 mm ×7 mm (width × length × thickness), and is suitable for making a headband. It consists of a high-performance, low-powered microcontroller, and a class 2 Bluetooth module. The Bluetooth module has a range of 10 m, a default transmission rate of 115 k baud. The microcontroller reads the data from the accelerometer, gyroscope and ThinkGear chip. After getting the left prefrontal EGG and the resultant acceleration and angular velocity, it transmits them to the mobile smart device via Bluetooth. The tri-axial accelerometer has a range of ±16 g. The tri-axial gyroscope has a full-scale range of ±2000°/sec.

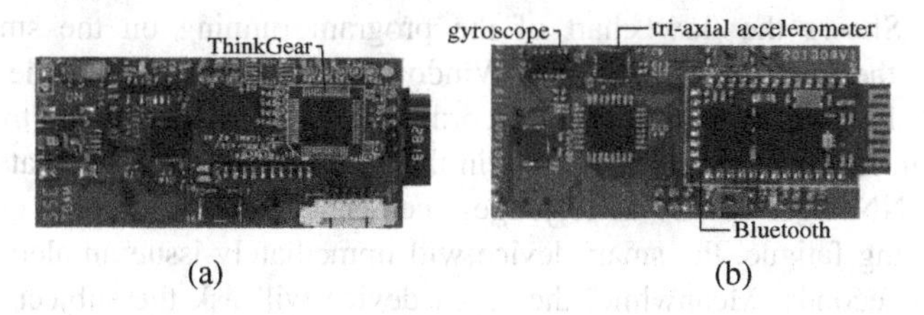

Fig. 7. Sensor board. (a) Front of the sensor board with ThinkGear (b) Back of the sensor board with tri-axial accelerometer, gyroscope and Bluetooth

4.3 Software Design

The software includes the on-chip program running on the sensor board and the driving fatigue detection application downloaded onto a mobile smart device. The key step of the on-chip program is as follows:

1. Initializes the tri-axial accelerator, gyroscope and ThinkGear. Sets the baud rate for Bluetooth, and the frequency for viewing the tri-axial acceleration, angular velocity and left prefrontal EEG.
2. Monitors the accelerations and angular velocities from tri-axial accelerometer and gyroscope in intervals of 0.1 s. And read the Meditation and Attention EEG data from the ThinkGear chip in intervals of 1 s.
3. Calculates the resultant acceleration (namely α) and angular velocity (namely ω) according to the data from tri-axial accelerator and gyroscope.
4. Sends the Meditation and Attention EEG, and the resultant acceleration and angular velocity to the mobile smart device via Bluetooth.

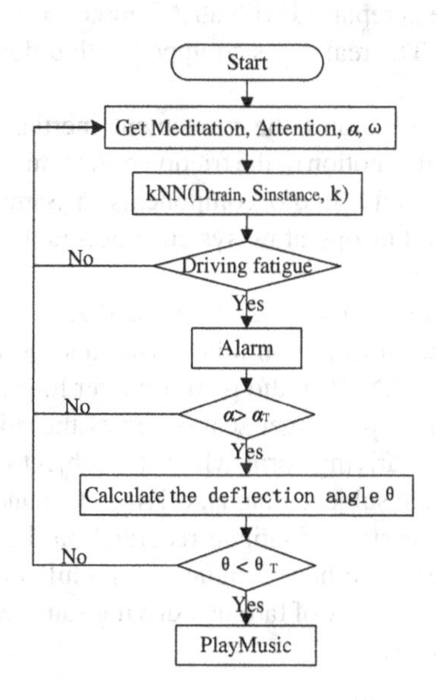

Fig. 8. Flow chart of the program running on Android

Figure 8 Shows the flow chart of the program running on the smart device. According to the instance of a Sliding Window of the input stream, the correlation coefficient of the instance is calculated according to the formula (1), and the similarity between the instance and training sample in the training dataset is calculated using by calling the k-NN algorithm so as to judge whether it is driving fatigue or not. After detecting driving fatigue, the smart device will immediately issue an alert which will continue five seconds. Meanwhile, the smart device will ask the subject whether he wants to play music or not. The smart device will automatically play music to awake the subject when it detects the subject's nod with agreement, of which means the resultant acceleration α is greater than the threshold α_T, and the deflection angle θ is less than the threshold θ_T.

5 Experimental and Data Analysis

5.1 Experimental Setup

Due to the serious danger of driving fatigue, the driving simulation platform (SCANeR Studio) is selected as the experimental platform. SCANeR Studio not only provides a real-time driving environment, but also keeps the subject safe. SCANeR Studio is comprised of the following parts:

Computer - in real-time tracks the movement of the real car (the driving simulator) as the subject drives, and produces a real life 3D traffic environment for the experiment.

Surrounding projection screens - consist of four projection screens locating at the front and back, left and right around the real car. Each screen is 3 meters wide, and 2 meters high. The real life driving environment produced by the computer is projected on those screens. Those screens give the driver a real driving experience.

The real car - its engine is replaced with an AC motor, and all wheels are supported by a pair of friction rollers. The real car is equipped with different kinds of sensors that collect data for the computer.

Mechanics simulator - generates the movement inertia, friction, and turbulence characteristics according to the motion of the friction roller while the subject drives the car.

In addition to the SCANeR Studio components, a Samsung Galaxy Note 10.1, which integrates with 3G and its operating system Android 4.2, is introduced to run the software to do the experiment.

Twelve subjects who are aged from 22 to 43, and each of who has at least 1 year driving experience are selected to put on a headband and do one hour experiments to get training samples from SCANeR studio platform after lunch. During the experiment, an observer sitting at the front passenger seat observes the driving state of the subject and records the state on the driving form while the subject drives. The driving form shown in Table 1 includes the subject's name, driving state and time. After driving, the subjects should confirm their state of fatigue recorded on the driving form. There are only six subjects who appeared to have fatigue states while driving. Hence, ten sets of normal driving samples and ten sets of fatigued driving samples are selected from those 6 subjects as training samples.

5.2 Data Analysis

In order to test the driving fatigue detection system, 12 males and 6 females who are aged from 18 to 43, and have at least one year driving experience are selected to drive at Visual SCAN platform after lunch for half an hour. During the test, an observer sitting at the front passenger seat records the subject's state of driving fatigue on the driving form as long as the system detects driving fatigue and issues an alert. The subjects should then confirm their fatigue states detected by the software after driving. In total there are 345 sets of normal driving samples and 142 sets of fatigued driving samples in the test. Table 2 shows the results of the test. Among the 345 sets of normal driving samples, 312 samples are correctly detected, and the other 33 samples were wrongly classified as fatigued driving. Among the 142 sets of fatigued driving samples, 115 samples were correctly detected, and the other 27 samples are wrongly classified as normal driving.

Sensitivity and specificity are introduced to evaluate the driving fatigue detection system. Sensitivity is the capacity to detect a fatigued driver, and specificity is the capacity to detect only a fatigued driving [17]. According to Table 1, the sensitivity and specificity of the system are 80.98 % and 90.43 %. The reason why the sensitivity is only 80.98 % is possiblely because some subjects could not make sure the detected state of driving fatigue after their driving. It perhaps will be better that subjects make sure their state soon as the system detects their driving fatigue.

Table 1. An example of the driving form

Name	Driving State	Date	Start Time	End Time
Zuohao Jia	Normal	2013-10-08	12:05	12:30
Zuohao Jia	Fatigue	2013-10-08	12:31	12:33

Table 2. Test results

Detecting results	Total	Correct	Wrong
Normal driving	345	312	33
Fatigued driving	142	115	27

During the test, the system takes only ten seconds to sample Attention and Meditation EEG so as to construct the sliding windows, and the response delay time to detect driving fatigue is less than 0.1 ms (millisecond) once the Sansung Galaxy Note receives the Attention and Meditation EEG data. Meanwhile, the software is also installed on a Moto 525 (a smart phone) whose operating system is Android 2.2, and the response delay time to detect driving fatigue is also less than 0.1 ms. It proves that the system has portable and real-time advantages.

Tables 3 and 4 show the experimental results of male subjects and female subjects. The male Sensitivity and Specificity are 82.93 % and 87.61 %, while the female Sensitivity and Specificity are 78.33 % and 95.28 %. One can see that the sensitivity of males is 4.6 higher than that of females, and the specificity of female is 7.67 % higher than that of male.

Table 3. Test results of male subjects

Detecting results	Total	Correct	Wrong
Normal driving	218	191	27
Fatigued driving	82	68	14

Table 4. Test results of female subjects

Detecting results	Total	Correct	Wrong
Normal driving	127	121	6
Fatigued driving	60	47	13

Fig. 9. An emergent message with GPS location

Each of the detected driving fatigue not only triggers an alert in 5 s, but also makes a call to monitor, or sends an emergent message with GPS location to bus schedule center immediately, so as to provide a fast and accurate intervention. Figure 9 shows an example of an emergent message with GPS location when He is in state of driving fatigue.

6 Conclusion

According to portable and real-time problems for its practical application of EEG based driving fatigue detection, a headband integrated with Thinkgear EEG chip, tri-axial accelerometer, gyroscope and Bluetooth is developed to collect the subject's left pre-frontal Attention, Meditation EEG and head movement data. And a new method for detecting driving fatigue is proposed based on k-NN and the correlation coefficient of subject's Attention and Meditation. Meanwhile, a driving fatigue detection system based on the above technologies is implemented on an Android device. The experiment proves that it has noninvasive and real-time advantages, as well as its sensitivity and specificity are at high levels of 80.98 % and 90.43 % respectively. In the future, new algorithms (such as artificial neural network, Bayesian) will be studied in order to improve the accuracy of system detection.

Acknowledgment. This work was supported by the Beijing Natural Science Foundation under grant No. 4102005, and partly supported by the National Nature Science Foundation of China (No. 61040039). The authors would also like to thank Dr. Xiaohua Zhao who opens the Beijing Transportation Engineering key Lab to do experiments.

References

1. Idogawa, K.: On the brain wave activity of professional drivers during monotonous work. Behav. Metrika **30**, 23–34 (1991)
2. Yaxuan, W., Wenshu, L., Guosheng, S., et al.: Survey on fatigue driving detection method research. Chinese J. Ergon. (2003). (in Chinese)
3. Khushaba, R.N., Kodagoda, S., Lal, S., Dissanayake, G.: Driver drowsiness classification using fuzzy wavelet-packet-based feature-extraction algorithm. IEEE Trans. Biomed. Eng. **58**(1), 121–131 (2011)
4. Lin, C.-T., Ko, L.-W., Chung, I.-F., Huang, T.-Y., Chen, Y.-C., Jung, T.-P., Liang, S.-F.: Adaptive EEG-based alertness estimation system by using ICA-based fuzzy neural networks. IEEE Trans. Circ. Syst. I Regul. Pap. **53**(11), 2469–2476 (2006)
5. Saccade velocity characteristics Intrinsic variability and fatigue, Schmidt, D., Abel, L.A., Dell'Osso, L.F., Daroff, R.B.: Saccade velocity characteristics: Intrinsic variability and fatigue. Aviat. Space Environ. Med. **50**, 393–395 (1979)
6. Lal, S.K.L., Craig, A.: Psychological effects associated with drowsiness: Driver fatigue and electroencephalography. Int. J. Psychophysiol. **15**(3), 183–189 (2001)
7. Ji, Q., Zhu, Z., Lan, P.: Real-time nonintrusive monitoring and prediction of driver fatigue. IEEE Trans. Veh. Technol. **53**(4), 1052–1068 (2004)
8. Jap, B.T., Lal, S., Fischer, P., Bekiaris, E.: Using EEG spectral components to assess algorithms for detecting fatigue. Expert Syst. Appl. **36**(2), 2352–2359 (2009)
9. Tsuchida, A., Bhuiyan, M.S., Oguri, K.: Estimation of drowsiness level based on eyelid closure and heart rate variability. In: proceedings of the 31st Annual International Conference IEEE Engineering in Medicine and Biology Society (EMBS), pp. 2543–2546 (2009)
10. Van Orden, K., Jung, T.P., Makeig, S.: Combined eye activity measures accurately estimate changes in sustained visual task performance. Biol. Psychol. **52**(3), 221–240 (2000)
11. Jung, T.P., Makeig, S., Stensmo, M., Sejnowski, T.J.: Estimating alertness from the EEG power spectrum. IEEE Trans. Biomed. Eng. **44**(1), 60–69 (1997)
12. Fu, J.W., Li, M., Lu, B.L.: Detecting drowsiness in driving simulation based on EEG. In: Proceedings of the 8th International Workshop Autonomic System-Self-organization, Manage., Control, pp. 21–28, October 6–7 2008
13. Kandel, E.R., Schwartz, J.H., Jessell, T.M.: Principles of Neural Science Fourth Edition. McGraw-Hill, United State of America, pp. 324 (2000)
14. American Electroencephalographic Society: Guideline thirteen: Guidelines for standard electrode position nomenclature. J. Clin. Neurophysiol. **11**, 111–113 (1994)
15. Erdogan, S.Z., Bilgin, T.T.: A data mining approach for fall detection by using k-nearest neighbour algorithm on wireless sensor network data. IET Commun. **6**(18), 3281–3287 (2012)
16. He, J., Hu, C., Li, Y.: An autonomous fall detection and alerting system based on mobile and ubiquitous computing. In: 10th International Conference on Ubiquitous Intelligence and Computing, pp. 539–543 (2013)
17. Kangas, M., Vikman, I., Wiklander, J., Lindgren, P., Nyberg, L., Jämsä, T.: Sensitivity and specificity of fall detection in people aged 40 years and over. Gait Posture **29**(4), 571–574 (2009)

Efficient Regional Traffic Signal Control Scheme Based on a SVR Traffic State Forecasting Model

Ziyi You[1,2](✉) and Pu Chen[1,2]

[1] Department of Physics & Electronic Science,
Guizhou Normal University, Guiyang 550001, China
357534271@qq.com

[2] Key Laboratory of Special Automotive Electronics Technology of the Education Department of Guizhou Province, Guiyang 550001, Guizhou, China
740461512@qq.com

Abstract. Intelligent control of urban traffic signal takes a very important role in the Internet of Vehicle (IOV). This paper proposes an intelligent control method to satisfy the real-time and accuracy of the regional traffic signal control (ICMRT). On the basis of the existing wireless sensor network structure, ICMRT adopts the unequal clustering strategy to create a model of discrete switched system for the regional traffic system. Furthermore, taking the network delay and packet loss rate in data transmission into consideration, the state observer of the discrete switched system uses the improved ε-SVR theory to realize the online prediction of the multi-source data based traffic state, so that the overall traffic signal can be coordinately controlled. The asymptotic stability of discrete switched system is proved using Lyapunov function. Finally, simulation results show that ICMRT has better performance in the intersection average delay time compared with ordinary fuzzy neural prediction and ordinary ε-SVR method. So we can get the conclusion that this scheme is feasible and effective.

Keywords: Regional traffic signal system · Internet of vehicle · Dynamic feedback system · Traffic state prediction · Asymptotic stability

1 Introduction

Intelligent transportation system (ITS) is one of the most effective methods to solve the traffic problems. So far, traffic signal control is the important research topic of its subsystem "advanced traffic management system", which has become one of the hot spots in the field of intelligent transportation [1].

As the development of traffic signal control technology, so far, traffic control systems have three leading types, .i.e. regional coordinated signal control, centralized control and hierarchical control [2–4]. These three kinds of systems above play a very important role in alleviating the urban problems, but still could not absolutely adapt to the development of urban transportation system in China. It is given for three reasons. Firstly, the anticipation of operating characteristics

C.-H. Hsu et al. (Eds.): IOV 2015, LNCS 9502, pp. 198–209, 2015.
DOI: 10.1007/978-3-319-27293-1_18

of traffic flow in China looks inadequate. Secondly, based on the mathematical model, original systems above cannot meet the requirements of real time control. Finally, there are limitations in signal timing at intersection. Therefore, the development of traffic signal control system in china should be further combined with the characteristics of Chinese traffic environment.

Due to the complexity of the road traffic system and the real-time of the dynamic allocation, dynamic feedback mechanisms should be used in the urban traffic control. However, the nonlinearity and randomness of the traffic system make it so difficult to describe the control algorithm based on the dynamic feedback system. Therefore, the application of artificial intelligence methods is highly valued by the academic community. In the existing reports, a traffic signal fuzzy control method is proposed by M. Bani Yassein et al. [5] for the four phase single intersection. The control mode of the fuzzy controller in [5] is directly determined by the length of vehicle queue from each phase. But the control rules and membership functions which come from original fuzzy controllers are usually based on the experts' experiences, so it is difficult to obtain the excellent performances for original fuzzy controllers. Subsequently, a genetic algorithm based optimization scheme of fuzzy membership degree is proposed [6], in which the optimization of fuzzy rules of fuzzy controller is realized. Considering the impact of traffic signal timing schemes on the traffic flow prediction, Lin De-gang et al. [7] designed a scheme based on Propagation Back (BP) neural network to predict the traffic flow at intersection. Although the neural network has very strong nonlinearity, it also has some disadvantages such as slow convergence and so on. As an improvement for the scheme above in [7], Liu Yan [8] combines the advantages of fuzzy control theory and neural network, and constructs the fuzzy rough neural network to predict and control the traffic flow in the intersection. In addition, [9] presents a short-time traffic flow combined prediction model based on constrained Kalman filter ,in order to overcome the instability of the performance of a single traffic flow prediction model. However, Kalman filter algorithm needs a lot of matrix and vector operations, it is difficult to real-time online prediction using Kalman filter. A short-time traffic flow prediction method based on Support Vector Regression (SVR) is proposed by GAO Xuehui et al. [10], and the prediction results show its effectiveness. But this method is based on the single point prediction model of vehicle detectors, hence, the optimization of related parameters, the improvement of prediction accuracy and the application of the model still need to be further studied.

In a word, all the methods above just focus on a certain location of the traffic section or a certain kind of parameters to predict and control traffic flow. Combining with the existing unequal clustering strategy [11], this paper presents a ε-SVR based intelligent control method, used for regional traffic signal coordination control (ICMRT).

The rest part of the paper is organized as follows. Following section discusses the ε-SVR theory. The ICMRT scheme is described in detail in Sect. 3. The observer design is described in Sect. 4 and followed by analysis of stability and scheduling in Sect. 5. Section 6 focuses on the performance simulation analysis of the scheme. Finally, in Sect. 7, the conclusion is given.

2 Support Vector Regression ε-SVR

Support vector machines (SVMs) [12] have wide application areas such as pattern recognition, bio-informatics etc. Particularly, along with the introduction of ε-insensitive loss function, SVMs also have been extended to solve nonlinear regression estimation problems, which are so-called support vector regression (SVR) [13].

It could be said that the prediction accuracy and efficiency of the SVR model are directly influenced by the size of the training data set. Furthermore, both accuracy and efficiency usually are mutual restricted. The original SVR theory carries out a large number of matrix operations, resulting in a slow convergence rate of the algorithm. Therefore, SVR training needs to be improved in following two aspects, i.e. the convergence speed of the training algorithm and the training problem of large sample set [14]. In this paper, we adopt the training method of fixed working variable set, so that the size of the working sample set can be fixed in the limit of algorithm efficiency. When the computation load of an online SVR algorithm increasing, the accelerated decremental strategy could be used to reduce the computation amount and cut online sample set size [15].

3 The ICMRT Scheme

3.1 Network Model

The network structure of regional traffic signal could be a multi-tiered and WSN based cluster structure, which is obtained based on the Reference [13]. Each cluster is constituted of a collector (the cluster head) installed on the roadside and one or more passing by vehicles (the cluster members). Due to the mobility of vehicles, each cluster head is designated by the sink node as an infrastructure in the corresponding cluster. In a cluster region, the cluster head obtains other members' sampling data, which will be integrated into the non-redundant data set. AS the sink node receives plenty of flow information from multiple clusters, it can perform classifications based on the accurate timestamps and geographic locations, and then separately upload them to the application layer. After analyzing the aggregation results from each cluster, the application layer sends timely decision messages back to each actuator of the network control system in order to achieve regional traffic coordinated control.

In the process of collecting traffic information, ICMRT uses data-based gridding urban traffic system Assume that all the sensor nodes (vehicles and infrastructures) randomly distribute in a two-dimensional plane rectangular region: $Z^2 = \{(x, y), 0 \leq x, 0 \leq y\}$. This region can be a grid model where principles and methods [16] are proposed to deploy the collection equipment for flow data Every square grid is defined as $\alpha \times \alpha$, the value of α depends on the accuracy of certain application task. For example, the location of node can be recorded as $L_p(x_i, y_i)$. The detail can be described in [11].

3.2 *l* The State Feedback Model

Figure 1 shows a dynamic feedback model of the traffic signal control system which has communication constraints and time delay. As can be seen in Fig. 1, the remote control terminal contains two components .i.e. the state observer and the controller. After receiving the aggregated data from the sink node, the state observer could obtain the predictive value of the controlled object state according to these data. And then, the controller sends the corresponding control signals back to actuators.

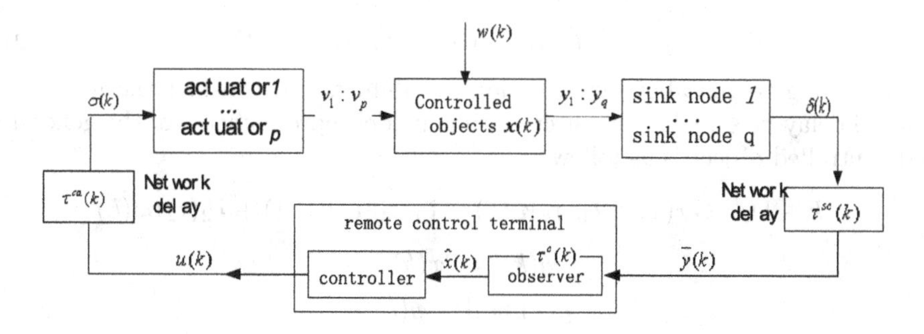

Fig. 1. System state feedback model

As shown in Fig. 1, $\boldsymbol{x}$ denotes the state variable of controlled objects and $\boldsymbol{x} = [x_1, ..., x_n]^T \in R^n$, $\boldsymbol{v}$ denotes the controlled output which actuators receive and $\boldsymbol{v} = [v_1, ..., v_p]^T \in R^p$, $\boldsymbol{y}$ denotes the control input from sink nodes and $\boldsymbol{y} = [y_1, ..., y_q]^T \in R^q$, $\boldsymbol{w}$ denotes the energy limited interference signals and $\boldsymbol{w} = [w_1, ..., w_n]^T \in R^n$. $\tau^{sc}(k)$ stands for the network delay which from the sensors to the state observer of the remote control terminal, $\tau^{ca}(k)$ stands for the network delay which from the controller of the remote control terminal to the actuators, and $\tau^{c}(k)$ stands for the computation delay of the state observer. For convenience of investigation, we make the following assumptions.

1. The actuator is the event-driven and the sampling period of the cluster collector is T.
2. The total delay of closed-loop system $\tau(k) = \tau^{sc}(k) + \tau^{ca}(k) + \tau^{c}(k)$, and $0 \leq \tau^{c}(k) \leq T$.
3. Due to sensor data or control signals are subject to constraints in network transmission, a sink node can only transfer $d_s (d_s \leq n)$ state vectors at the same time, while a controller can only transfer $d_c (d_c \leq p)$ control signals. All signals are transmitted in the form of data packets.

Based on the assumptions above, the packet loss may also appear due to interference or network congestion. The random variable $\{\sigma_i | i = 1, ...q\}$ describes the case of data packet loss sent from the sink node i to the observer. When $i \neq j$, σ_i and σ_j are mutual independent. If $\boldsymbol{\Lambda}_\delta(k) = diag(\boldsymbol{\delta}(k))$, it means that the observer successfully receives $y_i(k)$, otherwise the $y_i(k)$ transmission fails.

Define $\boldsymbol{\Lambda}_\delta(k) = diag(\boldsymbol{\delta}(k))$. Taking into account the observer's compensation function, but also to simplify the design, the input for the observer is as Eq. (1).

$$\overline{\boldsymbol{y}}(k) = \boldsymbol{\Lambda}_\delta(k)\boldsymbol{y}(k) \tag{1}$$

Similarly, random variable $\{\sigma_j | j = 1, ..., p\}$ describes the case of control packet loss sent from the controller to the actuator . During the sampling period k, if $\sigma_j(k) = 1 (j = 1, ..., p)$, it means that $u_j(k)$ is successfully transferred to the controlled object, otherwise $u_j(k)$ is discarded. Define $\boldsymbol{M}_\sigma(k) = diag(\boldsymbol{\sigma}(k))$, the input for the controlled objects are as Eq. (2).

$$v(k) = \boldsymbol{M}_\sigma(k)\boldsymbol{u}(k) + (I - \boldsymbol{M}_\sigma))\boldsymbol{u}(k-1) \tag{2}$$

According to the above information scheduling, and consider the network-induced delay $\tau_k \in [0, \mathrm{T}]$, the discrete time model representation of the generalized controlled object is as follows.

$$\boldsymbol{x}(k+1) = \boldsymbol{G}x(k) + \boldsymbol{\Gamma}_0(\tau_k)\boldsymbol{u}(k) + \boldsymbol{\Gamma}_1(\tau_k)\boldsymbol{u}(k-1) + \boldsymbol{\Gamma}_2(\tau_2)\boldsymbol{w}(k)$$

$$\boldsymbol{y}(k) = \boldsymbol{C}\boldsymbol{x}(k)$$

$$\overline{\boldsymbol{y}}(k) = \boldsymbol{\Lambda}(k)\boldsymbol{y}(k)$$

$$\boldsymbol{v}(k) = \boldsymbol{M}_\sigma(k)\boldsymbol{u}(k) + (I - \boldsymbol{M}_\sigma(k))\boldsymbol{u}(k-1) \tag{3}$$

where $\boldsymbol{G} = e^{AT}, \boldsymbol{\Gamma}_0(\tau_k) = \int_0^{T-\tau_k} e^{At}B_1 dt$, $\boldsymbol{\Gamma}_1(\tau_k) = \int_{T-\tau_k}^{T} e^{At}B_1 dt$, $\boldsymbol{\Gamma}_2(\tau_k) = \int_0^T e^{At}B_2 dt$, A, B_1, B_2, C are appropriate dimensions matrices. Depending on mathematical transformation [17], $\boldsymbol{\Gamma}_0(\tau_k)$ can be expressed as $\boldsymbol{\Gamma}_0(\tau_k) = H_0 + D\boldsymbol{F}(\tau_k')E$, and $\boldsymbol{\Gamma}_1(\tau_k)$ can be expressed as $\boldsymbol{\Gamma}_1(\tau_k) = H_1 + D\boldsymbol{F}(\tau_k')E$,where, H_0, H_1, D, E are constant matrices. Along with τ_k changes, $\boldsymbol{F}(\tau_k')$ changes, and satisfies $\boldsymbol{F}^T(\tau_k')\boldsymbol{F}(\tau_k') \leq I, \tau_k' = \tau_k - T/2$.

Using the same time period, the state observer is designed as following Eq. (4).

$$\hat{\boldsymbol{x}}(k+1) = \boldsymbol{\Phi}^+\overline{\boldsymbol{y}}(k)$$

$$\hat{\boldsymbol{y}}(k) = C\hat{\boldsymbol{x}}(k) \tag{4}$$

and the controller is as Eq. (5).

$$\boldsymbol{u}(k) = S_k\hat{\boldsymbol{x}}(k) \tag{5}$$

In Eq. (4), $\boldsymbol{\Phi}^+$ is the prediction function for the observer. S_k is relative to the scheduling policies $s_k(k = 0, ..., N-1)$, N is corresponding to the communication sequence period. Set the augmenting vectors $\boldsymbol{Z}(k) = [\boldsymbol{x}^T(k), \boldsymbol{x}^T(k-1), \boldsymbol{v}^T(k-1)]^T$. The closed-loop system which is based on the state observer and controller is as Eq. (6).

$$\boldsymbol{Z}(k+1) = \boldsymbol{\Psi}^*\boldsymbol{Z}(k) + \begin{bmatrix} \boldsymbol{\Gamma}_2(\tau_k) \\ 0 \\ 0 \end{bmatrix} \boldsymbol{w}(k) \tag{6}$$

$$\begin{aligned}\Psi^* &= \begin{bmatrix} G & \Gamma_0(\tau_k)M_\sigma S_k \Phi^+ \Lambda_\delta C & \Gamma_0(\tau_k)(I - M_\sigma) + \Gamma_1(\tau_k) \\ I & 0 & 0 \\ 0 & M_\sigma S_k \Phi^+ \Lambda_\delta C & (1 - M_\sigma) \end{bmatrix} \\ &= \overline{G} + (\overline{H}_0 + I_2)\overline{M}K_0 + I_1(I_2^T + \overline{M}K_0)DF(\tau_k')E \end{aligned} \tag{7}$$

In Eq. (11): $\overline{G} = \begin{bmatrix} G\,0\,H_1 \\ I\;0\;0 \\ 0\;0\;0 \end{bmatrix}$; $\overline{H}_0 = \begin{bmatrix} H_0 \\ 0 \\ 0 \end{bmatrix}$; $I_1 = \begin{bmatrix} I \\ 0 \\ 0 \end{bmatrix}$; $I_2 = \begin{bmatrix} 0 \\ 0 \\ I \end{bmatrix}$;

$\overline{M} = \begin{bmatrix} 0 \\ M_\sigma \Lambda_\delta \\ 1 - M_\delta \end{bmatrix}^T$; $K_0 = \begin{bmatrix} 0 & 0 & 0 \\ 0 & S_k \Phi^+ C & 0 \\ 0 & 0 & I \end{bmatrix}$.

4 The Observer Design

In the upper section, the output of the traffic signal is recorded, corrected and parameterized. According to the traffic distribution law of a city section, during a week, the monitoring time is divided into several different periods, and ε-SVR prediction function could be established respectively. In each period of time, the sample training method is firstly used to set the fixed work variable set. In the process of collecting the output variable, $\overline{y}(k)$ is corresponding to the current data at different moments$k$. The sampling sequence is divided into blocks on fixed length, the initial length of the block is l, and the step length of each sampling period is 1. In addition, the monitoring system cannot always save all of the sampled data, the data is so much that not be conducive to real-time computing. Thus, the observer starts to select a sample from the current moment S, and a collection S is established.

$$S = \begin{bmatrix} S_1 \\ S_2 \\ \vdots \\ S_{m-l} \end{bmatrix} = \begin{bmatrix} (\overline{y}(k-m+1), ..., \overline{y}(k-m+l))^{\mathrm{T}} \\ (\overline{y}(k-m+2), ..., \overline{y}(k-m+l+1))^{\mathrm{T}} \\ \vdots \\ (\overline{y}(k-l), ..., \overline{y}(k-l))^{\mathrm{T}} \end{bmatrix} \tag{8}$$

Each vector S of the set is used as the training sample, and the expected value of each training sample P is as Eq. (9).

$$P = \begin{bmatrix} \hat{y}(k-m+l+1) \\ \hat{y}(k-m+l+2) \\ \vdots \\ \hat{y}(k) \end{bmatrix} \tag{9}$$

The steps of the dynamic prediction algorithm in the observer are described as follows:

Step1: According to the feedback model of the network control system, the corresponding method of traffic data collection is adopted.

Step 2: Extract $m$samples from the front data as the benchmark data. Set the sample collection $\boldsymbol{S}$ and get the expected value $\Pr e$ through the set $\boldsymbol{S}$.

Step 3: Take the collection $\boldsymbol{S}$ and the expected value $\Pr e$ as the ε-SVR prediction training samples. The Gauss radial basis function is selected as the training tool, where the penalty factor $\overline{\boldsymbol{\alpha}} = (\overline{\alpha}_1, \overline{\alpha}_1^*, ..., \overline{\alpha}_l, \overline{\alpha}_l^*)^T$ could be 50, the parameter ε could be 0.0240, and the parameters w and b are adjusted.

Step 4: The last set of samples is extracted as the input value, and the predicted value of the controlled object is obtained by using ε-SVR function.

Step 5: Delete the samples which need to be forgot by online training, update the support vector set and error support vector set.

Step 6: Carry on one signal sampling, and record the current sampling time k and data $\overline{\boldsymbol{y}}(k)$. Assuming that the sampling period of training samples is $2 \sim 5s$

Step 7: When the computation amount of the processor resource exceeds a certain proportion such as 70% (this value is obtained by a large number of experimental data), the length of the block could be adjusted. Update the set $\boldsymbol{S}$ and expected values $\Pr e$. Repeat from step 2 to step 5.

The prediction algorithm is based on the Gauss radial basis kernel function, and the decision function are constructed as follows

$$\hat{\boldsymbol{x}}(k+1) = \sum_{i=1}^{l} (\overline{\alpha_i}^* - \overline{\alpha}_i) \exp(\frac{-\|\overline{\boldsymbol{y}}(k) - \overline{\boldsymbol{y}}_i\|^2}{2\sigma^2}) + \overline{b} \tag{10}$$

where, the optimal solution is $\overline{\boldsymbol{\alpha}} = (\overline{\alpha}_1, \overline{\alpha}_1^*, ..., \overline{\alpha}_l, \overline{\alpha}_l^*)^T$.

5 Analysis of Stability and Scheduling

Assuming that the state of the system can be measured, take the stabilization problem of the system (6) into consideration in the absence of external disturbances (i.e. $w(t) \equiv 0$). The closed-loop system is designed with no disturbance input as Eq. (11).

$$\boldsymbol{Z}(k+1) = \boldsymbol{\Psi}^* \boldsymbol{Z}(k) \tag{11}$$

Lemma 1. [18] For a given symmetric matrix $S = \begin{bmatrix} S_{11} & S_{12} \\ S_{21} & S_{22} \end{bmatrix}$ S_{11} is the $r \times r$ matrix The following 3 conditions are equivalent:

1. $S < 0$.
2. $S_{11} < 0, S_{22} - S_{12}^T S_{11}^{-1} S_{12} < 0$.
3. $S_{22} < 0, S_{11} - S_{22}^{-1} S_{12}^T < 0$.

Lemma 2. [18] Given matrices $W, M, N, \boldsymbol{F}(k)$ of compatible dimensions $\boldsymbol{F}(k)$ with symmetric , then inequality:

$$W + M\boldsymbol{F}(k)N + N^{\mathrm{T}}\boldsymbol{F}^{\mathrm{T}}(k)M^{\mathrm{T}} < 0 \text{ holds}$$

If and only if $\varepsilon > 0$,there exists a constant

$$W{+}\varepsilon MM^{T}{+}\varepsilon^{-1}N^{T}N < 0$$

Theorem 1. For the system (11), if there is a symmetric positive definite matrix $\boldsymbol{X}$, scalar $\varepsilon_1 > 0$ and $\boldsymbol{Y}$, The following matrix inequalities are established:

$$\begin{bmatrix} -P^{-1} + \varepsilon_1[I_1(I_2^T + \overline{\boldsymbol{M}}\boldsymbol{K}_0)D][I_1(I_2^T + \overline{\boldsymbol{M}}\boldsymbol{K}_0)D]^T & \overline{\boldsymbol{G}}\boldsymbol{X} + (\overline{\boldsymbol{H}}_0 + I_2)\overline{\boldsymbol{M}}\boldsymbol{Y} & 0 \\ * & -\boldsymbol{X} & \boldsymbol{X}E^T \\ * & * & -\varepsilon_1 I \end{bmatrix} \tag{12}$$

Then, the closed-loop system (11) is asymptotically stable, and the state feedback gain matrix is $\boldsymbol{K}_0 = \boldsymbol{Y}\boldsymbol{X}^{-1}$.

Proof: Take Lyapunov function $\boldsymbol{V}(k) = \boldsymbol{z}^T(k)P\boldsymbol{z}(k)$, then

$$\begin{aligned} \Delta &= \boldsymbol{V}(k+1) - \boldsymbol{V}(k) \\ &= \boldsymbol{z}^T(k+1)P\boldsymbol{z}(k+1) - \boldsymbol{z}^{\mathrm{T}}(k)P\boldsymbol{z}(k) \\ &= [(\overline{\boldsymbol{G}} + (\overline{\boldsymbol{H}}_0 + I_2)\overline{\boldsymbol{M}}\boldsymbol{K}_0 + I_1(I_2^T + \overline{\boldsymbol{M}}\boldsymbol{K}_0) \\ &\quad D\boldsymbol{F}(\tau_k')E)\boldsymbol{z}(k)]^{\mathrm{T}}P[(\overline{\boldsymbol{G}} + (\overline{\boldsymbol{H}}_0 + I_2)\overline{\boldsymbol{M}} \\ &\quad \boldsymbol{K}_0 + I_1(I_2^{\mathrm{T}} + \overline{\boldsymbol{M}}\boldsymbol{K}_0)D\boldsymbol{F}(\tau_k')E)\boldsymbol{z}(k)] - \boldsymbol{z}^{\mathrm{T}}(k)P\boldsymbol{z}(k) \\ &= \boldsymbol{z}^{\mathrm{T}}(k)([\overline{\boldsymbol{G}} + (\overline{\boldsymbol{H}}_0 + I_2)\overline{\boldsymbol{M}}\boldsymbol{K}_0 + I_1(I_2^{\mathrm{T}} + \overline{\boldsymbol{M}} \\ &\quad \boldsymbol{K}_0)D\boldsymbol{F}(\tau_k')E)^{\mathrm{T}}P[\overline{\boldsymbol{G}} + (\overline{\boldsymbol{H}}_0 + I_2)\overline{\boldsymbol{M}}\boldsymbol{K}_0 + \\ &\quad \boldsymbol{K}_0 + I_1(I_2^{\mathrm{T}} + \overline{\boldsymbol{M}}\boldsymbol{K}_0)D\boldsymbol{F}(\tau_k')E)\boldsymbol{z}(k)] - \boldsymbol{z}^{\mathrm{T}}(k)P\boldsymbol{z}(k) \\ &= \boldsymbol{z}^{\mathrm{T}}(k)([\overline{\boldsymbol{G}} + (\overline{\boldsymbol{H}}_0 + I_2)\overline{\boldsymbol{M}}\boldsymbol{K}_0 + I_1(I_2^{\mathrm{T}} + \overline{\boldsymbol{M}} \\ &\quad \boldsymbol{K}_0)D\boldsymbol{F}(\tau_k')E]^{\mathrm{T}}P[\overline{\boldsymbol{G}} + (\overline{\boldsymbol{H}}_0 + I_2)\overline{\boldsymbol{M}}\boldsymbol{K}_0 + \\ &\quad I_1(I_2^{\mathrm{T}} + \overline{\boldsymbol{M}}\boldsymbol{K}_0)D\boldsymbol{F}(\tau_k')E] - P)\boldsymbol{z}(k) \\ &= \boldsymbol{z}^{\mathrm{T}}(k)(\boldsymbol{W}^{\mathrm{T}}P\boldsymbol{W} - P)\boldsymbol{z}(k) \end{aligned} \tag{13}$$

In above Eq. (13), $\boldsymbol{W} = \overline{\boldsymbol{G}}{+}(\overline{\boldsymbol{H}}_0 + I_2)\overline{\boldsymbol{M}}\boldsymbol{K}_0 + I_1(I_2^T + \overline{\boldsymbol{M}}\boldsymbol{K}_0)D\boldsymbol{F}(\tau_k')E$. If $\boldsymbol{W}^{\mathrm{T}}P\boldsymbol{W}{-}P < 0$, then $\Delta\boldsymbol{V}(k) < 0$, the closed-loop system (11) is asymptotically stable. By $W^{\mathrm{T}}PW - P < 0$, according to Lemma 1, transform is carried out, and W could be got into and obtained as Eq. (14).

$$\begin{bmatrix} -P^{-1} & \overline{\boldsymbol{G}} + (\overline{\boldsymbol{H}}_0 + I_2)\overline{\boldsymbol{M}}\boldsymbol{K}_0 + I_1(I_2^{\mathrm{T}} + \overline{\boldsymbol{M}}\boldsymbol{K}_0)D\boldsymbol{F}(\tau_k')E \\ * & -P \end{bmatrix} \tag{14}$$

Eq. (13) can be re written as

$$\begin{aligned} &\begin{bmatrix} -P^{-1} & \overline{\boldsymbol{G}} + (\overline{\boldsymbol{H}}_0 + I_2)\overline{\boldsymbol{M}}\boldsymbol{K}_0 \\ * & -P \end{bmatrix} + \begin{bmatrix} I_1(I_2^{\mathrm{T}} + \overline{\boldsymbol{M}}\boldsymbol{K}_0)D \\ 0 \end{bmatrix} \boldsymbol{F}(\tau_k') \\ &[0\ E] + [0\ E]^{\mathrm{T}}\, \boldsymbol{F}^{\mathrm{T}}(\tau_k') \begin{bmatrix} I_1(I_2^{\mathrm{T}} + \overline{\boldsymbol{M}}\boldsymbol{K}_0)D \\ 0 \end{bmatrix}^{\mathrm{T}} < 0 \end{aligned} \tag{15}$$

By the previous analysis, $\boldsymbol{F}^T(\tau_k')\boldsymbol{F}(\tau_k') \leq I$. According to Lemma 2, set $\varepsilon_1 > 0$, and Eq. (14) is equivalent to following Eq. (16).

$$\begin{bmatrix} -P^{-1} & \overline{\boldsymbol{G}} + (\overline{\boldsymbol{H}}_0 + I_2)\overline{\boldsymbol{M}}\boldsymbol{K}_0 \\ * & -P \end{bmatrix} + \varepsilon_1 \begin{bmatrix} I_1(I_2^{\mathrm{T}} + \overline{\boldsymbol{M}}\boldsymbol{K}_0)D \\ 0 \end{bmatrix} \begin{bmatrix} I_1(I_2^{\mathrm{T}} + \overline{\boldsymbol{M}}\boldsymbol{K}_0)D \\ 0 \end{bmatrix}^{\mathrm{T}} + \varepsilon_1^{-1} \begin{bmatrix} 0 & E \end{bmatrix}^{\mathrm{T}} \begin{bmatrix} 0 & E \end{bmatrix} < 0 \tag{16}$$

According to Lemma 1, Eq. (14) is equivalent to following Eq. (17).

$$\begin{bmatrix} -P^{-1} + \varepsilon_1[I_1(I_2^{\mathrm{T}} + \overline{\boldsymbol{M}}\boldsymbol{K}_0)D][I_1(I_2^{\mathrm{T}} + \overline{\boldsymbol{M}}\boldsymbol{K}_0)D]^{\mathrm{T}} & \overline{\boldsymbol{G}} + (\overline{\boldsymbol{H}}_0 + I_2)\overline{\boldsymbol{M}}\boldsymbol{K}_0 & 0 \\ * & -P & E^{\mathrm{T}} \\ * & * & -\varepsilon_1 I \end{bmatrix} \tag{17}$$

Equation (17) respectively multiply $diag[I, P^{-1}, I]$ by the left and the right, and set $\boldsymbol{X} = P^{-1}$, $\boldsymbol{Y} = \boldsymbol{K}_0 P^{-1}$, then Eq. (12) can be obtained. Therefore, Eq. (12) is equivalent to $\boldsymbol{W}^{\mathrm{T}} P \boldsymbol{W} - P < 0$. According to $\boldsymbol{Y} = \boldsymbol{K}_0 P^{-1} = \boldsymbol{K}_0 \boldsymbol{X}$, $\boldsymbol{K}_0 = \boldsymbol{Y}\boldsymbol{X}^{-1}$ can be obtained easily. End of Proof.

In the case of interference, the anti-interference ability of the system (6) can be further discussed by defining 2, and the verification method is similar to the case without interference.

Theorem 2. The communication sequence period for network control system is defined as N. In a certain kind of scheduling policy, if there is a synthesis method to make the closed-loop system stable, and then the scheduling strategy of the system is available.

Proof: The traffic signal control system is usually a discrete periodic system in which the time period is 5. Assuming its communication sequence is s_k, $s_{k+N} = s_k (k = 0, 1, ..., N_1)$, $s_k \in \{\Delta(1), \Delta(2), ...\Delta(n)\}$. $\Delta_j(i)(i = 1, 2, ..., j = 1, 2, ..., p)$ represents the scheduling parameter which transmitted to the actuator j within the sampling period k. If Eq. (18).

$$\mathrm{rank}[s_0 s_1 ... s_{N+1}] = n \tag{18}$$

holds, then the scheduling strategy of the system is allowed.

6 Simulations

In order to verify the effectiveness of the ICMRT, simulate the traffic conditions [19], and analyse the ordinary fuzzy neural network control [17], ordinary ε-SVR predictive control [18] and ICMRT method [20].

At first, set the state controller to control the traffic lights of a road intersection. The method used is described as follows: The travel time of straight lane

Table 1. Network simulation environment parameter settings

Number of the nodes	50
Total simulation time	300 s
Area size	800m × 400m
Packet size	1024 bytes
MAC protocol	IEEE 802.11b
Vehicle moving speed	10 km/h–40 km/h
Vehicle mobility model	Random waypoint
Link bandwidth	128 kbps
Radio range	100 m
protocols	fuzzy neural, ordinary ε-SVR, ICMRT

last for 80 s, the travel time of left lane is 25 s, while the right lane without time limit. If a vehicle at a certain phase still not go through the intersection on the last second of the green traffic light duration, then the green light is extended for 3 s. When there is no vehicle to reach the phase of intersection, or the extending time of the green signal has reached the maximum value, then the green light will switch the phase in the switching sequence. When the phase switches, the lasting time of the colour lamp is 5 s.

Assuming the vehicle arrivals at each direction of the intersection with random, when the arrival speed of traffic flow at one phase is less than 800PCU/h, traffic flow is subject to the Poisson distribution, on the contrary, it is subject to binomial distribution. The saturated traffic flow of the straight lane is 3500 PCU/h, the same to the right lane, while the left lane is 2650 PCU/h. The network environment parameters are shown in Table 1.

Table 2. Simulation results (average delay time: s/PCU)

Control method / Traffic flow	Ordinary fuzzy neural network control	Ordinary ε-SVR	ICMRT control	The improvement compared to ordinary fuzzy neural network	The improvement compared to ordinary ε-SVR
200	5.82	5.84	5.66	2.75%	3.08%
400	5.94	5. 98	5.73	3.54%	4.18%
600	6.09	6.16	5.84	4.11%	5.19%
800	6.67	6.81	6.38	4.35%	6.31%
1000	9.58	9.83	9.16	4.38%	6.82%
1200	14.89	15.08	13.17	11.55%	12.67%
1400	24.94	26.03	22.68	9.06%	12.87%
1600	36.07	38.98	32.21	10.70%	17.37%
1800	50.38	53.43	44.32	12.03%	17.05%
2000	67.06	71.23	58.64	12.56%	17.68%

Table 2 shows the simulation results of the traffic flow in the range of 2000PCU/h $\sim$ 200PCU/h. As can be seen from Table 2, when the vehicle arrival rates reaches a low value, the average delay time of these 3 methods is very close to the average delay time. However, as the increase of vehicle arrival rate, ICMRT control has better performance in the average delay compared to ordinary fuzzy neural network and ordinary ε-SVR. When the traffic flow reaches 1800PCU, the improvement rate of ICMRT is 12.03% compared with the ordinary fuzzy neural network, while compared with ordinary ε-SVR is 17.5%. When the traffic flow reaches 2000 PCU, the improvement rate of ICMRT is 12.56% compared with the ordinary fuzzy neural network, while compared with ordinary ε-SVR is 17.68%. When the vehicle flow saturation is higher, the improvement effect is more obvious, which shows that the ICMRT control is real-time and effective.

7 Conclusions

Taking the complexity of urban road scale and the real-time feedback of dynamic traffic flow into consideration, this paper presents a regional intelligent traffic signal control strategy (ICMRT) based on ε-SVR theory. Depending on a clustering method, ICMRT can divide the complex traffic signal system into small and mutual coordinated communication subsystems. On the basis of this, considering two information scheduling modules, i.e. from the sensors to the observers and from the controllers to the actuators, the network control system is modeled as a discrete switched system with uncertain parameters. In the discrete switched system, the state observer is used to predict the state of the traffic signal, which is based on support vector regression theory. ε-SVR model can take the system's output value and the actual value as the training samples. By improving the training method adjust the regression function, so as to improve the accuracy of the controlled object prediction. Next, the asymptotic stability and the scheduling performance of discrete switched system are verified by using the piecewise Lyapunov function method. Finally, the real-time and effectiveness of ICMRT are analyzed by network simulation.

The research results can be theoretically instructive for the intelligent control of urban traffic signal. But there are some shortcomings in this paper, such as the observability and robustness of the system will be further studied in the future.

Acknowledgement. The work presented in this paper was partially financed by the-Science and Technology Foundation of Guizhou Province No. J[2013]2222, and PHD Research Project Funding of Guizhou Normal University named "The control of network topology and secure information transmission in WSN based intelligent transportation system".

References

1. Wu, Y.: Application research of wireless sensor network in intelligent transportation system. Adv. Mater. Res. **108—-111**, 1170–1175 (2010)

2. Triguero, I., Derrac, J.: A taxonomy and experimental study on prototype generation for nearest neighbor classification. IEEE Trans. Syst. Man Cybern.-Part C: Appl. Rev. **42**(1), 86–100 (2012)
3. Liying, S., Wei, L., Jianjun, S., Guangxia, C., Deshen, S.: The analysis and application actuality of traffic signal control system-SCOOT and ACTRA in Beijing. Road Traffic Saf. **7**(2), 10–13 (2007)
4. Liang, C., Fan, B.Q., Han, Y.: Coordination control method of regional traffic flow. J. Traffic Transp. Eng. **11**(3), 112–117 (2011)
5. Nair, B.M., Cai, J.: A fuzzy logic controller for isolated signalized intersection with traffic abnormality considered. In: IEEE Intelligent Vehicles Symposium, pp. 1229–1233 (2007)
6. Gao, J., Li, J., Chen, Y., Zhao, X.: Optimize design and simulation of traffic signal two level fuzzy control system. J. Beijing Univ. Technol. **35**(1), 19–24 (2009)
7. Glin, D., Zheng, C., Chen, S.: Study of traffic delay predicting at signalized intersection based on BP neural network. Chin. J. Dalian Jiaotong Univ. **34**(4), 53–56 (2013)
8. Min, H.A.N., Ya-Nan, W.A.N.G.: Multivariate time series online predictor with kalman filter trained reservoir. Acta Autom. Sinica **36**(1), 169–173 (2010)
9. Gao, X., Liu, Y., Wang, Q., Jia, S., Sun, H.: Prediction of short-term traffic flow with on-line support vector regression algorithm. J. Shandong Univ. Sci. Technol. (Nat. Sci.) **30**(1), 78–82 (2011)
10. Alpaydin, E.: Introduction to Machine Learning, FAN Ming, translated. Beijing: China Machine Press, pp. 9–20 (2009)
11. You, Z., Cao, X., Wang, Y.: An unequal clustering strategy for WSNs based urban intelligent transportation system. J. Inf. Comput. Sci. **12**(10), 4001–4012 (2015)
12. Jiqin, W.X.W.: A survey on support vector machines training and testing algorithms. Comput. Eng. Appl. **21**(13), 75–78 (2004)
13. He, K., Li, X., Schick, B., Qiao, C., Sudhaakar, R., Addepalli, S., Chen, X.: On-road video delivery with integrated heterogeneous wireless networks. Ad Hoc Netw. **11**(7), 1992–2001 (2013)
14. Wang, X., Jiqin, W.: A survey on support vector machines training and testing algorithms. Chin. J. Comput. Eng. Appl. **21**(13), 75–78 (2004)
15. Shan, B., Zijian, J., Tong, W.: Application of online SVR on the dynamic liquid level soft sensing. In: Proceedings of 25th Chinese Control and Decision Conference, pp. 3003–3007. IEEE, Piscataway (2013)
16. Ossenbruggen, P., Laflamme, E.: Time series analysis and models of freeway performance. J. Transp. Eng. **138**(8), 1030–1039 (2012)
17. Huang, Y., Cui, B.: Stochastic Stability and stabilization of networked control systems with uncertain data losses and long delay. Chin. J. Huaihai Inst. Technol. (Nat. Sci. Ed.) **22**(4), 22–27 (2013)
18. Chen, X.: linear System Theory. Machinery Industry Press, New York (2011)
19. Xiao-ming, L.O.U.: Urban road network signal design optimization considering the impact of intersections. J. Transp. Eng. Inf. **11**(1), 108–110 (2013)
20. Li, X.: The Traffic Flow and the Signal Control of Intersections Research with MATLAB. Central China Normal University, pp. 17–39 (2008)

Managing Trust for Intelligence Vehicles: A Cluster Consensus Approach

Shu Yang(✉), Jinglin Li, Zhihan Liu, and Shangguang Wang

State Key Laboratory of Networking and Switching Technology, University of Posts and Telecommunications, Beijing, China
{assureys, jlli, zhihan, sgwang}@bupt.edu.cn

Abstract. Managing trust is critical for Intelligence Vehicles (IVs) collaboration. Forming clusters/platoons, IVs can work together to accomplish complex jobs that they are unable to perform individually. To improve safety and efficiency of collaboration, IVs' trust management has been extensively studied and a number of approaches have been proposed. However, most of these proposals either pay little attention on consensus forming among vehicles or ignore the utility of networked Road-Side-Units(RSUs). In this paper, we introduce a cluster consensus-based trust management scheme for IVs, where some malicious or incapable vehicles are existing on roads. The proposed scheme works by allowing IVs to evaluate each other, communicate and finally converge to a consensus where some trustworthy Cluster Head (CH) are generated. Periodically, the CHs take responsibility for intra-cluster trust management. Moreover, the scheme is enhanced with a distributed supervising mechanism and a central reputation arbitrator. The simulation results show that our scheme can achieve a high robustness to malicious/incapable IVs by limiting failure rate below 1 %.

Keywords: Trust management · Cluster consensus · Affinity Propagation · Intelligence Vehicles

1 Introduction

Internet of Vehicles (IoV) is an open converged network system supporting human-vehicles-environment cooperation [1]. Fusing multiple advanced terms, such as VANET, autonomous driving, cloud computing, multiagent system (MAS), this hybrid term plays a fundamental role towards a cooperative and effective intelligence transport system. This paper aims to manage trust for intelligence driving in an IOV environment. As commercial IVs are drawing near, vehicles are more and more intelligent. Meanwhile, cyber vehicles are unprecedented vulnerable when supported by an uncertain and dynamic network. Malicious attacks, information tampering, along with system failures, will directly threaten human lives and properties. A trust management scheme is desirable in an environment where filled up with uncertainty.

In this work we investigate how IVs can reach a consensus of authoritative nodes and how to build an integral trust management system upon these nodes and extra infrastructures. To highlight our motivations, we present following two illustrative scenarios. *Scenario 1*: In cluster/platoon-based driving, IVs frequently communicate

C.-H. Hsu et al. (Eds.): IOV 2015, LNCS 9502, pp. 210–220, 2015.
DOI: 10.1007/978-3-319-27293-1_19

with each other to maintain lateral/longitudinal control. Vehicles with incapability or malicious intention may join the cluster/platoon. Their behaviors are very likely to temper/disturb collaboration. In this safety-oriented case, local vehicles should be able to maintain robust intra-cluster trust to wipe out unqualified vehicles. *Scenario 2*: In efficiency-oriented case, where IVs need to collaborate in a wilder area, they exchange message to pre-sense traffic conditions, request parking lots information through VANET, even negotiate routes to prevent traffic congestions. None of these applications would be efficient without trustworthy collaboration. Above two scenarios suggests that a trust management scheme is in urgent need. We thus propose an approach called Cluster Consensus Trust Management (CCTM). Figure 1 describes the framework. The Cluster-based trust component and Central Reputation component work together to maintain trust management. The former is responsible for local trust evaluation and the latter is designed for global reputation.

The major contribution of this paper lies in the following two aspects:

- We identify cluster-based trust and reputation as two major components of dynamic trust. To reach a consensus of cluster-based trust, we proposed a *Modified Affinity Propagation Clustering* to generate the most trustworthy cluster head. The algorithm runs in a distributed manner and shows robustness to malicious/incapable vehicles.
- We enhance cluster-based trust with two extra mechanisms: mutual supervision and central reputation. Both of them are embedded into clustering process and show improvements for trust management system.

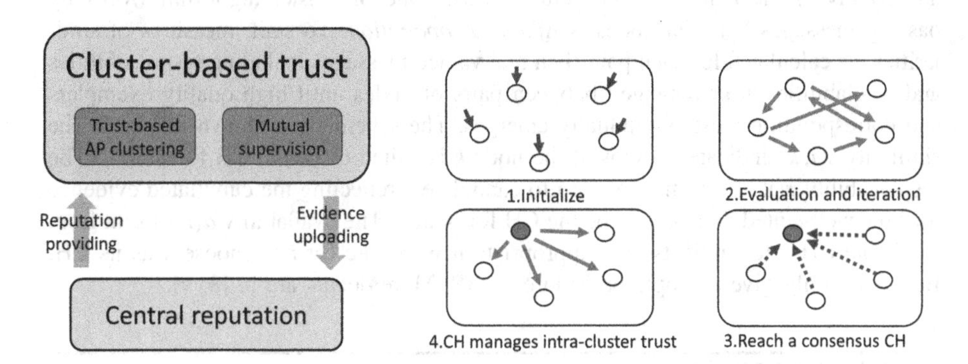

Fig. 1. Framework of CCTM

Fig. 2. Steps of Cluster-based trust establishment

2 Related Work

Most existing trust management systems in VANETs use distributed approach. Raya et al. [2] argues that the trust should be attributed to data per se in ephemeral ad hoc networks and proposed a framework for data-centric trust establishment. Their scheme shows high resilient to attackers and could converges to stable right decision. However, Raya's trust mechanism could not reduce attacker numbers in system level, since there is no punishment for cheating. Chen [3] present a decentralized framework combined with message propagation and trust evaluation in VANET. Specifically, trust measurement consists of role-based trust and experience-based trust. It is a good attempt to synthesize static priori trust (role-based trust) with dynamic situational trust (experience-based trust). Nonetheless, they didn't take historical reputation into consideration.

Recently, the RSU deployment is promoted by Intelligent Transport System group. With the help of RSUs, centralized trust management is not an ambiguous goal. Some works have emerged centralized trend. Wang [4] proposed a vertical handoff method, which improves availability of network access. His method therefore makes contributions to build centralized trust management system. Machado [5] aims to aggregate advantages of both centralized and distributed trust computation. In his work, RSUs play a central authority who keep track of messages and accordingly maintain a global trust-score for each vehicle. Their central grading system could efficiently distinguish dishonest nodes in real-life scenarios. Huang [6] et al. utilize identity-based cryptography to integrate entity-based trust and social trust in proxy server. The email interactions among individuals are mined to obtain social trust. The author then propose a situation-aware trust architecture for VANETs [7].

3 Trust Establishment and Reputation Evaluation

We illustrate the process of reaching a trust consensus. Cluster and its head are generated after several rounds of iteration. The generated exemplar, called CH, is an authoritative node managing intra-cluster trust. One of cluster algorithm works by passing messages between nodes is *Affinity Propagation*. To start, measures of similarities are calculated for each pair, then real valued messages including responsibilities and availabilities are exchanged between pairs of nodes until high quality exemplars and corresponding clusters gradually emerges. The schematic is shown in Fig. 2 The similarity $s(i,k)$ indicates how well the node k is suited to be the CH for node i. The responsibility $r(i,k)$ sent from node i to candidate k reflecting the cumulated evidence for how well-suited node k is to be the CH for node i. The availability $a(i,k)$ sent from candidate CH k to i reflects how appropriate it would be for i to choose k as its CH. Below we only give a simple procedure of AP. More details are in [8, 9].

Primitive AP iteration process

$$r(i,j) \leftarrow s(i,j) - \max_{k \ s.t. k \neq j} \{a(i,k) + s(i,k)\} \quad (1)$$

$$a(i,j) \leftarrow \min\{0, r(j,j) + \sum_{\forall k \neq i,j} \max\{0, r(k,j)\}\} \quad (2)$$

$$a(j,j) \leftarrow \sum_{k \ s.t. k \neq j} \max\{0, r(k,j)\} \quad (3)$$

To make real-valued message converge, messages are damped by λ, $Message_{new} = \lambda Message_{old} + (1 - \lambda)Message_{new}$, where λ is a weighing factor ranges from 0 to 1. When messages converged, a cluster head is generated:

$$CH_i = \max_j \{a(i,j) + r(i,j)\} \tag{4}$$

3.1 Untrust Degree

We design an *UntrustDegree* function as "distance measurement" for AP algorithm to find "the most trustworthy node", i.e. to reach a consensus on which node is suitable for CH. The function, $UntrustDegree(i,j)$ is automatically calculated by IV:

$$UntrustDegree(i,j) = F_i\left(Identity, \overrightarrow{Situtation}, \overrightarrow{Behavior}\right) \in [0,1] \tag{5}$$

Identity is one item from set $Id = \{bus, taxi, police, private, \ldots\}$ denoting real identity of one car and could be realized by a unique digital number. $\overrightarrow{Situation}$ is a vector pre-defined as some basic values which gives environmental context (e.g. the weather). $\overrightarrow{Behavior_j}$ is a vector recording basic actions that node j has done recently, with the help of behavior detection technologies [10]. The value is primarily positive but set to negative to fit AP algorithm. We reasonably assume that IV is intelligent enough to evaluate each other based on three ingredients.

The self UntrustDegrees, $UntrustDegree(i,i)$, are initialized to the same value, which is set to minimize the number of clusters produced. It should be noted that a higher self-trust degree make one node more likely to become the cluster head. In our final model, valid self-trust (also called *preference*, will be discussed in Sect. 4.2) is set a value which balances *IV's evaluation* and *historical reputation. Historical reputation* can only be legally announced by Central Arbitrator. When a group of IVs pass by a RSU, RSU will proactively download/broadcast reputations to IVs.

3.2 Mutual Supervision

A malicious/incapable node can cheat/mistake in message passing process by broadcasting a false $a(i,j)$. For example, if IV_i broadcast very high $a(i,j)$ to other nodes, it is more likely to be elected CH according to the AP algorithm. We need a mechanism to prevent nodes broadcast false availability or responsibility. The core of mutual supervision model is to match IV_i a supervisor IV_j. The integral mechanism of supervisor model is illustrated in Fig. 3.

To assure a stable and honesty supervisor, we apply below algorithm. From this algorithm, a IV_i has possibility to supervise another IV_j only when (1) IV_i doesn't tend to believe IV_j (2) IV_i and IV_j have small relative mobility. i.e. They are stable driving companions. This mechanism builds up mutual supervision relationship between two

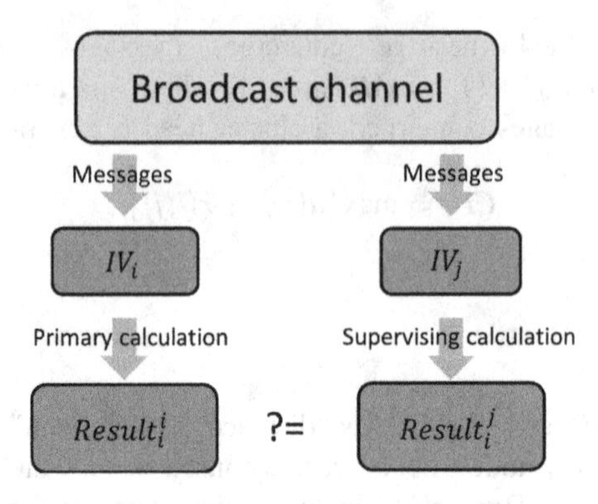

Fig. 3. Mutual supervision model

adversary nodes so that supervisor and supervisee are not likely to collude. More important, it can identify cheating nodes in message passing process.

Supervisor Matching algorithm

Input: a supervisor IV_i, nearby node' states (position, speed, behavior)
Output: *pair(supervisor, supervisee)*
1. For *Node j in DSRC(Dedicated Short Range Communication) range*
2. $M_{i,j} = MobMetr((Pos_i, Pos_j),(Speed_i, Speed_j))$ *//calculate mobility metric*
3. If *j has no supervisor* and $UntrustDegree(i,j)$ *is low*
4. Then *Add j in Supervisee Candidate List*
5. EndFor
6. For *j in Supervisee Candidate List*
If $M_{i,j} < M_{max}$ Then *k=j* *//find the most stable supervisee*
7. EndFor
8. Return *matched pair(i, k)* *//i supervises k*

3.3 Reaching a Consensus by Message Passing

We try to use a distributed algorithm to reach a consensus among large amounts of opinions. Each IV_i maintains a neighbor list N_i. As Table 1 shows, the list consists N_i^j

Table 1. Neighbor and Supervision filed

Neighbor field		Supervision field	
$(x,y)_j$	position of IV_j	$a'(k,j)$	IV_k's last availability received from IV_j
$(v_x, v_y)_j$	speed of IV_j		
$UntrustDegree(i,j)$	UntrustDegree from IV_i to IV_j	$a'(j,k)$	IV_k's last availability send to IV_j

(*Continued*)

Table 1. (*Continued*)

Neighbor field		Supervision field	
$a(i,j)$	last availability received from IV_j		
$a(j,i)$	last availability send to IV_j	$r'(k,j)$	IV_k's last responsibility received from IV_j
$r(i,j)$	last responsibility received from IV_j		
$r(j,i)$	last responsibility send to IV_j	$r'(j,k)$	IV_k's last responsibility send to IV_j
$CH_{cnvg,j}$	cluster head converge flag for IV_j		
$SUPVE_j$	IV_j's supervisee		

for each neighbor IV_j. Additional, IV_i also maintains a supervision fields for a supervisee IV_k.

Hello beacons are broadcast and received to maintain local awareness.

Broadcast and Receive Hello Beacons process

1. Every T_{hello}, each IV_j broadcast *hello beacon*:

$$\left\langle j,(x,y)_j,(v_x,v_y)_j,CH_j,SUPVE_j \right\rangle$$

2. Each receiving neighbor IV_i, calculates $UntrustDegree(i,j)$ if they are traveling in the same direction
3. IV_i adds/updates N_i^j in its neighbor list:

$$\left\langle j,(x,y)_j,(v_x,v_y)_j,UntrustDegree(i,j),CH_j,SUPVE_j \right\rangle$$

Availability and Responsibility messages should be broadcast periodically. We define this period as $T_{message}$. Each IV_i will calculate and broadcast $a(j,i)$ and $r(j,i)$ for each neighbor IV_j.

According to mutual supervision model, the process of calculating $a(j,i)$ and $r(j,i)$ should be supervised. Each IV automatically choose a supervisee by *Supervisor Matching algorithm*. Supervisor will check supervisee's calculation result, and release *alert* on condition that supervisee's message is suspicious. The process enhanced by mutual supervision model is illustrated below.

An integral message passing process period is T_{round} consisted of *several* $T_{message}$. $T_{message}$ must be small enough to allow algorithm converged within a T_{round}. We have injected a supervision mechanism into clustering process.

Supervising and Message passing process

Every $T_{message}$, each IV_i will:

1. Find a matching supervisee IV_k prepare for the next T_{round}'s iteration. If found in this $T_{message}$, claims out by $SUPVE_j$. If failed, try next $T_{message}$.
2. If hear an *alert* about IV_n, Each IV_i will ignore IV_n's messages in this T_{round}.
3. Calculate responsibility, $r(i,j)$ for each neighbor IV_j
4. Update with damping factor and stored: $r(i,j)=(1-\lambda)r(i,j)_{new}+\lambda r(i,j)_{old}$
5. Calculate availability, $a(j,i)$ for each neighbor IV_j
6. Update with damping factor and stored: $a(j,i)=(1-\lambda)a(j,i)_{new}+\lambda a(j,i)_{old}$
7. Determine if itself converged to CH: If $r(i,i)+a(i,i)>0$, then set $CH_{cnvg,j}$
8. Broadcast Responsibility and Availability array, $r(i,j)$ and $a(j,i)$
9. Supervising IV_k : IV_i updates and calculates $r'(j,k)$ and $a'(j,k)$ for IV_k
10. IV_i listens IV_k's messages: $r(k,j)$ and $a(j,k)$, if $|r(k,j)-r'(k,j)|>Threshold$ or $|a(k,j)-a'(k,j)|>Threshold$ then broadcast *alert* about IV_k.

In any $round_r$, there is a CH_{r-1} generated from $round_{r-1}$. Empowered by cluster consensus, CH_{r-1} will claim its role and broadcast *Authoritative Message*, which represent CH's final evaluation to each cluster member IV_i.

$$AuthoritativeMessage = \{UntrustDegree(CH_{r-1}, i)\} \tag{6}$$

Authoritative Message is trustworthy since it is sent from CH, which are elected as "the most trustworthy node" by cluster consensus. Build upon *Authoritative Message*, intra-cluster trust management is relatively reliable to support IVs' collaborations.

3.4 Evidence Evaluation

IVs will observe and evaluate qualities of each other. Moreover, they form *evidences* and report them to Central Arbitrator. A global reputation is established for on-the-road IVs to choose potential collaborators, and to incent good behaviors as well as to punish bad.

IVs leverage "store-upload" mechanism in delivering evidences to a CA. Since RSUs are sparsely deployed, each IV would store evidences in its storage firstly, then upload them when moving into a RSU's service range. CA should be able to make a conclusion on certain behavior by evaluating and merging different pieces of evidences from different individual IVs. Note that not all evidences are consistent, nor all evidences are trustworthy.

To mathematically model evidence evaluation, assume CA has to decide among several basic behavior $\beta_i \in \Omega$, based on K pieces of evidences $\{e_k^j\}$ to IV_j uploaded from different k IVs. Let B^j denotes the final judgement on behavior type of IV_j. Majority voting is leveraged, with the ability to filter false evidences. The final evaluation accords with the majority. Given counts of each types of observed behaviors, $count_i$, the behavior type of IV_j is defined by:

$$B^j = \beta_{argmax(count_i)} \tag{7}$$

Corresponding change will be reflected on *reputation* based on evaluated behavior.

4 Performance Evaluation

To evaluate performance of our scheme, we ran an extensive simulation in TransModeler with real map and high fidelity data. We use a map of urban area of San Antonio, US. We feed real macroscopic traffic data, which are measured in critical roads and sections, to reconstruct real traffic scenario. We believe that macroscopic data could reflect traffic dynamic to a high extent. We do not simulate the wireless medium in this case since it is orthogonal to our evaluation. All simulations were performed with approximate 400 vehicles on a 6 miles expressway. Five RSUs are sparsely and equally deployed along the expressway with an interval of 1.5 miles. The DSRC range is set 300 m. Each simulation ran for 600 s, however only the last 400 s were used for performance metric calculations.

Three *Basic Behaviors* are modeled for unqualified node: (1) Life-critical behavior (2) Inefficient behavior (3) Normality behavior. Each type of behavior causes different evaluated trust value. Moreover, to depict complex malicious/incapable behavior patterns, which are often mixed with different basic behaviors, we design several *behavior pattern*. Each pattern could trigger several basic behaviors with different probability. In simulation, a complex behavior of one unqualified node is triggered in every T_{round}.

4.1 The Effect of AP Algorithm

Ideally, AP clustering would generate a CH for every on-the-road vehicles. However, a small portion of vehicles, N_{left}, could be left alone when iterations are finished. There are two major reasons for these nodes: (1) the node could not find a converged CH candidate in its neighborhood, (2) the node itself is the CH but is the only member of cluster. Beside N_{left}, there are N_{in} nodes which forms M normal clusters. In unqualified node-free simulation, several results are shown below in Table 2. *Covered ratio* is a parameter describing how much the clustering results could cover the whole participants:

$$CoverRatio = \frac{N_{in}}{N_{in} + N_{left}} \tag{8}$$

The simulations are ran several times so figures below are averaged.

Table 2. The results of primary AP clustering

Damping Factor	Iteration Cycle	DSRC Range	Average Covered Ratio	Average Normal Cluster Number	Average Member Number	State
0.6	6	300 m	85.8 %	26.5	13.2	Under-Damping
0.7	5	300 m	38.9 %	9.3	17.8	Under-Iteration
0.7	6	200 m	91.5 %	46.7	8.5	Ok
0.7	6	300 m	94.3 %	39.9	10.0	Ok
0.7	7	300 m	95.5 %	85.5	4.8	Over-Iteration
0.8	6	300 m	74.3 %	18.5	22.5	Over-Damping

A trade off between *Covered Ratio* and *Cluster Member Number* could be found through Table 2. In general, the higher *Covered Ratio,* the lower *Cluster Member Number*. Several parameters should be meticulously adjusted for real deployment, among them the most important ones are *Damping Factor* and *Iteration Cycle*.

We use two metrics to measure effectiveness of modified AP algorithm:

(1) *Direct Influence*. We define one *Failure* is a malicious/incapable node elected to be CH, *Failure Rate* is to measure direct influence of one unqualified node:

$$Failure\,Rate = \frac{Number_{mal-CH}}{Number_{mal-node}} \tag{9}$$

(2) *Indirect Influence*. We define *Risk Degree* to feature how much potential influence an unqualified VI_i has when it is in one cluster.

$$RiskDegree = (1 - UntrustDegree(CH, i)) \times Number_{clustermembers} \tag{10}$$

Risk Degree could feature the indirect influence of an unqualified according to its role (CH or memeber) and UntrustDegree.

4.2 Comparison of Four Models

Our performance evaluation is based on four models: Primary AP model (PAP), Tempering AP model (TAP), Tempering & Supervising AP model (TSAP), and Converged AP model (CAP). PAP is directly derived from AP clustering algorithm. TAP models the clustering scenario where unqualified nodes could temper/disturb message passing process. In short, TAP considers tempering/disturb behaviors over PAP. To alleviate influence of tempering/disturbing, TSAP model injects *Mutual Supervision Model* into PAP to identify unqualified nodes. Finally, CAP is a converged model which enhanced TSAP with historical reputation.

For CAP, CA uses Majority Voting to fuse evidences. We assume that unqualified nodes use a random reporting strategy, which means generate evidences randomly regardless of what other nodes really have done. Normal nodes will always report true evidences. In PAP/TAP/TSAP, IV_n's preference is set median of $UntrustDegree(i,n)$. However, in CAP where historical reputation is considered by algorithm, IV_n's preference is calculated by:

$$Preference_n = PunishFactor_n \times Median(UntrustDegree(i,n)) \tag{11}$$

When IV_n's reputation is low, $PunishFactor_n \in [1,2]$ is big, preference therefore becomes small (preference is a negative real number), indicating IV_n is not suitable to be CH.

Figure 4 shows *Unqualified Percentage-Failure rate* curve of four models. Generally, when unqualified nodes percentage is low ($\leq 5\%$), *Failure Rate* is 0 %. As percentage goes up, *Failure Rate* also goes higher. TAP is a model with tempering/disturbing and no supervision mechanism, it performs worse than PAP (no tempering/disturbing) and TSAP (tempering/disturbing, supervision model). In contrast, CAP is a converged model (tempering/disturbing, supervision model, reputation) with a strong defense to unqualified nodes, so it shows the highest robustness among four models. Furthermore, either model could limit failure rate below 1 % even when unqualified nodes percentage is up to 25 %. This results demonstrates that our algorithms could effectively filter out unqualified nodes by reaching a trust consensus.

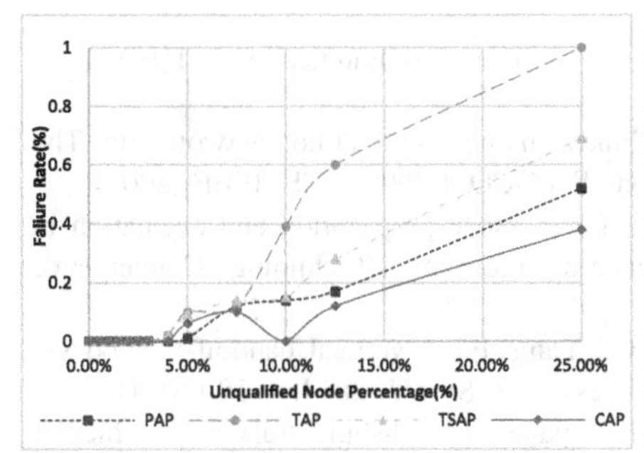

Fig. 4. Unqualified Percentage-Failure Rate curve

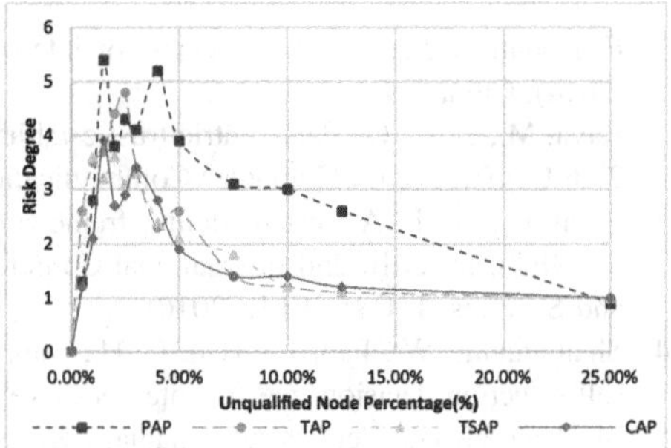

Fig. 5. Unqualified percentage-Risk Degree

Figure 5 shows the *Unqualified Percentage-Risk Degree* curve for four models. Risk Degree is firstly low when *unqualified node Percentage* is low, and markedly goes to peak when *Percentage* slightly increases. Finally, it stably declines with increasing *percentage*. The explanation is: (1) when *Percentage* is very low ($\leq 1\%$), tempering/disturbing is few, unqualified nodes therefore are easily distinguished by normal nodes. As a result, unqualified nodes are very likely to be left alone. So the overall *Risk Degree* is low. (2) When *Percentage* goes higher ($\leq 5\%$), this percentage still indicates a "safe environment", IVs tend to form "big clusters". However, with increase of unqualified percentage, more unqualified nodes have chances to join big clusters by more tempering/disturbing. According to formula (10), even one unqualified node in a big cluster leads a big risk degree. (3) When *Percentage* increases over 5 %, our algorithms tend to be conservative and form “small clusters”. Fewer members leads to lower *Risk Degree*. Overall, CAP could limit Risk Degree under 4, demonstrating our trust management is effective on risk control.

5 Conclusion and Future Work

Our work aims to build a fundamental trust management platform to handle IVs scenarios. To this end, we modified Affinity Propagation (AP) to elect a most trustworthy node, called CH. We also considered that AP is executed in attack/disturbance environment. Lastly, we integrate another component, CA, into our system.

In the future, great efforts are needed on both the in-vehicular system and RSUs to strengthen our system. These efforts include deploying mobile and local CA using cloud computing techniques, and reducing overhead of consensus-reaching process.

Acknowledgment. This work is supported by the National High-tech Research and Development Program (863) of China under Grant No. 2012AA111601, and 2015 Construction of key discipline under No. 700200253.

References

1. Fangchun, Y., et al.: An overview of internet of vehicles. Communications **11**(10), 1–15 (2014). China
2. Raya, M., et al.: On data-centric trust establishment in ephemeral ad hoc networks. In: The 27th Conference on Computer Communications INFOCOM 2008 IEEE. IEEE (2008)
3. Chen, C., et al.: A trust modeling framework for message propagation and evaluation in VANETs. In: 2010 2nd International Conference on Information Technology Convergence and Services (ITCS), IEEE (2010)
4. Shangguang, W., Fan, C., Hsu, C.-H., Sun, Q., Yang, F.: A vertical handoff method via self-selection decision tree for internet of vehicles. IEEE Syst. J. **2014**, 1–10 (2014)
5. Machado, R.G., Venkatasubramanian, K.: Short paper: Establishing trust in a vehicular network. In: 2013 IEEE Vehicular Networking Conference (VNC), IEEE (2013)
6. Huang, D., et al.: Establishing email-based social network trust for vehicular networks. In: 2010 7th IEEE Consumer Communications and Networking Conference (CCNC), IEEE (2010)
7. Huang, D., Hong, X., Gerla, M.: Situation-aware trust architecture for vehicular networks. Commun. Mag. IEEE **48**(11), 128–135 (2010)
8. Frey, B.J., Dueck, D.: Clustering by passing messages between data points. Science **315** (5814), 972–976 (2007)
9. Shea, C., Hassanabadi, B., Valaee, S.: Mobility-based clustering in VANETs using affinity propagation. In: Global Telecommunications Conference, GLOBECOM 2009 IEEE. IEEE, (2009)
10. Sivaraman, S., Trivedi, M.M.: Looking at vehicles on the road: A survey of vision-based vehicle detection, tracking, and behavior analysis. Intell. Transp. Syst. IEEE Trans. **14**(4), 1773–1795 (2013)

Design and Evaluation of a Smartphone-based Alarming System for Pedestrian Safety in Vehicular Networks

Jiyoon Kim[1], Younghwa Jo[2], and Jaehoon (Paul) Jeong[1](✉)

[1] Department of Interaction Science, Sungkyunkwan University, Suwon, Gyeonggi-Do 440-746, Republic of Korea
{yoonssl,pauljeong}@skku.edu

[2] Department of Software, Sungkyunkwan University, Suwon, Gyeonggi-Do 440-746, Republic of Korea
jyh1127@skku.edu

Abstract. This paper proposes a depth-based alarming service for pedestrian's safety in streets. This service is based on attentional network in cognitive neuroscience field. For the detailed study, we developed an Android alarming App, letting a smartphone user be informed of risk in advance. The alarming App was evaluated with six types of alarm along with a remote control car. During experiment, four metrics were measured, such as response time, collision, disturbance, and satisfaction by recording and questionnaire. The results show that, among six sorts of alarm, providing pre-warning with colorful background was useful to a pedestrian's response time to a main warning, but transparency for background color was not useful. These results demonstrate that offering pre-warning alarm makes the users promptly avoid the collision with cars. On the part of preference, going against our assumption, people prefer an apparent warning to a transparent background-color warning for less disturbance. Participants expressed that the less clear notification message is provided with transparent background, the more disturbed they are. Through experiments, it is shown that the proposed two-level depth-based alarming service can significantly reduce a pedestrian's reaction to a main warning message, leading to the provisioning of better safety for pedestrians.

Keywords: Smartphone · Alarming · Pre-warning · Distracted walking · Texting

1 Introduction

These days, when we walk around the street, we can often see people using their smartphone while walking. This scene means two aspects. One is how advanced our technology is and the other is the high possibility of collision beween a pedestrian and a car. Due to the highly advanced technology of smartphones, people can easily access the Internet for a variety of purposes (e.g., entertainment and

C.-H. Hsu et al. (Eds.): IOV 2015, LNCS 9502, pp. 221–233, 2015.
DOI: 10.1007/978-3-319-27293-1_20

email checking) anytime and anywhere. Now, this accessibility of a smartphone, however, puts people in danger. People lose their attention by smartphone and are exposed to the collision with a car. In the United States, 200,000 of people received hospital treatment, and approximately 4,000 people were killed because of distracted walking by smartphones in only 2009 [1]. According to the report of the US NSC (National Safety Council) in 2011, at least 1.6 million people suffered from accidents related to smartphones, and this trend has been increasing constantly [2]. Especially, in case of South Korea, the research of the impact of smartphone use on pedestrian safety risk was conducted by Transportation Safety Authority in 2013. This research reports that traffic accidents associated with smartphones have been steadily rising for the last four years. Korea has 1.7 times higher increase rate in smartphone-related accidents than the OECD [3]. Thus, the increase of smartphone users causes more traffic accidents related to smartphones, which is a global issue.

To the best of our knowledge, this paper is the first work to study a mobile alarm service for smartphone users, which is Smartphone-Assisted Navigation Application (SANA). The proposed alarm service is designed to ensure the safety of pedestrians with minimal disturbance. In particular, we choose a way of providing effective visual alarm for texting users while walking, which may encounter dangerous situations. The proposed alarm service takes a two-stage alarming, such as pre-warning and warning (called alarm). A pre-warning is an additional alarm generated before a warning so that a pedestrian can react to the warning within safety margin time to avoid an accident. After the delivery of the pre-warning, a warning is generated to let the pedestrian avoid an accident. In this paper, we design this two-stage alarming and evaluate our design in terms of user experience. Note that this paper is the enhanced version of our early Korean paper [4]. The contributions of the paper are as follows:

- A design of a depth-based alarming service. The proposed alarming service consists of two-level alarms for effective user reaction, such as pre-warning and warning.
- The evaluation of the depth-based alarming service. The alarming service is implemented by Android Application and the performance is statistically evaluated.

The remaining structure of the paper is as follows. Section 2 summarizes related work. Section 3 describes our design of SANA. Section 4 specifies our experiment method to evaluate our SANA. In Sect. 5, we conclude the paper along with future work.

2 Related Work

Nowadays, distracted walking has become an issue as one of major causes in road accidents. Smartphone users have been dramatically increasing for several years, and it is expected that one third of the worldwide population will use it by 2018 [5]. Distracted driving is now reducing due to regulation and education, but distracted walking is increasing on the contrary. Some research indicates that distracted walking could be more dangerous than distracted driving [6]. In comparison to feature phone, smartphone would also be more dangerous. That is because various functions of smartphone make people spend their time more than before. Many people are using their smartphone for texting or music-listening while walking. Therefore, distracted walking needs to be studied as well as distracted driving for safety.

Visually distracted people tend to walk slower about 2 s in average than the undistracted person on the intersection, and also they tend to look around less [7]. In order to avoid distracted walking accident, smartphone users need to do less texting than other activities with smartphones.

In order to avoid distracted walking, both the methods in both technology and education are explored. BumpFree service for preventing collision with obstacles is in progress [8]. BumpFree is a service to prevent the indoor/outdoor collision caused by distraction from happening through a smartphone alarm in a passage or street. On the other hand, in this paper, we have more interest in collision avoidance between a pedestrian and a vehicle rather than collision avoidance from stationary obstacles. Some movements to avoid distracted walking also have been started. There is an international movement towards improvement for the pedestrians safety. The Washington DC constructed a campaign for smartphone users in sidewalk lane [9], and in Korea, *distracted walking prevention campaign* was held for children [10]. However, even though these are good approaches, it seems still insufficient to avoid or prevent countless accidents throughout the world. To reduce distracted walking, therefore, needs not only the methods of education , but also that of technology. As a technology for accident prevention, we propose a smartphone-based alarming system, which informs pedestrians of the important moment in advance when they could be in danger because of distraction. Moreover, we investigate an effective design of alarm with minimal disturbance from the alarm. In this context, one natural research question is *how to provide pedestrians using smartphones with safety alarm effectively with minimal disturbance.*

In order to study a proper method for testing the effectiveness of an alarm on distracted walking, attentional network in cognitive neuroscience has been researched. Attentional network has three separable functional components, that is, alerting, orienting, and executive control [11]. *Alerting* can be described as a change in the internal state in preparation for perceiving a stimulus. *Orienting* can be defined as the aspects of attention that support the selection of specific information from numerous sensory inputs. *Executive control* can be meant as more complex, mental operations in detecting and resolving conflicts between computations occurring in different brain areas [12]. Among them, especially, orienting is affected if different conditions are given for an experiment, because giving an alarm while texting includes orienting as well as alerting. This is what we want to verify in this study, such as the effectiveness of visual alarm with depth when the alarm was given. In behavioral studies, orienting is often manipulated by presenting a *cue*, indicating where a subsequent *target* will (or will not) appear [12,13]. Therefore, for experiments in this paper, the revised attention network test (called ANT-R) is used, which is for testing the effects of alarms, including behavior using cue and target [12].

3 Design of SANA

In this section, we introduce the architecture, scenario, and implementation of our SANA.

3.1 Architecture

Our framework is based on wireless networks between a vehicle and pedestrian. Above all, the connection issue was on the table first. During the experiment of this study, we assume that a wireless networks from each of their devices such as a vehicle and a smartphone is available. That is, either Dedicated Short-Range Communications (DSRC) [14] or 4G-LTE [15] is used for the communications between the vehicle and the pedestrian, as shown in Fig. 1. Actually, the vehicle and pedestrian can communicate with each other via Road-Side Unit (RSU) in DSRC or Evolved Node B (eNodeB) in 4G-LTE [16]. When the people set this application up at the smartphone, the network connection could be automatically built up and supports data sharing.

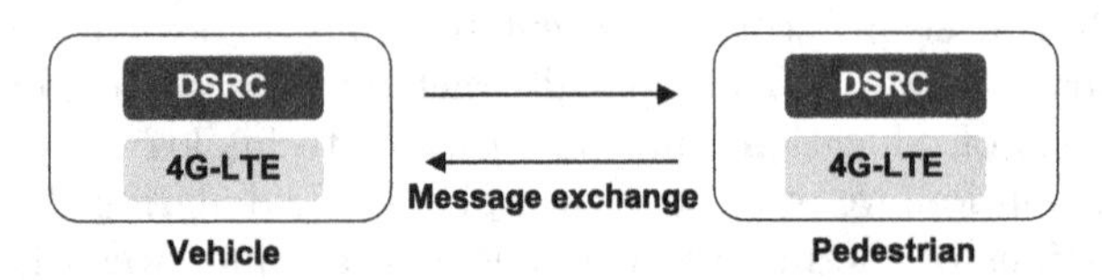

Fig. 1. Wireless communications for interaction between vehicle and pedestrian

In Fig. 2 shows the framework for user experience test in SANA. In Step (a) of the figure, Server determines time instants for alarm, and in Step (b), it sends an alarm to the smartphone Applications of both Participant (i.e., pedestrian) and Experimenter (i.e., vehicle driver). The figure also shows how Participant reacts to the alarm. In Step (c), Participant gets the alarm from the Application and Experimenter provides stimulus (i.e., remote control car) to Participant at the same time. Finally, in Step (d), Participant reacts to the alarm.

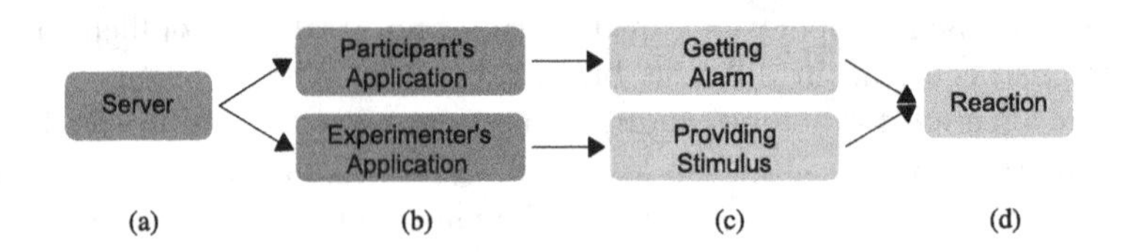

Fig. 2. The framework for user experience test in SANA

3.2 Scenario

Along with the basic steps for providing an alarm, we consider a useful method for alarming under two scenarios in a target environment (e.g., street) and the ways to use smartphones.

Basic Principle. Figure 3 shows the basic idea of providing alarming service. Two circles in the figure are used as the boundary for detecting a vehicle. When the vehicle is detected at the outer circle, pre-warning is delivered, and when it hits the inner circle, warning is provided. Our early work in [16] explains the details for detecting vehicles, filtering vehicles, and energy efficiency.

Surrounding Environments of Pedestrian. Pedestrian's surrounding environments include roadside, alley, and crosswalk. Roadside has a roadway and a sidewalk, and they can be easily located by using database for traffic information. When a pedestrian is in sidewalk, and a vehicle is in roadway, even though the vehicle and pedestrian are close to each other, this may not be a dangerous circumstance. Through using this information, false positive alarms in short distance can be filtered out to reduce the disturbance of pedestrians. Vector information of user's smartphone is also a good way to increase the accuracy of the alarm. In Fig. 3(a), vehicle goes straight, and, in Fig. 3(b), it turns left for different direction. When the pedestrian crosses the street, he may encounter a collision. But if the directions of the vehicle and pedestrian are heading to each other, and they keep their position (pedestrian in sidewalk and vehicle in roadway), it is not dangerous at all. The combination of traffic information and direction of smartphone will block such false positive alarms. However the street is narrow or the distance between the vehicle and the pedestrian is too close, it could be a dangerous case. Especially, a street like alley has many blind spots. In this case, warning should be generated accurately for safety.

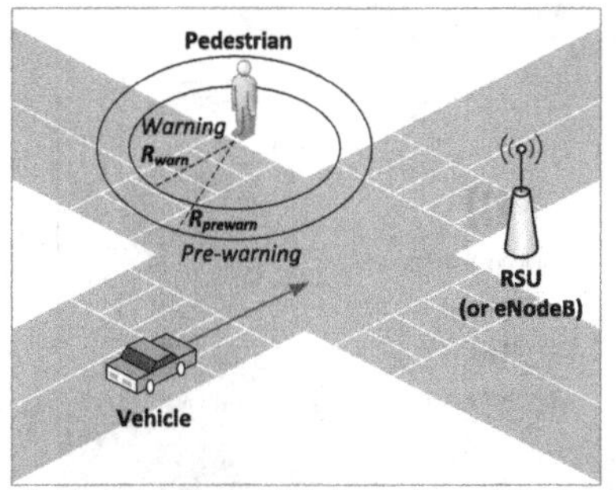

(a) No Overlap of Trajectories of Pedestrian and Vehicle

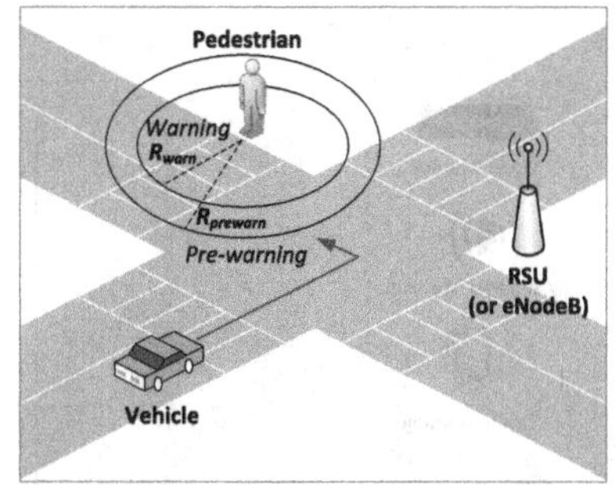

(b) Overlap of Trajectories of Pedestrian and Vehicle

Fig. 3. Scenarios in target environment

Figure 4 shows two ways of using smartphone while walking. Those may cause visual and cognitive distraction, and cognitive distraction only [1]. The first case in Fig. 4(a) is related to screen, and the second case in Fig. 4(b) is not. When a pedestrian is watching screen while walking, the alarm could be given in visual, auditory, or tactile way. On the other hand, when a pedestrian walk with talking and listen only, visual warning could not be an option. Also, in both cases, we can consider to give alarm not only in one way, but also in multiple ways. In this paper, we investigate the first case of the visual and cognitive distraction which is the most dangerous case in distracted walking, and we use visual warning among the three warning ways.

3.3 SANA Design

This section explains the design of user interface and user experience test for our SANA experiments, and the synchronization of devices used for the experiments.

Design of User Interface. This subsection explains our user interface design in SANA. For alarming, SANA provides a pop-up window with a warning message (e.g., Watch out) in red background, as shown in Fig. 5(a). SANA provides a pedestrian with depth-based alarming that consists of pre-warning (as cue) and alarm (as target) [12,13], as shown in Fig. 5(b). The background color for target is always red to draw attention from the pedestrian. On the other hand, the background color for cue is either yellow or red. Also, the transparency levels of the background is investigated, such as 0 % (as completely transparent color, that is, no background color), 50 % (as an half transparent color for less disturbing the pedestrian), and 100 % (as the solid color for strong attention). The interval between cue and target is 2 s [17]. The target has no interval for alarming. In this paper, we investigate which combination is the most effective for alarm in terms of the pedestrian's response time and disturbance for the alarm.

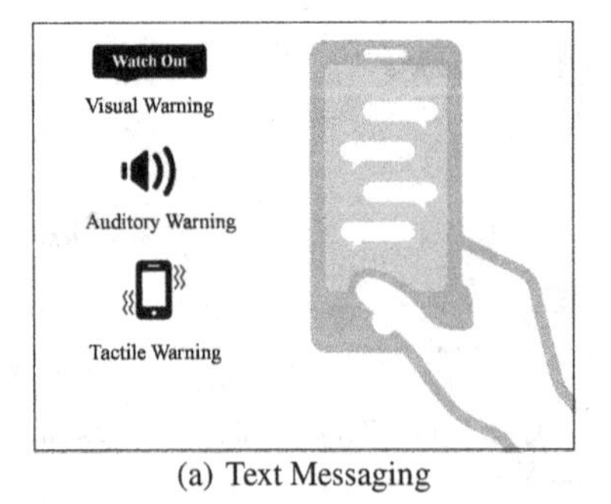

(a) Text Messaging

(b) Music Listening

Fig. 4. Scenarios of smartphone usage

(a) User Interface for Participant's Smartphone

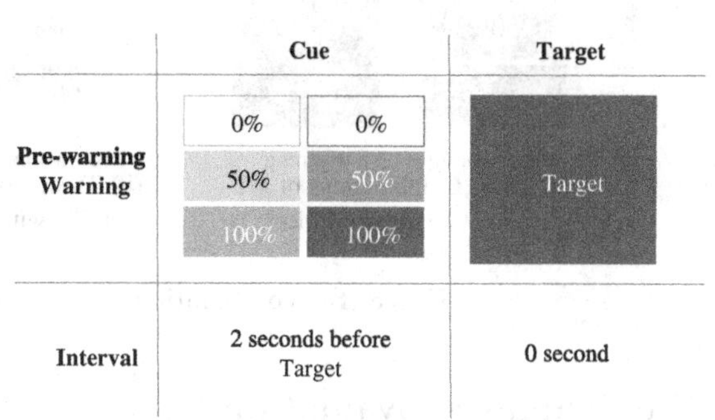

Cue
Target
Pre-warning
Warning
0%
0%
50%
50%
100%
100%
Target
Interval
2 seconds before Target
0 second

(b) Depth-based Alarming with Cue and Interval

Fig. 5. Design of SANA user interface

Design of User Experience Test. There are three components for user experience test, as shown in Fig. 6. Those are experimenter, participant and server. Experimenter controls all of experiments. Experimenter starts experiment, stops experiment, and records log file. Participant is the subject of the experiment. Participant gets warning alarm from server on time. Server waits for the command from experimenter. When experimenter orders to start an experiment, server generates a random time instant for collision event and sends it to both experimenter and participant. The devices of experimenter and participant know when an remote-controlled vehicle will come. But the difference is that experimenter can see all time instants server generates and participant cannot. Thus, experimenter can control the vehicle to collide with participant, as shown in Fig. 7.

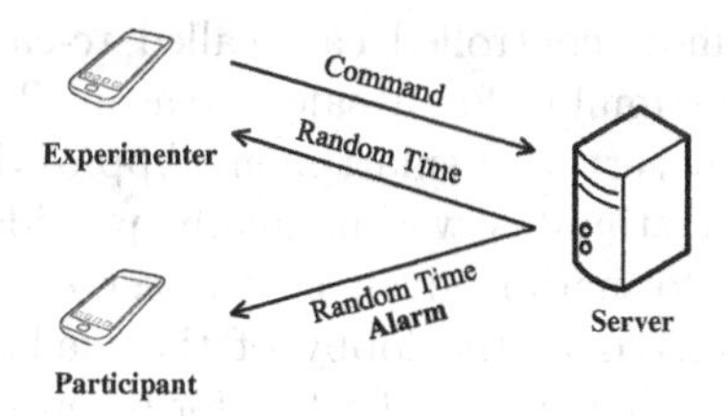

Fig. 6. Three components for experiment

Fig. 7. Experiment with remote-control car

Synchronization. For an accurate experiment, all devices and server must be synchronized. An easy way to synchronize is to use device time. All devices have different device time from a millisecond to more than 10 seconds at first. To correct this error, Server sends servers device time to other devices. Then experimenter sets the time of other devices to be the same with the received time from server.

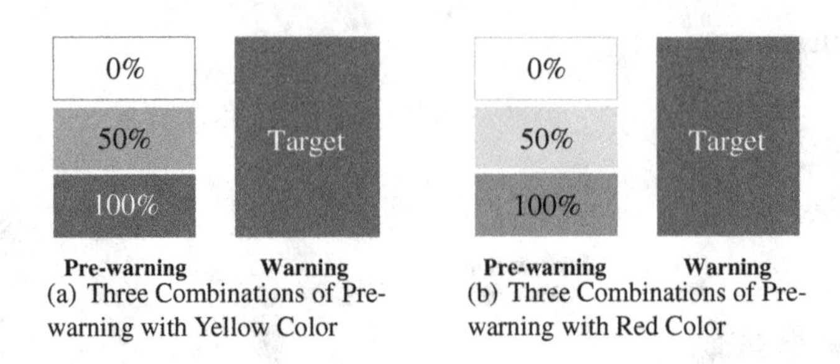

Fig. 8. Six combinations of pre-warning

4 Performance Evaluation

This section explains our experiment method to evaluate our SANA. This study conducts experiments to verify our research questions and hypotheses. The experiments use six combinations with different background color and transparency for alarm message. We categorize them into two kinds of experiments, such as (i) within-subject experiment with three options in transparency and (ii) between-subject experiment with two options in color. Participants were randomly assigned to have transparency options and were exposed to a warning stimulus more than 10 times. During the experiments, every single reaction of participant was recorded, and the results were calculated under operational definitions, specified in Sect. 4.1. We categorized the tests into three different types of transparency, 0 %, 50 % and 100 %. After every single experiment, individuals answered questionnaires regarding to their disturbance and satisfaction through online survey. Each experiment was repeated ten times and then questionnaire was given, as shown in Fig. 8. Total 29 (21 Males and 8 Females) students using smartphone daily were recruited from the Natural Sciences Campus at Sungkyunkwan University (SKKU), Korea.

The experiments were performed by remote-controlled car (called rc-car), such as M40S Volkswagen Touareg3, as a stimulus for a safety reason. The alarm stimuli were automatically provided by server to smartphone App of the participants, as shown in Fig. 6. The rc-car stimulus was manually provided from experimenter according to the time information given by the server, as shown in Fig. 7. The experiment was conducted in the lobby of the medical school auditorium at SKKU using a camcorder (i.e., Canon FS200) for recording the students' physical reaction, a laptop computer (i.e., LG ZD360-GD70K) for answering questionnaire, and mobile devices (i.e., Samsung SHV-250, SHV-160, and SHW-M380S) for recording experiments and operating rc-car.

4.1 Operational Definitions for Measurement

The operational definitions for this study are as follows.

Warning: Warning is an on-time alarm which indicates a probability of collision with rc-car within a second.

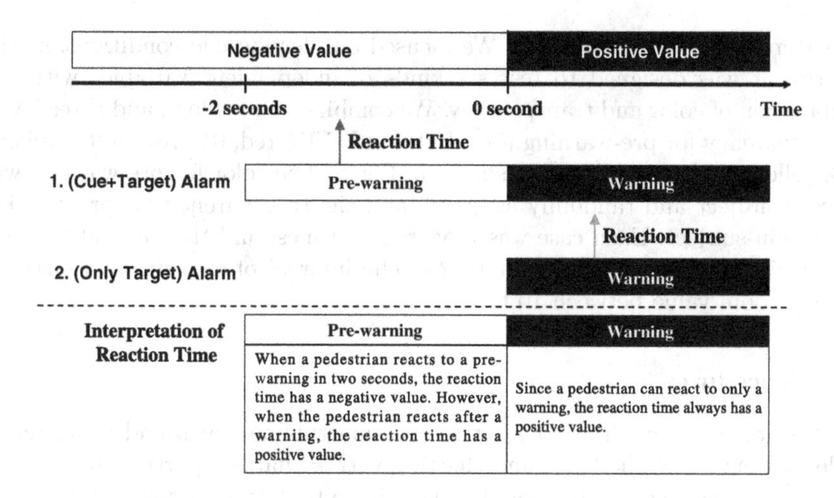

Fig. 9. Operational definitions on pre-warning and warning

Pre-warning: Pre-warning is a preceding warning which is provided two seconds earlier than a warning.

Reaction Time: Reaction time is relative time interval between the generating time of a warning and the response time of a pedestrian to a rc-car, as shown in Fig. 9. In the figure, for *(Cue+Target) Alarm* with both pre-warning and warning, the response time is negative, but, for *(Only Target) Alarm* with only warning, the response time is positive.

Collision: Collision is either a physical collision of a pedestrian and a rc-car or the distance between them is less than 20 centimeters. That is, when a pedestrian is within 20 centimeters of the rc-car, it is counted as a collision.

Disturbance: Disturbance is the value of disturbance questionnaire which is customized from *the development of the noise sensitivity questionnaire* [18]. A major factor is divided into two disturbances. One is the daily disturbance of the smartphone alarm, and the other is the disturbance of the smartphone alarm during the experiment. Participants responded to each item on 7-point Likert scale, ranging from 1= "strongly disagree" to 7= "strongly agree". The value of satisfaction questionnaire is customized from *the assessment of client/patient satisfaction* [19]. Satisfaction also uses the same Likert scale as the disturbance questionnaire.

4.2 Experimental Treatment Conditions

In the ANT-R experimental frame which was designed by Jin Fan, he considered interval as a cue-target onset asynchrony (ctoa) and alerting cue conditions (no-cue, temporal cue, and spatial cue) [12]. We considered one interval of 2 s in this study [17], because of the difference of individual response time which would be better to be considered later. We focused on alerting cue conditions, so the experiment was designed to test six kinds of independent variables with the combination of color and transparency. We combined two colors, and three levels of transparency for pre-warning (i.e., 100 % red, 50 % red, 0 % red, 100 % yellow, 50 % yellow, and 0 % yellow), as shown in Fig. 8. The color for pre-warning was between-subject and randomly assigned, and the transparency for pre-warning was within-subject. Each case was repeated 10 times, and the interval between pre-warning and warning was set to 2 s. The interval of experiment repetition was a random value between 10 to 30 s.

4.3 Procedure

The task of participants is to react to the smartphone alarm and avoid rc-car while walking along the track and chatting with a chatting partner in the same time using a message application in smartphone that was provided by us. Also, the participants were asked to wear earplug and not to react to any noise of a remote control car or anything else during the experiment. In order to establish a similar situation as the real case, we hired a chatting partner. The chatting topic was fixed to Korean thanksgiving holidays that will come soon. The experiment was processed for each individual and conducted 30 times (i.e., every 10 times for one of three random combinations). Right after each experiment, participants marked their disturbance and satisfaction on questionnaire. After the experiment, response time and collision were measured through the analysis of recording data. Reaction time was measured in millisecond. When participants reacted not to alarm but to a noise, the data were excluded in statistical analysis for more accurate measurement.

4.4 Data Analysis

SPSS 21 was adopted for whole testing. ANOVA was conducted to investigate the effect of pre-warning and color. Generalized Linear Model (GLM) was used to test the effect of transparency and gender on pre-warning. Correlation analysis was conducted to test the correlations of response time and collision, and of disturbance and satisfaction. Among the collected data, the data of a person who does not react to the given alarm or overreact to the stimulus were excluded from statistical analysis as an outlier.

4.5 Experiment Results

This subsection summarizes the results of user experience in the SANA alarming service. Table 1 shows the results of the statistical analysis about whether hypotheses were supported or not. Providing pre-warning has significant effect on response time and collision (see H1). Providing color pre-warning has significant effect on collision and satisfaction (see H2). Providing transparent pre-warning has significant effect on response time and collision (see H3).

Table 1. Results of hypothesis test

Hypotheses			Supported
H1a	Providing pre-warning alarm will have positive effect on	(a) response time	**yes**
		(b) collision	**yes**
		(c) disturbance	**no**
H1b	Providing pre-warning alarm will have negative effect on	(d) satisfaction	**no**
H2a	In case of providing pre-warning alarm, different background-color alarm would have positive effect on	c) disturbance	**no**
		(d) satisfaction	**yes**
H2b	In case of providing pre-warning alarm, different background-color alarm would have no effect on	(a) response time	**yes**
		(b) collision	**no**
H3a	In case of providing pre-warning alarm, providing transparency to alarm will have positive effect on	c) disturbance	**no**
		(d) satisfaction	**no**
H3b	In case of providing pre-warning alarm, providing transparency to alarm will have no effect on	(a) response time	**no**
		(b) collision	**no**

Manipulation Check. A reliability analysis was conducted to check the consistency of the questionnaire. Disturbance questionnaire was reliable (Cronbachs $\alpha = 0.761$). Satisfaction questionnaire was also reliable (Cronbachs $\alpha = 0.894$). In the case of two factors of disturbance, the questionnaire about participant's daily disturbance was reliable (Cronbachs $\alpha = 0.629$). The disturbance questionnaire during experiment was also reliable (Cronbachs $\alpha = 0.716$).

Statistical Validation. We performed one-way ANOVA to show statistical validation of the effectiveness of our alarming system with pre-warning and warning.

To find out the effect of a pre-warning, one-way ANOVA was conducted. The result shows that there was a significant effect on response time with pre-warning (Mean(M) = −0.601, Standard Error(SE) = 0.1398) compared with no pre-warning (M = 0.8573, SE = 0.3962), and degree of freedom (1, 658) = 53.33, p-value < 0.5. The result also confirms that there was a significant effect on collision with pre-warning (M = 1.83, SE = 0.018) compared with no pre-warning (M = 1.48, SE = 0.034), $F(1, 658) = 104.18$, $p < 0.5$. However, there was no significant effect on disturbance with pre-warning and as well as on satisfaction. The difference on response time and collision with pre-warning and without pre-warning is shown in Fig. 10. Hypotheses 1a were partially supported, however, Hypothesis 1b was not supported.

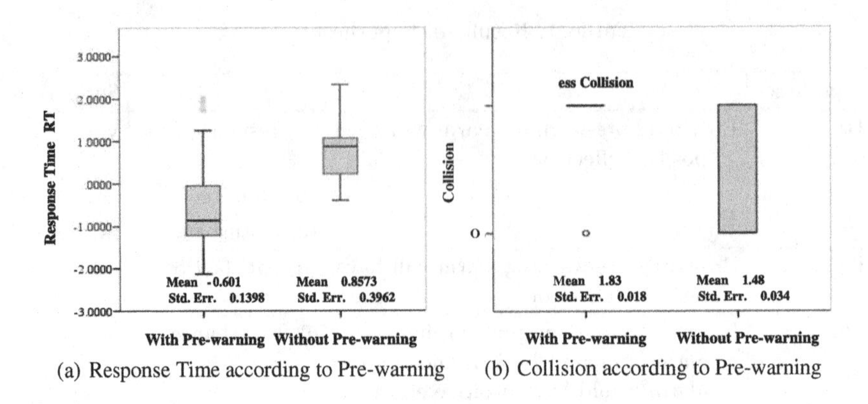

(a) Response Time according to Pre-warning (b) Collision according to Pre-warning

Fig. 10. The impact of pre-warning on reaction time and collision

To investigate the effect of the color and transparency for pre-warning, we performed one-way ANOVA in a similar way with the previous ANOVA test. For the color, the result shows that red pre-warning is more effective for both response time and collision avoidance than yellow pre-warning. Also, for the transparency, the result shows that 100 % or 50 % tranparent pre-warning is more effective for both response time and collision avoidance compared with 0 % transparent pre-warning.

5 Conclusion

There are several fundamental outcomes in this research. First, providing alarm for distraced-walking pedestrians can improve safety. Second, notification with different depth of color and transparency provided to pedestrians can bring different reactions to the moving vehicle, then it can be helpful for collision avoidance. Third, the warning message with solid-background color allows collision avoidance with less disturbance for pedestrians. Finally, an alarming App for smartphone users seems a simple and effective method to bring safety in streets. As future work, we can design an audio-based alarming for the case where pedestrians are listening to music or having phone-call.

Acknowledgment. This research was supported by Basic Science Research Program through the National Research Foundation of Korea (NRF) funded by the Ministry of Science, ICT &Future Planning (2014006438). This research was supported in part by Global Research Laboratory Program (2013K1A1A2A02078326) through NRF, and the ICT R&D program of MSIP/IITP (14-824-09-013, Resilient Cyber-Physical Systems Research) and the DGIST Research and Development Program (CPS Global Center) funded by the Ministry of Science, ICT &Future Planning. Note that Jaehoon (Paul) Jeong is the corresponding author.

References

1. Byington, K.W., Schwebel, D.C.: Effects of mobile Internet use on college student pedestrian injury risk. Accid. Anal. Prev. **51**, 78–83 (2013)
2. National Safety Council. National safety council estimates that at least 1.6million crashes each year involve drivers using cell phones and texting. http://www.nsc.org/learn/NSC-Initiatives/Pages/distracted-driving-problem-of-cell-phone-distracted-driving.aspx
3. Korea Transportation Safety Authority. Report on the behavior of smartphone users in walk. http://www.ts2020.kr/tsk/rck/InqDetPTRTrafficSafety.do?ctgCd=1&searchCtgCd=+&bbsSn=5275&pageIndex=3+&searchCnd=&searchWrd
4. Kim, J.Y., Jo, Y.-H., Jeong, J.P.: A study on alarm service for providing safety for pedestrian using smartphone: focused on the effectiveness of depth-based alarming. In: KIPS 2014-Fall, November 2014
5. eMarketer. 2 Billion Consumers Worldwide to Get Smartphones by 2016. http://www.emarketer.com/Article/2-Billion-Consumers-Worldwide-Smartphones-by-2016/1011694
6. Nasar, J.L., Troyer, D.: Pedestrian injuries due to mobile phone use in public places. Accid. Anal. Prev. **57**, 91–95 (2013)
7. Thompson, L.L., Rivara, F.P., Ayyagari, R.C., Ebel, B.E.: Impact of social and technological distraction on pedestrian crossing behaviour: an observational study. Inj. prev. **19**(4), 232–237 (2013)
8. Shin, K.G.: BumpFree: Real-time warning for distracted pedestrians with smartphones. In: The 9th International Symposium on Embedded Technology, May 2014
9. Yahoo. Cellphone Talkers Get Their Own Sidewalk Lane in D.C. https://www.yahoo.com/tech/cellphone-talkers-get-their-own-sidewalk-lane-in-d-c-92080566744.html
10. Safe Kids Korea. FedEx Korea, Held Distraction Walking Prevention Campaign for children. http://news.mk.co.kr/newsRead.php?year=2013&no=402088
11. Fan, J., Posner, M.: Human attentional networks. Psychiatr. Prax. **31**(2), S210–S214 (2004)
12. Fan, J., Gu, X., Guise, K.G., Liu, X., Fossella, J., Wang, H., Posner, M.I.: Testing the behavioral interaction and integration of attentional networks. Brain Cogn. **70**(2), 209–220 (2009)
13. Posner, M.I.: Orienting of attention. Q. J. Exp. Psychol. **32**(1), 3–25 (1980)
14. Xu, Q., Mak, T., Ko, J., Sengupta, R.: Vehicle-to-vehicle safety messaging in DSRC. In: Proceedings of the 1st ACM International Workshop on Vehicular Ad Hoc Networks, pp. 19–28. ACM (2004)
15. de Monfreid, C.: The lte network architecture - a comprehensive tutorial. Alcatel-Lucent White Paper (2009)
16. Hwang, T., Jeong, J.P., Lee, E.: SANA: Safety-aware navigationapp for pedestrian protection in vehicular networks. In: Proceeding of International Conferenceon ICT Convergence, ICTC 2014. IEEE, October 2014
17. Fingelkurts, A.A., Fingelkurts, A.A., Krause, C.M., Sams, M.: Probability interrelations between pre-/post-stimulus intervals and ERD/ERS during a memory task. Clin. Neurophysiol. **113**(6), 826–843 (2002)
18. Schutte, M., Marks, A., Wenning, E., Griefahn, B.: The development of the noise sensitivity questionnaire. Noise Health **9**(34), 15 (2007)
19. Larsen, D.L., Attkisson, C.C., Hargreaves, W.A., Nguyen, T.D.: Assessment of client/patient satisfaction: development of a general scale. Eval. Program Plann. **2**(3), 197–207 (1979)

An Approach for Map-Matching Strategy of GPS-Trajectories Based on the Locality of Road Networks

Aftab Ahmed Chandio[1,2,3](✉), Nikos Tziritas[1], Fan Zhang[1], and Cheng-Zhong Xu[1,4]

[1] Shenzhen Institutes of Advanced Technology, Chinese Academy of Sciences, Shenzhen 518055, China
{aftabac,nikolaos,zhangfan,cz.xu}@siat.ac.cn
[2] University of Chinese Academy of Sciences, Beijing 100049, China
[3] Institute of Mathematics and Computer Science, University of Sindh, Jamshoro 70680, Pakistan
[4] Department of Electrical and Computer Engineering, Wayne State University, Detroit 48202, USA

Abstract. A map-matching process plays a pivotal role in ascertaining the quality of many location based services (LBS). Map-matching process is to determine the accurate path of a vehicle onto road network in a form of digital map. Most of the current map-matching strategies are based on the shortest path queries (SPQs) providing best performance in terms of accuracy. Unfortunately, the execution of the SPQs is the most expensive part of the map-matching process in terms of computational cost, which may be unaffordable for real-time processing. This paper introduces LB-MM (i.e., Locality Based Map-Matching), a novel approach for map-matching strategy that is based on locality of road network. LB-MM approach addresses a key challenge of SPQs in map-matching strategies by adaptively tuning the interior parameters of the map-matching process. The interior parameters, i.e., a number of candidate points (CP) and error circle radius (ECR) are fine-tuned based on different classes of locality of road network for each GPS sampling point. We characterize the locality of road network in different classes which result by splitting road network into small grids. In that way, a set of interior parameters is chosen based on locality that drastically reduces a number of SPQs and the overall computation time of map-matching process. The evaluation of proposed strategy against the SPQ-based ST-MM (i.e., Spatio-Tempo Map-Matching) strategy found in the literature is performed through simulation results based on both synthetic and real-world datasets. In LB-MM strategy, the total number of SPQs is counted as less than 27 % against those of ST-MM.

Keywords: Location based services · Map-matching · GPS data · Spatiotemporal data mining · GIS · Locality of road network

C.-H. Hsu et al. (Eds.): IOV 2015, LNCS 9502, pp. 234–246, 2015.
DOI: 10.1007/978-3-319-27293-1_21

1 Introduction

Recently urban computing becomes mature due to the advent and growth of information and communication technologies (ICT) in the developed and developing countries. Particularly, in urban computing, vehicles, buildings, devices, sensors, roads, and people are accessed as a component to probe city dynamics [1]. The data represented in the aforementioned components is usually obtainable in the form of global positioning system (GPS) data. Information available in the GPS data is accessed in many GPS-based applications to achieve better quality of services (QoS). Such applications include location based services (LBS) and intelligent transportation systems (ITS) (i.e., traffic flow analysis [2], the hot route finder [3], geographical social network [4], and the route planner [5]).

The process of map-matching plays an essential role in navigation systems (such as road guidance, driving directions, traffic flow analysis, and moving object management) of GPS-based vehicles. Basically, map-matching process is to determine the correct path of a vehicle onto road network by aligning observed GPS positions of a vehicle on that path [6,7]. Most of the current map-matching strategies (such as [6–10]) suggested that the best performance of a map-matching process in terms of accuracy is based on the transition probability which is incorporated with the shortest path between two candidate points of consecutive GPS points. On the other hand, the execution of the shortest path queries (SPQs) needs high computational cost [6]. Moreover, the noisy and imprecise natures (i.e., sampling error and measurement error) of GPS data degrade the performance of render map-matching strategy in both accuracy and time complexity [11]. The high-sampling-rate of GPS data may hold unnecessary data (such as at the time of traffic jam, traffic signal, moving slowly, etc.). Next, the imprecise data is commonly produced because of the limitations of the GPS technology (i.e., measurement error).

Furthermore, the current map-matching strategies use fixed settings for interior pa-rameters i.e., candidate points (CP) and error circle radius (ECR) for considering most likely candidate road segments. Using fixed interior settings to map-match the noisy and imprecise GPS data may incur either extra number of SPQs (i.e., high computa-tional cost) or uncertainty of identifying zero or less candidate road segments than fixed values (i.e., low-accuracy).

Due to the above facts, unluckily, the map-matching strategies found in the current literature are unaffordable for real-time processing. Particularly, dynamic traffic management and control systems are taken input of real-time traffic information by real-time processing [9,12]. Therefore, it is of paramount importance for the real-time map-matching to handle the executions of the SPQs in an effective and flexible way. In this paper, we introduce LB-MM (**L**ocality **B**ased **M**ap-**M**atching), a novel approach for map-matching of GPS trajectories based on locality of road network. LB-MM strategy addresses a key challenge of SPQs by adaptively tuning the interior parameters of the map-matching process. Specifically the interior parameters considered in this paper i.e., candidate points (CP) and error circle radius (ECR) are fine-tuned based on different classes of locality of road network for each GPS sampling point. We characterize the locality of road network in different classes which result by splitting road network into small grids. In that way, the interior parameters are chosen in a way, such that to drastically reduce a number of SPQs and the overall computation time of map matching process. By employing parameter tuning of map-matching process based on locality of road network, we propose a viable solution to the issues when performing SPQs. The above approach is empirically evaluated using real-world and synthetic datasets against the SPQ-based map-matching strategy. The real-world dataset is collected from the Shenzhen Transportation Systems. In LB-MM strategy, the total number of SPQs is counted as less than 27 % against those of ST-MM.

The rest of the paper is organized as follows. Section 2 sets forth the state-of-the-art of map-matching process and expresses major definitions of terms used in this paper. In Sect. 3, we present the proposed LB-MM strategy that is based on locality of road network. In the sequel, in Sect. 4, we describe our experimental settings, the results evaluation, and discussion, while Sect. 5 concludes the paper.

2 State-of-the-Art of Map-Matching

Particularly, the map-matching process of all of the current map-matching strategies found in the literature can be categorized into three major steps as shown in Fig. 1. We explain each of them as follow.

1. The *initialization* step prepares a number of CPs that is projected on the road segments within an ECR.
2. The *weight calculation* step calculates a weight function of a link between two CPs of two consecutive observed GPS sampling point.
3. The *weight aggregation* step in a map-matching strategy aggregates the weight scores calculated in the previous step.

A complete map-matching strategy is generally classified as a/an: (a) incremental [13,14], (b) global [6,8,15], or (c) statistical [9,10,16,17] method. In the incremental method, a map-matching strategy firstly finds a local match of geometries for each given GPS point. A small region of a road network that is close to the point is considered. Then this map-matching strategy accumulates the weight scores with the results of the previous GPS sampling point. In terms of accuracy, the incremental method provides better results under the high-sampling-rate (i.e., 2–5 s) of the GPS points.

On the other hand, in the global method, a map-matching strategy matches an entire trajectory with a digital road network graph. In terms of accuracy, the global map-matching strategy produces better results against incremental method especially when applying low-sampling-rate of GPS points. The reason of the low accuracy provided by incremental method is that it performs matching GPS points one by one in incremental way. In addition, matching GPS points one by one also enables the map-matching strategy as much as fast. Unfortunately, such a strategy working in an incremental way breaks the continuity of

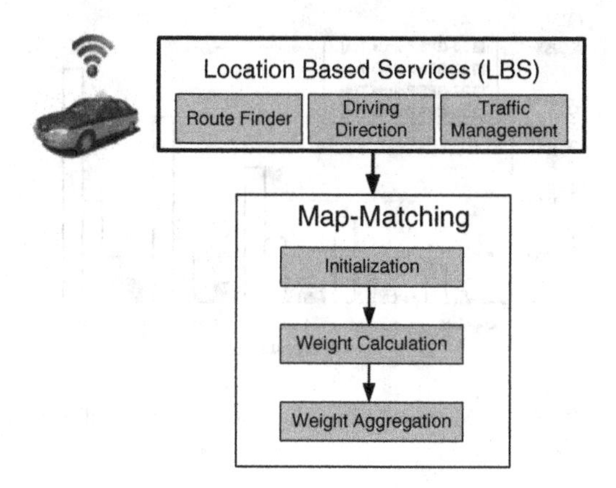

Fig. 1. Basic steps in a Map-Matching

trajectory, which degrades the accuracy. More precisely, a map-matching strategy based on global method is suffered by the increased computational cost while producing more accurate results. The map-matching strategy in the last category performs statistical methods (such as, Bayesian classifiers, Hidden Markov model [17], Kalman filter, and cubic spline interpolation [16]) to match GPS points with a road network. These methods are used to handle the GPS errors (i.e., measurement and sampling errors).

The proposed approach for map-matching, LB-MM, is motivated from the ST-MM [6] global map-matching strategy that provides high accuracy. The ST-MM is incorporated with the spatial (i.e., geometric structure) and temporal (i.e., speed) constraints to solve the problem of low-sampling-rate GPS points. Specifically, the ST-MM calculates the weight function with respect to two consecutive GPS sampling points and their candidate points. The weight function considered in the ST-MM is completely based on the spatial and temporal constraints. The ST-MM then generates a true path by following the largest summation of weight scores.

In the experimental evaluation [6,18], we found that the ST-MM violates the fast response time, QoS. The weight function in the ST-MM strategy uses the SPQ between two consecutive GPS points which is the most expensive part of map-matching process. Figure 2 shows a total number of SPQs processed by the ST-MM with respect to different candidate points and GPS sampling points. The figure reveals that when increasing the number of CPs and GPS points in trajectories, then the total number of SPQs increases. Moreover, the reason of high computation cost is attributed to the fact that a fixed number of CP and ECR in the map-matching strategy results in a large number of SPQs.

Another remark in this regard is that a map-matching process using a fixed number of interior parameters may result in low accurate results. Especially, when the respective part of a road network (called ECR) does not contain a suitable number of candidate road segments for CPs [19]. For example, in Fig. 3.

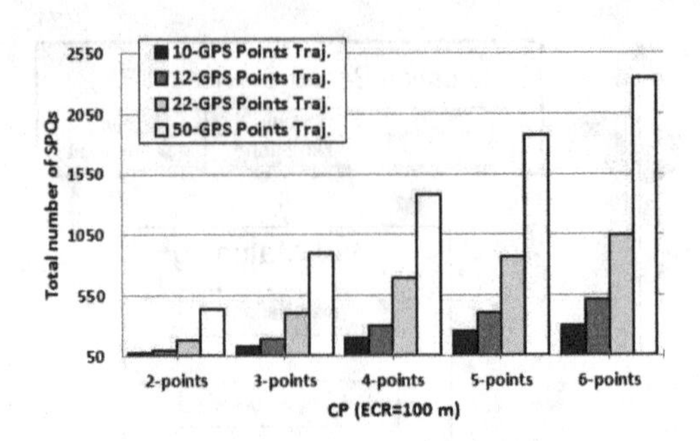

Fig. 2. ST-MM processing the total number of SPQs

we show a case of two fixed values for a part of a road network (i.e., error circle radius r_1 = small and r_2 = large). In that figure, when r_1 = small, then we result in two candidate points, while when r_2 = large, then we result in four candidate points. To address the aforementioned problems, we propose an approach for map-matching by applying an adaptive tuning strategy. Our proposed strategy is based on locality of road network that adaptively fine-tunes the interior parameters of map-matching (i.e., CPs and ECR) for each GPS sampling point. Subsequently, a reasonable number of SPQs reduces the overall running time in the map-matching process.

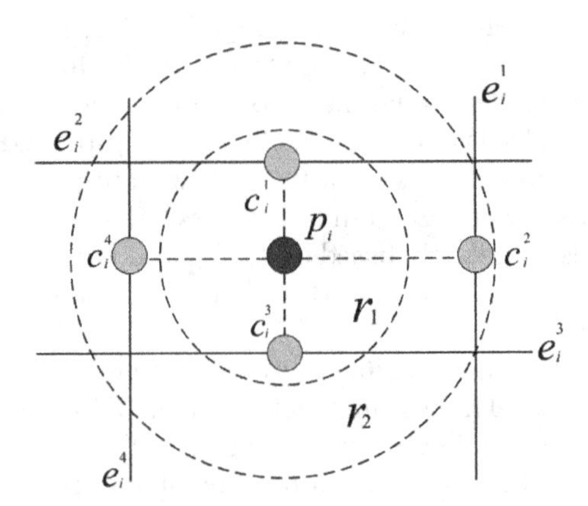

Fig. 3. The interior setting for considering candidate points for a sampling point

Definition 1. GPS Trajectory*: A route completed by a taxi is called trajectory T and consists of a sequence of GPS sample points i.e., $T : p_1 \rightarrow p_2 \rightarrow \cdots \rightarrow p_n$, where $0 < p_{(i+1)}.t - p_i.t < \Delta T (1 \leq i < n)$ and T is the time spent between two consecutive GPS points. Each GPS point $p_i \in T$ contains information including GPS position in latitude $p_i.lat$, longitude $p_i.lon$, and timestamp $p_i.t$*

Definition 2. Road graph*: A directed road network graph $G(V, E)$ is called a road graph. Let V denote a set of points intersecting the road segments, called vertices. E signifies a set of road segments, called edges. A directed edge e is associated with: (a) a unique id e.gid, (b) the average travel speed e.v, (c) the road length e.l, (d) the starting point e.start, (e) the ending point e.end, (f) and the intermediate points comprising the road polyline.*

Definition 3. CP*: A candidate point c is a point projected on a candidate road segment e .*

Definition 4. ECR*: An error circle radius r is a space of road network graph which is used for choosing most likely candidate road segments.*

Definition 5. Path*: A path P is a list of connected road segments between two given vertices (V_i, V_j) in a road network G , i.e., $P : e_1 \rightarrow e_2 \rightarrow \cdots \rightarrow e_n$, where $e_1.start = V_i, e_n.end = V_j, e_k.end = e_{k+1}.start, 1 \leq k < n$*

3 Map-Matching Strategy Based on Locality of Road Network

This section addresses the map-matching strategy that is based on locality of the given road network (i.e., LB-MM). LB-MM approach adaptively fine-tunes the interior parameters of the map-matching process. In this paper, we use two interior parameters: (a) candidate points (CP) and (b) error circle radius (ECR). In order to achieve high accuracy, we incorporate a LB-MM approach with the spatio-tempo strategy (i.e., ST-MM [6]). In the following section, we first explain the major components of the map-matching strategy. We then introduce the approach for tuning the interior parameters of map-matching process, which is based on locality of road network.

3.1 Map-Matching Components

The map-matching strategy is consisted of four major components: (a) tuning, (b) candidate preparation, (c) spatio- and tempo-analysis, and (d) matching score. The description of them is mentioned as follow.

The tuning phase is a major component of our proposed LB-MM. Specifically; the tuning phase is a responsible for tuning the interior parameters, i.e., CPs and ECR that are used for mapping the GPS sampling point to the road that is most probable to be correct. The interior parameters are tuned based on the locality of the road network. As the characterization and analysis of workloads help the overall performance of a system [20], we also characterized the locality of the road network. To characterize the locality of road network, we split the road network graph into a grid format. The number of grids can be varied according to a memory capacity of a system, since all grids tuples are stored in the memory during map-matching process (e.g., in our experiment, we have set out 200 X 200 grids). The following information for each grid is stored in a separate tuple.

1. *gId* is for the grid identification.
2. *coordinate* keeps information of both start and end of longitude and start and end of latitude for the grid.
3. *density* stores the total number of road segments intersected to or located into the gird.
4. *rIds* stores all unique ids of all road segments into the grid.

Table 1. Parameters based on locality of road network

Parameters	Class-I	Class-II	Class-III	Class-IV	Class-V
Density($\sum e$)	≥ 21	≥ 16 to < 21	≥ 12 to < 16	≥ 9 to < 12	< 9
ECR (r)	60	80	100	120	150
CP (c)	3	4	5	6	7

After characterizing the locality of the road network, we categorize the grids into five different classes (as mentioned in Table 1), where each class considers different CPs and ECR. If a grid has a high density (i.e., several road segments) then map-matching strategy will have maximum chance to consider the number of candidate road segments under the small ECR and fewer CP. In that way, the map-matching strategy can adaptively choose the interior parameters for each GPS sampling point. Our observation is that by adaptively choosing the interior parameters for each GPS sampling point, we can drastically reduce the number of SPQs and the overall computation time of map matching process. Mainly in this phase, the map-matching strategy initially finds the grids for each GPS point and then returns the best number of CP and ECR.

Algorithm 1 (Fig. 4.) describes the mechanism of proposed tuning approach for map-matching strategy based on locality of road network. Initially, by the latitude and longitude of the GPS sampling point, the algorithm finds a grid cell on which the observed GPS sampling point is located as well as its neighbor grid cells (line 1). Then it finds the maximum road density of the grid cells (line 2). The interior parame-ters are chosen based on the road density as shown in Table 1 (line 3–4). Lastly, it returns the total rIds of the road segments located in the grid cells based on the interior parameters values (line 5–8). Consequently, the interior parameters are adjusted for each observed GPS sampling point.

The candidate preparation phase is a next step that first retrieves a set of possible candidate road segments e within an ECR r , then it finds a set of candidate points CP on the retrieved candidate road segments by using geometry projection. A CP c is the closest point to the GPS point with respect to the Euclidean distance such that $c = argmin_{(\forall c_i \in e)} dist(c_i, p)$. Figure 3 illustrates the strategy.

We assume that the GPS error follows a normal distribution $N(\mu, \sigma^2)$ of the distance between GPS sampling point and CP, i.e., $x_i^j = dist(c_i^j, p)$. As the above assumption is common practice in developing map-matching strategies, we also

Algorithm 1: The proposed approach for map-matching strategy based on locality of road network

```
Input:  A GPS point
Output: CPs
1: gc=findGridCells (p.long,p.lat); //find neighboring grid cells
2: ge=findMaxEdges (gc); //find a maximum number of roads in gc
3: cp=bestCPNumber (ge); //find a best number for candidate points
4: ecr=bestECRNumber (ge); //find a best number for circle radius
5:   for i in gc do
6:      gids[] = getGids (i);//collect ids of all roads in the grid
        cells
7:   end for loop
8: return c[]= getCPs (gids, cp, ecr); //get candidates points
```

Fig. 4. Algorithm

assume $\mu = 0$ and $\sigma = 20$ meters. For the above, the reader is referred to [6] for further details. The equation for the normal distribution of a candidate point is given by (1).

$$N(c_i^j) = \frac{1}{\sqrt{2\pi}\sigma} e^{\frac{(x_i^j - \mu)^2}{2\sigma^2}} \tag{1}$$

Spatial and temporal analysis phase is responsible to generate the CP graph $G'_T(V'_T, E'_T)$. We adopt the spatial and temporal analysis function from ST-MM [6] in order to maintain the accuracy of the proposed approach, denoted by:

$$F(c_{i-1}^t \rightarrow c_i^s) = F_s(c_{i-1}^t \rightarrow c_i^s) \mathrm{X} F_t(c_{i-1}^t \rightarrow c_i^s), 2 \leq i \leq n \tag{2}$$

where F_s is a spatial analysis function calculated by:

$$F_s(c_{i-1}^t \rightarrow c_i^s) = N(c_i^s) \mathrm{X} V(c_{i-1}^t \rightarrow c_i^s), 2 \leq i \leq n \tag{3}$$

where $V = \frac{d_{(i-1 \rightarrow i)}}{w_{((i-1,t) \rightarrow (i,s))}}$ is a likelihood method in transmission probability, which defines a true path followed by the SPQ. On the other side, F_t is a temporal analysis function, defined by:

$$F_t(c_{i-1}^t \rightarrow c_i^s) = \frac{\sum_{u=1}^{k} (e'_u.v \mathrm{X} \overline{V}_{(i-1,t) \rightarrow (i,s)})}{\sqrt{\sum_{u=1}^{k} (e'_u.v)^2} \mathrm{X} \sqrt{\sum_{u=1}^{k} (\overline{V}_{(i-1,t) \rightarrow (i,s)})^2}} \tag{4}$$

where $e'_u.v_1 \rightarrow e'_u.v_2 \rightarrow \cdots \rightarrow e'_u.v_k$ describes the total speed of the road segments in the shortest path. Particularly, temporal analysis function calculates a score of each CP by considering the average speed. Further details of spatial and temporal analysis are described in the paper [6].

The last phase evaluates the CP graph $G'_T(V'_T, E'_T)$, wherein V'_T in G'_T is a set of CPs associated with the probability of GPS error $N(c_i^j)$ and E'_T in G'_T is a set of road segments associated with the score of spatial $F_s(c_{i-1}^t \rightarrow c_i^s)$ and temporal $F_t(c_{i-1}^t \rightarrow c_i^s)$ analysis function. A sequence of CPs is then chosen from a CP graph $G'_T(V'_T, E'_T)$ by considering the maximum summation of the score.

4 Experimental Settings, Results, and Discussions

4.1 Computation Environment

In our experiment, a custom-based simulation setup is designed based on Java envi-ronment (version 1.7). All of the datasets used for evaluating the studied map-matching strategies are stored in the PostgreSQL (version 9.1). We also used the PostGIS (version 2.0) as a spatial tool and pgRouting for executing SPQs.

4.2 Datasets

A real-world road network graph is used in our experiments, which is taken from Shenzhen transportation department. The graph is comprised of a total of 86335 vertices and 133397 road segments [18]. On the other hand, we also used a real-world trajectory dataset to evaluate the studied map-matching strategies. The dataset is contained of a group of real-world trajectories collected from a taxi traveled in Shenzhen city on a day of 10th October 2013.

A synthetic trajectory dataset is also used in our experiments. In our simulation, we randomly generate synthetic trajectories by executing a SPQ between two random vertices in the road network graph. Then the returned edges in the shortest path between two random vertices are considered as a ground truth. The simulator generates GPS sampling points on the respective edges according to the given sampling-rate. Specifically, each GPS sampling point is simulated by normal distribution with μ being zero, while σ being 20. The aforementioned method is adopted and the same values for normal distribution are chosen by [6] to create a synthetic trajectory dataset.

4.3 Performance Metrics

The running time and accuracy are chosen as the basic metrics to evaluate the studied map-matching strategies. The running time is the overall time captured during map-matching processing. Alternatively, the percentage of total correct matching points (CMP) is considered as the accuracy performance metric. We calculate the percentage of CMP in the following way as defined in [8].

$$\text{CMP} = \frac{\text{correct matched points}}{\text{total number of points to be matched}} \text{ X } 100\,\% \qquad (5)$$

Additionally, two other performance metrics (i.e., total number of CPs and SPQs) are considered to investigate the impact of the map-matching parameters. We benchmark the proposed strategy with the ST-MM. The ST-MM chooses a fixed number of CPs equaling 5, and a fixed number of ECR equaling 100 m.

4.4 Results and Discussions

Firstly, we analyze the proposed approach against ST-MM strategy in terms of the running time and accuracy metrics. Figure 5 shows the results of the aforementioned primary performance metrics for map-matching of synthetic trajectories with respect to different GPS sampling rate. Specifically, the aforementioned figure shows the average results for matching a set of synthetic trajectories, with the low-sampling-rate ranging between 2 and 6 min. Because our proposed approach incorporates the spatial and temporal function of the ST-MM, it achieves almost similar results with ST-MM in terms of accuracy, as shown in Fig. 5(a). Since LB-MM approach is incorporated with the spatio-tempo functions it provides high accuracy for low-sampling-rate of GPS trajectories as compare to high-sampling-rate of GPS trajectories. On the other hand, the results of running time in Fig. 5(b) clearly reveal that the proposed approach provides faster response time as compared to ST-MM strategy.

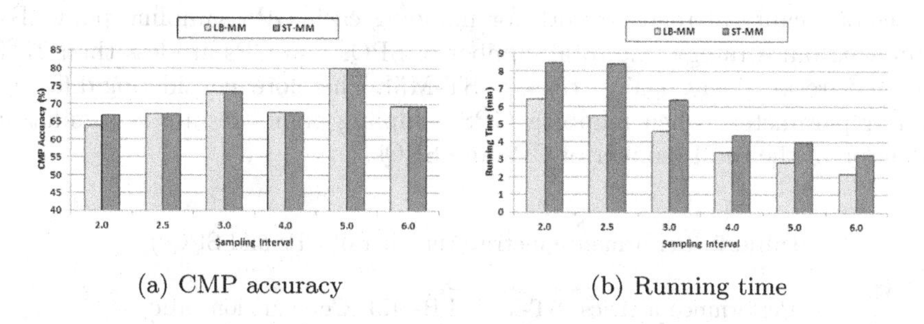

(a) CMP accuracy (b) Running time

Fig. 5. Primary performance metrics

Another remark is that the proposed approach, LB-MM outperforms the ST-MM when the trajectories consist of more than 40 GPS points. The results of the above statement are clearly shown in Fig. 6(a), wherein we used real-world trajectories randomly selected from a full-day trajectory. After analyzing the studied map-matching strategies with respect to the primary metrics, we also evaluate them in terms of the distribution of GPS sample data in each class that is based on locality of road network (see Table 1). Particularly, we examine the impact of interior parameters tuning by considering the percentage of GPS sampling points in each class. Their distributions are shown in Fig. 6(b). The results show that by adaptively changing the interior parameters for each class, we can achieve better execution time and more accurate results against a static strategy like ST-MM. In the figure, the distributions of GPS sampling data in the Class-I, II, and III is dominant against the rest classes. Because of choosing small values for the CPs and ECR parameters, the proposed strategy significantly reduces the number of SPQs and provides fast running time.

Furthermore, we evaluate the performance of our proposed strategy against the ST-MM strategy in terms of two more performance metrics: (a) a total

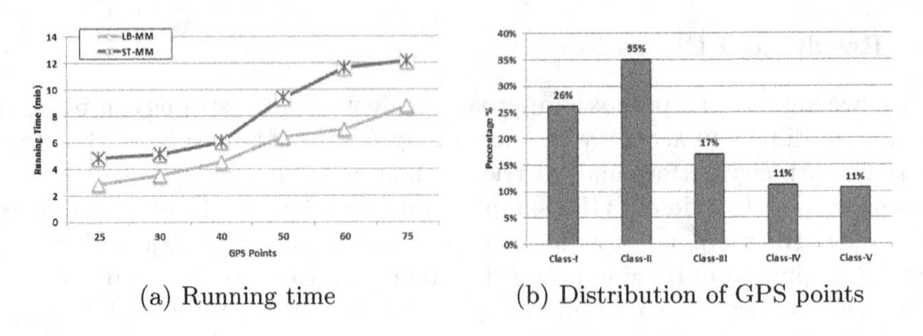

(a) Running time (b) Distribution of GPS points

Fig. 6. Distribution of GPS points in map-matching process

number of SPQs and (b) a total number of considered CPs. The results of the metrics are shown in Table 2. The results in the table reveal that the proposed approach (i.e., LB-MM) outperforms the SPQ-based ST-MM strategy in terms of both performance metrics. The LB-MM strategy adaptively chooses different classes of locality of road network for mapping each GPS sampling point. By employing our strategy, the total number of SPQs and CPs are less than 27 % and 15 %, respectively, against those of ST-MM. Therefore, considering different interior parameters when mapping GPS sampling points results in significant reduction of the total number of CPs and SPQs.

Table 2. Performance metrics (number of CPs and SPQs)

Performnce metrics	ST-MM	LB-MM	Comparision ratio
Total CPs	1636	1397	1.17
Total SPQs	7692	5636	1.36

5 Conclusions

Because map-matching process uses shortest path queries (SPQs) to ensure accuracy, the navigation systems of land vehicles become more and more computation hungry. The above may be interpreted as that the map-matching process becomes unaffordable for real-time processing. This paper proposed an approach based on locality of road network for map-matching of GPS trajectories. By adaptively fine-tuning the interior parameters, the proposed strategy outperformed the state-of-the-art approach (ST-MM) found in the literature. Specifically, our proposed strategy aims at reducing the number of SPQs, which is the most expensive part of map-matching process, while at the same time to maintain the accuracy. We analyzed its performance in terms of (a) running time, (b) accuracy, and (c) total number of SPQs and (d) candidate points (CP) on real-world GPS data as well as synthetic data. By employing LB-MM approach, the total number of SPQs in the map-matching process is counted as less than 27 % against those of ST-MM. In future work, unlike the traditional approaches, we intend to propose a strategy that pre-computes the shortest path distances by exploiting the parallel computing paradigm in cloud environments. Furthermore, we plan to provide a way for tuning the GPS sampling rate of real-time GPS trajectories and use of different parameters.

Acknowledgments. The work is funded in part by the grant of the National Natural Science Foundation of China (No. 61100220 and U1401258), U.S. NSF (CCF-1016966) and National Basic Research Program (973 Program, No. 2015CB352400). AA. Chandioś work was partly supported for his PhD studies in Shenzhen Institutes of Advanced Technology, Chinese Academy of Sciences, Shenzhen, China.

References

1. Zheng, Y., Capra, L., Wolfson, O., Yang, H.: Urban computing: concepts, methodologies, and applications. ACM Trans. Intell. Syst. Technol. **5**, 1–55 (2014)
2. Kuehne, R., Schfer, R., Mikat, J., Thiessenhusen, K., Boettger, U., Lorkowski, S.: New approaches for traffic management in metropolitan areas. In: Proceedings of IFAC CTS Symposium (2003)
3. Li, X., Han, J., Lee, J.-G., Gonzalez, H.: Traffic density-based discovery of hot routes in road networks. In: Papadias, D., Zhang, D., Kollios, G. (eds.) SSTD 2007. LNCS, vol. 4605, pp. 441–459. Springer, Heidelberg (2007)
4. Zheng, Y., Wang, L., Zhang, R., Xie, X., Ma, W.Y.: Geolife: Managing and understanding your past life over maps. In: 9th International Conference on Mobile Data Management, MDM 2008, pp. 211–212. IEEE (2008)
5. Gonzalez, H., Han, J., Li, X., Myslinska, M., Sondag, J.P.: Adaptive fastest path computation on a road network: a traffic mining approach. In: Proceedings of the 33rd International Conference on Very Large Data Bases, VLDB Endowment, pp. 794–805 (2007)
6. Lou, Y., Zhang, C., Zheng, Y., Xie, X., Wang, W., Huang, Y.: Map-matching for low-sampling-rate gps trajectories. In: Proceedings of the 17th ACM SIGSPATIAL International Conference on Advances in Geographic Information Systems, pp. 352–361. ACM (2009)
7. Quddus, M., Washington, S.: Shortest path and vehicle trajectory aided map-matching for low frequency GPS data. Transp. Res. Part C: Emerg. Technol. **55**, 328–339 (2015)
8. Yuan, J., Zheng, Y., Zhang, C., Xie, X., Sun, G.Z.: An interactive-voting based map matching algorithm. In: 2010 Eleventh International Conference on Mobile Data Management (MDM), pp. 43–52. IEEE (2010)
9. Goh, C.Y., Dauwels, J., Mitrovic, N., Asif, M., Oran, A., Jaillet, P.: Online map-matching based on hidden markov model for real-time traffic sensing applications. In: 2012 15th International IEEE Conference on Intelligent Transportation Systems (ITSC), pp. 776–781. IEEE (2012)
10. Newson, P., Krumm, J.: Hidden markov map matching through noise and sparseness. In: Proceedings of the 17th ACM SIGSPATIAL International Conference on Advances in Geographic Information Systems, pp. 336–343. ACM (2009)

11. Fang, S., Zimmermann, R.: Enacq: energy-efficient GPS trajectory data acquisition based on improved map matching. In: Proceedings of the 19th ACM SIGSPATIAL International Conference on Advances in Geographic Information Systems, pp. 221–230. ACM (2011)
12. Chen, B.Y., Yuan, H., Li, Q., Lam, W.H., Shaw, S.L., Yan, K.: Map-matching algorithm for large-scale low-frequency floating car data. Int. J. Geogr. Inf. Sci. **28**, 22–38 (2014)
13. Wenk, C., Salas, R., Pfoser, D.: Addressing the need for map-matching speed: Localizing global curve-matching algorithms. In: 18th International Conference on Scientific and Statistical Database Management, pp. 379–388. IEEE (2006)
14. Greenfeld, J.S.: Matching GPS observations to locations on a digital map. In: Transportation Research Board 81st Annual Meeting (2002)
15. Brakatsoulas, S., Pfoser, D., Salas, R., Wenk, C.: On map-matching vehicle tracking data. In: Proceedings of the 31st International Conference on Very Large Data Bases, VLDB Endowment, pp. 853–864 (2005)
16. Hummel, B., Tischler, K.: Robust, GPS-only map matching: exploiting vehicle position history, driving restriction information and road network topology in a statistical framework. In: GIS Research UK Conference (GISRUK), pp. 68–77 (2005)
17. Pink, O., Hummel, B.: A statistical approach to map matching using road network geometry, topology and vehicular motion constraints. In: 11th International IEEE Conference on Intelligent Transportation Systems, ITSC 2008, pp. 862–867 (2008)
18. Chandio, A.A., Zhang, F., Memon, T.D.: Study on LBS for characterization and analysis of big data benchmarks. Mehran Univ. Res. J. Eng. Technol. **33**, 432–440 (2014)
19. Quddus, M.A., Ochieng, W.Y., Noland, R.B.: Current map-matching algorithms for transport applications: state-of-the art and future research directions. Transp. Res. Part C: Emerg. Technol. **15**, 312–328 (2007)
20. Chandio, A.A., Bilal, K., Tziritas, N., Yu, Z., Jiang, Q., Khan, S.U., Xu, C.Z.: A comparative study on resource allocation and energy efficient job scheduling strategies in large-scale parallel computing systems. Cluster Comput. **17**, 1349–1367 (2014)

V2V and M2M Communications

An Improved GNSS Receiver Autonomous Integrity Monitoring Method Using Vehicle-to-Vehicle Communication

Liu Jiang[1(✉)], Wu Xi[2], and Cai Bai-gen[1]

[1] School of Electronic and Information Engineering, Beijing Jiaotong University, Beijing 100044, China
jiangliu@bjtu.edu.cn

[2] Beijing Engineering Research Center of EMC and GNSS Technology for Rail Transportation, Beijing 100044, China

Abstract. The GNSSs (Global Navigation Satellite Systems) positioning-based applications are of great significance to support many IoV (Internet of Vehicles) services and implementations. Conventional satellite-based vehicle positioning solutions mainly concern the improvement of precision, continuity and service availability under different scenarios. The integrity monitoring of the satellites should also be considered for many safety-related applications, even under the low-visibility conditions with assistance from additional sensors. In this paper, the cooperative vehicle positioning using DSRC-based vehicle-to-vehicle wireless communication is concentrated to establish a GNSS/DSRC integrated solution within the IoV scheme. And a virtual satellite-based approach is presented to make it possible to realize an improved RAIM (Receiver Autonomous Integrity Monitoring) calculation using redundant DSRC measurements through the cooperation among the vehicles. Thus, results of the improved integrity monitoring can enhance the capability of safe positioning. Simulations with step and ramp pseudo-range measurement errors are carried out to validate performance of the proposed solution, and the results of RAIM availability and the detection identification latency demonstrate the potential of enabling field IoV implementations with critical performance requirements of safety.

Keywords: Cooperative positioning · Satellite navigation · Integrity monitoring · Vehicular communication · Dedicated Short Range Communication

1 Introduction

In the past few years, the Internet of Vehicles (IoV) has been a new emerging technology, by which a large number of devices/objects can be connected to the vehicles through Internet, and thus many advanced applications will be enabled for the future Intelligent Transportation Systems (ITSs) [1, 2]. Owing to the rapid development of the information and communication technologies, the vehicles are connected with the wireless communication capabilities so that they will be proactive, cooperative, well informed, and coordinated to support the advanced functions and applications, even in the safety-related

C.-H. Hsu et al. (Eds.): IOV 2015, LNCS 9502, pp. 249–260, 2015.
DOI: 10.1007/978-3-319-27293-1_22

fields [3]. As low-cost GNSS (Global Navigation Satellite System) receivers are assumed to be standard in the commercial vehicles, satellite positioning seems to be the most promising technology to enable the location-based applications under an IoV scheme, where the space-time information of the vehicles is an essential foundation for both the self-perception and environmental-awareness [4, 5]. Additionally, the rapid development of the GNSS constellations and technologies is providing more opportunities to achieve a higher cost-efficiency for field implementations.

In order to support specific IoV applications that might have a critical requirement for safety, i.e. the active safety to avoid collisions, it is highly required that the vehicle positioning module can real-timely monitor the quality of measurements and its calculation process, by which the safety of the output can be effectively identified. When using the satellite navigation systems, the concept of integrity has been widely used in many application fields. Integrity is defined as an indication of the capability to warn the users when it is operating out of the specified performance limits [6]. The integrity monitoring at the user level, Receiver Autonomous Integrity Monitoring (RAIM), has been widely applied in many areas to satisfy the safety requirements [7, 8]. However, there is a requirement for satellite visibility in conventional RAIM solutions, in which more than 5 satellites should be available in navigation calculation and thus the potential fault can be detected and identified. In the urban operation environment, the satellite signals may be influenced by surrounding geography conditions, and that leads to the failures for satisfying the visibility criterion. Under this circumstance, the available satellite-based measurements are considered to be assisted by additional sensors, i.e. inertial sensors [9], odometer [10] or vision sensors [11], to achieve the continuity performance. However, the loss of the RAIM function may result in the safety risks since the potential faults or failures in GNSS measurements cannot be detected with 4 or less satellites. Under the IoV scheme, vehicular communication using DSRC (Dedicated Short Range Communication) is enabling technique to bring benefits from the VANETs (Vehicular Ad Hoc Networks). It has also been proved with a capability of cooperative positioning with ranging or ranging-rate-based methods [12]. When using an integrated architecture for GNSS/DSRC cooperative positioning, it is necessary to evaluate the quality of the range measurements of the satellites, even there are only 4 or less satellites available and the conventional RAIM cannot be realized.

In this paper, we propose an improved receiver integrity monitoring method, which uses redundant DSRC measurements from cooperative neighborhoods to guarantee the safety of vehicle positioning. It is beneficial to release the requirements of satellite visibility. Section 2 gives the principles of DSRC-assisted GNSS positioning, and the improved RAIM method is detailed introduced in Sect. 3 with simulation results in the following part. Section 5 concludes the paper and indicates the future plans.

2 DSRC-Assisted GNSS Positioning

The range or range-rate between a vehicle and its cooperative neighborhoods can be applied to estimate the position and the corresponding dynamic states. With a DSRC-based cooperative localization architecture, there are three main approaches for location estimation: RSS (Received Signal Strength) based ranging, time-based ranging and CFO (Carrier Frequency Offset) based range-rating, where the CFO-based method is identified as a potential solution for cooperative localization with a fair level of precision [13]. Based on the principle of Doppler effect, the Doppler shift or the integrated carrier phase difference between the DSRC transmitter (equipped on one of the connected neighborhood vehicles) and a DSRC receiver (equipped on the target vehicle) can be used to estimate the range-rate, where [13]

$$f_{\mathrm{r}} = F_{\mathrm{t},j}(1 - \frac{1}{c}\dot{r}_j) \tag{1}$$

where f_{r} and $F_{\mathrm{t},j}$ indicate the carrier frequency of the received and transmitted signals from the jth neighborhood respectively, c is the speed of light, and $\dot{r}_j = dr_j/dt$ denotes the range-rate between the two vehicles which implies a distance r_j.

Due to existence of clock errors between the receiver and transmitters, the Doppler shifts cannot be extracted by the CFO results directly. We have estimate CFO with an assumed clock error condition and thus the range-rating function as (1) can be applied to build an observation model for the vehicle location determination. Considering the frequency offsets in f_{r} and $f_{\mathrm{t},j}$, the range-rate function can be written as

$$\dot{r}_j = c(1 - \frac{f_x}{F_{\mathrm{t},j}}) = c(1 - \frac{F_r + \delta_r}{f_{\mathrm{t},j} + \delta_{\mathrm{t},j}}) = -c \bullet \frac{F_r - f_{\mathrm{t},j} + \delta_r - \delta_{\mathrm{t},j}}{f_{\mathrm{t},j} + \delta_{\mathrm{t},j}} \tag{2}$$

where δ_{r} denotes the offset between the actual received frequency f_{r} and the measured frequency F_{r}, and $\delta_{\mathrm{t},j}$ is the offset between the known actual transmit frequency $F_{\mathrm{t},j}$ from the jth neighborhood and the nominal value $f_{\mathrm{t},j}$.

Therefore, if we utilize the CFO measurement as $\varphi_j = F_{\mathrm{r}} - f_{\mathrm{t},j}$ with the clock error $\Delta_j = \delta_{\mathrm{r}} - \delta_{\mathrm{t},j}$, the range-rate can be calculated as

$$\dot{r}_j = -c_j \bullet \frac{\varphi_j + \Delta_j}{f_{\mathrm{t},j}} \tag{3}$$

where the offset $\delta_{\mathrm{t},j}$ in the denominator can be ignored as $f_{\mathrm{t},j} + \delta_{\mathrm{t},j} \approx f_{\mathrm{t},j}$ owing to the great disparity of the order of magnitude.

Based on the analysis of the CFO-based range-rate estimation model, an observation model for DSRC-based vehicle positioning can be established with m_k available neighborhood vehicles at instant k, which means

$$\begin{aligned} \dot{r}_{kj} &= -c \bullet \frac{\varphi_{kj}}{f_{\mathrm{t},j}} - c \bullet \frac{\Delta_{kj}}{f_{\mathrm{t},j}} + \varepsilon_{kj} \\ r_{kj} &= \sqrt{(e_k - e_{k,j})^2 + (n_k - n_{k,j})^2} \end{aligned} \qquad j = 1, 2, \cdots, m_k \tag{4}$$

where the subscript k denotes the discrete time instant, $\varepsilon_{k,j}$ is the equivalent random measurement error, (e_k, n_k) and $(e_{k,j}, n_{k,j})$ are two-dimensional position of the target vehicle and the jth neighborhood in the Gauss plane coordinate system.

When the condition $m_k \geq 3$ is fulfilled, the unknown position can be estimated and identified using specific nonlinear estimation algorithms, i.e. the Bayesian filters.

3 Improved Integrity Monitoring Method

In the conventional RAIM algorithms with the single-fault assumption, the navigation calculation with pseudo-range measurements from visible and available satellites (the value of DOP (Dilution of Precision) and satellite elevation criterion are fulfilled) is carried out with additional embedded logics to estimate the error bound and the probable fault/ failure. The LS estimation method is usually employed in a standard navigation calculation process, and the derived SSE (Sum Square Error) of residual error can be involved to build the statistic for fault detection. In the presented DSRC-aided GNSS vehicle positioning, a non-linear Bayesian filter is applied to achieve the integrity monitoring in advance, after which the location determination for a target vehicle would be realized using sufficient fault-free measurement data. There are usually two steps for integrity monitoring:

(1) Pre-calculation for a navigation result.
(2) HPL (Horizontal Protection Level) estimation and fault identification.

For fault detection and identification, more than 5 satellites should be available for providing pseudo-range measurements due to the characters of adopted residual-based test statistic. If the number of available satellites is less than 5, despite the additional observations from DSRC can contribute to a continuous location computation, it is of great necessity to identify the quality of the adopted measurements from the satellites. The redundant range-rate observations from DSRC make it possible of enhancing the satellite visibility and fulfilling the requirements for RAIM calculation, where measurements from the Virtual Satellites (VSs) are introduced.

When we assume that n_k satellites are available and u_k virtual satellites could be created through m_k neighborhood vehicles, the condition for RAIM can be written as

$$n_k^+ = n_k + u_k = n_k + [\tfrac{m_k}{3}] \geq 5, \quad u_k \geq 1, m_k \geq 3 \tag{5}$$

where the symbol $[*]$ indicates the integral part.

By using the range-rate measurements from m_k neighborhoods, a pre-calculation can be performed according to the observation model presented in Sect. 2, and the DSRC-based result $(\hat{e}_k^+, \hat{n}_k^+)$ is transformed into three-dimensional coordinate system (x_k^+, y_k^+, z_k^+) with specific coordinate transformation logic and the altitude information. Considering the requirement for a sufficient n_k^+, multiple DSRC pre-calculations with independent CFO measurements are used to get results $(x_{k,v}^+, y_{k,v}^+, z_{k,v}^+)$, $v = 1, 2, \cdots, u_k$. With this basis, the VSs pseudo-range measurements can be derived according to the ephemeris information from the receivers, which means

$$\rho_{k,v}^+ = \sqrt{(x_{k,v}^+ - x_{k,v})^2 + (y_{k,v}^+ - y_{k,v})^2 + (z_{k,v}^+ - z_{k,v})^2} \tag{6}$$

where $(x_{k,v}, y_{k,v}, z_{k,v})$ denote the instantaneous 3D position of the vth satellite extracted by the ephemeris, despite the satellite is not practically observed by the receiver.

Based on the extended pseudo-range measurements, a measurement vector can be defined as $\mathbf{z}_k = [\rho_{k,1}, \rho_{k,2}, \cdots, \rho_{k,n_k^+}]^{\mathrm{T}}$, and an enhanced observation model can be

$$\begin{cases} \cdots\cdots \\ \rho_{k,i'} = \sqrt{(x_k - x_{k,i'})^2 + (y_k - y_{k,i'})^2 + (z_k - z_{k,i'})^2} + c\alpha_k + \xi_{k,i'} \\ \cdots\cdots \\ \rho_{k,v}^+ = \sqrt{(x_k - x_{k,v})^2 + (y_k - y_{k,v})^2 + (z_k - z_{k,v})^2} + c\alpha_k + \xi_{k,v} \\ \cdots\cdots \end{cases} \tag{7}$$

$$i' = 1, 2, \cdots, n_k$$
$$v = 1, 2, \cdots, u_k$$

where α_k is the receiver clock error, $\xi_{k,i'}$ and $\xi_{k,v}$ denote the observation noise of the i' th visible satellite and the corresponding equivalent observation noise from the v th simulated virtual satellite respectively.

Thus, the above mentioned two steps for the enhanced integrity monitoring can be achieved as the following procedures.

First, the nonlinear filter-based navigation calculation is carried out by the standard cubature Kalman filter (CKF) [14]. The state vector of the target vehicle is defined as $\mathbf{x}_k^+ = [x_k, y_k, z_k, \alpha_k]^{\mathrm{T}}$, and the one for the state estimation is expanded with high-order components as $\mathbf{x}_k = [x_k, \dot{x}_k, \ddot{x}_k, y_k, \dot{y}_k, \ddot{y}_k, z_k, \dot{z}_k, \ddot{z}_k, \alpha_k]^{\mathrm{T}}$, where the location, velocity, acceleration and the clock error are involved. The system and measurement model are

$$\begin{cases} \mathbf{x}_{k+1} = f(\mathbf{x}_k, k) + \boldsymbol{\beta}_k \\ \mathbf{z}_k = h(\mathbf{x}_k, k) + \boldsymbol{\lambda}_k \end{cases} \tag{8}$$

where $f(*)$ and $h(*)$ are nonlinear functions representing the state transition process and the observation model as depicted in (7), $\boldsymbol{\beta}_k$ and $\boldsymbol{\lambda}_k$ are system and measurement errors that are assumed as Gaussian noises with variances $\mathbf{Q}_k$ and $\mathbf{R}_k$.

The final estimation of the state vector is derived by an integration of the model-based state prediction $\hat{\mathbf{x}}_{k|k-1}$ and the observation innovation as

$$\hat{\mathbf{x}}_k = \hat{\mathbf{x}}_{k|k-1} + \mathbf{W}_k \mathbf{r}_k = \hat{\mathbf{x}}_{k|k-1} + \mathbf{W}_k[\mathbf{z}_k - h(\hat{\mathbf{x}}_{k|k-1})] \tag{9}$$

$$\mathbf{P}_k = \mathbf{P}_{k|k-1} - \mathbf{W}_k \mathbf{P}_{zz,k|k-1} \mathbf{W}_k^{\mathrm{T}} \tag{10}$$

where $\mathbf{P}_k$ is the error covariance of estimation, $\mathbf{W}_k$ indicates the Kalman gain matrix derived by the innovation covariance $\mathbf{P}_{zz,k|k-1}$ and cross-covariance $\mathbf{P}_{xz,k|k-1}$ as

$$\mathbf{W}_k = \mathbf{P}_{xz,k|k-1} \mathbf{P}_{zz,k|k-1}^{\mathrm{T}} \tag{11}$$

$$\mathbf{P}_{zz,k|k-1} = \frac{1}{2N}\sum_{i=1}^{2N} \mathbf{Z}_{i,k|k-1}\mathbf{Z}_{i,k|k-1}^{\mathrm{T}} - \hat{\mathbf{z}}_{k|k-1}\hat{\mathbf{z}}_{k|k-1}^{\mathrm{T}} + \mathbf{R}_k \tag{12}$$

$$\mathbf{P}_{xz,k|k-1} = \frac{1}{2N}\sum_{i=1}^{2N} \mathbf{X}_{i,k|k-1}\mathbf{Z}_{i,k|k-1}^{\mathrm{T}} - \hat{\mathbf{x}}_{k|k-1}\hat{\mathbf{z}}_{k|k-1}^{\mathrm{T}} \tag{13}$$

The CKF algorithm uses a set of sigma points to solve the nonlinear transformation problem and realize the estimation with the form as (9). The detailed procedure of the involved standard CKF can be found in [13].

Second, we may use the intermediate results of the filter-based navigation calculation to estimate the HPL for fault detection. The RAIM availability should be investigated in advance to determine a permitted RAIM calculation. The value of HPL can be updated according to the given principle as

$$\begin{aligned} HPL_k &= \sigma_0^2\sqrt{\eta}\bullet[\max(\sqrt{\frac{A_{k,lq}^2 + A_{k,2q}^2}{S_{k,qq}}})] \\ A_k &= \left(\Sigma_k^T \Sigma_k^T\right)^{-1} \Sigma_k^T \\ S_k &= I - \Sigma_k (\Sigma_k^T \Sigma_k)^{-1} \Sigma_k^T \\ \Sigma_k &= H_k D^{-1} \end{aligned} \tag{14}$$

where $\mathbf{x}_k^+ = \mathbf{D}\mathbf{x}_k$, η denotes the parameter of the non-central χ^2 distribution under the faulty hypothesis that is derived with a given missed detection probability p_{md}.

Only if the RAIM availability is identified by the criterion HPL<HAL (Horizontal Alert Limit), the following procedures would be carried out.

The residual error of the state estimation can be written as

$$\mathbf{e}_k = \mathbf{z}_k - \hat{\mathbf{z}}_k = \mathbf{z}_k - h(\hat{\mathbf{x}}_k) \tag{15}$$

When we use the equivalent measuring matrix $\mathbf{H}_k$ to represent the relationship of the estimated measurement $\hat{\mathbf{z}}_k$ and the truth $\mathbf{z}_k$, the right part of (15) can be changed as a linear form $\mathbf{e}_k = \mathbf{z}_k - \mathbf{H}_k\hat{\mathbf{x}}_k$ with

$$\mathbf{H}_k = (\mathbf{P}_{k|k-1}^{-1}\mathbf{P}_{xz,k|k-1})^{\mathrm{T}} \tag{16}$$

Considering the estimation formula (9) with an equivalent linear observation model, the transformed residual error corresponding to $\mathbf{x}_k^+$ can be derived as

$$\mathbf{e}_k^+ = (\mathbf{I} - \mathbf{H}_k\mathbf{W}_k)\mathbf{r}_k \tag{17}$$

The test statistic will be built as

$$\hat{T} = \sqrt{\frac{(\mathbf{e}_k^+)^{\mathrm{T}}\mathbf{e}_k^+}{n_k^+ - 4}} \tag{18}$$

It is assumed that the test statistic fulfills the central χ^2 distribution under the fault-free hypothesis. According to the pre-defined false-alarm probability p_{fa}, the threshold for fault detection is derived as the following equation

$$p\left(\frac{(\mathbf{e}_k^+)^{\mathrm{T}}\mathbf{e}_k^+}{\sigma_0} < \mathrm{T}_{\mathrm{d}}^2\right) = \int_0^{\mathrm{T}_{\mathrm{d}}^2} f_{\chi^2(n_k^+-4)}(\tau)d\tau = 1 - p_{\mathrm{fa}} \tag{19}$$

Therefore, the criterion for fault detection and identification is

$$\hat{T} \geq \frac{\sigma_0 \square \mathrm{T}_{\mathrm{d}}}{\sqrt{n_k^+ - 4}} \tag{20}$$

which means the fulfillment of criterion (20) will indicate the existing of fault corresponding to a certain satellite or its measurement.

When the available enhanced measurements fulfill the condition $n_k^+ \geq 6$, the fault diagnosis is performed to identify which satellite among the n_k visible ones is faulty and should be isolated in the final location calculation process.

4 Simulation and Analysis

According to the theoretical analysis of the improved RAIM method, simulations are carried out to validate the capability in potential field implementations due to difficulties of large-scale practical experiment under the connected environment. A rectangle area from (116.7210, 39.5802) to (116.7337, 39.5740) is selected for test. The traffic flow of the test area is simulated by Paramics and used for generating DSRC observations corresponding to a focused target vehicle. Then, the trajectory of the target vehicle is loaded into a GNSS simulator to obtain the satellite measurements. In the simulations, 4 BeiDou satellites (#2, #5, #7, #9) are involved to establish a low-visibility scenario for vehicle positioning, and DSRC localization result is employed to build a virtual satellite observation component using the ephemeris data of the #6 BeiDou satellite. To investigate the capability of fault detection using DSRC assistance data, the constant step noise e_1 from $t = 100s$ is firstly coupled into the original pseudo-range measurement of #5 satellite. When we define $p_{\mathrm{fa}} = 0.00001$ and $p_{\mathrm{md}} = 0.001$, Fig. 1 shows the derived HPL value with $e_1 = 50$, where a high RAIM availability is achieved as 98.8%. To demonstrate the fault detection performance under different noise magnitudes, Figs. 2 and 3 depict the comparison results of the test statistics and the threshold, and Table 1 summarizes the detection rates.

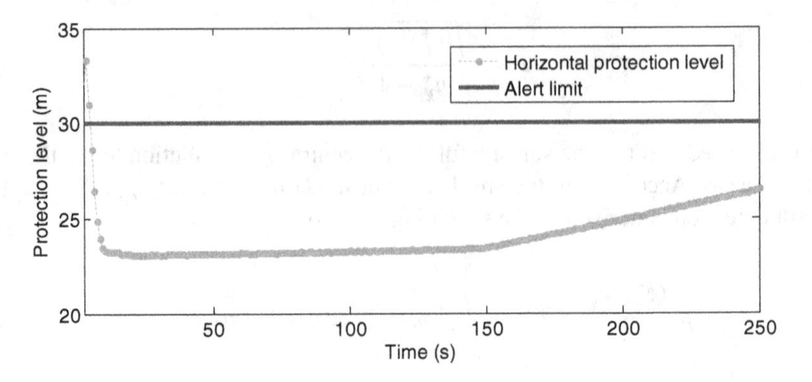

Fig. 1. Horizontal protection level under a constant step noise condition

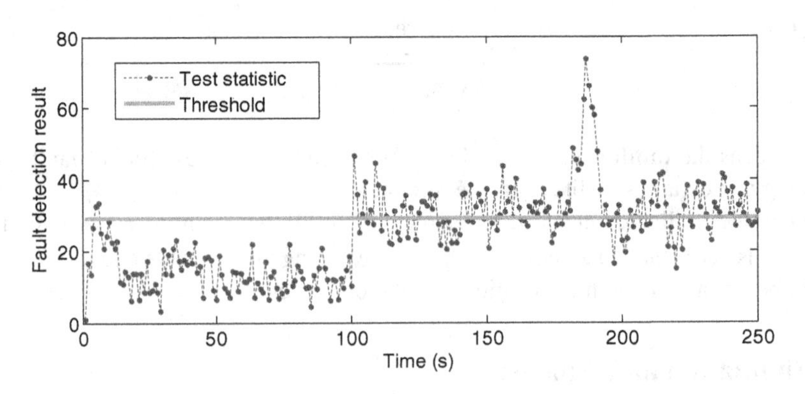

Fig. 2. Fault detection result under a constant noise magnitude (75 m)

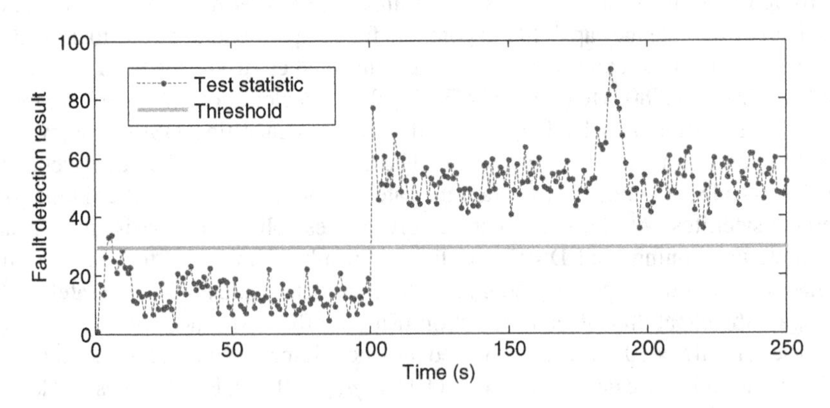

Fig. 3. Fault detection result under a constant noise magnitude (125 m)

Table 1. Comparison of detection performance with different noise magnitude conditions

Noise magnitude (m)	50	75	100	125
Detection rate	15.33%	62.67%	98.67%	100%
Missed-detection rate	84.67%	37.33%	1.33%	0

To further investigate the capability of fault detection under random conditions, the variance magnification events are simulated and coupled into original pseudo-range measurements of the faulty satellite after $t = 100s$. The coupled noise is described by the variance var(e_2). Figures 4 and 5 show the added noise and fault detection result under the noise condition var(e_2) = 30.0, where a detection rate 54.0% is achieved.

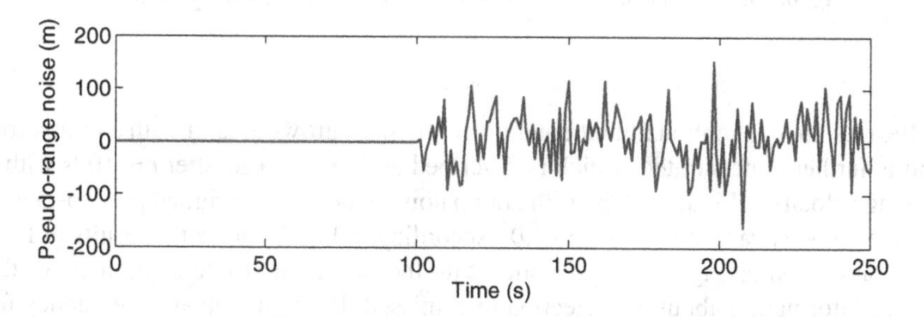

Fig. 4. The combined noise for generating the variance magnification event

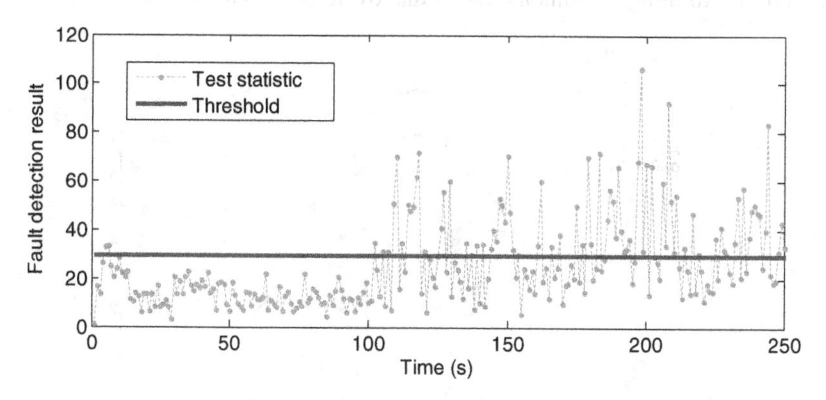

Fig. 5. Fault detection result under a variance magnification event (var(e_2) = 30.0)

Due to the detection capability may be influenced by the randomness features of the events, 30 Monte Carlo simulations are carried out separately for var(e_2) = 30.0 and var(e_2) = 50.0 to get statistical results for an evaluation purpose. Figures 6 and 7 give the results of detection rates, from which it can be found that a relatively low detection rate is presented due to the randomness of noises. However, the detection rate is proportionate to the increased magnitude of the noises, and that performs with the same character as the step noise scenarios.

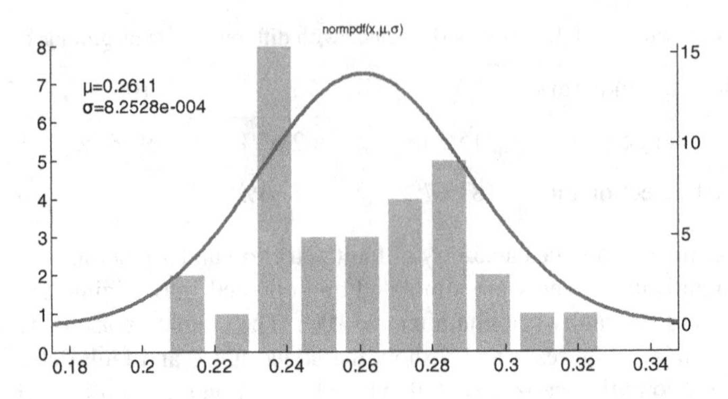

Fig. 6. Statistical result of detection rate under the event $\mathrm{var}(e_2) = 30.0$

Besides the conventional fault conditions, the slow growing fault with a ramp error form is further concentrated, which is described as $e_3(t) = \bar{v} \bullet \Delta t$ after $t = 100s$ with a growing velocity $\bar{v}$. Figure 8 depicts the ramp noise added to the original pseudo-ranges with the velocity ranges from 1.0 to 3.0. Accordingly, Fig. 9 shows the results of fault detection by comparing the test statistics with the threshold. Table 2 summarizes the detailed information about the detection rate, missed-detection rate and the latency for the identification of the ramp fault. The results illustrate the effectiveness of the proposed solution for performing continuous RAIM that overcomes visibility limitations.

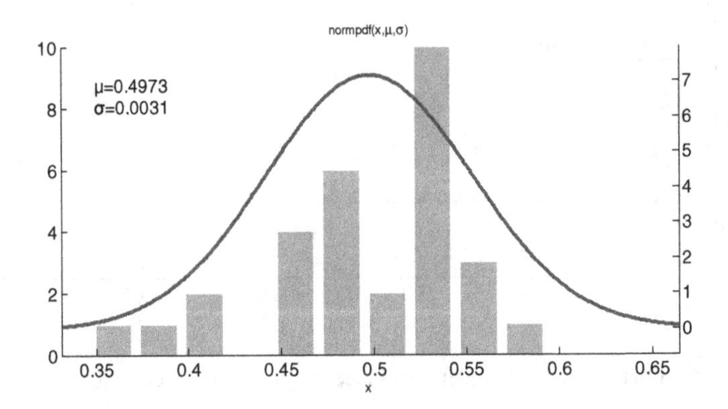

Fig. 7. Statistical result of detection rate under the event $\mathrm{var}(e_2) = 50.0$

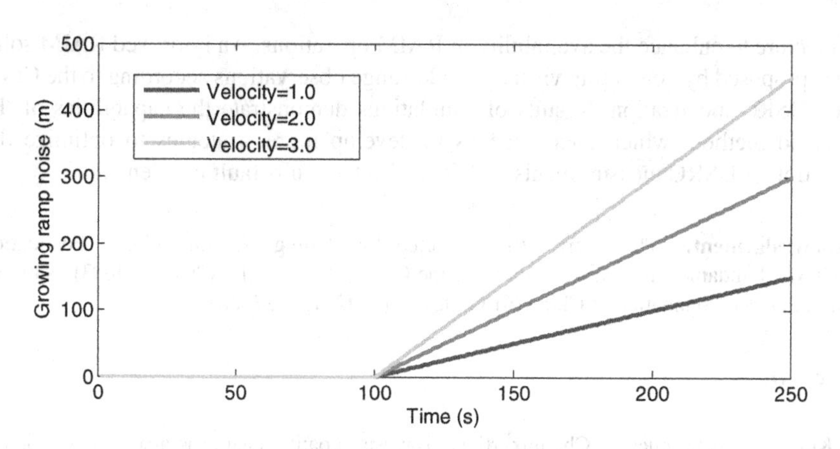

Fig. 8. Simulated ramp noises with different growing velocities

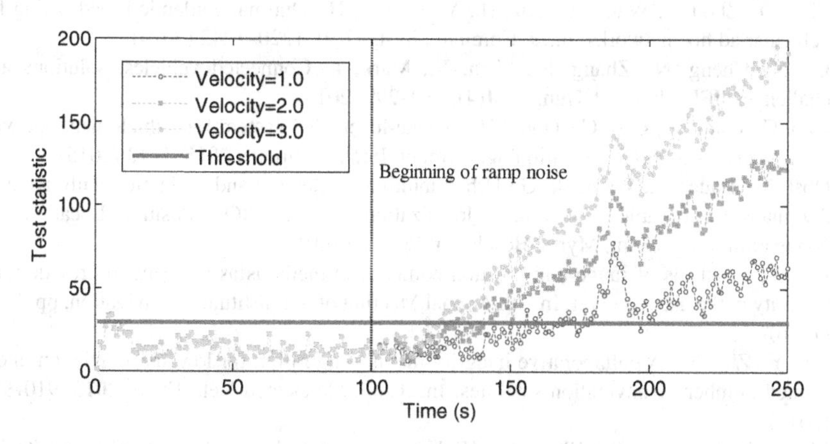

Fig. 9. Fault detection results under different growing velocity conditions

Table 2. Statistical results under different growing velocity conditions ranging from 0.5 to 3.0

Velocity	0.5	1.0	1.5	2.0	2.5	3.0
Detection rate	19.33%	54.00%	70.00%	74.67%	82.67%	85.33%
Missed-detection rate	80.67%	46.00%	30.00%	25.33%	17.33%	14.67%
Latency (s)	82	56	41	31	26	20

5 Conclusion and Future Plans

In this paper, the integrity monitoring of vehicles under the IoV application scheme is concerned. Based on the limitation of satellite visibility by the conventional RAIM methods, the DSRC-enabled cooperative positioning is coupled in the GNSS positioning architecture to enhance the availability of RAIM operations. An improved RAIM solution is proposed by generating virtual pseudo-range observations according to the CFO-based DSRC localization. Results of simulations demonstrate the capabilities of the presented method, which encourage us to develop more strategies to optimize the utilization of DSRC measurements and deal with the multi-fault problems.

Acknowledgement. This work was supported by Beijing Natural Science Foundation (4144081), Fundamental Research Funds for the Central Universities (2014JBM003), National Natural Science Foundation of China (61403021, U1334211, 61273089).

References

1. Kumar, N., Rodrigues, J., Chilamkurti, N.: Bayesian coalition game as-a-service for content distribution in Internet of Vehicles. IEEE Int. Things J. **1**(6), 544–555 (2014)
2. Tian, D., Zhou, J., Wang, Y., Xia, H., Yi, Z., Liu, H.: Optimal epidemic broadcasting for vehicular ad hoc networks. Int. J. Commun. Syst. **27**(9), 1220–1242 (2014)
3. Lu, N., Cheng, N., Zhang, N., Shen, X., Mark, J.: Connected vehicles: solutions and challenges. IEEE Internet Things J. **1**(4), 289–299 (2014)
4. Guo, C., Guo, W., Cao, G., Dong, H.: A lane-level LBS system for vehicle network with high-precision BDS/GPS positioning. Comput. Intell. Neurosci. **2015**, 1–13 (2015)
5. Obst, M., Bauer, S., Wanielik, G.: Urban multipath detection and mitigation with dynamic 3D maps for reliable land vehicle localization. In: IEEE/ION Position Location and Navigation Symposium, Myrtle Beach, pp. 685–691 (2012)
6. Lee, Y.: Analysis of range and position comparison methods as a means to provide GPS integrity in the user receiver. In: 2th Annual Meeting of the Institute of Navigation, pp. 1–17 (1986)
7. Liu, Y., Zhu, Y.: A collaborative integrity monitor algorithm for low space aviation under limited number of navigation satellites. Int. Conf. Connected Veh. Expo. **2013**, 910–915 (2013)
8. Li, L., Mohammed, Q., Zhao, L.: High accuracy tightly-coupled integrity monitoring algorithm for map-matching. Transp. Res. Part C: Emerg. Technol. **36**, 13–26 (2013)
9. Alam, N., Kealy, A., Dempster, A.: Cooperative inertial navigation for GNSS-Challenged vehicular environments. IEEE Trans. Intell. Transp. Syst. **14**(3), 1370–1379 (2013)
10. Wieser, A.: Development of a GNSS odometer. In: 20th International Technical Meeting of the Satellite Division of The Institute of Navigation, pp. 1466–1476 (2007)
11. Won, D., Lee, E., Heo, M., Sung, S., Lee, J., Lee, Y.: GNSS integration with vision-based navigation for low GNSS visibility conditions. GPS Solutions **18**(2), 177–187 (2014)
12. Alam, N., Dempster, A.: Cooperative positioning for vehicular networks: facts and future. IEEE Trans. Intell. Transp. Syst. **14**(4), 1708–1717 (2013)
13. Alam, N.: Vehicular positioning enhancement using DSRC. Univ. New South Wales, Sydney, Australia (2012)
14. Arasaratnam, I., Haykin, S.: Cubature Kalman filters. IEEE Trans. Autom. Control **54**(6), 1254–1269 (2009)

An Adaptive Data Traffic Offloading Model for Cellular Machine-to-Machine Networks

Tao Lei(✉), Shang-guang Wang, and Fang-chun Yang

State Key Laboratory of Networking and Switching Technology,
Beijing University of Posts and Telecommunications, No.10 Xitucheng Road,
Haidian District, Beijing, China
{leitao,sgwang,fcyang}@bupt.edu.cn

Abstract. With the emergence of a large number of businesses and applications based cellular Machine-to-Machine (M2M) communication such as telematics, smart metering, point-of-sale terminals, and home security, a heavy data traffic which need through the cellular network has been produced. Although many schemes have been proposed to reduce data traffic, they are inefficient in practical application due to poor adaption. In this paper, we focus on how to adaptively offload data traffic for cellular M2M networks. To this end, we propose an adaptive data traffic offloading model (AOM). This model can decide whether to adopt opportunistic communications or communicate via cellular networks adaptively. In the AOM, we introduce traffic offloading rate (called TOR) and local resource consumption rate (called LRCR), and analyze them based on continue time Markov chain (CTMC). Theory proof and extensive simulations demonstrate that our model is accurate and effective, and can adaptively offload data traffic of cellular M2M networks.

Keywords: Cellular M2M networks · Opportunistic communication · Continue time markov chain

1 Introduction

Machine-to-machine (M2M) networks are composed of large numbers of nodes, since the main subject participating in M2M communication is a machine or object, or indeed it can be everything around us. Because of machine can sense itself or its surrounding physical environment, the traffic per machine is very small. However, a large number of objects generate a large quantity of data with different format. These various data traffics from a large number of sensors are gathered through cellular network, an unpredictable pattern is created at the cellular M2M networks. The accumulated traffic in the cellular networks may cause network congestion and outage of network resources [1]. Although M2M devices have different traffic patterns from smartphones, they are generally competing with smartphones for shared network resources [2].

The easiest way to solve this problem is to add more infrastructures (e.g. picocell, femtocell) [3], but imposed a serious challenge as the increase in infrastructure capacity does not scale accordingly. Therefore, the most popular research issue is how to offload the data traffic via WiFi. But, in fact, some nodes receive data from other nodes when

C.-H. Hsu et al. (Eds.): IOV 2015, LNCS 9502, pp. 261–272, 2015.
DOI: 10.1007/978-3-319-27293-1_23

they are moving such as vehicles. Then, some research used the opportunistic communication into mobile networks to reduce the data traffic [4–8]. A solution [4] has been proposed to offload the amount of mobile data, which exploits opportunistic communications to facilitate information dissemination in the emerging Mobile Social Networks (MoSoNets). This study only focuses on offloading mobile data on MoSoNets, which cannot be applied cellular M2M networks. Then, in order to improve practical application, an integrated architecture has been proposed by Dimatteo et al. [5], which migrates data traffic from cellular M2M networks to metropolitan WiFi Access Points by exploiting the opportunistic networking paradigm. Bulk file transfer and video streaming have been quantified the benefits of their architecture. Their research shows that it is very significantly to improve the delivery performance even with a sparse WiFi network. Meanwhile, iCAR [9] has been proposed as an alternative scheme, which is an integrated cellular and ad hoc relaying scheme, to divert traffic from one (possibly congested) cell to another (non-congested) cell. And the basic idea is to lay up some Ad hoc Relaying Stations (ARSs) at the key locations. ARSs can be used to forward signals between Mobile Hosts and Base Transceiver Stations. However, the iCAR cannot offload the data traffic. Instead, it diverts traffic from one (possibly congested) cell to another (non-congested) cell.

More recently, several schemes have been designed to offload data traffic [6–8]. For example, HAN et al. [7] proposed a scheme to reduce cellular data traffic, and they intentionally delay the delivery of information and use the opportunistic communications to offload the cellular data traffic. Soon, Whitbeck et al. [8] proposed a content dissemination framework, push-and-track, which minimized the load on the wireless infrastructure by exploiting ad hoc communication opportunities while guaranteeing tight delivery delays. If new copies need to be re-injected into the network through the 3G interface, Push-and-track will use a control loop to collect user-sent acknowledgements to determine it. The scenarios of Periodic message flooding and floating data can use the Push-and-track. Xiaofeng Lu et al. [6] proposed a Subscribe-and-Send architecture and an opportunistic forwarding protocol. In this approach, users can receive subscribed contents from other users who have the contents through WiFi, instead of downloading from the Content Service Provider. This scheme can offload data traffic, but it has poor adaption in the process of data forwarding since it cannot obtain offloaded data traffic and fails in deciding when to use opportunistic communications.

In this paper, we propose an adaptive data traffic offloading model for cellular M2M networks. This model can adaptive select different data transfer mechanisms, such as opportunistic communications or through cellular M2M networks, to offload data traffic of cellular networks. To describe our model more clearly, we define and analyze data traffic offloading rate (called TOR) based on continue time Markov chain in cellular M2M networks, and local resource consumption rate (called LRCR). Besides, we extend the ONE[1] by adding infection probability and multi-destination nodes setting, and make it to simulate data transfer in the opportunistic networks. Extensive simulations demonstrate that our derivations is accurate and effective.

[1] https://www.netlab.tkk.fi/tutkimus/dtn/theone/

The rest of this paper is organized as follows. In Sect. 2, we define TOR, LRCR and minimum completion time, and propose our model. The TOR and LRCR for data transfer mechanism are analyzed in Sect. 3. Extensive simulations are given in Sect. 4. Finally, we conclude in Sect. 5.

2 Proposed Model

In order to differentiate nodes with different function better in this paper, we define four types of mobile nodes, i.e., subscription nodes, un-subscription nodes, services nodes and seeds. Subscription nodes denote the mobile nodes which need obtain the data from cellular M2M networks. In contrast, the un-subscription nodes need not obtain the data. And services nodes denote the intermediate nodes to transfer data. Seeds denote the nodes which have the data initially. In other word, the subscription nodes need seeds transfer the data to them. To better understand our model, we define the notions of this paper as shown in Table 1.

Table 1. Notations

Symbol	Meaning
t	Time
j	Number of subscription nodes that have received data, $j \in [0, N]$
K	Number of groups, $K \leq s$
k	The k-th group, $k \in [1, K]$
ψ	Infection probability
T_{d_k}	The deadline in the k-th group
M_k	Number of mobile nodes in the k-th group, $M = \sum_k M_k$
N_k	Number of subscription nodes in the k-th group, $N = \sum_k N_k$
s_k	Number of seeds in the k-th group, $s = \sum_k s_k$
$C_k(t)$	Number of service nodes in the k-th group at time t, $C(t) = \sum_k C_k(t)$
$D_k(t)$	Number of subscription nodes among the service nodes in the k-th group at time t, $D(t) = \sum_k D_k(t)$
$\Omega_k(t)$	Number of un-subscription nodes among the service nodes in the k-th group at time t, $\Omega(t) = \sum_k \Omega_k(t)$
T_h	Minimum interval of time for $C(t)$ from i (i = 1, 2, ..., M-s) nodes to (i + 1) nodes
t_i	$t_i = \sum_{h=0}^{i} T_h$, i (i = 1, 2, ..., $M-s$)

2.1 Related Definitions

To measure how much data traffic can be offloaded, we define *traffic offloading rate* (called TOR) as a performance metric of our model as follows:

Definition 1 (TOR). *At time* t*, traffic offloading rate (* $\gamma(t)$*) is the ratio of* $D(t)$ *and* N*, denoted by* $\gamma(t) \in [0, 1]$*, is given by the following*:

$$\gamma(t) \triangleq \frac{D(t)}{N} \quad (1)$$

According to the Definition 1, $\gamma(T_d)$ (T_d indicates deadline) is the maximum value of TOR in our model. However, if we only pay attention on TOR, the effectiveness of the model may be low. For example, one subscription node may need amount of un-subscription nodes to transfer the data, which not only cause communication mechanism inefficient, but also consume much of local resource of un-subscription nodes.

Therefore, we define local resources consumption rate (called LRCR) as another performance metric of our model to measure the resource consume of un-subscription nodes by the following:

Definition 2 (LRCR). *At time t, local resource consumption rate ($\chi(t)$) is a radio of the local resource consumption of un-subscription nodes and the data traffic offloading as follows:*

$$\chi(t) \triangleq f(\Omega(t), D(t)) \quad (2)$$

where $f()$ indicates the resource consumption function.

In addition, the minimum time of all subscription nodes receive data is one of key time to in our model. Thus, we introduce a new metric of minimum completion time, as defined next:

Definition 3 (*minimum completion time*). *In the process of data transfer, the time required for transfer to all subscription nodes at first time, denoted by $T_{\min}$, is defined by the following*:

$$T_{\min} \triangleq \inf\{t : D(t) = N\}. \quad (3)$$

We call $T_{\min}$ the minimum completion time throughout this paper.

Because of T_d is determined by subscription nodes when they send the data request to cellular M2M networks. If $T_{\min} \leq T_d$, $\gamma(T_d) = 1$, this means all subscription nodes could obtain data before deadline.

2.2 Our Model

Generally, the cellular M2M networks transfer data to subscription nodes via cellular networks directly. The advantage of this communication mechanism is that all subscription nodes can obtain the data immediately, but it could increase the communication cost of the cellular networks at the same time. To solve this problem, a novel communication mechanism has been proposed [6–8]. In this communication mechanism, the cellular networks only transfer the data to some nodes. Then these nodes transfer the data to other nodes through opportunistic communication. The advantage of this communication mechanism is that it can offload data traffic through opportunistic communication. However, this mechanism may be inefficient when the subscription nodes distribution is sparse.

Different from traditional offloading data traffic models, we consider the two communication mechanisms above, and propose adaptive offloading data traffic (AOM) model for cellular M2M networks. As shown in Fig. 1, AOM model contains two phrases, computation phrase and selection phrase. The TOR and LRCR are calculated in the computation phrase. And different communication mechanisms could be adopted in accordance with TOR and LRCR in the selection phrase.

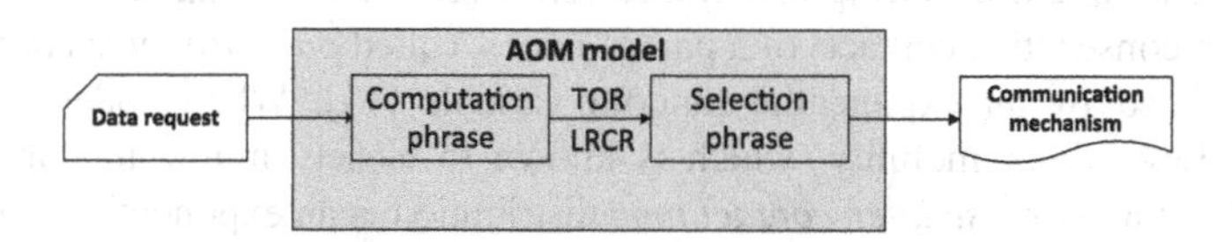

Fig. 1. Adaptive offloading data traffic model

In computation phrase, when subscription nodes need data from others, the seeds could transfer the data to the cellular networks or other nodes. And the cellular M2M networks calculates TOR and LRCR instead of transfer data to subscription nodes immediately. Firstly, according to the distribution of subscription nodes, the cellular M2M networks should determine whether transfer data to some nodes so that they become seeds. Then, the deadline of the data transfer should be determined in accordance with the subscription nodes. Finally, cellular M2M networks calculates the TOR and LRCR based on the CTMC with the seeds and deadline under the opportunistic communication mechanism.

In selection phrase, the cellular M2M networks first initialize the thresholds of TOR and LRCR. Then, compared the values of TOR and LRCR with the thresholds. Finally, determined the communication mechanism depending on the results of the comparison. If the TOR and LRCR greater than the thresholds, the cellular M2M networks could adopt opportunistic communication to transfer data before deadline, otherwise transfer the data to the subscription nodes directly.

3 Analysis TOR and LRCR

In this section, we first outline the basics of the opportunistic communication mechanism. Then, we describe the derivation of the expectations of the number of service nodes and the number of subscription nodes among the service nodes. Thirdly, we examine the parameters of TOR and LRCR, such as population sizes, infection probability, and multiple groups. Finally, we derive formulas for TOR and LRCR.

3.1 The Basic of Opportunistic Communication

In classic information dissemination scheme, researchers always use the epidemiology to simulate information transfer between mobile nodes [10]. Without loss of generality, we also use the epidemiology in our model. In the data transfer mechanism, a services node (say node α), meets a mobile node not having the data yet (say node β), it should transfer data to node β with probability $\psi_\alpha \in (0, 1]$. Similarly, node β receive data successfully with probability $\psi_\beta \in (0, 1]$. In our opportunistic communication mechanism, a mobile node becomes a service node once it receives data successfully. Therefore, node β becomes a service node with the infection probability $\psi_\alpha \psi_\beta$, we use ψ indicates the infection probability $\psi_\alpha \psi_\beta$.

In the pairwise meeting process of mobile networks, the temporal behavior of the data transferring process is determined by its stochastic characteristic which is a critical factor for the mobile transferring process. Recently studies show that the time duration between two consecutive contacts of a pair of nodes, called *pairwise inter-contact time*, can be modeled by an exponential random variable [11]. When nodes movement follow the Lévy flight mobility, which is known to closely mimic human mobility patterns [11], the *pairwise inter-contact time* distribution is an exponential distribution. The work [12] assumed that the *pairwise inter-contact time* of node α and node β, denoted by $M_{\alpha,\beta}$, follows an exponential distribution with rate $\lambda_{\alpha,\beta}(>0)$, i.e.,

$$P\{M_{\alpha,\beta} > t\} = \exp(-\lambda_{\alpha,\beta} t), \quad t \geq 0 \tag{4}$$

Suppose that node α is a service node and node β is an un-service node. According to the meeting process between them, we can obtain the interval of data transferring successful. Then assuming $\lambda^{suc}_{\alpha,\beta} = \lambda_{\alpha,\beta}\psi$, from (1), we have [12]:

$$P\{M^{suc}_{\alpha,\beta} > t\} = \exp(-\lambda^{suc}_{\alpha,\beta} t), \quad t \geq 0 \tag{5}$$

where $M^{suc}_{\alpha,\beta}$ denotes the interval of data transfer successful from α to β.

3.2 Derivation for TOR and LRCR

To derive TOR, we need know the temporal distribution of total number in services nodes $C(t)$ according to Definition 1. To this end, we first divide M nodes into K (K = *1, 2,…, s*) mutual independence groups. Then, at time t, we have $C(t) = \sum_K C_k(t)$ and $s = \sum_K s_k$. Finally, we use the multi-dimensional CTMC to identify the distribution of $C(t)$ as shown in Lemma 1.

Lemma 1 (Joint temporal distribution of $C(t)$). *For K = 1, 2,…, s, let*

$$\mathbb{C}(t) \triangleq (C_1(t), C_2(t), \ldots, C_K(t)).$$

Then, the process $\{\mathbb{C}(t) : t \geq 0\}$ is a K-dimensional CTMC. And the state space is given by:

$$\varepsilon \triangleq \prod\nolimits_{k=1}^{K} \{s, s+1, \ldots, M-s\} \backslash 0.$$

Proof: Assume that the minimum time for the number of service nodes from i nodes to $(i + 1)$ nodes is T_i in group k ($k = 1, 2, \ldots, K$), and s_k indicate the number of seeds. And it is easy to proof that $C_k(t)$ is a CTMC. Because each group is mutually independent, each group is a CTMC. Consequently, process $\{\mathbb{C}(t) : t \geq 0\}$ is a K-dimensional CTMC. □

According to Definition 1, $D(t)$ determines TOR. To obtain $D(t)$, the probability distribution of $D(t)$ can be computed via finding the relationship between $D(t)$ and $C(t)$, and we have the follow lemma:

Lemma 2 (the probability of $D(t)$): *Assuming that the minimum interval time of $C(t)$ from i ($i = 1, 2\ldots, M-s$) nodes to $(i + 1)$ nodes is T_i, and $t_i = \sum_{h=0}^{i} T_h$, we have $C(t_i) = s + i$. At time t_i, assuming $D(t_i) = j$ ($j = 0, 1, \ldots, N-1$), then we give the probability of $D(t) = j + 1$ at time $t(> t_i)$ as follows:*

$$P\{D(t) = j + 1\} = (N - j) \sum_{i=j}^{M-s-N+j} \frac{P\{C(t) = s + i + 1\}}{M - s - i}. \tag{6}$$

where, i denotes the services nodes number, j denotes the number of subscription of services nodes.

Proof: Using the Bayes formula can be obtained the result of (6). Due to the limitation of the length we omit the rest of the proof. □

Opportunistic communication usually relies on free shortwave transmission (such as Bluetooth), therefore we mainly consider the number of un-subscription nodes to participate data transfer. To derive LRCR, we define the resource consumption function $f(\Omega(t), D(t))$ is the ratio of the $D(t)$ and $(\Omega(t) + s)$ according to Definition 2, and LRCR $(\chi(t))$ is denoted as follows:

$$\chi(t) = f(\Omega(t), \mathrm{D}(t)) \triangleq \frac{\mathrm{D}(t)}{C(t) - D(t)} = \frac{\mathrm{D}(t)}{\Omega(t) + s}. \tag{7}$$

From Lemmas 1 and 2, we can derive the expectations of $C(t)$ and $D(t)$.

Let $\pi(t) = (\pi_1(t), \pi_2(t), \ldots, \pi_{M-s}(t))$ denote the probability distribution of $C(t)$ at time t. And the expectation of $C(t)$ and $D(t)$ can be represent as follows.

$$E[C(t)] = \sum_{i=0}^{i=M-s} (i + s)\pi_i(t). \tag{8}$$

$$E[D(t)] = \sum_{j=0}^{N} \sum_{i=j}^{M-s-N+j} \frac{\mathrm{j} \cdot (N - j)}{M - s - i} \pi_i(t). \tag{9}$$

3.3 Parameters Analysis

In opportunistic communication mechanism, each non-seed node is considered as a workload to complete. However, once the node becomes a service node, it operates in a similar manner as the seed and is involved in transfer of the data. Hence, it is not straightforward whether the population sizes (*M*, *N*, *s*) accelerates or slows down the speed of the data transfer. Assume that the minimum time for $C(t)$ from *i* (*i* = *1*, *2*, ..., *M*−s) nodes to (*i* + *1*) nodes is T_i. Let $t_i = \sum_{h=0}^{i} T_h$, then $P\{T_{\min} < t_N\} = 0$, according to Lemma 2, we can obtain the expectation of $T_{\min}$:

$$E[T_{\min}] = \sum_{i=N}^{M-s} \varphi(t_i) \cdot \frac{1}{M\psi} t_i. \tag{10}$$

where $\varphi(t_i) = \frac{P\{C(t)=s+i|C(t_{N-1})=s+i-1\}}{M-s-i+1}$. Similarly, $E[t_i] = \sum_i E[T_i]$. Hence, one has

$$E[t_i] = \sum_i E[T_i] \approx \frac{1}{\psi M} \int_1^i \left(\frac{1}{s+t} + \frac{1}{M-s-t}\right) \mathrm{dt} = \frac{1}{\psi M} \ln\left(\frac{M-s-1}{M-s-i} \cdot \frac{s+i}{s+1}\right). \tag{11}$$

As shown in (10), the infection probability can have an effect on $T_{\min}$. Furthermore, according to (5), the infection probability is one of the factors that determine the data transfer time. The following theorem describes how the infection probability affects $T_{\min}$:

Theorem 1 *(Effect of infection probability). Suppose that the infection probability* ψ *is scaled* ω *(> 0) times for all* α, β, *denoted by* $\widehat{\psi}$, *i.e.,* $\widehat{\psi} = \omega\psi$. *Let* $\hat{T}_{\min}$ *be the correspondences of* $T_{\min}$, *then their relationship is given as*

$$\hat{T}_{\min} \triangleq \omega^{-1} T_{\min}. \tag{12}$$

where $\triangleq$ *denotes "equal in distribution."*

Proof: According to (10) and (11), we know that $\hat{T}_{\min}$ and $T_{\min}$ have same probability at time *t*, and one has,

$$Et_i = \frac{1}{\psi M} \ln\left(\frac{M-s-1}{M-s-i} \cdot \frac{s+i}{s+1}\right).$$

Further, due to $\hat{\lambda}_{\alpha,\beta}^{suc} = \omega\lambda_{\alpha,\beta}^{suc}$, the expectation of $\hat{t}_i$ is denoted as follows:

$$E\hat{t}_i = \frac{1}{\hat{\psi} M} \ln\left(\frac{M-s-1}{M-s-i} \cdot \frac{s+i}{s+1}\right) = \frac{1}{\omega\psi M} \ln\left(\frac{M-s-1}{M-s-i} \cdot \frac{s+i}{s+1}\right) = \omega^{-1} Et_i.$$

In other words, $P\{\hat{T}_{\min} < t\} = P\{\omega^{-1} T_{\min} < t\}$, then, $\hat{T}_{\min} \triangleq \omega^{-1} T_{\min}$. □

In our model, the cellular M2M networks divides all mobile nodes into $k(k = 1, 2, \ldots, K)$ mutually independent groups. Suppose that the corresponding population sizes are M_k, N_k, and s_k in the k-th group. For different groups, the minimum completion time may be different. We calculate the minimum completion time of the k-th group, $T_{\min_k}$, based on (10), where $T_{\min} = \max\{T_{\min_1}, T_{\min_2}, \ldots, T_{\min_K}\}$. Similarly, we can set the deadline of the k-th group, T_{d_k}, according to $T_{\min_k}$, without loss of generality, to arrange as $T_{d_1} \leq T_{d_2} \leq \ldots \leq T_{_K}$. On the basis of the above analysis, we derive the formula for TOR as (13), and the formula for LRCR as (14).

$$\gamma(T_d) = \frac{\sum_K D_k(T_{d_k})}{N}. \tag{13}$$

$$\chi(T_d) = \frac{\sum_K D_k(T_{d_k})}{\sum_K (C_K(T_{d_k}) - D_K(T_{d_k}))}. \tag{14}$$

3.4 Formulation of TOR and LRCR

According to Definition 1, TOR is restricted by time t. Then, according to (10), we can calculate the expectation of $T_{\min}$, and the value of TOR is one when the deadline is greater than $T_{\min}$. If we set $T_d = \mathrm{T}_{\min}$, then (14) can be rewritten by combining (7) and (8) as follows:

$$E[\chi(T_{\min})] = \frac{\sum\limits_{k=1}^{K} \sum\limits_{i=0}^{M_k - s_k} (i + s_k)\pi_{k_i}(T_{\min_k})}{\sum\limits_{k=1}^{K} (\sum\limits_{i=0}^{M_k - s_k} (i + s_k)\pi_{k_i}(T_{\min_k}) - \sum\limits_{j=0}^{N_k} \sum\limits_{i=j}^{M_k - s_k - N_k + j} \frac{N_k - j + 1}{M_k - s_k - i} \pi_{k_i}(T_{\min_k}))}. \tag{15}$$

where $\pi_{k_i}(t)$ indicates the probability distribution of the k-th group.

Similarly, when $\mathrm{T}_{\min} > T_d$, the formula for TOR is (16) and the formula for LRCR is given by (17).

$$E[\gamma(T_d)] = \frac{1}{N} \sum_{j=0}^{N_k} \sum_{i=j}^{M_k - s_k - N_k + j} \frac{j \cdot (N_k - j + 1)}{M_k - s_k - i} \pi_{k_i}(T_{d_k}). \tag{16}$$

$$E[\chi(T_d)] = \frac{\sum\limits_{k=1}^{K} \sum\limits_{i=0}^{i=M_k - s_k} (i + s_k)\pi_{k_i}(T_{d_k})}{\sum\limits_{k=1}^{K} (\sum\limits_{i=0}^{i=M_k - s_k} (i + s_k)\pi_{k_i}(T_{d_k}) - \sum\limits_{j=0}^{N_k} \sum\limits_{i=j}^{M_k - s_k - N_k + j} \frac{j \cdot (N_k - j + 1)}{M_k - s_k - i} \pi_{k_i}(T_{d_k}))}. \tag{17}$$

where, $\pi_{k_i}(t)$ indicates the probability distribution of the k-th group.

Using (16) and (17), the cellular M2M networks can obtain TOR and LRCR. On the basis of TOR and LRCR, our model can decide whether to adopt opportunistic communications to transfer data or transfer data from the cellular M2M networks.

4 Simulations

To verify TOR and LRCR, we first extended the ONE simulator by adding multi-destination node settings and infection probability. Then, we used the extended ONE simulator to simulate data transfer in opportunistic networks.

As shown in Fig. 2, the minimum completion time ($T_{\min}$) decreases with increasing infection probability. In accordance with Theorem 1, we calculated $T_{\min}$ with different

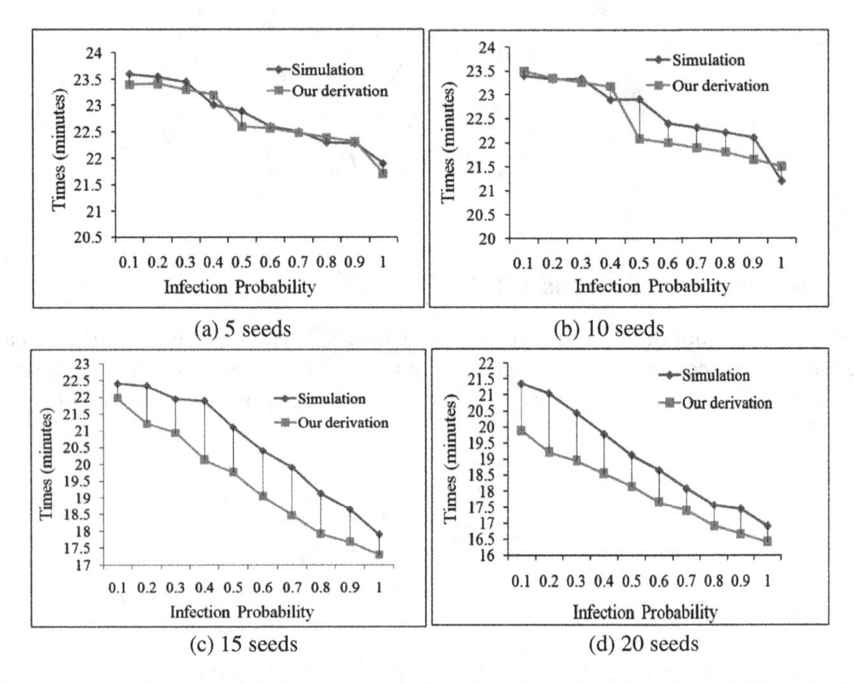

(a) 5 seeds (b) 10 seeds

(c) 15 seeds (d) 20 seeds

Fig. 2. The minimum completion time with 5 seeds, 10 seeds, 15 seeds, and 20 seeds. The simulation results show Theorem 1 is correct.

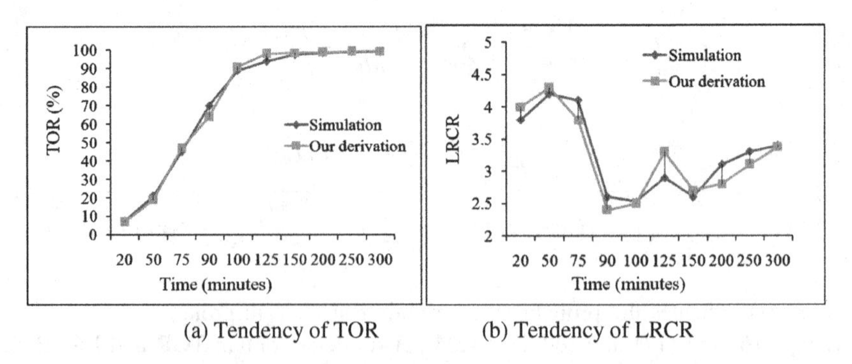

(a) Tendency of TOR (b) Tendency of LRCR

Fig. 3. The tendency of TOR and LRCR. And the results confirm that our derivation for TOR and LRCR is correct.

infection probabilities. Although the results of the simulation are irregular because of the random movement of the nodes, the tendency of $T_{\min}$ is consistent with our derivation. Thus, the simulation results also confirm that Theorem 1 is correct. Finally, we investigated the formulas for TOR and LRCR. As shown in Fig. 3(a), the value of TOR increases with increasing time t. Further, as shown in Fig. 3(b), the value of LRCR changes irregularly. Thus, the simulation results confirm that our derivation for TOR and LRCR is correct.

5 Conclusion

In this paper, we first proposed an adaptive offloading data traffic model for cellular M2M networks. This model contains two data transfer mechanisms: transfer the data via the cellular networks and transfer of data through opportunistic communication. Then, we derived TOR and LRCR based on continuous time Markov chain. Finally, the results of extensive simulations conducted demonstrate that the derivations in our model is accurate and effective.

Acknowledgments. This work is supported by the Natural Science Foundation of Beijing under Grant No. 4132048, National Natural Science Foundation of China under Grant No. 61202435 and 61272521

References

1. Kim, J., Lee, J., Kim, J., Yun, J.: M2M service platforms: survey, issues, and enabling technologies. IEEE Commun. Surv. Tutorials **16**, 61–76 (2014)
2. Shafiq, M.Z., Ji, L., Liu, A.X., Pang, J., Wang, J.: A first look at cellular machine-to-machine traffic: large scale measurement and characterization. In: ACM SIGMETRICS Performance Evaluation Review, pp. 65–76 (2012)
3. Beckman, R., Channakeshava, K., Fei, H., Vullikanti, V.S.A., Marathe, A., Marathe, M.V., et al.: Implications of dynamic spectrum access on the efficiency of primary wireless market. In: 2010 IEEE Symposium on New Frontiers in Dynamic Spectrum, pp. 1–12, April 2010
4. Bo, H., Pan, H., Kumar, V.S.A., Marathe, M.V., Jianhua, S., Srinivasan, A.: Mobile data offloading through opportunistic communications and social participation. IEEE Trans. Mob. Comput. **11**, 821–834 (2012)
5. Dimatteo, S., Pan, H., Bo, H., Li, V.O.K.: Cellular traffic offloading through WiFi networks. In: 2011 IEEE 8th International Conference on Mobile Adhoc and Sensor Systems (MASS), pp. 192–201 (2011)
6. Lu, X., Hui, P., Lio, P.: Offloading mobile data from cellular networks through peer-to-peer WiFi communication: a subscribe-and-send architecture. Commun. China **10**, 35–46 (2013)
7. Han, B., Hui, P., Kumar, V.S.A., Marathe, M.V., Pei, G., Srinivasan, A.: Cellular traffic offloading through opportunistic communications: a case study. In: Proceedings of the 5th ACM Workshop on Challenged Networks (2010)
8. Whitbeck, J., Lopez, Y., Leguay, M., Conan, V.: Fast track article: push-and-track: saving infrastructure bandwidth through opportunistic forwarding. Pervasive Mob. Comput. **8**, 682–697 (2012)

9. Wu, H., Chunming, Q., De, S., Tonguz, O.: Integrated cellular and ad hoc relaying systems: iCAR. IEEE J. Sel. Areas Commun. **19**, 2105–2115 (2001)
10. Andersson, H., Britton, T.: Stochastic epidemic models and their statistical analysis. Springer Science & Business Media (2012)
11. Yoora Kim, K.L., Shroff, N.B., Rhee, I., Chong, S.: On the generalized delay-capacity tradeoff of mobile networks with lévy flight mobility. In: *arXiv*, vol. The Ohio State Universi, Tech, July 2012
12. Yoora, K., Kyunghan, L., Shroff, N.B., Injong, R.: Providing probabilistic guarantees on the time of information spread in opportunistic networks. In: 2013 IEEE 32th International Conference on Computer Communication (INFOCOM 2013), pp. 2067–2075, April 2013

Similarity-Based Trust Management System for Detecting Fake Safety Messages in VANETs

Hind Al Falasi(✉) and Nader Mohamed

College of Information Technology, United Arab Emirates University,
Al Ain, United Arab Emirates
{hindalfalasi,nader.m}@uaeu.ac.ae

Abstract. VANETs (Vehicular Ad hoc Networks) are a special case of ad hoc networks with highly mobile nodes (onboard vehicles). In VANETs, vehicles form trust relationships to achieve many goals. Cooperative driving is one of the goals of VANETs. In this paper, we present a scheme that uses similarity-based trust relationships to detect false safety event messages from abnormal vehicles in VANETs. Our scheme reacts to safety events claims made by a vehicle. The scheme predicts that the source vehicle will react to a truthful safety event report. We perform simulations to study the effectiveness of the proposed scheme in uncovering false safety event messages sent by abnormal vehicles.

Keywords: Similarity · Trust · False data injection · VANET · Safety system

1 Introduction

Vehicular Ad hoc Network (VANET) is a special class of Mobile Ad hoc Networks (MANETs) [1]; it is comprised of moving nodes. The nodes in the network are vehicles that are self-organized. The mobility pattern of VANET nodes is much faster. The topology of the network is dynamic; changes in the networks are very frequent due to the speed by which vehicles travel and the changes in the underlying road infrastructure. The resources available for the nodes in VANET are greater than the available resources for the nodes in MANET.

VANETs have a variety of applications; some of them are safety applications. Safety applications include: emergency breaking, lane changing warning messages and collision avoidance applications [2]. They are the most critical applications in VANETs because of the severity of loss associated with them. In these applications, vehicles exchange messages to communicate their current speed and location. In addition, vehicles send safety event messages to warn other vehicles of an incident in the road. The security of the safety event messages is critical, similarly the correctness of these messages is as critical. The vehicles in VANET need to trust safety event messages sent by their neighbors in the network; a few false messages from misbehaving vehicles can disturb the performance of the network. In safety applications, false messages can cause serious accidents. Due to the nature of the applications, any disturbance to the normal operation of the network can threaten the lives of road users' and inflect losses to their properties.

C.-H. Hsu et al. (Eds.): IOV 2015, LNCS 9502, pp. 273–284, 2015.
DOI: 10.1007/978-3-319-27293-1_24

In this paper we present a similarity-based trust management system. Vehicles use similarity to assign trust ratings for their one hop neighbors in the network. The preliminary results of the effectiveness of this similarity-base management system are presented in [3]. In addition, we develop a scheme that utilizes the trust ratings of the vehicles to determine whether the safety event reported by a vehicle is truthful or not. We present the performance of the scheme that is based on trust relationships derived from similarity. Our ultimate goal is to enhance the decision making process using trust; we want to study whether the vehicle's reaction to a message reporting an event in the network is the right one. Did it believe a false report or did it ignore a genuine report.

The paper is organized in the following way. In Sect. 2, the related work is presented. We provide a brief background of VANET environment, adversary model and the trust management system in Sect. 3. In Sect. 4, the scheme overview is presented. Simulation and results are presented in Sect. 5. Finally, Sect. 6 discusses future directions of research and concludes the paper.

2 Related Work

Data verification is an important element of security in VANETs; in addition to the standard security services, vehicles need to ensure that the information they receive is correct and truthful i.e. trustworthy. Without data verification, an adversary would be able to inject false data into the network to alter the behavior of the participating vehicles. In order to preserve the security of data in VANETs, researchers proposed the use of reputation systems. In reputation systems, the trustworthiness of the data is derived from the trustworthiness of the source of the data i.e. entity-centric trust. In [4], the authors propose a scheme to assess the reliability of a reported warning message using reputation. The vehicle's decision to whether accept a warning message or not is taken by considering three sources of information: the reputation of the sender, the recommendation of the neighbors, and the reputation of the sender provided by a central authority if existed.

The problem with this scheme is that a vehicle has little time to make a decision about a safety event in the network; it doesn't have time to query its neighbors about the reliability of the reported safety event.

The authors in [5] use an event-based reputation model to identify false data. Upon receipt of a safety event message, a vehicle will observe the behavior of the sender to determine the truthfulness of the reported event; if the behavior of the sender matches the behavior expected in the presence of such events, then the receiver will conclude that the event is true. Otherwise, the receiver will assign a low reputation value for the reported message. Although this seems like a good solution, it is susceptible to attacks; the malicious vehicle injecting false data can modify its behavior during the observation period to trick the receiver. Another event-based reputation system is presented in [6], VARS is an event-based reputation system. In the system, the vehicles make the decision to forward an event based on the reputation of the event; when an event is sent, an opinion is attached to it. A vehicle calculates the forwarding probability of an event using three attributes: direct experience of the event, trust value of the reporting vehicle and aggregated partial opinions. The problem with this system is the use of aggregated opinions; a vehicle will take into account an opinion from a stranger that it had never interacted with.

In order to prevent false data injection attacks, the authors in [7] suggest that every vehicle models its neighborhood, and then compares the received information from its neighbors to the locally built model. The premise of the research is that a misbehaving vehicle is likely to be an outlier in the local model. Such approach is hard to implement; this is mainly due to the nature of the neighbors list as it changes frequently. In [8], the authors use the speed, density and flow to build a model to identify malicious vehicles. Flow is calculated from speed and density, and then every vehicle compares the locally calculated flow and density with the flow and density calculated by the sender. If the information doesn't match, then the sender is assumed malicious. The model fails to detect false data injection attack if the data sent by the malicious vehicle conforms to the locally built model. Consensus is used by the authors in [9] to prove that a vehicle is relevant to the event it has reported; the burden of proof is on the sender. The reporting vehicle must collect endorsement messages from witnessing vehicles in the area of the detected event to serve as proof. This scheme is prone to failure in areas where there is low traffic density. The authors in [10] investigate the consensus problem; the authors investigate the best threshold value needed to react to a warning message.

In this paper, we use echoing to observe the behavior of the abnormal vehicle. When a vehicle reports a safety event in the network, the receiver of the report will echo that message. If the originator of the safety event message reacts to its own report, then they are assumed truthful; otherwise, they are assumed dishonest, and their trust rating is demoted as a consequence.

3 Background

Our scheme is proposed to protect vehicles from the false data injection attack. It reacts to safety event claims made by a vehicle. The scheme predicts that the source vehicle will react to a truthful safety event report. Figure 1 shows the state transition diagram.

3.1 Goals

For a given partition of a VANET, we would like to detect a false data injection attack, i.e. an attempt by an adversary to disseminate false information to disturb the behavior of other vehicles. We would like to detect this attack using a distributed solution due to the issues inherent in centralized solutions [6, 11]. We evaluate the scheme by examining the probability of detecting an attack given the adversary's trust rating. We also evaluate the efficiency of the scheme; our aim is to detect the attack using the minimum number of messages to minimize the communication overhead on the network as a whole and on the individual vehicles.

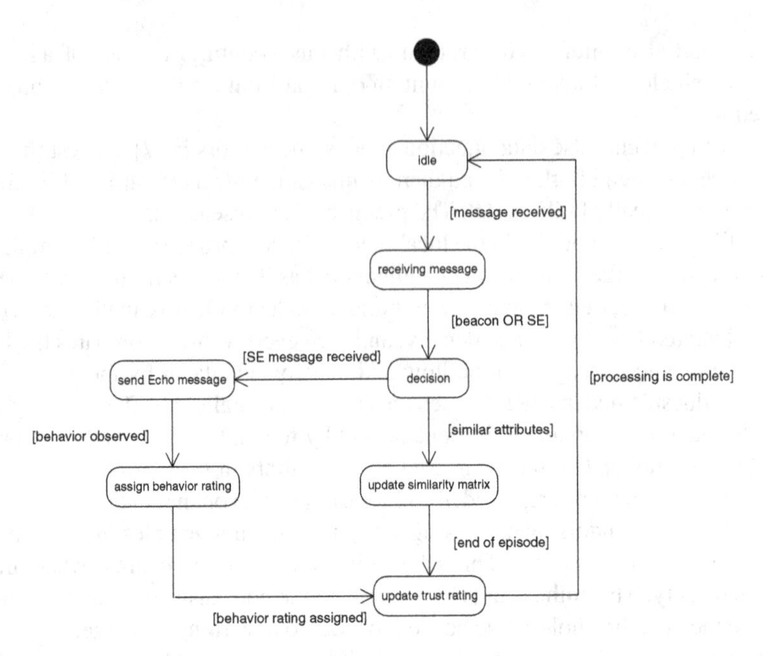

Fig. 1. State transition diagram of the OBU

3.2 Collaborative Vehicles

Vehicular Ad hoc Network (VANET) is comprised of vehicles that cooperate to achieve some advantages. The vehicles in VANET work towards some individual goals, and other collective goals of the members of the network. For example, an individual goal would be for a vehicle to ensure security of its communications. On the other hand, an example of a collective goal would be to reduce the network communications overhead, which in turn helps the vehicles achieve another individual goal of a better quality of service in the network.

Cooperation in VANETs means that vehicles rely on reports generated by other vehicles on the road to react to safety events. We proposed a Trust Management System architecture for VANETs in [3]. Using similarity, vehicles construct local views of their surroundings, and form opinions about their neighbors. Assuming trust is established, vehicles rely on their calculation of trust rating of their peers to validate event generated reports.

3.3 Adversary Model

In examining the security of VANET environment, we take a conservative approach by assuming that the adversary is working alone. We consider a type of attack that is localized; it only affects the immediate neighbors of the adversary. If most of the adversary's neighbors are in collision with them, then no scheme running in the network can withstand an attack carried out by the majority of the network's participants.

The adversary sends a broadcast to its one hop neighbors to falsely warn about a safety event in the network for their selfish objectives. The adversary is assumed to exhibit an abnormal behavior i.e. the vehicle is abnormal. Abnormal vehicles are vehicles that exhibit unpredictable behaviors, such as: irregular rates of acceleration and deceleration, and failure to maintain safety distance; we expect these vehicles to have low trust ratings as their actions cannot be counted on i.e. untrustworthy.

3.4 Similarity

In our model, trust is achieved through similarity. A vehicle calculates the similarity rating between itself and the other vehicles it encounters throughout its journey. Vehicles in VANETs use periodically broadcast beacons to communicate information related to their location and speed. Every vehicle listens for beacons sent from its one hop neighbors. Additional information might be available in the beacon depending on the application(s) running onboard. The received information is processed and stored in the receiving vehicle for handling later on at the end of the listening period. The listening period is limited. At the end of the listening period, calculations begin in the receiving vehicle to compute the degree of similarity between itself and its neighbors.

Each vehicle will store its own view of the network; the degree of similarity between the vehicle and its one hop neighbors.

In order to discover associations among the vehicles, we mine for frequent item sets to build a set of vehicles that frequently exhibit similar speeds within a limited area of communication range, and at a certain window of time.

Therefore, their behavior is more predictable. During the listening period, whenever a vehicle and its one hop neighbor have similar speeds, the frequency of meeting (which we call the met value) is incremented every second in the Similarity Matrix (SMij) in the vehicles. For example, Table 1 shows SM2j of vehicle V2. V2 observed the network for a listening period of 5 s.

Table 1. Sample similarity matrix in vehicle V_2.

V_0	V_1	V_2	V_3	V_4	V_5	V_6
1	?	1	?	?	?	?
1	1	1	?	?	?	?
1	?	1	?	?	?	?
1	?	1	?	?	1	?
?	?	1	?	?	1	1

During the listening period, V_2 encountered the below vehicles driving at a similar speed:

V_0: 4 times
V_1: 1 time
V_5: 2 times
V_6: 1 time

In order to investigate the existence of relationships between the vehicles, we use Apriori which is a data mining algorithm that is used to mine frequent item sets and develop association rules [12]. When we feed this dataset to Apriori, the below association rules are derived:

$V_2 = 1 ==> V_0 = 1$, confidence: (0.8)
$V_2 = 1 ==> V_1 = 1$, confidence: (0.2)
$V_2 = 1 ==> V_5 = 1$, confidence: (0.4)
$V_2 = 1 ==> V_6 = 1$, confidence: (0.2)

Where the confidence value of each association rule represents its correctness; for example, V_2 and V_0 will travel at the same speed at the same space in time with probability = 80 %, and V_2 and V_5 will travel at the same speed at the same space in time with probability = 40 %.

We use the association rules to compute the similarity rating. At the end of the listening period, the similarity rating S_{ij} is calculated using the following equation:

$$S_{ij} = \mathrm{Freq}_{ij}/X \tag{1}$$

Where $Freq_{ij}$ is the *met* value for vehicles i and j, and x = the duration (in seconds) of the observation (listening period). Three measures are considered when calculating the pair-wise similarity degree: Location, speed and time. The vehicle keeps listening for beacons from its one hop neighbors i.e. when their locations are within the communication range of the vehicle. Only neighbor vehicles travelling at a similar speed as the listening vehicles cause the increment of the *met* value.

Finally, a vehicle V listens for other vehicles that have similar speeds as itself during the listening period. The resulting similarity rating is then used to calculate the trust rating.

3.5 Trust

Each vehicle listens to the beacons sent from its neighbors throughout the journey. The duration of the journey is divided into equal length time-intervals (e.g. 10 s), which we call *periods*. Calculating similarity and consequently trust ratings is done over these periods of time. Trust is a cumulative value where at the end of each listening period, the trust rating is updated by adding the current similarity rating to the previous trust rating. VANETs are constantly changing as they comprise of highly mobile nodes. In order to capture this characteristic of VANETs, we use an exponential decay function to assign a weight to the old and new values of trust in our calculation. We derive trust from similarity; we look for vehicles that exhibit similar behaviors in terms of acceleration and deceleration rates. We use the similarity rating calculated at the end of each listening period to compute a trust rating.

Below is the equation we use to calculate T_{ij}, the trust rating between vehicles i and j:

$$T_{ij}^{n} = \left[(1-\alpha)\,T_{ij}^{n-1} + \alpha S_{ij}^{n}\right] B_{ij}^{n}\ ,\ T_{ij}^{0} = \varphi \tag{2}$$

Where α is the rate of decay, S_{ij}^{n} is the similarity rating between vehicles i and j at the current period n, $T_{ij\text{-}}^{n-1}$ is the trust value in the previous period, n-1, and φ is the initial trust value. α is a predefined value which can be increased or decreased depending on the application or vehicle preference. B_{ij}^{n} is the behavior rating assigned by the receiver to the source of the safety event message:

$$B_{ij}^{n} = \begin{cases} y, & v_i \textit{ is dishonest} \\ 1, & \textit{otherwise} \end{cases} \tag{3}$$

As shown in (3), y is the penalty given to the source vehicle for reporting a false safety event.

3.6 Scalability

Each vehicle listens to the beacons sent from its neighbors throughout the journey. If a vehicle stops receiving beacons from once a neighbor for a while, the neighbor is then removed from the neighborhood list. At every given moment, each vehicle has a finite number of neighbors which allows this model to be implemented in big networks.

4 Scheme Overview

In our verification scheme we have two participants depending on the role they play in the network:

1. **Safety Event Reporter (SER):** A vehicle is designated as SER if it is the originator of the safety event message.
2. **Safety Event Evaluator (SEE):** The one hop neighbors of the Safety Event Reporter are designated as Safety Event Evaluators.

Our aim is to use the trust rating calculated by every vehicle in the network to validate a safety event reported by a *SER*. A *SEE* has the responsibility of identifying true from false messages from a *SER*. When a *SEE* receives a safety event message from *SER*, it has t time to verify the event before it must make a decision.

Given the fact that *SER* and *SEE* are one hop neighbors, they have already established a trust relationship between each other following the equations presented in the previous section. The *SEE* can verify the truthfulness of the received safety event message even though it hasn't experienced it directly; when a *SEE* receives a safety event message from a *SER*, it will react by sending the same message to the *SER*.

Intuitively, the *SER* will react to its own message and the *SEE* will observe the *SER*'s reaction. If the behavior of the *SER* matches the typical behavior expected by the *SEE*, it will consider the message as trusted. For example, if a *SER* sends a safety event message about a road deadlock ahead, the *SEE* will send the same message back to the *SER* expecting that the *SER*'s behavior would be to slow down or change route. If the observed behavior of the *SER* doesn't match the behavior the *SEE*'s expectation, it will conclude that the safety event message is false. *SEE*'s trust rating of the *SER* will be updated accordingly to reflect its misbehavior.

We factor the trust rating of the *SER* calculated by the *SEE* in the decision making process; trust rating is calculated from attribute similarity, our aim is to improve the precision of the calculated trust ratings by factoring a behavioral element in the calculation.

4.1 Echo Protocol

As presented earlier, we have two participants in the Echo protocol; the *SER* and the *SEE*. The *SER* sends the Safety Event (*SE*) message, and reacts to an echo message; the *SER* reacts to the echo message in one of the following actions:

- **Brake:** The SE is a genuine safety event message, and therefore the SER is honest.
- **Do Nothing:** The SE is a fake safety event message, and therefore the SER is a dishonest vehicle.

The receiver of *SE* sends the echo message to the *SER* upon receipt of the *SE*. The echo message contains the original safety event message and the hash of the message for authentication. The protocol message exchange is shown in Fig. 2. The receiver will continue to receive updates in the form of beacons from the source vehicle and will use these updates to observe which reaction the source vehicle is exhibiting upon receipt of the Echo message. The reaction of the source vehicle will help the receiver to draw conclusions about the behavior of the source vehicle; therefore, determine the appropriate behavior rating to use in updating the trust rating of the source vehicle.

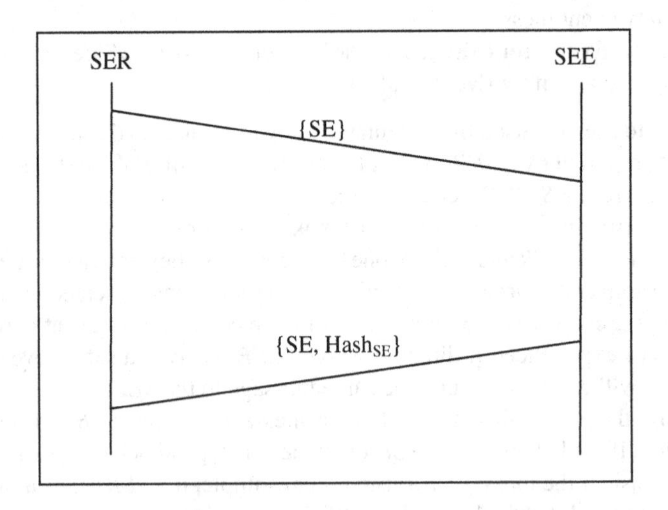

Fig. 2. The echo protocol

5 Simulation and Results

In this section we will describe the conducted simulation to demonstrate the validity of the Echo protocol using our similarity-based trust management system. We use SUMO [13] to simulate the traffic in a network of vehicles.

SUMO is a microscopic simulator. It generates realistic traffic traces of vehicles' movements. The trace files generated from SUMO indicate the speed and location of every vehicle in the network every step of the simulation. Each step of the simulation represents of 1 s of simulation time. Our simulation is a simplified highway topology; one-way highway with three lanes. The length of the highway segment is 5000 m.

In all of the simulation runs we have 100 vehicles. We use the built-in Sigma parameter to define the driver imperfection in the simulation; the driver's ability to adapt to the desired safe speed. P is the percentage of vehicles with Sigma = 1. We have three simulation runs with different values of P: 20, 50 and 80. The listening period used to calculate similarity and trust ratings is set to 60 s in all the runs. On top of the traffic generated from SUMO we build an application that assigns trust ratings to all the vehicles in the network.

We read the trace file generated by SUMO for further processing. In addition, the application is used to validate the Echo scheme by incorporating the behavior rate in the calculation of trust.

We use formula (2) to compute the average trust rating between normal vehicles, and abnormal vehicles when a vehicle reports a safety event.

The average trust rating is continuously updated throughout the simulation, either through re-calculating similarity or through the behavior rating of the source vehicle. We simulate a safety event in the network, and then calculate how many vehicles believed the reporting vehicle. The reporting vehicle can either be truthful or dishonest. We use the trust rating of the reporting vehicle to assist the receiving vehicles in making their decision about the safety event message. We validate the system by calculating the percentage of vehicles that believed a true report of a safety event in the network vs. the percentage of vehicles that believed a false report of a safety event. A vehicle that receives a report about a safety event uses the calculated trust rating of the sender to make a decision on whether to believe the report or not.

The simulation was run in three network setups where the percentage of abnormal vehicles or vehicles with unpredictable behaviors was 20 %, 50 % and 80 %.

Figure 3 shows that the number of vehicles that trusted a true safety event message increases as the number of observation episodes increases. This is to be expected as the longer a vehicle communicates with another vehicle in the network, the more similarities there are between them; therefore, the better the trust rating is calculated.

In the first and second network, we can see that the average percentage of vehicles that believed a safety event message becomes steady as the number of observation episodes increase. In the third network; however, the number of vehicles fluctuates. This is understandable given the fact that the majority of the network participants are abnormal vehicles; therefore, it's difficult for the normal vehicles to find similarities with their neighbors to use to establish trust relationships. Therefore, the vehicles cannot use the calculated trust rating to identify truthful safety event messages.

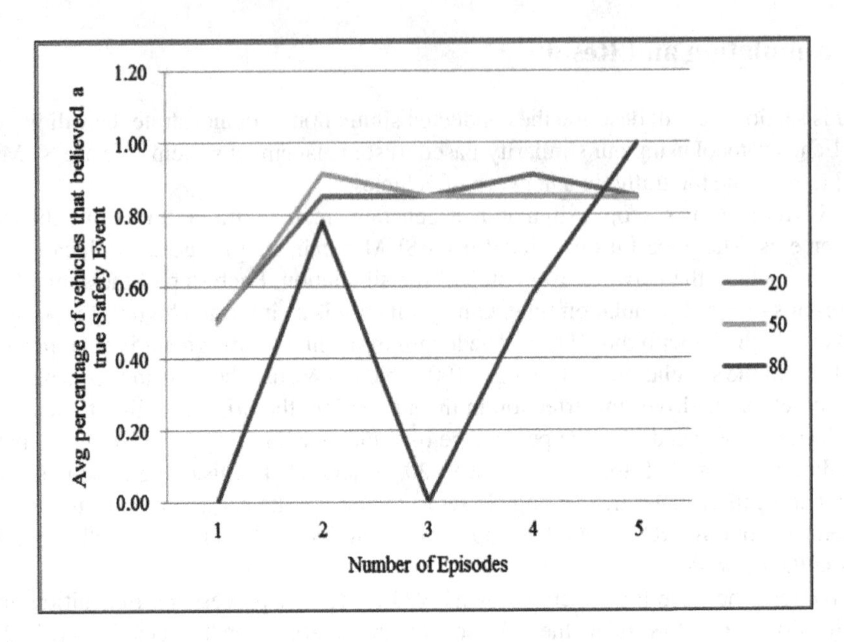

Fig. 3. Average percentage of vehicles that believed a true safety event report

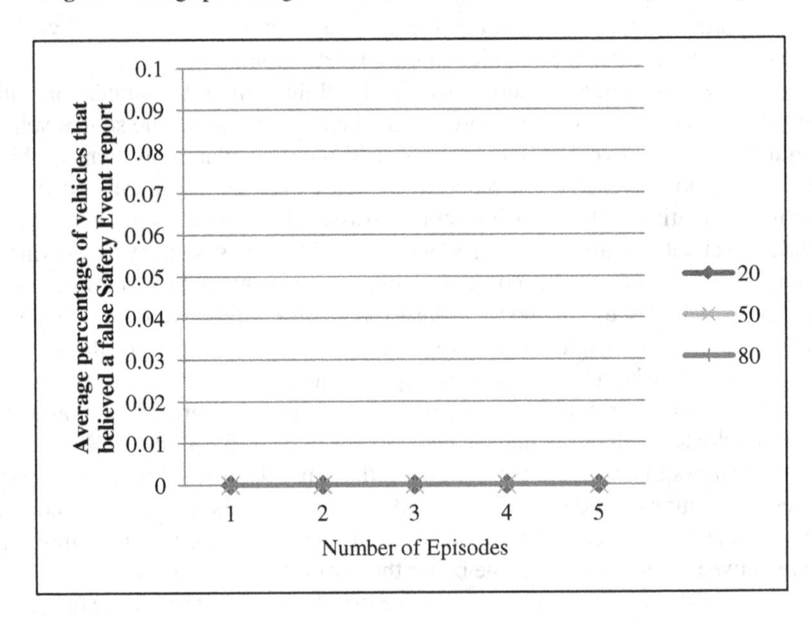

Fig. 4. Average percentage of vehicles that believed a false safety event report

Figure 4 on the other hand shows the number of vehicles that believed a false safety event message using information collected from different number of episodes. In all network simulations, the number of vehicles is zero. The simulation results show that our proposed system for using similarity to achieve trust is proof against false data injection attack.

6 Conclusion

Our proposed system has shown that similarity can be used to compute trust among vehicles; using the information from the beacons we were able to build association rules that predict the probability of correspondence in driving behavior i.e. rate of acceleration, rate of deceleration and the preservation of safety distance. The proposed system filters the data in the network in order to isolate useful information. The isolated bits of information are then used to calculate the similarity degree between a vehicle and all of its one hop neighbors. Finally, the generated similarity ratings are used to compute the trust ratings of the vehicles. The trust rating is updated using observations of other vehicles' behavior in the network.

The system is designed to use data mining techniques to find high value information in a highly dynamic network. In the system, similarity between vehicles is mined using association rule mining; each vehicle looks for other vehicles that frequently exhibit similar speeds as themselves in a given location and a specific timeframe. Additionally, in this paper we designed and evaluated an adversary model in order to study the effect of trust on the accuracy of the decision taken in the presence of false data in the network.

In the future, we intend to make the decision making process dynamic to adjust to the time constraints present in VANETs safety applications. Moreover, we plan to use more attributes to compute similarity rating among vehicles. We would like to investigate their effect on the average trust rating in the network for other type of VANET applications.

References

1. Hadim, S., Al-Jaroodi, J., Mohamed, N.: Middleware issues and approaches for mobile ad hoc networks. In: The IEEE Consumer Communications and Networking conference (CCNC 2006) (2006)
2. Jawhar, I., Mohamed, N., Zhang, L.: Inter-vehicular communication systems, protocols and middleware. In: 2010 IEEE Fifth International Conference on Networking, Architecture and Storage (NAS), IEEE (2010)
3. Al Falasi, H., Masud, M.M., Mohamed, N.: Trusting the same: using similarity to establish trust among vehicles. In: The 2015 International Conference on Collaboration Technologies and Systems 2015: Atlanta
4. Mármol, F.G., Pérez, G.M.: TRIP, a trust and reputation infrastructure-based proposal for vehicular ad hoc networks. J. Netw. Comput. Appl. **35**(3), 934–941 (2012)
5. Ding, Q., et al.: Reputation-based trust model in vehicular ad hoc networks. In: 2010 International Conference on Wireless Communications and Signal Processing (WCSP). IEEE (2010)
6. Dötzer, F., Fischer, L., Magiera, P.: Vars: A vehicle ad-hoc network reputation system. In: Sixth IEEE International Symposium on a World of Wireless Mobile and Multimedia Networks. IEEE (2005)

7. Ghaleb, F.A., Zainal, A., Rassam, M.A.: Data verification and misbehavior detection in vehicular ad-hoc networks. J. Teknologi **73**(2), 37–44 (2015)
8. Zaidi, K., et al.: Data-centric rogue node detection in VANETs. In: 2014 IEEE 13th International Conference on Trust, Security and Privacy in Computing and Communications (TrustCom). IEEE (2014)
9. Cao, Z., et al.: Proof-of-relevance: Filtering false data via authentic consensus in vehicle ad-hoc networks. In: INFOCOM Workshops 2008. IEEE (2008)
10. Petit, J., Mammeri, Z.: Dynamic consensus for secured vehicular ad hoc networks. In: 2011 IEEE 7th International Conference on Wireless and Mobile Computing, Networking and Communications (WiMob). IEEE (2011)
11. Raya, M.: Data-centric trust in ephemeral networks. In: École Polytechnique Fédérale De Lausanne 2009
12. Wu, X., et al.: Top 10 algorithms in data mining. Knowl. Inf. Syst. **14**(1), 1–37 (2008)
13. Krajzewicz, D., Bonert, M., Wagner, P.: The open source traffic simulation package SUMO. In: RoboCup 2006 Infrastructure Simulation Competition, vol. 1, pp. 1–5 (2006)

A Modified Time Domain Least Square Channel Estimation for the Vehicular Communications

Chi-Min Li(✉) and Feng-Ming Wu

Department of Communications,
Navigation and Control Engineering,
National Taiwan Ocean University,
Keelung, Taiwan
cmli@ntou.edu.tw

Abstract. Modern Intelligent Transportation System (ITS) uses the wireless communication technologies to increase the traffic safety and the transportation efficiency. However, the Doppler and the multipath fading problems severely attenuate the received signal in the vehicle communication environments. Therefore, channel estimation is a serious issue should be solved properly to improve the Bit Error Rate (BER) performance at the receiver. Besides, ITS has adopted the IEEE802.11p standard as its protocol and IEEE 802.11p defines the preamble with specific format to conduct the channel estimation. The well-known Least Square (LS) channel estimation method is commonly used with the help of preamble due to the simple and easy implementation advantages. The LS methods can be conducted either in the Time Domain (TD) or in the Frequency Domain (FD). In this paper, a modified TDLS channel estimation is proposed for the IEEE802.11p channel estimation. The proposed method has the channel length prediction ability to improve the conventional LS channel estimation performance at both the time domain and frequency domain.

Keywords: Channel estimation · Least square · IEEE802.11p

1 Introduction

Intelligent Transportation System (ITS) is a smart system that integrates the telecommunication, information, electronics technologies, and traffic management to efficiently manipulate public transportation in many modern metropolises. It adopts the mobile communication techniques to increase the traffic and the transportation efficiency in many applications such as the vehicle-to-vehicle communications (V2V) and the vehicle-to-road communications (V2R). For both the V2V and V2R communications, the signal transmitted at the Transmitter (Tx) will suffer the intensely channel distortion due to the Doppler effect and the multipath fading channel. In order to decode the transmitted symbols correctly, Receiver (Rx) has to conduct the channel estimation properly to improve the Bit Error Rate (BER) performance for further information processing.

C.-H. Hsu et al. (Eds.): IOV 2015, LNCS 9502, pp. 285–293, 2015.
DOI: 10.1007/978-3-319-27293-1_25

ITS has adopted the IEEE802.11p standard as its protocol [1]. Besides, IEEE 802.11p also known as the Wireless Access in the Vehicular Environment (WAVE) that can be applied in the Dedicated Short Range Communications (DSRC). Basically, IEEEE 802.11p is an Orthogonal Frequency Division Multiplexing (OFDM) system with the frequency band ranges from 5.850 GHz to 5.925 GHz. The data rate can up to 3 Mbps to 27 Mbps within the coverage 300 m to 1000 m. Six data channels and one control channel have been defined among the IEEE 802.11p. Each channel consists of 64 sub-carriers within the 10 MHz frequency band. In fact, 802.11p is an extension from the conventional IEEE 802.11a Wi-Fi system. However, the Guard Interval (GI) for the 802.11p is twice as

Table 1. Comparisons of the IEEE 802.11a and IEEE802.11p

Parameters	IEEE802.11a	IEEE802.11p	Changes
Bit rate (Mbit/s)	6, 9, 12, 18, 24, 36, 48, 54	3, 4, 5, 6, 9, 12, 18, 24, 27	Half
Modulation mode	BPSK, QPSK, 16QAM, 64QAM	BPSK, QPSK, 16QAM, 64QAM	No change
Code rate	1/2, 2/3, 3/4	1/2, 2/3, 3/4	No change
Number of subcarriers	52	52	No change
Guard Time	0.8 μs	1.6 μs	Double
Symbol duration	4 μs	8 μs	Double
FFT period	3.2 μs	6.4 μs	Double
Preamble duration	3.2 μs	6.4 μs	Double
Subcarrier spacing	0.3125 MHz	0.15625 MHz	Half

the IEEE 802.11a that can be very helpful in the Inter-symbol Interference (ISI) avoidance under the vehicular communication environments. Table 1 lists the comparisons between the 802.11a and the 802.11p systems.

In the literatures, many channel estimation methods for OFDM system have been proposed to conduct the channel compensation at the Rx. Channel estimation can be achieved in the frequency domain or in the time domain [2]. For example, the frequency domain estimations include the Inverse Fast Fourier Transform (IFFT) [3] and the Minimized Mean Square Error (MMSE) method [4] while the Finite and Infinite Length MMSE estimation [5] and the Linear Interpolation estimation [5] can be performed in the time domain. Among the literary methods, the well-known Least Square (LS) channel estimation method is commonly used with the help of preamble due to the simple and easy implementation advantages. The LS methods can be conducted in the time domain or in the frequency domain.

In IEEE 802.11p, a pre-defined preamble with specific format is provided to conduct the channel estimation (Fig. 1). In this paper, a modified TDLS channel estimation is proposed for the IEEE802.11p channel estimation. The main difference for the proposed method and the conventional LS method is that it can have the channel length prediction ability and improve the conventional time domain and frequency domain LS channel estimation. The paper is organized as follows, Sect. 2 describes the time domain and frequency domain LS channel estimation methods and the proposed

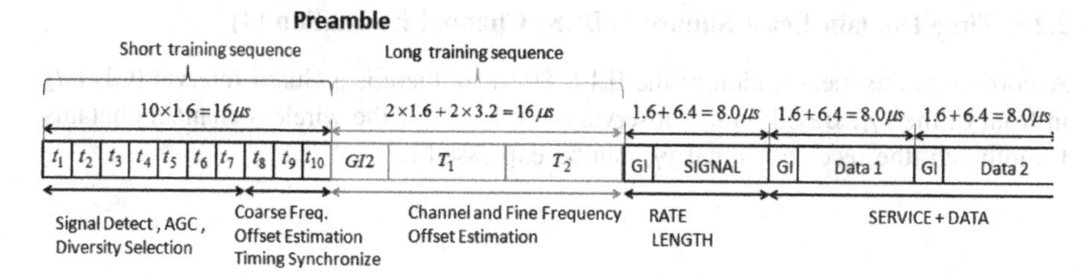

Fig. 1. Signal format of the IEEE 802.11p preamble [1]

modified scheme. Computer simulations based on the multipath channel model are given in Sect. 3. Finally, some conclusions are summarized in Sect. 4.

2 Methods Descriptions

In this section, the conventional Frequency Domain Least Square (FDLS), Time Domain Least Square (TDLS) and the proposed modified TDLS are provided. All these methods use the preamble defined in Fig. 1 to compensate the channel distortion for the OFDM signal.

2.1 Frequency Domain Least Square (FDLS) Channel Estimation [3]

Let the first and second long training sequence be denoted as x_{T_1} and x_{T_2} and the corresponding received signals denoted as y_{T_1} and y_{T_2} respectively. In IEEE 802.11p these two long training sequences are identical. The N-point Discrete Fourier Transform (DFT) of the y_{T_1}, y_{T_2} are

$$Y_{T_1}(k) = \sum_{n=0}^{N-1} y_{T_1}[n] e^{-j\frac{2\pi kn}{N}} \tag{1}$$

$$Y_{T_2}(k) = \sum_{n=0}^{N-1} y_{T_2}[n] e^{-j\frac{2\pi kn}{N}} \tag{2}$$

Then, the estimated frequency response of the fading channel is

$$\hat{H}_{FDLS}(k) = \frac{Y_{T_1}(k) + Y_{T_2}(k)}{2X_{T_1}(k)} \tag{3}$$

Where

$$X_{T_1}(k) = \sum_{n=0}^{N-1} x_{T_1}[n] e^{-j\frac{2\pi kn}{N}} \tag{4}$$

2.2 Time Domain Least Square (TDLS) Channel Estimation [4]

According to the specification of the IEEE 802.11p, there is a Guard Interval (GI) GI_2 in front of the x_{T_1}. Based on the observation in Fig. 2, if the wireless channel contains L-multipath, the received signal y_{T_1} can be expressed as

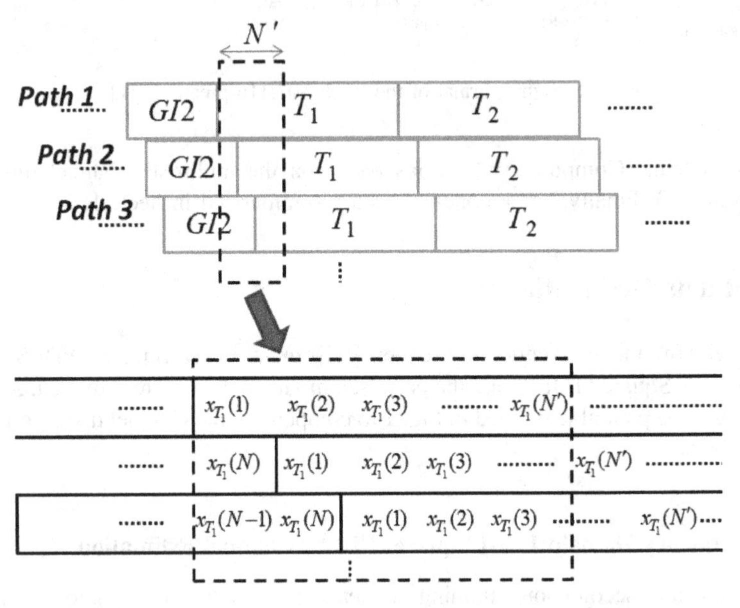

Fig. 2. Received signal through a multipath fading channel

$$y'_{T_1} = \begin{bmatrix} x_{T_1}(1) & x_{T_1}(N) & x_{T_1}(N-1) & \cdots & \cdots & x_{T_1}(N-(L+1)) \\ x_{T_1}(2) & x_{T_1}(1) & x_{T_1}(N) & & & x_{T_1}(N-(L+2)) \\ \vdots & x_{T_1}(2) & x_{T_1}(1) & & & x_{T_1}(N-(L+3)) \\ x_{T_1}(N'-2) & \vdots & x_{T_1}(2) & \ddots & & \vdots \\ x_{T_1}(N'-1) & x_{T_1}(N'-2) & \vdots & & \ddots & x_{T_1}(N'-(L-1)) \\ x_{T_1}(N') & x_{T_1}(N'-1) & x_{T_1}(N'-2) & \cdots & \cdots & x_{T_1}(N'-L) \end{bmatrix} \begin{bmatrix} h_1 \\ h_2 \\ \vdots \\ \vdots \\ \vdots \\ h_L \end{bmatrix} + \begin{bmatrix} w_1 \\ w_2 \\ \vdots \\ \vdots \\ \vdots \\ w_L \end{bmatrix} = M_{T_1}\mathbf{h} + \mathbf{w} \tag{5}$$

Where w_i is the Additive White Gaussian Noise (AWGN). The delay profile can be estimated as

$$\hat{h}_{T_1,TDLS} = (M_{T_1}^H M_{T_1})^{-1} M_{T_1}^H y'_{T_1} \tag{6}$$

Besides, after the N-N' zero padding, i.e., $\hat{h}'_{T_1,TDLS} = [\hat{h}^T_{T_1,TDLS} \quad 0_{N-N'}]$, the estimated frequency response of the channel can be

$$\hat{H}_{TDLS}(k) = \sum_{n=0}^{N-1} \hat{h}'_{T_1}[n] e^{-j\frac{2\pi kn}{N}} \tag{7}$$

2.3 The Proposed Modified Time Domain Least Square (MTDLS) Channel Estimation

In the TDLS, the define matrix M_{T_1} is a $N' \times N'$ matrix. Basically, the purpose of the calculation of the delay profile in Eq. (6) is simply trying to compensate the distortion due to the multipath. Therefore, if the Receiver (Rx) can have the information of channel length L, it will be more efficient and accurate to construct the matrix M_{T_1} for the channel estimation. Therefore, the modified TDLS method is illustrated as follows, First of all, a $N' \times L'$ matrix M'_{T_1} with L' range from 1 to N' is constructed to estimate the channel delay profile and frequency response as Eqs. (6) and (7).

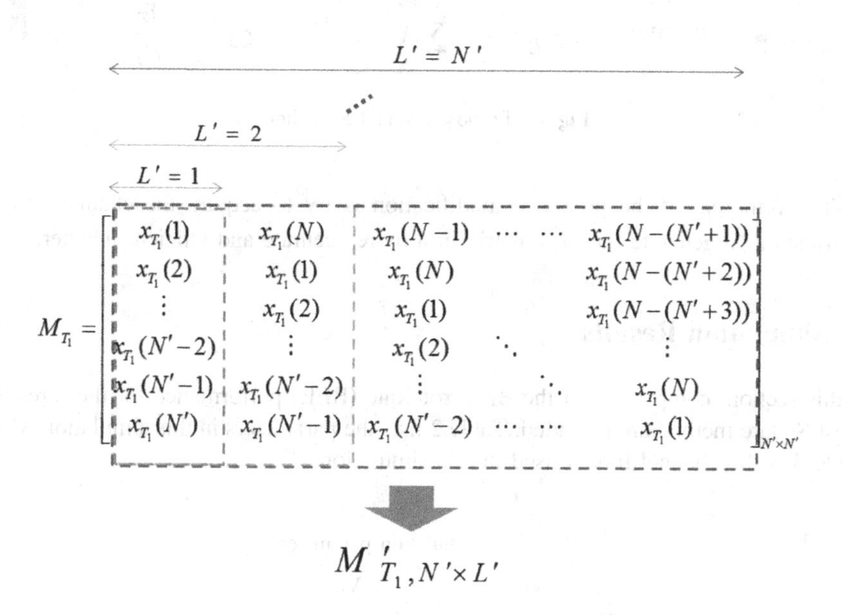

Fig. 3. Generation of the M'_{T_1}

Since the transmitted training sequence is known at both the Tx and Rx, Rx can predict the transmitted preamble by using Eq. (8).

$$\hat{X}_{T_1} = \frac{Y_{T_1}}{\hat{H}_{T_1}} \tag{8}$$

$$\hat{L} = \arg\min_{L'}\left(\frac{1}{N}\sum |X_{T_1} - \hat{X}_{T_1}|\right) \tag{9}$$

Then, the channel length information can be acquired via Eq. (9). After the channel length prediction, the delay profile and the channel frequency response can be re-calculated with the appropriate $N' \times \hat{L}$ matrix M'_{T_1}. The whole procedure can be illustrated by Fig. 4.

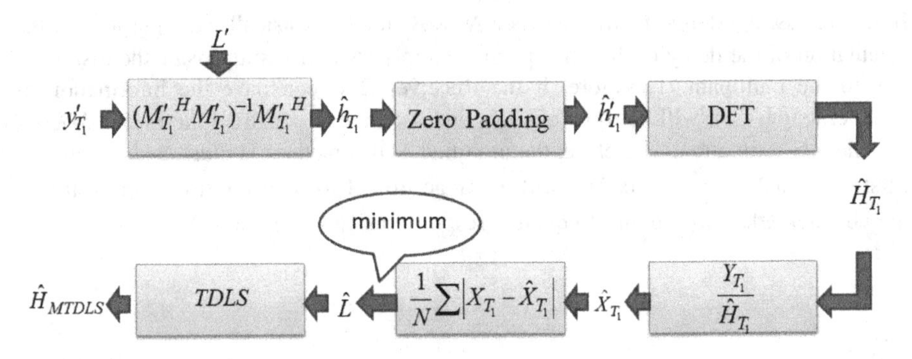

Fig. 4. Proposed MTDLS method

The principle of the proposed modification is try to acquire the channel length information to generate the M'_{T_1} matrix in a more accurate and efficient manner.

3 Simulation Results

In this section, comparison of the Bit Error Rate (BER) performance for these related Least Square method are provided. Table 2 lists the parameters in this simulation while Table 3 is the channel model used in this simulation.

Table 2. Simulation parameters

Parameters	Value
Number of multipath (L)	8
Modulation	QPSK
Number of Subcarrier(N)	52
Speed of the Mobile	300 km/hr
Guard Time	0.8 μs
Symbol duration	4 μs
FFT period	3.2 μs
Preamble duration	3.2 μs
Subcarrier spacing	0.3125 MHz

Table 3. Multipath fading channel used in the simulation.

Tap	Delay (ns)	Power (dB)
1	0	0, Rician, K = 3.3 dB
2	100	?9.3
3	200	?14.0
4	300	?18.0
5	400	?19.4
6	500	?24.9
7	600	?27.5
8	700	?29.8

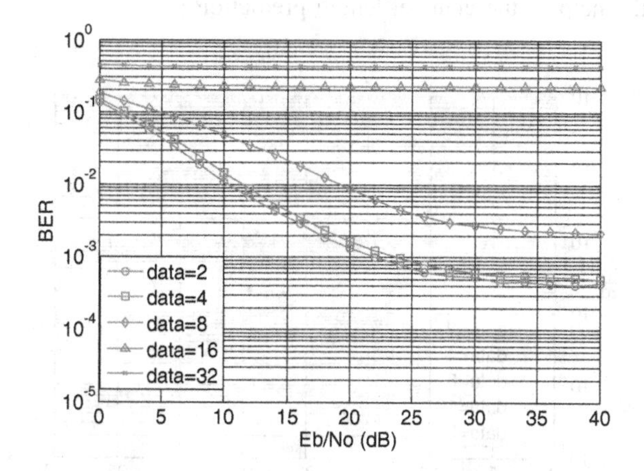

Fig. 5. BER performance of FDLS [3]

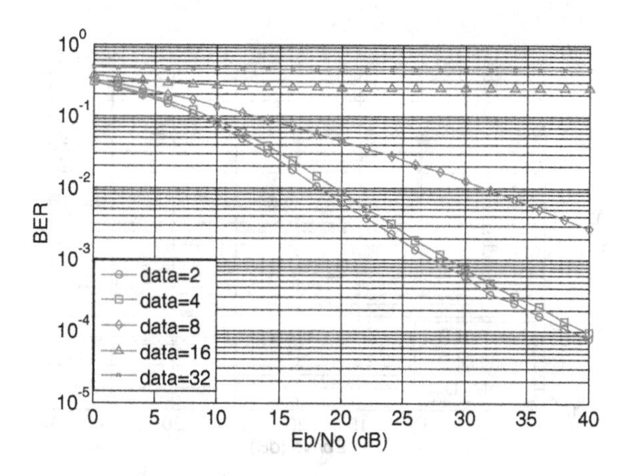

Fig. 6. BER performance of TDLS [4]

Figures 5, 6 and 7 are the BER performance of the TDLS, FDLS and the proposed MTDLS. Results show that for all these method, if the number of data between two successive preamble is greater than 8, their performance will be very poor. Figure 8 is the BER comparison if we re-plot for these method in the case of the number of data is four between two successive preamble. Results show that the FDLS and the proposed method have quit similar performance at the low-SNR region. However, the FDLS has its limitation of the BER improvement even the SNR is increased. This comes from the fact that the FDLS estimates the frequency response of the fading channel by an average sense via Eq. (3), if the speed of the mobile becomes high, the average value will become more and more inaccuracy than the actual response. Nevertheless, according to the simulation results, the performance of the TDLS and the proposed MTDLS can be improved if the SNR is increased. Besides, the proposed MTDLS method has the best performance at the high-SNR region compared with the other methods with the help of the channel length prediction.

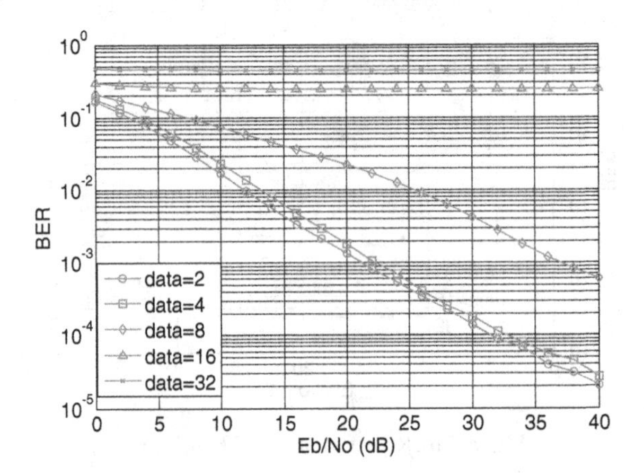

Fig. 7. BER performance of the proposed MTDLS

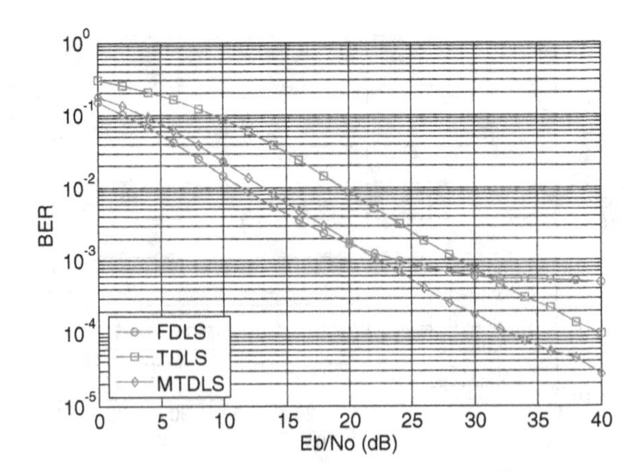

Fig. 8. BER comparison of the FDLS TDLS and the proposed MTDLS

4 Conclusions

In this paper, a modified TDLS channel estimation is proposed based on the IEEE802.11p vehicular communications. The proposed modification estimates the effective channel length before conducting the TDLS channel estimation. BER performance is evaluated at the high speed vehicle scenarios and results show that the proposed method can improve performance of the conventional LS channel estimation performance at both the time domain and frequency domain.

References

1. IEEE Standard for Information technology– Local and metropolitan area networks– Specific requirements– Part 11: Wireless LAN Medium Access Control (MAC) and Physical Layer (PHY) Specifications Amendment 6: Wireless Access in Vehicular Environments, "IEEE Std 802.11p-2010 (Amendment to IEEE Std 802.11-2007 as amended by IEEE Std 802.11 k-2008, IEEE Std 802.11r-2008, IEEE Std 802.11y-2008, IEEE Std 802.11n-2009, and IEEE Std 802.11w-2009), pp. 1, 51, July 15 2010
2. Supplement to IEEE Standard for Information Technology - Telecommunications and Information Exchange Between Systems - Local and Metropolitan Area Networks - Specific Requirements. Part 11: Wireless LAN Medium Access Control (MAC) and Physical Layer (PHY) Specifications: High-Speed Physical Layer in the 5 GHz Band," IEEE Std 802.11a-1999 (1999)
3. Fernandez, J.A., Stancil, D.D., Bai, F.: Dynamic channel equalization for IEEE 802.11p waveforms in the vehicle-to-vehicle channel. In: 2010 48th Annual Conf. on Communication, Control, and Computing, Allerton, pp. 542–551, September 29 2010–October 1 2010
4. Lin, C.-S., Lin, J.-C.: Improved time-domain channel estimation techniques. In: IEEE 802.11p environments. In: 2010 7th International Sym. on Communication Systems Networks and Digital Signal Processing (CSNDSP), pp. 437, 442, 21–23
5. Syafei, W.A., Nishijo, K., Nagao, Y., Kurosaki, M., Ochi, H.: Adaptive channel estimation using cyclic prefix for single carrier wireless system with FDE. In: 10th International Conference on Advanced Communication Technology, vol. 2, pp. 1032, 1035, 17–20, Febuary 2008

On the Performance Assessment of the OLSR Protocol in Mobile Ad hoc Networks Using TCP and UDP

Mushtaq Ahmad[1](✉), Zahid Khan[1], Qingchun Chen[1], and Muhammad Najam-ul-Islam[2]

[1] School of Information Science and Technology, Southwest Jiaotong University, Chengdu, Sichuan, People's Republic of China
mushtaq.ahmad91@gmail.com, Zahid.icp@gmail.com, qcchen@swjtu.edu.cn

[2] Department of Electrical Engineering, Bahria University Islamabad, Islamabad, Pakistan
najam@bahria.edu.pk

Abstract. In mobile Ad-hoc networks (MANETs) and Vehicle Ad-hoc networks (VANETs) various reactive and proactive protocols may be used for the transmission of packets from one node to another. Amongst all routing protocols, Optimized Link State Routing (OLSR) protocol is the most prominent and widely used protocol for MANETs. This paper presents the performance evaluation of OLSR protocol for TCP and UDP traffic patterns by (This work was partly supported by the NSFC under grant No. 61271246.) varying parameters like node density, node speed and pause time. Under different network scenarios, the performance of OLSR has been assessed in terms of the most widely used performance metrics, such as packet loss, end-to-end delay and throughput. The results prove that TCP performs considerably well in terms of throughput, end to end delay and packet loss in different node density and mobility scenarios, while considering pause time UDP turns out to be a better choice.

Keywords: OLSR · TCP · UDP · Performance assessment

1 Introduction

Mobile Ad-hoc Networks (MANETs) have attracted much research interest. MANETs have found use in different applications, such as: Members of a rescue team communicate with each other in a disaster recovery sight using mobile devices and Impromptu communications among groups of people like secure business discussions, government secret discussions in the parliament house. Areas where no existing communication infrastructure is available like battlefield is also suitable for such type of communication [1]. In MANETs, the nodes can be connected dynamically in a random fashion, and the infrastructure of the network is not stationary in such networks [2]. Each node works as a router and participates in maintenance and discovery of paths to the other nodes in the network. Generally MANETs can be visualized as a set of mobile nodes which are capable of communicating and networking. These nodes don't need any intervention or an infrastructure to establish and maintain a network. As nodes move around, they make and break links between them. Compared to a wired network, the connections between nodes in MANET break more frequently and dynamically. Consequently, the network

C.-H. Hsu et al. (Eds.): IOV 2015, LNCS 9502, pp. 294–306, 2015.
DOI: 10.1007/978-3-319-27293-1_26

topology changes with time. The routing protocols considered for wired networks (for instance, the Internet) generally make use of either link state or distance vector routing protocols. In distance vector routing, each router broadcasts its complete routing table information to each neighbor router at predetermined intervals. In link state routing protocols, updates arise when any change occurs in the network topology [3]. Due to the repeated changes of network topology and link status, these routing protocols will have high data and routing overhead to update and keep the route information between any couple of nodes in the MANET. So that's why wired networks routing protocols are quite wasteful if employed in MANETs.

The MANETs routing protocol designs are complicated in nature and we have to consider different design issues like dynamic topology, bandwidth constraint, error prone broadcast channel, hidden and exposed terminal problems, resource limitations, Quality of Services (QoS) limitations, security and so on [4]. An ample amount of effort has been made to tackle the issues associated with the routing protocols of the MANETs. As a consequence, different types of routing protocols have been developed, which can be briefly classified into position-based routing protocols and topology-based routing protocols [5]. The topology based routing protocols use the concept of conventional routing, such as distributing link state information or maintaining a routing table, while geographical routing or Position-based routing protocol relies on geographical position of the mobile stations to route the desired data packets to the destination. The topology based routing protocols can be further subdivided into two groups, i.e., Proactive (Table driven) and Reactive (On-demand) protocols [6]. As the name indicates, Proactive protocols try to maintain fresh and regular information within the system according to their destinations and routes. In Proactive protocols, the route must be established before the data transmission, the source node has predefined route to the destination node. In Reactive protocols, the nodes don't need to search for a predefined route because they establish route on demand. In the nutshell, we can also refer it to as the on-demand routing protocols.

In MANETs, it's very important to find whether TCP or UDP performs well in terms of the QoS metrics over various scenarios by using standardized parameters. TCP and UDP are well known traffics for the underlying routing protocols, and the specific choice corresponds to different overall performance requirement. For instance, in a critical situation where packet loss cannot be compromised, TCP should be employed. On the other hand, UDP find applications in real time communication scenarios. Previous research assessed TCP and UDP over different routing protocols with various mobility models. The primary focus of this paper is the comparative analysis of TCP and UDP over Optimized link state routing (OLSR) protocol under different involved factors, such as the mobility, varying pause time, node speed, and node density. Our work represents an effort to evaluate the performance of OLSR in terms of different QoS metrics over different environments. It is known that, the OLSR protocol is widely employed in wireless environment [7]. In static networks, the OLSR protocol outperforms the reactive routing protocols (e.g., the AODV, DSR), which send out additional messages at each discovery of path [8], in particular for short communications where the communication paths are accessible instantly [8]. The random waypoint mobility model is particularly suitable for TCP and UDP as its performance affected by the basic initialization of other mobility models, such as Reference Point Group mobility models (RPGM), Freeway model, Manhattan Model (MM) [9]. In this paper, Random Waypoint model is assumed since we want to assess the performance that is independent of temporal dependency, spatial dependency and geographic dependency.

In short our main input is as follows: For the first time, we evaluate the performance of OLSR for transport layer Protocols in NS-2. We examine the performance measuring in OLSR with node density, node speed and pause time, which are observed in actual traces. Our research may give several novel approaches for the design of node density, node speed and pause time based mobile infrastructure-less networks. The simulation results corroborate that in highly congested environment, OLSR outperforms in TCP connection, while UDP is slightly perform well in case of high pause time, where the loss rate is decreased.

The remainder of this paper is organized as follows: Sect. 2 provides an overview of the OLSR protocol and the related analysis. Section 3 will illustrate the simulation environments and simulation setup. The simulation results are presented in Sect. 4. Some discussion and concluding remarks are presented in Sect. 5 to summarize the paper.

2 Optimized Link State Routing Protocol (OLSR) and the Related Work

Optimized Link State Routing Protocol (OLSR) is a link state proactive/table-driven protocol developed by INRIA [7], which employs two control messages (HELLO and TC) to find out and then distribute link state information throughout the MANET. OLSR is a modified version of Link State Routing (LSR) protocol designed for MANETs and VANETs. In fact, OLSR is an optimized variant of LSR by using special nodes called MPRs (Multi-Point Relays). The neighboring nodes select their MPRs. In LSR protocol, every node will advertise its links while in OLSR only MPRs advertise their links. Secondly, each node forwards messages for its neighbors in LSR, while in OLSR only MPRs forward messages to their neighbor nodes, which pick them as their MPRs. In OLSR, the routing selection is localized. In the construction of routing table, OLSR keeps MPRs as the last hop to destination.

It is suitable to use OLSR in densely-populated mobile networks, as the optimization obtained in MPRs goes well in this context, because MPRs minimize the broadcast of HELLO messages among the nodes. The degree of optimization achieved is proportional to the size and density of the network. Thus it is more efficient when compared to the classic link state algorithm for large dense networks. OLSR makes use of hop-by-hop routing, i.e., each node routes packets using its local information. OLSR is suitable for networks, where the traffic is haphazard and irregular between a large number of nodes rather than between a small numbers of specific nodes. Being a proactive protocol, OLSR is also suitable for networks, where the communicating pairs change with the passage of time. As routes are retained for all known nodes at all times, and no extra control traffic is produced in this state.

To assess the performance of OLSR, a separate analysis is mandatory for the UDP and TCP transport protocols in the existence of several paths due to their heterogeneities. Due to the dynamic nature of OLSR, UDP streams are scarcely affected by unstable conditions of route. On the other hand, TCP is reasonably more sensitive to downlink. When the problem of down link packet losses occurs, which cause TCP to diminish the transmission window and progressively increase the Retransmission Time-Out (RTO) [10].

In literature, many performance evaluations have performed over different routing protocols under different scenarios [11–13]. They used different performance parameters to evaluate the performance of selected protocols. In our work, we not only evaluate OLSR under different scenarios, but also study the transport layer characteristics of OLSR as well. The inter-comparison of different protocols is also performed under different simulation environments in [14–17]. The authors in [18, 19] analyzed MANETs protocols through TCP connection and evaluate their effects on packet loss and delay. They comprehensively address all the flavors of TCP and check their reliabilities. Nonetheless, they did not deal with CBR or UDP effects; it is about the TCP variants in both static and mobile environment. The effect of node density and packet size has also been evaluated to assess the performance of reactive and proactive protocols in [20–22].

In nutshell, the performance of OLSR has been extensively evaluated using different parameters. However, a comprehensive schema is missing. In this paper, we aim at assessing the performance parameters like throughput, delay, packet loss, PDF for both TCP and UDP connections over a variety of simulation environments. And the related analysis may be useful to reveal the robustness and optimality of the OLSR protocol in various environments.

3 Simulation Environments

3.1 Simulation Setup

The Simulations of OLSR traffic has been carried out by using NS-2.34. NS-2 is one of the prominent simulators used to simulate MANETs routing protocols in different scenarios [23]. Random waypoint model is adopted in simulation for the movement of nodes. The simulation parameters are given in Table 1. The parameters used are sufficient to measure the performance of a protocol.

Our simulations are applied over a variety of scenarios implemented in Setdest tool [24]. The varying parameters in each of the scenario are node density, node speed, pause time. To analyze the impact of each of the parameters, one of the parameters is varied over a wide range while the other two are kept constant in each scenario. The performance metrics in our analysis consists of Throughput, End to End delay, and Packet loss, which are measured for every scenario described as below.

Table 1. Simulation settings

Parameters	Values
Simulator	NS-2.34
Mobility Model	Random Waypoint Model
Nodes	5,10,15,20,25,30,35,40
Protocol	OLSR
Simulation Time	500 s
Simulation Area	500×500 m^2
Traffic Type	UDP,TCP
Transmission Time	500 s
Data Payload	0.01 Mbps
No. Of Connections	8 connections

3.2 Scenarios

- **Node Density**
  TCP and UDP connections are evaluated using different number of nodes. For the analysis of node density, the number of nodes used is 10, 15, 20, 25, 30, 35 and 40. The positions of all these nodes have been selected through Setdest tools. To achieve precision and high confidence factor, the average values of 5 simulations for each parameter in each scenario are used in the analysis. In case of node density, node speed, and pause time are fixed to be 20 m/s and 4 s, respectively.
- **Node Speed**
  OLSR performance is evaluated under different mobility conditions. In our simulation, the node speeds are 10, 20, 30, 40, and 50 m/s. 20 nodes and 2 s pause time are assumed in case of the node speed. Each scenario is generated 5 times and its average value is considered for analysis.
- **Pause Time**
  OLSR traffic is analyzed with different pause times. Various scenarios are generated with different pause time of 0, 50, 100, 150, 250, 300, 350, and 400 ms. 20 m/s node speed and 20 nodes are assumed in case of the pause time.

3.3 Performance Parameters

The performance of OLSR is analyzed by different parameters. Our selected parameters are the most well known metrics used by many researchers. The selected parameters clearly describe the performance of networks and are the best candidates for comparing the OLSR in TCP and UDP connections. Our adopted parameters are given below.

- **Throughput**
  Throughput is the source to measure the threshold of a network. It refers to packets received by a node in unit time.
- **End to End Delay**
  End to end delay refers to the time span a packet takes to reach the receiver moving from the sender.
- **Packet Loss**
  Packet loss refers to the drop packets on receiving side, which are generated by TCP or UDP traffic agent.

4 Simulation Results and Evaluation

All the parameters are measured in a fixed simulation environment using two different traffic sources; TCP and UDP traffic sources. The OLSR performance is analyzed under the above mentioned traffic sources, but each traffic source is further analyzed with different traffic scenarios e.g. Node Density, Node Speed, Pause time. Each of the scenarios is evaluated using different parameters, which are Throughput, Delay and Packet loss.

In our simulation, we configure OLSR TCL (Tool Command Language) files for both TCP and CBR traffic sources. Each TCL file is further evaluated for different traffic scenarios (Node density scenario, Speed scenarios, Pause time scenarios). All the results and their comparative discussions are given in the following paragraphs.

4.1 OLSR Traffic Analysis with Respect to Node Density

Node density refers to the various numbers of nodes passing through TCP and UDP traffic sources. In our case, different number of nodes in both TCP and UDP traffic sources are evaluated for throughput, delay and packet loss as given below.

4.1.1 OLSR Throughput Analysis in Case of Node Density

OLSR protocol normally works very well in denser environments, which means that node density has a very slight impact on throughput both in TCP and UDP traffic sources as shown in Fig. 1. The line graph in Fig. 1 depicts that, node density, has almost no impact on Throughput in both TCP and UDP connections. The throughput remains relatively static from node size 15 to 40. Although TCP's throughput is more than UDP traffic source owing to the fact that TCP default window size for the given data rate. In short in denser environment having data rate varying from 100 Kbps to 1 Mbps, we should adopt TCP traffic source to maintain higher throughput.

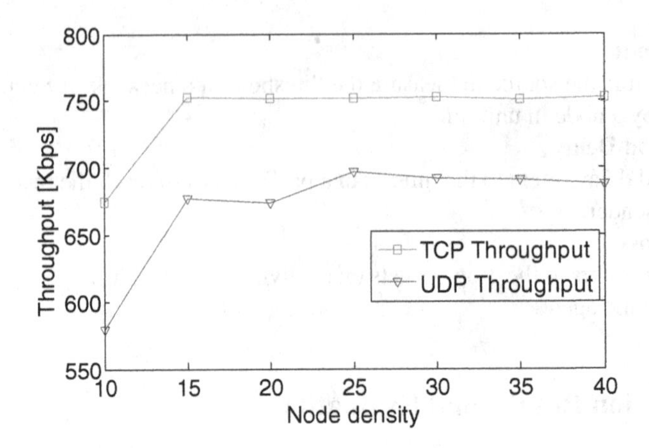

Fig. 1. Throughput comparison of OLSR over TCP & UDP connection in case of node density

4.1.2 OLSR Delay Analysis in Case of Node Density

Transport layer of OSI model dictates that UDP should be used in real time applications where delay is negligible [25]. In our experiment UDP delay is too low with increasing node density, while on the other hand TCP delay is higher throughout the simulation as shown in below Fig. 2. In case of TCP traffic source, the node density has a very good impact on delay. The delay drastically decreases with increasing number of nodes. UDP delay is almost static during the simulation. The statistics of the given graph show that in highly denser environment TCP connection acts like UDP in the context of delay.

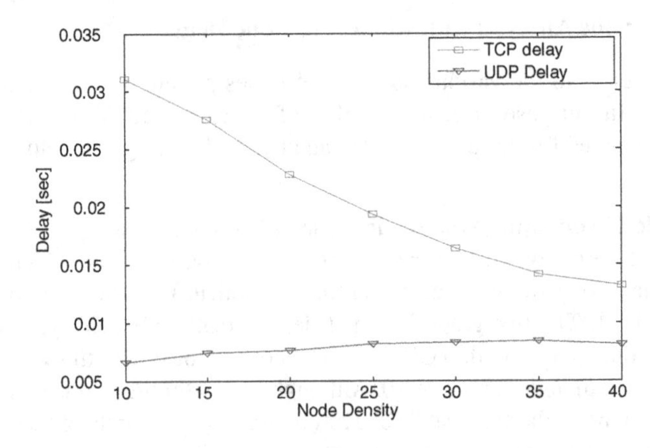

Fig. 2. Delay comparison of OLSR over TCP & UDP connection in case of node density

4.1.3 OLSR Packet Loss Analysis in Case of Node Density

UDP transmission compromises on packet loss as compared to packet's delay, that's why UDP traffic source can be adopted in live transmission where retransmission of packets has no meaning. The line graph of Fig. 3 reveals that UDP connection is unreliable as compared to TCP connection with respect to packet loss, because packet loss ratio of UDP connection is too much as compared to TCP, but on the other angle in case of UDP, we can analyze that node density has a good impact on packet loss. In Fig. 3, lost packets in case of 40 nodes are quite less than 10 nodes in UDP traffic source, because with very sparsely occupied environment the number of feasible paths between any two nodes is very less and hence the performance in term of packet loss is poor. Node Density has a very minor impact on packet loss in case of TCP connection.

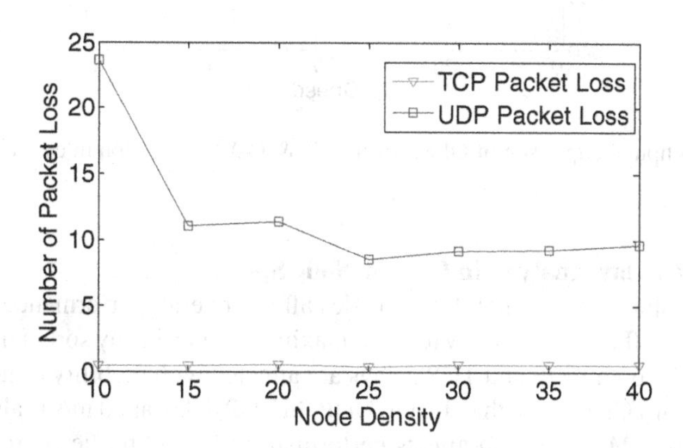

Fig. 3. Packet loss comparison of OLSR over TCP & UDP connection in case of node density

4.2 OLSR Traffic Analysis with Respect to Node Speed

As MANETs are concerned with mobility, hence node speed is an important metric to be evaluated. OLSR performance is analyzed with varying speeds in both TCP and UDP connections. The same QoS metrics are evaluated in case of node speed as that of node density.

4.2.1 OLSR Throughput Analysis in Case of Node Speed

The node speed greatly impacts the performance of wireless communication. The simulation results show that as mobility increases, the overall performance of MANETs affects badly, because when mobility of node increase the connection failures occur more quickly, ensuing in packet losses, frequent convergent and frequent path discoveries. Therefore, spontaneously, UDP throughput should continually decreases as the mobility is increased. As in TCP a virtual connection is established during transmission, hence its throughput does not affect as far as its connection maintains either node speed increase or decrease. Figure 4 shows that, in TCP connection throughput doesn't decrease with mobility of nodes. On the other hand UDP performance degrades with mobility, greater the mobility lower will be the throughput. In short in our simulation setup, TCP works well in dynamic topologies as compared to UDP connections.

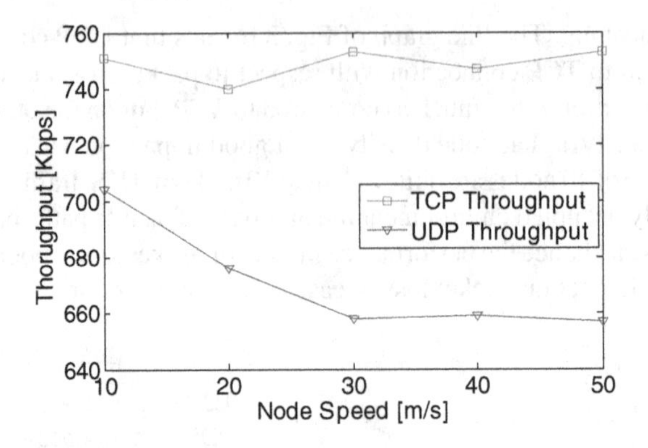

Fig. 4. Throughput comparison of OLSR over TCP & UDP connection in case of node speed

4.2.2 OLSR Delay Analysis in Case of Node Speed

In wireless communication, mobility of nodes affects overall performance of network. Normally TCP traffic source mostly tends to maximum delay in any sort of transmission. In case of OLSR both UDP and TCP delays are affected with mobility of nodes but the rate of change in TCP is more than that in UDP. In TCP delay at 50 m/s is almost double of that at 10 m/s. Mobility also affects performance of UDP traffic source but not as much as that of TCP as shown in Fig. 5.

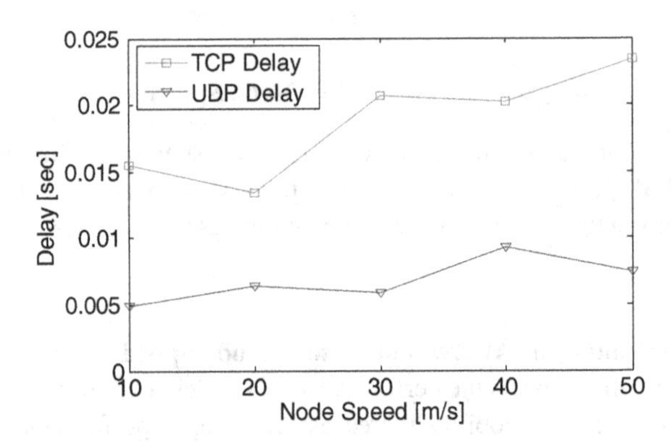

Fig. 5. Delay comparison of OLSR over TCP & UDP connection in case of node speed

4.2.3 OLSR Packet Loss Analysis in Case of Node Speed

Packet loss is one of the highly affected QoS metric regarding node speed. In Fig. 6, UDP performance degrades mostly with respect to speed as compared to TCP. In case of UDP traffic source packet loss raises to double by varying speed from 10 to 50 m/s. In short, in a highly dynamic environment we should use TCP traffic source.

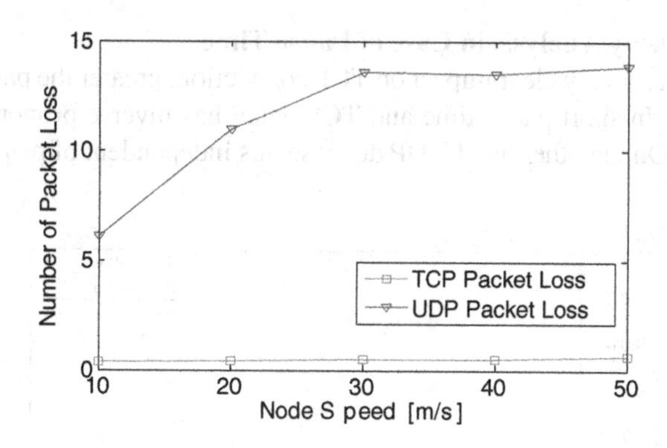

Fig. 6. Packet loss comparison of OLSR over TCP & UDP connection in case of node speed

4.3 OLSR Traffic Analysis with Respect to Pause Time

Pause time is also an important factor to be measured as performance metric in case of OLSR protocol. This section shows how throughput, delay, and packet loss behave with varying pause time.

4.3.1 OLSR Throughput Analysis in Case of Pause Time

TCP and UDP throughput comparison with respect to varying pause time is shown in Fig. 7. The figure depicts that for higher pause time throughput of UDP connection crosses TCP throughput. Hence we can say that in our given simulation environment, if the node's pause time exceeds 250 ms, then we have to use UDP connection for getting high throughput. The line graph shows that Pause time has no impact on TCP connection throughput.

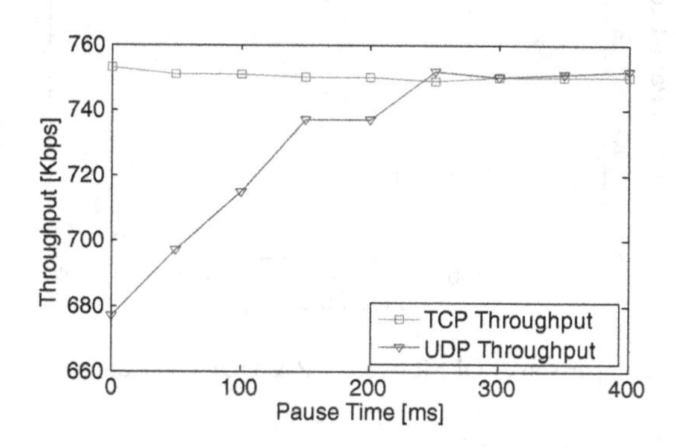

Fig. 7. Throughput comparison of OLSR over TCP & UDP connection in case of pause time

4.3.2 OLSR Delay Analysis in Case of Pause Time

The pause time has a very clear impact on TCP connection; greater the pause time lower will be its delay. In short pause time and TCP delay has inverse proportional relation with each other. On the other hand UDP delay seems independent of pause time as give in below Fig. 8.

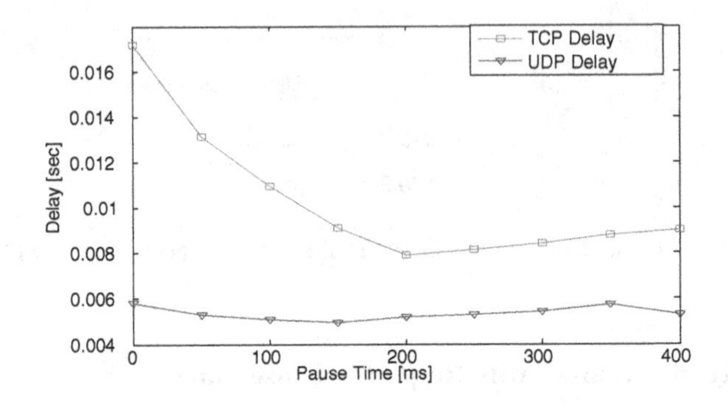

Fig. 8. Delay comparison of OLSR over TCP & UDP connection in case of pause time

4.3.3 OLSR Packet Loss Analysis in Case of Pause Time

In our given simulation setup pause time not only depends on UDP throughput but on packet loss as well. UDP packet loss and pause time has an inverse proportional relation with each other as shown in Fig. 9. TCP packet loss is independent of pause time as shown in above Fig. 9.

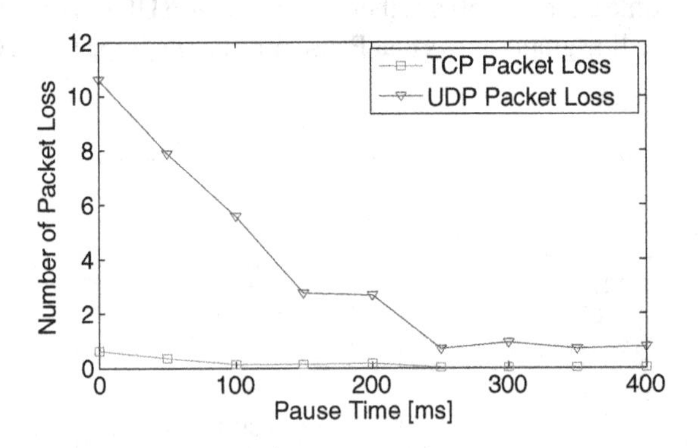

Fig. 9. Packet loss comparison of OLSR over TCP & UDP connection in case of pause Time

5 Conclusions

In this paper performance comparison of TCP and UDP has been presented over the OLSR routing protocol for MANETs. By using Network Simulator software, comparison in terms of packet loss, end to end delay and throughput has been conducted over different environments by varying parameters like node density, node speed and pause time. The simulation results show that TCP is much more suitable choice compared to UDP in highly dense networks except when real time transmissions are required. While considering the case of node mobility in MANETs, TCP performs better than UDP in terms of throughput and packet loss. Lastly, for higher pause time, UDP turns out to be a better choice due to its lower packet loss.

Acknowledgment. This work was partly supported by the NSFC under grant No. 61271246.

References

1. Guan, Q., Yu, F.R., Jiang, S., Leung, V.C.M.: Joint topology control and authentication design in mobile Ad Hoc networks with cooperative communications. IEEE Trans. Veh. Technol. **61**(6), 2674–2685 (2012)
2. Carofiglio, G., Chiasserini, C., Garetto, M., Leonardi, E.: Route stability in MANETs under the random direction mobility model. IEEE Trans. Mob. Comput. **8**(9), 1167–1179 (2009)
3. Wang, Z., Chen, Y., Li, C.: PSR: a light-weight proactive source routing protocol for mobile Ad Hoc networks. IEEE Trans. Veh. Technol. **PP**(99), 1 (2013)
4. Kilavuz, M.O., Soran, A., Yuksel, M.: Roadmap-based end-to-end traffic engineering for multi-hop wireless networks. In: 2014 IEEE 20th International Workshop on Local & Metropolitan Area Networks (LANMAN), pp.1–6, 21–23 May 2014
5. Condeiro, C.D.M., Agrawal, D.P.: Ad Hoc & Sensor Networks: Theory and Applications. World Scientific Publishing Co., Inc., River Edge (2006)
6. Altayeb, M., Mahgoub, I.: A survey of vehicular Ad hoc networks routing protocols. Int. J. Innov. Appl. Stud. **3**(3), 829–846 (2013)
7. Kaur, R., Rai, M.K.: A Novel review on routing protocols in MANETs. Undergraduate Acad. Res. J. (UARJ), ISSN, 1129–2278 (2012)
8. Clausen, T., Jacquet, P., Adjih, C., Laouiti, A., Minet, P., Muhlethaler, P., Qayyum, A., Viennot, L.: Optimized link state routing protocol (OLSR) (2003)
9. Saika, A., El Kouch, R., Najid, A., Himmi, M.M.: Implementation and performances of the protocols of routing AODV and OLSR in the Ad hoc networks. In: 2010 5th International Symposium on I/V Communications and Mobile Network (ISVC), pp.1–4, 30 September 2010–2 October 2010
10. Divecha, B., et al.: Impact of node mobility on MANET routing protocols models. J. Digit. Inf. Manage. **5**(1), 19–24 (2007)
11. Nácher, M., Calafate, C.T., Cano, J.-C., Manzoni, P.: Comparing TCP and UDP performance in manets using multipath enhanced versions of DSR and DYMO. In: Proceedings of the 4th ACM workshop on Performance Evaluation of Wireless Ad hoc, Sensor, and Ubiquitous Networks, pp. 39–45. ACM, Chania, October 2007
12. Kumari, S., et al.: Traffic pattern based performance comparison of AODV, DSDV & OLSR MANET routing protocols using freeway mobility model. Int. J. Comput. Sci. Inf. Technol. **2**(4), 1606–1611 (2011)
13. Tafti, V.A., Gandomi, A.: Performance of QoS parameters in MANET application traffics in large scale scenarios. World Acad. Sci. Eng. Technol. **72**, 857–860 (2011)
14. Vadhwani, D.N., Kulhare, D.: Traffic anslysis of DSR, AODV and OLSR using TCP and UDP. Int. J. Comput. Sci. Eng. (IJCSE) **5**(4), 221–224 (2013)

15. Nordstrom, E., et al.: A comprehensive comparison of MANET routing protocols in simulation, emulation and the real world. Uppsala University, pp. 1–12 (2006)
16. Kaur, R., Sharma, C.: Review paper on performance analysis of AODV, DSDV, OLSR on the basis of packet delivery. J. Comput. Eng. (IOSR-JCE) **11**(1), 51–55 (2013)
17. Sandhu, D.S., Sharma, S.: Performance evaluation of batman, DSR, OLSR routing protocols-A review. J. Inf. Oper. Manage. **3**(1), 225–227 (2012)
18. Baxla, S., Nema, R.: Performance analysis of AODV, OLSR, DSR and GRP routing protocol of Mobile Ad hoc Network- a Review. Int. J. Sci. Res. (IJSR) **2**(5), 275–278 (2013)
19. Oo, M.Z., Othman, M.: The effect of packet losses and delay on TCP traffic over wireless Ad Hoc networks. In: Theory and Applications of Ad Hoc Networks. INTECH Book Publication, January 2011
20. Seddik-Ghaleb, A.., Ghamri-Doudane, Y., Senouci, S.-M.: Effect of ad hoc routing protocols on TCP Performance within MANETs. In: 2006 3rd Annual IEEE Communications Society on Sensor and Ad Hoc Communications and Networks, SECON 2006, vol. 3, pp. 866-873, 28–28 September 2006
21. Tolani, M., Mishra, R.: Effect of packet size on various MANET routing protocols. Int. J. Appl. Inf. Syst. (IJAIS) **4**(9), 10–13 (2012)
22. Barolli, L., Ikeda, M., De Marco, G., Durresi, A., Xhafa, F.: Performance analysis of OLSR and BATMAN protocols considering link quality parameter. In: International Conference on Advanced Information Networking and Applications, AINA 2009, pp. 307–314, 26-29 May 2009
23. Johansson, P., Larsson, T., Hedman, N., Mielczarek, B., Degermark, M.: Scenario-based performance analysis of routing protocols for mobile Ad-hoc networks. In: Proceedings of the 5th Annual ACM/IEEE International Conference on Mobile Computing and Networking, pp. 195–206. ACM (1999)
24. The network simulator ns-2. www.isi.edu/nsnam/ns2
25. CMU Monarch Group. CMU monarch extensions to the NS-2 simulator. http://monarch.cs.cmu.edu/cmu-ns.htm
26. Ahn, G.S., Campbell, A.T., Veres, A., Sun, L.H.: Supporting service differentiation for real-time and best-effort traffic in stateless wireless Ad hoc networks (SWAN). IEEE Trans. Mob. Comput. **1**(3), 192–207 (2002)

Vehicle State Estimation Based on V2V System

Yong Pan[1], Ziqiang Tang[2](✉), Xianwu Gong[3], and Chao Tang[3]

[1] Key Laboratory of Intelligent Transportation Systems Technologies, Ministry of Communications, Beijing 100088, China

[2] School of Automobile, Chang'an University, Xi'an 710064, Shaanxi, China
tangzqa@126.com

[3] School of Electronic and Control Engineering, Chang'an University, Xi'an 710064, Shaanxi, China

Abstract. To improve the vehicle state estimation accuracy under V2V system, the vehicle state estimation method which uses the multiple information fusion is presented. Through the establishment of the automobile kinematics mode and state equation of vehicle state parameter estimation, the method uses extended kalman filter method to realize the accurately estimation of vehicle real-time state parameters. Based on Matlab/Simulink the vehicle state estimation simulation platform of constant speed circular motion is built, The simulation results show that this method can achieve the accurate state estimation of vehicle.

Keywords: V2V system · Vehicle state estimation · EKF · Vehicle kinematical

1 Introduction

Vehicle active safety technology and the development of intelligent vehicle provide an effective way to solve problem of reducing accident rates [1]. Based GPS literature [2] proposes the collision warning system coordinated under vehicle to vehicle (V2V). Literature [3] establishes collision avoidance support system based on communication of vehicle to vehicle. Literature [4] proposes the collision prediction algorithm based V2V. Above the study of vehicle collision warning safety system under V2V, accurate identification and state estimation of vehicle are the basis of collision warning safety system under V2V, which have significant impact on reliability of vehicle collision warning safety system.

Depending on the difference of the vehicle model, the current vehicle driving state estimation method can be divided into the kinematics estimation method and the dynamics estimation method [5]. For the dynamics estimation method, the tire dynamic model and the number of sensors have an important influence on the estimated accuracy and robustness [6]. For the kinematics estimation method, this method has better robustness and does not change with the change of vehicle dynamics state, but the flaw of the method is heavily dependent on the measurement accuracy of sensors, and it needs to timely eliminate the impact of cumulative integration errors, but through multi-information fusion method,

C.-H. Hsu et al. (Eds.): IOV 2015, LNCS 9502, pp. 307–314, 2015.
DOI: 10.1007/978-3-319-27293-1_27

the flaw can effectively be solved. Through the establishment of vehicle kinematics model, combined with vehicle sensors and on-board GPS, this paper uses the extended kalman filter method to accurately estimate real-time vehicle state under V2V. Based on Matlab/Simulink, the vehicle state estimation simulation platform of constant speed circular motion is built, the estimated situation of vehicle state parameters is simulated and analyzed.

2 Vehicle Kinematic Model

The position information which GPS receives is usually WGS-84 coordinates of the World Geodetic, the plane coordinates of a separate area coordinate is needed in the real navigation system. The common method is to get the latitude and longitude coordinates of the GPS receiver to WGS-84 reference ellipsoid as a benchmark for Gauss projection, then by plane coordinates conversion (such as similarity transformation affine transformation, complete quadratic polynomial transformation, etc.), the plane coordinate after gauss projection is unified to the national 54 countries coordinate, 80 coordinate system or any local coordinate system. The vehicle mounted GPS includes GPS coordinate, the vehicle coordinate system and the world coordinate system, because of the relationship between the vehicle coordinate system and the GPS coordinate system has been determined, when the GPS antenna is installed, GPS coordinate system can be unified into the vehicle coordinate system. Therefore, as shown in Fig. 1 the coordinate system can be simplified into world coordinate system and the vehicle coordinate system.

After ignoring vehicles roll, pitch and vertical movement, the vehicle state may be represented by the vehicle position, velocity, acceleration, heading angle and turned angle. Establishment of accurate vehicle kinematic model is the key factor to determine the vehicle trajectory(the position and heading angle). The vehicle position can not simply be modeled using the midpoint of the rear axle, the front axle or other reference point should also be taken to be modeled [7,8]. For the full description of vehicle position and orientation, this paper selects the rear axle midpoint $M_r(x_r, y_r)$ to be the origin of vehicle coordinate system, $M_f(x_f, y_f)$ to be the front axle midpoint coordinates. In the Fig. 1 the relationship between the front and rear axle midpoint coordinates is shown in Eq. (1).

$$x_f = x_r + l cos\theta \tag{1}$$

$$y_f = y_r + l sin\theta \tag{2}$$

In Eqs. (1) and (2), θ is the vehicle heading angle (the angle between the vehicle coordinate system and the world coordinate system); l is the vehicle wheelbase. So at t_i the vehicle state can be represented as the vector of X_{r-i} and X_{f-i}.

$$X_{r-i} = (x_{r-i}, y_{r-i}, v_i, a_i, \theta_i, \varphi_i) \tag{3}$$

$$X_{f-i} = (x_{r-i}, y_{r-i}) \tag{4}$$

Vehicle state from t_i to t_{i+1} can be expressed as Eqs. (5)–(10).

$$x_{r-i+1} = x_{r-i} + v_i(t_{i+1} - t_i)cos\theta_i + \frac{a_i cos\theta_i(t_{i+1} - t_i)^2}{2} \tag{5}$$

$$y_{r-i+1} = y_{r-i} + v_i(t_{i+1} - t_i)sin\theta_i + \frac{a_i sin\theta_i(t_{i+1} - t_i)^2}{2} \tag{6}$$

$$x_{f-i+1} = x_{f-i} + \frac{v_i(t_{i+1} - t_i)cos\theta_i}{cos\varphi_i} + \frac{a_i cos\theta_i(t_{i+1} - t_i)^2}{2cos\varphi_i} \tag{7}$$

$$y_{f-i+1} = y_{f-i} + \frac{v_i(t_{i+1} - t_i)sin\theta_i}{cos\varphi_i} + \frac{a_i sin\theta_i(t_{i+1} - t_i)^2}{2cos\varphi_i} \tag{8}$$

$$v_{i+1} = v_i + a_i(t_{i+1} - t_i) \tag{9}$$

$$\theta_{i+1} = \theta_i + \frac{v_i tan\varphi_i(t_{i+1} - t_i)}{l} \tag{10}$$

In Eqs. (4)–(10), x_{r-i} is the x-axis coordinate position of the rear axle midpoint; y_{r-i} is the y-axis coordinate position of the rear axle midpoint; v_i is longitudinal speed of the rear axle midpoint; θ_i is the heading angle of vehicle; x_{f-i} is the x-axis coordinate position of the front axle midpoint; y_{f-i} is the y-axis coordinate position of the front axle midpoint; $\varphi_i$is front wheel steering angle.

3 Method of Vehicle's State Estimation Based on EKF

Although the vehicle state can be estimated by Eq. (4), but this calculation method has time drift problem and be susceptible to on-board sensors. In order to obtain the more accurate vehicle state, this paper uses multi-information fusion method to accurately estimate vehicle real-time state. This paper uses extended kalman filter(EKF) to estimate vehicle real-time state. The Fig. 2 shows the flowchart of estimation. The specific estimation process is as follows:

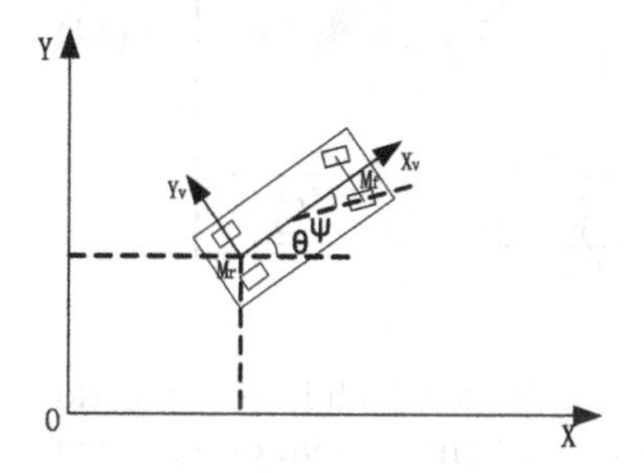

Fig. 1. Vehicle kinematic model

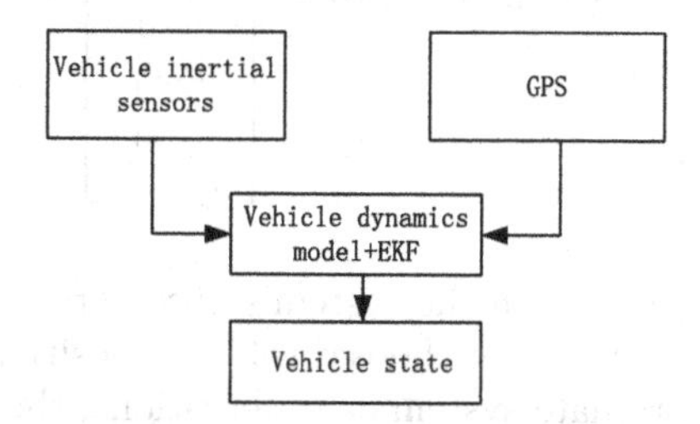

Fig. 2. The vehicle state estimation process based on EKF

(1) Establish the system's state equation

The vehicle state equation which is established includes eight state variables in this paper, they are respectively x-axis coordinate position of vehicle rear axle midpoint (x_r), the y-axis coordinate position of the rear axle midpoint (y_r), longitudinal speed of the rear axle midpoint (v), the heading angle of vehicle (θ), front wheel steering angle (φ),the x-axis coordinate position of the front axle midpoint (x_f) and the y-axis coordinate position of the front axle midpoint (y_f). when $\Delta t = t_{i+1} - t_i$ is comparatively small.

$$X_{i+1} = f(X_i\ U_i\ w_i) \tag{11}$$

$$X_{i+1} = \begin{bmatrix} x_{r-i+1} \\ y_{r-i+1} \\ v_{i+1} \\ a_{i+1} \\ \theta_{i+1} \\ \varphi_{i+1} \\ x_{f-i+1} \\ y_{f-i+1} \end{bmatrix} = \begin{bmatrix} x_{r-i} + v_i \Delta t cos\theta_i + \frac{(a_i+w_1)cos\theta_i(\Delta t)^2}{2} \\ y_{r-i} + v_i \Delta t sin\theta_i + \frac{(a_i+w_1)sin\theta_i(\Delta t)^2}{2} \\ v_i + (a_i + w_1)\Delta t \\ a_i \\ \theta_i + \frac{v_i tan\varphi_i}{l} + w_2 \\ \varphi_i \\ x_{f-i} + \frac{v_i \Delta t cos\theta_i}{cos\varphi_i} + \frac{(a_i+w_1)cos\theta_i(\Delta t)^2}{2cos\varphi_i} \\ y_{f-i} + \frac{v_i \Delta t sin\theta_i}{cos\varphi_i} + \frac{(a_i+w_1)sin\theta_i(\Delta t)^2}{2cos\varphi_i} \end{bmatrix} \tag{12}$$

In Eq. (12), w_1 and w_2 are input random noise vectors of the system state, and are the zero mean Gaussian white noise. Covariance Q of the system is shown in Eq. (13).

$$Q = \begin{bmatrix} w_1^2 & 0 \\ 0 & w_2^2 \end{bmatrix} \tag{13}$$

Transcendental state estimation of the system is shown in Eq. (14).

$$X_{i+1}^- = f(X_i\ U_i\ 0) = \begin{bmatrix} x_{r-i+1}^- \\ y_{r-i+1}^- \\ v_{i+1}^- \\ a_{i+1}^- \\ \theta_{i+1}^- \\ \varphi_{i+1}^- \\ x_{f-i+1}^- \\ y_{f-i+1}^- \end{bmatrix} = \begin{bmatrix} x_{r-i} + v_i \Delta t cos\theta_i + \frac{a_i cos\theta_i(\Delta t)^2}{2} \\ y_{r-i} + v_i \Delta t sin\theta_i + \frac{a_i sin\theta_i(\Delta t)^2}{2} \\ v_i + a_i \Delta t \\ a_i \\ \theta_i + \frac{v_i tan\varphi_i}{l} \\ \varphi_i \\ x_{f-i} + \frac{v_i \Delta t cos\theta_i}{cos\varphi_i} + \frac{a_i cos\theta_i(\Delta t)^2}{2cos\varphi_i} \\ y_{f-i} + \frac{v_i \Delta t sin\theta_i}{cos\varphi_i} + \frac{a_i sin\theta_i(\Delta t)^2}{2cos\varphi_i} \end{bmatrix} \tag{14}$$

(2) Establish the system's observation equation

This paper chooses vehicles' position and heading angle which is under world coordinate system as observations, the establishment of the system observation equation is Eq. (15). In Eq. (15), s_{i+1} is observation noise vector and are the zero mean white Gaussian noise, s_1 s_2 respectively represent the x and y direction observation noise of the midpoint of the vehicle's rear axle. s_3 represents measurement noise of vehicle heading angle, and covariance is Eq. (16).

$$Z_{i+1}^{-} = h(X_{i+1}^{-}\ s_{i+1}) = \begin{bmatrix} x_{r-i}^{-} + s_1 \\ y_{r-i}^{-} + s_2 \\ \theta_{i+1}^{-} + s_3 \end{bmatrix} \tag{15}$$

$$R = \begin{bmatrix} s_1^2 & 0 & 0 \\ 0 & s_2^2 & 0 \\ 0 & 0 & s_3^2 \end{bmatrix} \tag{16}$$

(3) Calculate partial differential matrix of Eqs. (17) and (18)

$$F_i = \frac{\partial f}{\partial x} = \begin{bmatrix} 1 & 0 & \Delta t \cos\theta_i & \frac{\cos\theta_i(\Delta t)^2}{2} & -v_i \Delta t \sin\theta_i - \frac{a_i \sin\theta_i (\Delta t)^2}{2} & 0 & 0 & 0 \\ 0 & 1 & \Delta t \sin\theta_i & \frac{\sin\theta_i(\Delta t)^2}{2} & v_i \Delta t \cos\theta_i + \frac{a_i \cos\theta_i (\Delta t)^2}{2} & 0 & 0 & 0 \\ 0 & 0 & 1 & \Delta t & 0 & 0 & 0 & 0 \\ 0 & 0 & 0 & 1 & 0 & 0 & 0 & 0 \\ 0 & 0 & \frac{\tan\varphi_i}{l} & 0 & 1 & \frac{v_i(\sec\varphi_i)^2}{l} & 0 & 0 \\ 0 & 0 & 0 & 0 & 0 & 1 & 0 & 0 \\ 0 & 0 & \frac{\Delta t \cos\theta_i}{\cos\varphi_i} & \frac{\cos\theta_i(\Delta t)^2}{2\cos\varphi_i} & \frac{-v_i \Delta t \sin\theta_i}{\cos\varphi_i} - \frac{a_i \sin\theta_i(\Delta t)^2}{2\cos\varphi_i} & \frac{-v_i \Delta t \cos\theta_i \sin\varphi_i}{(\cos\varphi_i)^2} - \frac{a_i \cos\theta_i \sin\varphi_i (\Delta t)^2}{2(\cos\varphi_i)^2} & 1 & 0 \\ 0 & 0 & \frac{\Delta t \sin\theta_i}{\cos\varphi_i} & \frac{\sin\theta_i(\Delta t)^2}{2\cos\varphi_i} & \frac{v_i \Delta t \cos\theta_i}{\cos\varphi_i} + \frac{a_i \cos\theta_i(\Delta t)^2}{2\cos\varphi_i} & \frac{-v_i \Delta t \sin\theta_i \sin\varphi_i}{(\cos\varphi_i)^2} - \frac{a_i \sin\theta_i \sin\varphi_i (\Delta t)^2}{2(\cos\varphi_i)^2} & 0 & 1 \end{bmatrix} \tag{17}$$

$$L_i = \frac{\partial f}{\partial w} = \begin{bmatrix} \frac{(\Delta t)^2 \cos\theta_i}{2} & 0 \\ \frac{(\Delta t)^2 \sin\theta_i}{2} & 0 \\ \Delta t & 0 \\ 0 & 0 \\ 0 & 1 \\ 0 & 0 \\ \frac{(\Delta t)^2 \cos\theta_i}{2\cos\varphi_i} & 0 \\ \frac{(\Delta t)^2 \sin\theta_i}{2\cos\varphi_i} & 0 \end{bmatrix} \tag{18}$$

(4) Initialize the initial state vehicle X_0 and the covariance P_0.And Time update of state priori estimation and the estimation error covariance of Eqs. (19) and (20)

$$H_{i+1} = \frac{\partial h_{i+1}}{\partial x} = \begin{bmatrix} 1 & 0 & 0 & 0 & 0 & 0 & 0 & 0 \\ 0 & 1 & 0 & 0 & 0 & 0 & 0 & 0 \\ 0 & 0 & 1 & 0 & 0 & 0 & 0 & 0 \end{bmatrix} \tag{19}$$

$$M_{i+1} = \frac{\partial h_{i+1}}{\partial s} = \begin{bmatrix} 1 & 0 & 0 \\ 0 & 1 & 0 \\ 0 & 0 & 1 \end{bmatrix} \tag{20}$$

(5) Update of state observation and error covariance and after state posteriori estimation

$$K_{i+1} = P_{i+1}^{-} H_{i+1}^{T} (H_{i+1} P_{i+1}^{-} H_{i+1}^{T} + M_{i+1} R M_{i+1}^{T})^{-1} \tag{21}$$

$$X_{i+1} = X_{i+1}^{-} + K_{i+1}(Z_{i+1} - Z_{i+1}^{-}) \tag{22}$$

$$P_{i+1} = (I - K_{i+1} H_{i+1}) P_{i+1}^{-} \tag{23}$$

In Eqs. (21)–(23), K_{i+1} is the kalman gain; $Z_{i+1} - Z_{i+1}^{-}$ is the difference between the actual measured values and the prediction measuring values; I is the 8-order unit matrix.

4 Simulation Analysis of State Parameter Estimation

In order to verify effectiveness of proposed state estimation algorithm, based on Matlab/Simulink, this paper builds the vehicle state estimation simulation platform of constant speed circular motion, and treat is as a theoretical validation tool to verify the validity of the algorithm. In the simulation vehicle is made regular circumferential moving around the coordinate point (2, 2). The Fig. 3

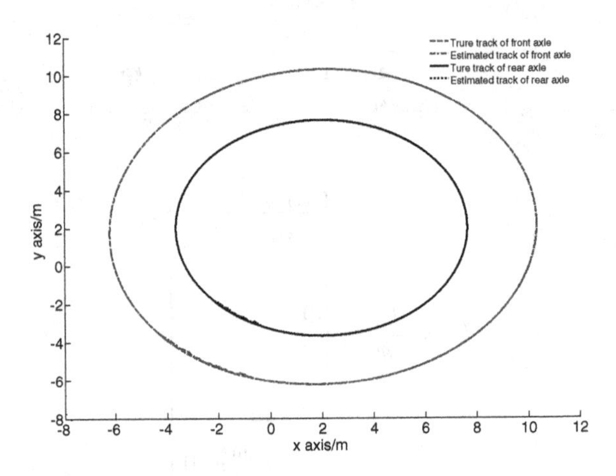

Fig. 3. Vehicle actual position and the estimation position of rear/front axle midpoint

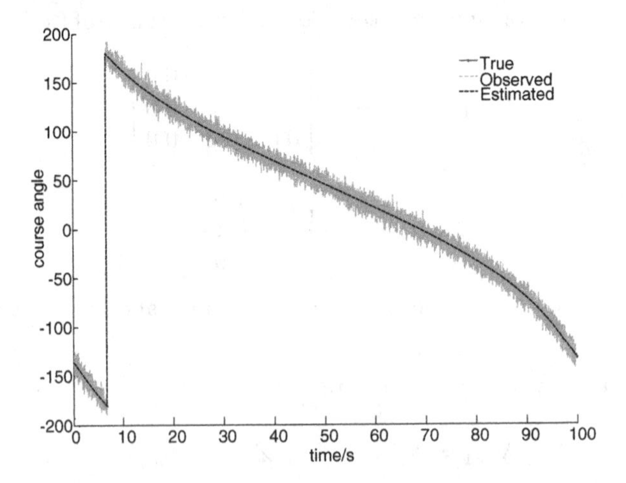

Fig. 4. Vehicle actual and estimation heading angle

shows the actual position of the vehicle rear axle midpoint and estimation position of the vehicle rear axle midpoint based on EKF, and the actual position of the vehicle front axle midpoint and estimation position of the vehicle front axle midpoint based on EKF. The actual heading angle of the vehicle and the vehicle estimation heading angle based on EKF are shown in Fig. 4. According to Figs. 3 and 4, it can be found that the estimation position of the rear and front axle midpoint and heading angle are good, and estimation can be able to achieve good accuracy.

5 Conclusion

In order to improve the estimation accuracy of the vehicle state under V2V, using extended kalman filter method to real-time accurately estimate vehicle state parameters, through building the vehicle state estimation simulation platform of constant speed circular motion, The simulation results show that the method of vehicle state estimation based on extended kalman filter is effective.

Acknowledgements. This paper is supported by the Opening Project of Key Laboratory of Intelligent Transportation Systems Technologies, Ministry of Communications, P.R.China and the special fund project of central university basic scientific research business expenses (No. 2014G1321040) and the national college students' innovative entrepreneurial training program(No. 201510710036).

References

1. Rieger, G., Scheef, J., Becker, H., Stanzel, M.: Activesafety systems change accident environment of vehicles significantly a challenge for vehicle design. In: International Technical Conference on the Enhanced Safety of Vehicles (ESV), pp. 181–184. IEEE Press, Washington DC (2005)
2. Huang, J., Tan, H.: DGPS-Based vehicle-to-vehicle cooperative collision warning: engineering feasibility viewpoint. J IEEE Trans. Intell. Transp. Syst. **7**, 415–428 (2006)
3. Sato, Y., Shimonaka, Y., Maruoka, T.: Vehicular collision avoidance support system v2 (VCASSv2) by GPS+INS hybrid vehicular positioning method. In: 2007 Telecommunication Networks and Applications Conference, pp. 29–34. IEEE Press, Christchurch (2007)
4. Ammoun, S., Nashashibi, F.: Real time trajectory prediction for collision risk estimation between vehicles. In: IEEE 5th International Conference on Intelligent Computer Communication and Processing, pp. 417–422. IEEE Press, Cluj-Napoca (2009)
5. Guo, H.Y., Chen, H., Zhao, H.Y.: State and Parameter estimation for running vehicle:recent developments and perspective. J. Automob. Eng. **36**, 968–973 (2014). (in Chinese)
6. Xie, B.Y., Wang, J.Q., Qin, X.H.: Vehicle state estimation based on V2I system. J. Control Theor. Appl. **30**, 661–672 (2013). (in Chinese)

7. Morette, N., Novales, C.: Inverse versus direct kinematics model based onflatness and escape lanes to control CyCab mobile robot. In: Vieyres, P., (eds.) Proceedings of 2008 IEEE International Conference on Robotics and Automation, pp. 2240–2246. IEEE Press, Pasadena (2008)
8. Wang, J.F., Li, P.: State estimation of vehicle based on information fusion. J. Comput. Simul. **30**, 131–136 (2013). (in Chinese)

A Compact Dual Band Antenna for IOV Applications

Malay Ranjan Tripathy[1(✉)], Priya Ranjan[1], Arun Kumar[1,2], and Sunil Kumar[1,3]

[1] Amity School of Engineering and Technology, Amity University Uttar Pradesh, Noida, India
mrtripathy@amity.edu

[2] Tata Teleservices, Main Mathura Road, New Delhi, India

[3] Department of Computer Science and Engineering, ABES Engineering College, Ghaziabad, U.P., India

Abstract. A compact dual band antenna is designed on FR4 substrate with permeability (ε_r) 4.4. The dimension of antenna is 30 X 35 X 1.5 mm^3. A wheel shaped antenna with slots and stubs is proposed. Coplanar wave guide feeding is made to make it simple and compact. Dual band is obtained in S11 vs. frequency plots. Lower band is obtained from 1.3 GHz to 2.9 GHz. The higher band is obtained from 5.3 GHz to 7.5 GHz. Both the bands are broad and highly useful for IoT and IoV applications. Apart from this, these bands are important for WLAN, WiFi and ISM applications. From radiation pattern plots for Φ, $\theta = 0$, omni directional patterns are obtained but for Φ, $\theta = 90$, bidirectional radiation pattern are obtained at 2.4 GHz. The current density distribution on CPW fed antenna is shown and is found to be interesting for further improvements of antenna parameters.

Keywords: CPW antenna · WLAN · ISM · WSN · Iot · Iov

1 Introduction

Innovative antenna design is critical to the success of communication between mobile platforms. Recent developments in antenna design due to their ubiquitous and pervasive nature and related miniaturization of the hardware inspires a new class of innovative antennas which elevate the physical layer performance by an order of magnitude and remain compatible with modern standards of IoT and IoV.

Looking at the earlier existing art of antenna design, radiation pattern of a Global Positioning System (GPS) antenna on a ground plane and five distinct vehicle platforms are studied where antenna was mounted under the instrument panel (IP) in all vehicle platforms and compared to the radiation pattern of the same antenna on a ground-plane [1]. Designing vehicle active antennas for weak signal performance while meeting the requirements of vehicle radiated immunity tests according to ISO 11451-2 and ISO 11452-2 remains one of the big challenges for the automobile and vehicle antenna manufacturers. Major factors addressed are technical parameters of the vehicle antenna designs like transfer functions and component set ups, parameters useful for the "Electromagnetic Compatibility" (EMC) and identifying the causes of the antenna amplifier failures (amplifier distortion and destruction) [2, 3].

C.-H. Hsu et al. (Eds.): IOV 2015, LNCS 9502, pp. 315–323, 2015.
DOI: 10.1007/978-3-319-27293-1_28

Drastic size reduction of axial-mode helical antenna based on hexaferrite-glass composite is demonstrated in the context of aviation. Axial-mode helical antenna was employed to provide reliable communication for unmanned aerial vehicle (UAV) applications. Presented hexaferrite helical antenna showed 82 % of volume reduction and good impedance matching compared to the air-core antenna in the axial-mode operation at 2.44 GHz with gain of 2.0 dBi [4].

Impact of topology and morphology of a particularly complex scenario on the deployment of ZigBee wireless sensor networks is analyzed. It is shown that ZigBee is a viable technology for successfully deploying intra-car wireless sensor networks [5]. Design of a highly efficient and compact micro-strip patch array antenna for wireless communication is presented. The antenna array is designed to transmit at 2.4 GHz frequency (ISM band). The main contribution of the paper is the integration of metamaterials with a 'U-shape' slotted rectangular micro-strip patch antenna, in order to achieve high gain and enhance the beam directivity [6].

An innovative circularly polarized dual band micro strip patch antenna for vehicular wireless communication at 2.45 GHz and 5.5 GHz is presented. Different radiation patterns, improved circular polarization, small dimensions and low cross polarization at both bands make this antenna particularly suitable for vehicle to vehicle communication using wireless as a medium [7]. Some cases SRR is loaded on feed line of antenna for 3G/Bluetooth/WiMAX and UWB applications [8, 9]. In other cases SRR loaded microstrip array antenna is used for obtaining low side lobe levels in radiation patterns.

Dual band antenna is becoming prominent because of their killer applicability in scientific and industrial wireless sensor networks specifically for vehicle to vehicle communication. Literature is replete with many case studies for such antenna [10–19], for example co-planar waveguide (CPW) fed antenna [11] seems to be apt for field applications and rugged due to simple design, planar profile, compact shape, light weight and ease of fabrication which is crucial to industrial success.

In this work CPW fed antenna is proposed for dual band operational capability. Basic objective is to integrate this antenna with existing industrial wireless sensor network platform provided by YUKTIX with next generation features in a new pico-powered platform and multiple communication options like BLE, 6LowPAN, Xbee, in a increased reception range and improved signal processing and overall leading to coin cell operated sensor end nodes. Some of the objectives to take this route to lab-grown design are as follows:

- Multi broadband antenna for reconfigurable radio terminals
- WiFi/WiMAX/WLAN antenna
- RADAR and Satellite communication
- Wireless Sensor Network antenna
- UWB scanning antenna

Generally speaking WSN and related antenna have critical applications in medical science for body sensors, tracking, medical telemetry, low resolution radar etc. Automotive applications include positioning for navigation in real time, proximity and convoying sensors, toll payments, security systems, baby-location services etc. Agriculture domains includes tractor/truck positioning for harvest and crop analysis, livestock monitoring, active cattle tagging and locationing etc. Major thrust of YUKTIX is targeting a host of environmental applications in the context of global warming and depleting water bodies. Information monitored using this platform can be fed to weather monitoring/predicting computational engines to improve their quality of predictions. Civil engineering applications include monitoring structures in real time, civil amenities like bridges, indoor climate control etc. Military has been a leader in using this technology in multiple contexts like intelligent battle theater design, hazard and calamity/loss data collection, situational awareness, targeting and mission tracking under given command and control etc.

This paper is organized as follow: The next section of this paper is discussing our basic WSN platform with which our proposed CPW fed antenna will be integrated. Design and performance result for our given antenna and discussions are made in Sect. 3 and conclusion is presented in Sect. 4.

2 Description of WSN Platform and Antenna Design

2.1 WSN Platform Under Integration

When we talk about hardware here, we mean end-sensor node hardware which consists of Micro-controller, small battery, external timer for synchronization if required, external sensor interface, local memory for some storage and transceiver and some other components. Lastly in our RWQM project we had used Yuktix IOT board which consists of on-board multiple communication options like WiMax/WLAN/WiFi, GPRS, Xbee, Ethernet and power options like Solar, AC and battery (Fig. 1).

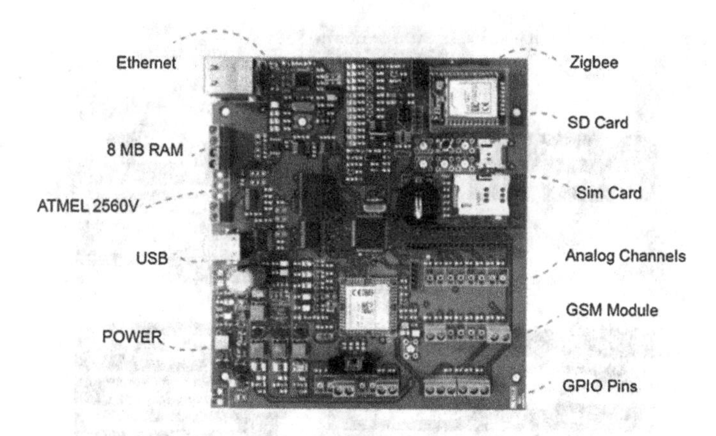

Fig. 1. Yuktix IOT platform with specifications.

Yuktix platform provides a high degree of processing power but lacks on-board data encryption features. Also power consumed by Yuktix factor makes it not the best option to be used in power optimised robust WSN node.

What we are proposing here is a hardware platform with ARM cortex M3 based CPU that is SimpleLink CC26xx / CC13xx. The ultra low power sensor controller has inbuilt

16 bit CPU with peripherals like ADC, analog comparators, SPI/I2C. This leads to not to wake up entire system when sampling sensor values and thus saving a lot of power. With a radio requiring peak current of 6 mA, it is only used when TX and RX is required and rest of time it is kept in sleep mode. In standby mode CC26xx consume only 1μ-A and thus with a 220mAh battery, life expectancy of node can reach upto 10 years with a condition that node consume average of 2.5 μ-Amp. The same processor is equipped with multiple communication options on-board like Bluetooth smart(BLE), Sub-1 MHz, 6LoWPAN, ZigBee, ZigBee RF4CE which provides us the properties of MIMO.

2.2 Firmware

Firmware also plays a major role in minimizing energy consumption by being in sleep for most of the time. As ContikiMAC RDC has a duty cycle of .6 %, that means, their nodes remain in sleep mode for 99.4 % of the total time. We, while ensuring that data is delivered to the sink node via using fall back data communication option, will try to reach the same duty cycle. Below is firmware design layout that we will try to apply in our firmware design. Also while dealing with fall-back option for data transmission, we need to ensure that the net duty cycle for our platform don't increase at more than 1 %. Also while transmitting data, we can put all other peripherals to sleep mode so as to achieve maximum efficiency in terms of power saving.

For security reasons we will use previous work done for energy efficient encryption algorithm for WSN. This algorithm uses exclusive OR and shifting and provides Energy Efficient Encryption Algorithm with 64-bit block length and 128 bit key length (Figs. 2 and 3).

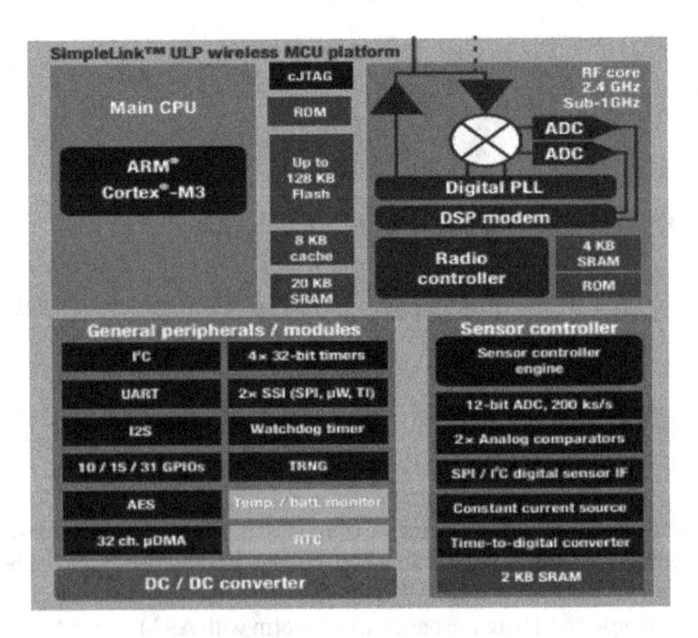

Fig. 2. SimpleLink CC26xx/ CC13xx platform

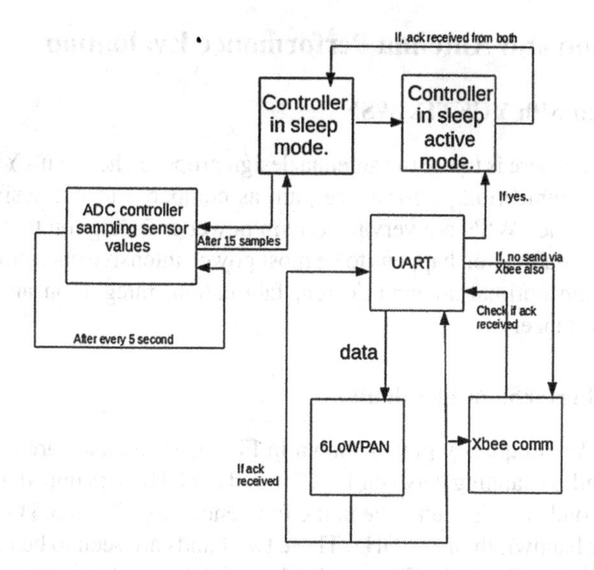

Fig. 3. Overall architecture

2.3 Antenna Design

A compact dual band CPW fed antenna is designed on FR4 substrate with permeability (ε_r) 4.4. The dimension of antenna is 30 X 35 X 1.5 mm^3. A wheel shaped antenna with slots and stubs is designed. This CPW fed antenna is simple, compact and miniaturized. This is shown in Fig. 4.

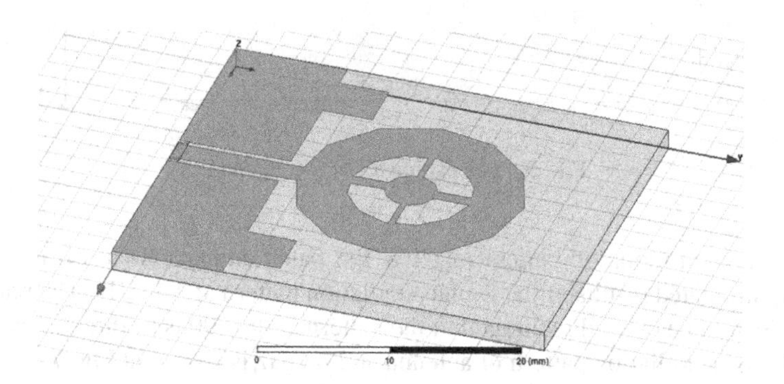

Fig. 4. CPW fed patch antenna

3 Integration and Antenna Performance Evaluation

3.1 Integration with YUKTIX WSN

Our main objective here is to use the antenna design proposed here with YUKTIX board and see the communication performance gain as compared to old designs. This is a challenging task since WSN are very miserly in power consumption to maximize their life time and communication happens to be most power-intensive operation draining the battery. Hence, an optimal antenna design, fabrication, integration and testing is of paramount importance.

3.2 Antenna Performance Evaluation

The return loss Vs. frequency plot is shown in Fig. 5. Dual bands are obtained in this design. First band is spanning between 1.3 GHz and 2.9 GHz with impedance bandwidth of 1.6 GHz. Second band is seen to be in the frequency range from 5.3 GHz to 7.5 GHz with impedance bandwidth of 2.2 GHz. These two bands are seen to be broad band and useful for IOT, VOT, WLAN, WiFi and ISM applications. The return loss at 1.9 GHz & 6.6 GHz are14.98 dB & 17.07 dB respectively. The electromagnetic simulator HFSS is used to simulate this design and the plots for return loss and radiation pattern of CPW fed patch antenna are obtained.

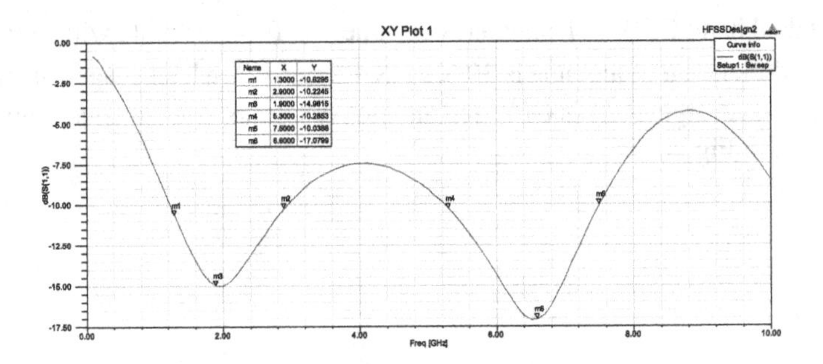

Fig. 5. S11 parameter of the antenna

The simulated radiation pattern of CPW fed patch antenna is shown in Fig. 6. It shows the simulated results of the E-plane radiation pattern at 2.4 GHz. The pattern for $\Phi = 0$ is seen to be omni-directional but the pattern at $\Phi = 90$ is bi-directional.

The H-plane radiation pattern of antenna at 2.4 GHz is shown in Fig. 7.

It is seen that the radiation pattern for $\theta = 0$ is omni-directional but the radiation pattern for $\theta = 90$ is bidirectional. The gain and characteristics of radiation pattern are seen to be similar to the E-plane radiation pattern.

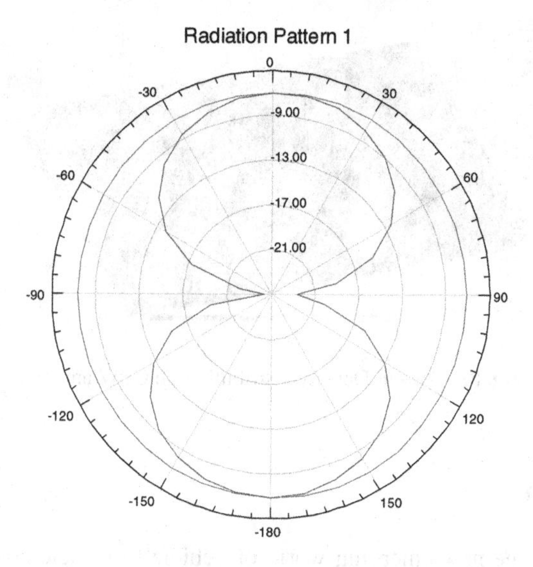

Fig. 6. E field radiation pattern

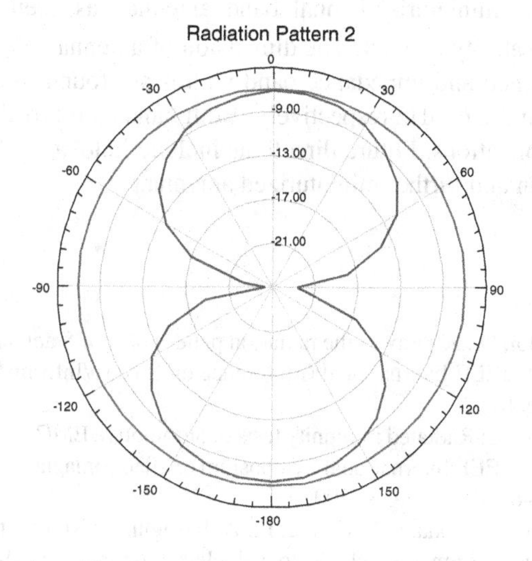

Fig. 7. H field pattern of the antenna

Figure 8 shows the surface current density distribution on radiating patch at 2.4 GHz frequency. It is seen that current density is distributed throughout the radiating patch. The coupling between ground and feed line is very strong. The gain and other parameters can be improved by adjusting design parameters of coplanar ground planes and feed line of the proposed compact, simple and miniaturized patch antenna.

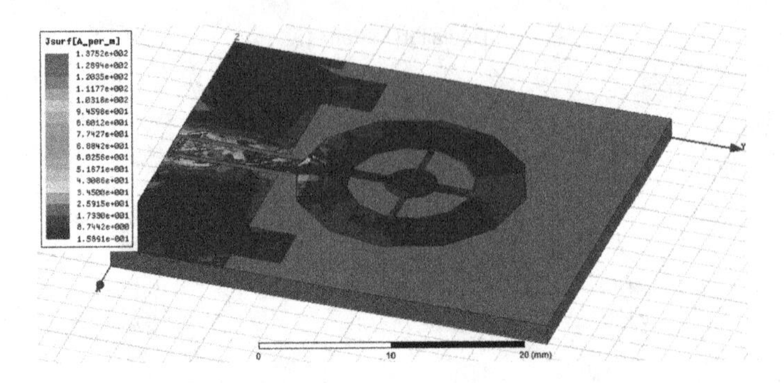

Fig. 8. Current Density distribution on CPW antenna

4 Conclusion

This work targets the new emerging world of vehicle to vehicle communication and designs a new class of antenna suitable for the purpose. In this regards, an innovative, compact, simple and miniaturized, dual band antenna has been designed on FR4 substrate with permeability (εr) 4.4. The dimension of antenna is 30 X 35 X 1.5 mm^3. Two bands are obtained and impedance band widths are found to be 1.6 GHz and 2. 2 GHz at 1.9 GHz and 6.6 GHz respectively. Both bands are broad and highly useful for IoT and IoV applications. Future directions include innovative design of metamaterial based high gain and further miniaturized antenna.

References

1. Alsliety, M.K., Aloi, D.: A study of the radiation pattern of a GPS antenna on several vehicle platforms. In: 2007 IEEE International Conference on Electro/Information Technology, pp. 1–5, 17–20 May 2007
2. Kuvedu-Libla, J.-R.K.: Radiated immunity tests of automotive EMC challenge vehicle active antenna designs. In: IEEE International Symposium on Electromagnetic Compatibility, 2008. EMC 2008, pp. 1–6, 18–22 August 2008
3. Toycan, M., Kaka, A., Ogala, C.J., Celebi, E., Abbasoglu, S.: Miniarutized and lightweight automotive antenna design for vehicle to vehicle communication. In: 2012 20th Signal Processing and Communications Applications Conference (SIU), pp. 1–4, 18–20 April 2012
4. Neveu, N., Hong, Y.-K., Lee, J., Park, J., Abo, G., Lee, W., Gillespie, D.: Miniature hexaferrite axial-mode helical antenna for unmanned aerial vehicle applications. IEEE Trans. Magn. **49**(7), 4265–4268 (2013)
5. Lopez Iturri, P., Aguirre, E., Azpilicueta, L., Garate, U., Falcone, F.: ZigBee radio channel analysis in a complex vehicular environment [Wireless Corner]. Antennas Propag. Mag. IEEE **56**(4), 232–245 (2014)
6. Baviskar, J., Singh, V., Jeyakumar, A.: Meta-material embedded designing of 2.4 GHz patch array antenna for wireless communication. In: 2015 International Conference on Communication, Information and Computing Technology (ICCICT), pp. 1–6, 15–17 January 2015

7. Ramya, R., Rao, T.R.: Circularly polarized dual band micro strip patch antenna for vehicular wireless network communications. In: 2014 IEEE International Conference on Advanced Networks and Telecommuncations Systems (ANTS), pp. 1–4, 14–17 December 2014
8. Behera, S.S., Singh, A., Sahu, S., Behera, P.: Compact tapered fed dual-band monopole antenna for WLAN and WiMAX application. In: 2014 International Conference for Convergence of Technology (I2CT), pp. 1–6, 6–8 April 2014
9. Singh, V., Kukde, A., Warty, C., Wagh, S.: Wireless Power Transfer using microstrip antenna assigned with graded index lenses for WSN. In: 2014 International Conference on Advances in Computing, Communications and Informatics (ICACCI, pp. 1664–1669, 24–27 September 2014
10. Niu, W., Li, J., Liu, S., Talty, T.: Intra-vehicle ultra-wideband communication testbed. In: Military Communications Conference, IEEE MILCOM, pp. 1–6 (2007)
11. Norimatsu, T., Fujiwara, R., Kokubo, M., Miyazaki, M., Maeki, A., Ogata, Y., Kobayashi, S., Koshizuka, N., Sakamura, K.: A UWB-IR transmitter with digitally controlled pulse generator. Solid-State Circuits, IEEE J. **42**(6), 1300–1309 (2007)
12. Oppermann, I., Stoica, L., Rabbachin, A., Shelby, Z., Haapola, J.: UWB wireless sensor networks: UWEN - a practical example. Commun. Mag. IEEE **42**(12), S27–S32 (2004)
13. Yong, L., Zhengli, H., Toshiyoshi, H., Fujita, H.: A novel J-Shape active MEMS Meta-material Antenna. In: 2014 IEEE Antennas and Propagation Society International Symposium (APSURSI), pp. 569–570, 6–11 July 2014
14. Abdalla, M., Abdelnaby, U., Mitkees, A.A.: Compact and triple band meta-material antenna for all WiMAX applications. In: 2012 International Symposium on Antennas and Propagation (ISAP), pp. 1176–1179, October 29 2012–November 2 2012
15. Abdalla, M.A., Hu, Z.: A compact dual band meta-material antenna for wireless applications. In: 2012 Loughborough Antennas and Propagation Conference (LAPC), pp. 1–4, 12–13 November 2012
16. Ying, P., Zhirun, H.: 60 GHz meta-material wideband antenna for FPGA Giga bit data transmission. In: Proceedings of 2010 IEEE International Symposium on Circuits and Systems (ISCAS), pp. 1440–1443, May 30 2010–June 2, 2010
17. Louertani, K., Ruifeng, H., Tan-Huat, C.: Wideband and low-profile unidirectionnal spiral antenna with Meta-material absorber. In: 2013 IEEE Antennas and Propagation Society International Symposium (APSURSI), pp. 682–683, 7–13 July 2013
18. Suenari, N., Murata, H., Okamura, Y.: 100 GHz-band wireless millimeter-wave-lightwave signal converter using electro-optic modulation with meta-material structure. In: 2014 International Topical Meeting on Microwave Photonics (MWP) and the 2014 9th Asia-Pacific Microwave Photonics Conference (APMP), pp. 75–78, 20–23 October 2014
19. Singh, V., Kukde, A., Warty, C., Wagh, S.: Wireless Power Transfer using microstrip antenna assigned with graded index lenses for WSN. In: 2014 International Conference on Advances in Computing, Communications and Informatics (ICACCI, pp. 1664–1669, 24–27 September 2014

A Beacon Rate Control Scheme Based on EMBMS in Cellular-VANETs Heterogeneous Networks

Weilong Wu(✉), Ping Wang, Chao Wang, and Fuqiang Liu

Broadband Wireless Communication and Multimedia Laboratory,
School of Electronics and Information Engineering,
Chengdu, People's Republic of China
wuweilong08@163.com,
{pwang, chaowang, liufuqiang}@tongji.edu.cn

Abstract. In Vehicle Ad-hoc Networks (VANETs), vehicles need to broadcast beacon messages periodically to ensure the cooperative awareness of vehicles in the vicinity. However, a constant beacon rate may result in channel congestion and severe deterioration of network performance because of limited channel resource. Since service provision technologies with point-to-multipoint (PtM) transmission mode, such as the Multimedia Broadcast Multicast Service (MBMS) and enhanced MBMS (eMBMS), have been well discussed and standardized, they can be considered as an additional solution appropriate for vehicular applications. In this paper, we propose an effective approach by combining services in VANETs and eMBMS in LTE to solve the distributed congestion control (DCC) problem for VANETs. Through simulation results we demonstrate that the proposal method is capable of combining the advantages of such two heterogeneous networks and attaining good integral performance.

Keywords: LTE · Embms · IEEE802.11p · Beacon rate control · Heterogeneous

1 Introduction

As the number of vehicles increases rapidly, traffic problems have attracted more and more attention. Intelligent transportation system (ITS) presents a new paradigm aiming to improve traffic safety and transportation efficiency via information and communication technologies (ICT). To meet pervasive appeal of vehicle communications, the IEEE Standards Association formed the 802.11p task group in 2004 and published the IEEE 802.11p protocol in 2010. The protocol is an approved amendment of the IEEE 802.11 standard in order to add wireless access in vehicular environments. In vehicular ad-hoc networks (VANETs) based on the 802.11p protocol, safety related messages are divided into two categories: event driven messages and periodical beacon messages. The former relates to emergency information to alert vehicles about emergency events. Beacon messages are disseminated to exchange vehicles' status (including position, speed, direction, etc.) among neighbor vehicles in a certain range. Both of these two

C.-H. Hsu et al. (Eds.): IOV 2015, LNCS 9502, pp. 324–335, 2015.
DOI: 10.1007/978-3-319-27293-1_29

types of messages are transmitted on a single channel with 10 MHz (generally referred to the CCH, Control Channel).

In VANETs, channel congestion is an important issue that affects the transmission of messages. When congestion occurs, the event driven messages can't be transmitted in time, which would be a threaten to the traffic safety. Considering that VANETs are distinguished from other wireless networks by their specific characteristics, such as limited channel resources, high speed of vehicles, varied channel conditions and frequent changing of network topology etc. So VANET's protocols have to cope with various traffic scenarios from sparse to dense situations, meeting the low delay and high reliability requirements of vehicle safety and transportation efficiency [1, 2].

Many approaches have been proposed to alleviate the channel congestion problem and improve the performance in VANETs. In general, most previous works focused on the control of two parameters: beacon transmit power and beacon transmit rate. Power control adjusts the message transmission power of vehicle nodes in order to control communication coverage, which changes the number of vehicles occupying the channel. Rate control is to reduce the channel occupying ratio by controlling vehicles' beacon message generation rate [3]. Researches have shown that a constant, high beacon transmit power and beacon transmit rate may result in channel congestion, especially in a crowded traffic state.

One generally accepted mechanism focus on power control is the D-FPTV [4]. The method calculates the max transmit power for each car to ensure the beacon load below the present channel threshold by collecting the status information of neighboring vehicles. While the algorithm may cause isolation of vehicles when the vehicle density decreases.

A dynamic beacon rate adaptation algorithm is introduced in [5]. In the algorithm, each vehicle node adjusts its beacon rate based on the vehicle density estimated by the number of received beacon messages from neighbors within a certain communication range. Such an approach of calculating local vehicle density is not always sufficiently accurate.

Given that in cellular networks, broadcast and multicast technologies implementing a point-to-multipoint (PtM) transmission mode is one efficient way to utilize the limited wireless resource for the same content reception, especially when the number of receivers is very large, it can be regarded as another solution to disseminate safety related messages to nearby vehicles. In recent years, several wireless and mobile multicast and broadcast solutions have been proposed and standardized. The 3rd Generation Partnership Project (3GPP) has carried out great efforts for these services in the evolution to future cellular networks, such the MBMS in release 6 and eMBMS in release 9 [6]. With the proliferation of smart phones, it's quite easy for drivers in vehicles to access the broadcast and multicast services, so it should be an effective solution to alleviate congestion problems.

Taleb and Benslimane [7] introduces an integration of Cellular-VANET heterogeneous wireless networks by using mobile gateways. The work concentrates on 3GPP unicast services, groups vehicles into clusters and selected the gateway vehicle which could serve as a relay between VANETs. However, the performance is not fully satisfactory in high dynamic vehicular communication environments. Reference [8] proposes a congestion control scheme in Cellular-VANET heterogeneous networks.

It combines the advantages of LTE network and 802.11p network. But the adjustment of LTE multicast update rate ignores the variation of local traffic density, which couldn't reflect the traffic status precisely.

Now that the successful transmission and reception of the beacon messages are crucial to VANETs application, we dedicate to perform an adaptive beacon rate scheme to reduce the possibility of 802.11p channel congestion and possible broadcast storm. To follow the above trend, our work concentrates on how to solve the DCC problem in VANETs with the support of eMBMS, based on an integrated Cellular-VANET heterogeneous network framework. In this paper, we propose an integral framework of Telematics service in the cellular network and 802.11p network, by combining the advantages of both networks.

According to the actual situation, the commercialization of 4G network has been lasted for a certain time and the coverage of the 4G is more and more widely, the eMBMS traffic safety service is accessible to drivers. However, the 802.11p device is less common. We assume that each 1 vehicle subscribes to eMBMS traffic safety service, and a part of vehicles are equipped with 802.11p devices. Adaptive beacon rate scheme based on LTE multicast service (ABRL) can only be applied in vehicles equipped with both of the two kinds of devices. We utilize the theory of traffic to calculate the local traffic density, which can be regarded as a factor to adjust the LTE uplink update rate. The proposed beacon rate adaptation scheme based on the LTE uplink update rate can guarantee the efficiency and reliability of safety-related messages' dissemination. In order to demonstrate the efficiency of the proposed scheme, we simulate the performance in terms of average packet delivery ratio and average packet loss rate by using the platform composed of VISSIM and NS3.

The remainder of this paper is organized as follows: A beacon rate adaptation scheme is presented in details in part 2. Simulations and results are shown in part 3. Finally, we conclude our work in part 4.

2 Adaptive Beacon Rate Scheme in Heterogeneous Networks

2.1 EMBMS in LTE

To utilize the limited network resource effectively, 3GPP has proposed MBMS in release 6. With MBMS, the same content is transmitted to multiple users in the MBMS service area, which always consists of multiple cells. In each cell participating in the transmission, a PtM radio resource is configured and all users subscribing to the MBMS service simultaneously receive the same transmitted signal. If the transmissions from different cells are time-synchronized, from a terminal point of view, the resulting signal will appear as a transmission from a single point over a time-dispersive channel [9]. In LTE Release-11, the service is called eMBMS, which is an evolution of MBMS. eMBMS is presented to support high quality of multimedia and timely multicast and broadcast service. In this paper, we consider using eMBMS to resolve the 802.11p CCH channel congestion which may be caused by large numbers of vehicles' beacons in a limited area when emergency event happens.

2.2 Adaptive Beacon Rate Scheme Based on LTE Multicast Service (ABRL)

In our scheme, emergency messages are broadcasted in both LTE network and 802.11p network, as shown in Fig. 1. But we only discuss the emergency messages transmitted in LTE network and beacon messages transmitted in 802.11p network. After all, the purpose of our scheme is to adjust the beacon rate and solve the congestion problem in 802.11p network, which can also guarantee higher reliability of emergency messages' dissemination. When an emergency event happens, the eNB within the communication range of the emergency vehicles will send the emergency messages to traffic safety multicast server through LTE uplink. The server asks telecom operator's LTE evolved package core (EPC) to broadcast these messages in the cell. If a vehicle in this cell has subscribed to eMBMS traffic safety service, it will be informed of the emergency event in advance. At the same time, the same information is broadcasted in the 802.11p network to allow nearby vehicles to attain more timely information of the emergency. If the vehicle is equipped with 802.11p device, its beacon rate will be adjusted based on the LTE update rate of multicast service in order to lower the possibility of CCH channel congestion and broadcast storm.

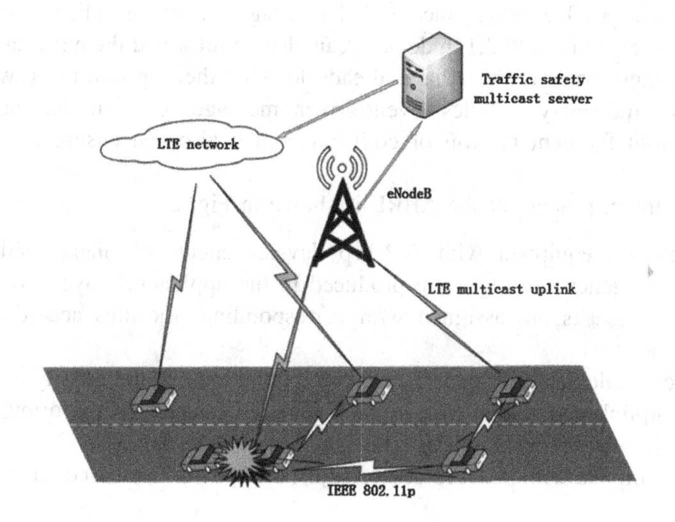

Fig. 1. Diagram of the emergent scenario

According to the number of LTE multicast messages vehicle A received in one observing time (*OT*), it will adjust its beacon rate adaptively. Though higher beacon rate means that we could get more accurate information about the vehicle' neighboring status, given that vehicle A has already known the emergency event through LTE network or 802.11p network ahead of time, decreasing the beacon rate properly has little effect on the safety of vehicle A. Besides, the mechanism can reserve more channel resources for event-driven messages in 802.11p network because of limited network resources. Therefore it can prevent possible channel congestion or alleviate

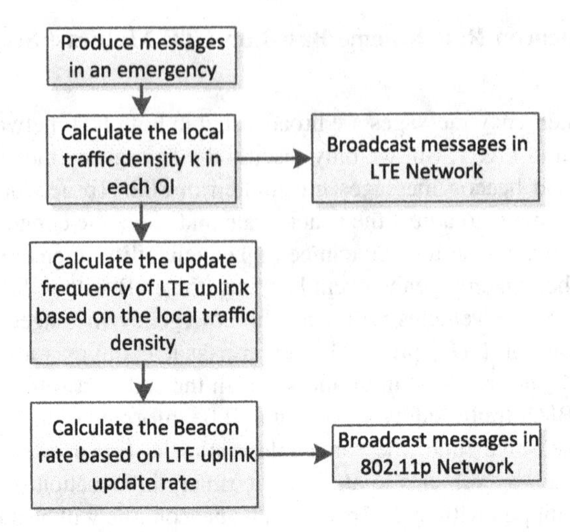

Fig. 2. The adjustment process of the ABRL

existing congestion effectively. Since eNB has a lager coverage radius than 802.11p node, the vehicles without 802.11p devices can also be informed the emergency ahead of time. Although many vehicles have already lowered their beacon rates, when they approach the emergency vehicles, event-driven messages can still be successfully received without frequent backoff or collision, which also can ensure the safety of them.

The adjustment process of the ABRL is shown in Fig. 2.

- When vehicles equipped with 802.11p devices encounter unexpected circumstances, emergency messages are produced in the application layer. At the same time, the messages are assigned with corresponding priorities according to the vehicle types.
- The vehicles calculate the space mean velocity and local traffic density in each *OT*.
- The LTE uplink update rate R_{ULi} of each vehicle is calculated by formula (5) and formula (6), which consist of the LTE multicast update rate R_{LM}.
- Vehicles equipped with 802.11p device adjust the beacon rate based on the value of R_{LM}.

The key criteria is presented in detail as follows.

(1) Observing Time (*OT*) & Priority Index (*Pri*)

OT, means a set of equal length intervals. In order to implement the mechanism during each *OT*, every vehicle maintains a neighboring table, each record of which consists of position, velocity and other information of vehicles within the communication range. The table is updated based on the 802.11p beacons it receives in IEEE 802.11p network. Different vehicles have different priority value. Actually, vehicle with high priority needs relative high beacon rate to ensure the traffic safety, such as truck loaded with oil or some other dangerous chemicals, ambulance and so on. We use (1) to record the priority index:

$$\text{Pri} = 1, 2\ldots H \tag{1}$$

Where the factor *Pri* is an integer and *H* stands for the highest level while 1 means the lowest priority.

(2) Space mean velocity and local traffic density k

Studies have shown that a constant and high beacon rate easily lead to channel congestion in crowded traffic state. The key to control congestion is sense the vehicle density nearby accurately, while previous works could hardly calculate the density accurately. In this part, we will use the record collected in the neighboring table introduced earlier to calculate the vehicle density. Assuming that during the ith *OT*, the number of the vehicles stored in vehicle A's neighbor table is M. v_j represents the jth neighbor vehicle's velocity. According to the theory of traffic [10], the space mean velocity of one section of a road in one *OT* is given below.

$$\overline{\mathrm{u}_s(t)} = \frac{1}{\frac{1}{M}\sum_{j=1}^{M}\frac{1}{v_j}} \tag{2}$$

The corresponding estimation density in one *OT* is as follow:

$$\lambda(t) = \frac{M}{T} \tag{3}$$

$$k(t) = \frac{\lambda(t)}{u_s(t)} \tag{4}$$

where $\lambda(t)$ stands for traffic flow and T equals the duration time of one *OT*. According to the number of vehicles in the neighboring table's record of beacons in each *OT*, each vehicle can calculate the current traffic flow λ. The current local traffic density k is subject to traffic flow λ and space mean velocity $\overline{u_s}$ at that period of time.

(3) LTE uplink update rate (R_{ULi}) and LTE multicast update rate(R_{LM}) in eMBMS

When an emergency event occurs, an emergency message is produced and transmitted to eNB through the LTE uplink. Then the message will be delivered to the traffic safety multicast server at a certain update rate. The update frequency is concerned with vehicle type, velocity, surround vehicle density and emergency degree of different events, such as crash, engine breakdown, and traffic jam etc. The final LTE uplink update rate is suggested by

$$R_{ULi} = h * \alpha \tag{5}$$

$$\alpha(i+1) = \begin{cases} \alpha(i) * \alpha & if((k(i) - k(i-1)/k(i-1) > thr1) \\ \alpha(i) & if(0 \leq (k(i) - k(i-1)/k(i-1) > thr1) \\ \alpha(i)/\alpha & else \end{cases} \quad (6)$$

where h is a coefficient subject to emergency event type,α is the increase factor which is concerned with the growth ratio of local traffic density $k.thr1$ is the increase threshold of the traffic density. Here, we take α as 2 and $thr1$ as 0.5 through simulation.

In our proposal, traffic safety multicast service has higher priority than usual eMBMS service. EPC will broadcast the messages in the cell after receiving the broadcast request by Internet server. The messages are the same as the LTE uplink content which was sent by the emergency vehicles several hundred milliseconds before. LTE multicast update rate R_{LM} is given below.

$$R_{LM} = \sum_{i=1}^{N} R_{ULi} \quad (7)$$

(4) Adaption of beacon rate in 802.11p network

Beacon messages are disseminated on a single channel with 10 MHz (generally referred to the Control Channel) in the IEEE 802.11p based vehicular networks. Due to the limited bandwidth, it is obvious that schemes with adaptive beacon rate have an advantage over those with fixed rate. In ABRL, beacon rate is concerned with its vehicle type, current and relative velocity, and the LTE multicast update rate R_{LM}. The calculated beacon rate in theory is determined as formulas (8) and (9).

$$R_{BEA} = Pri * c * \frac{u}{u_{\max}} * P, \ 0 < p \leq 1 \quad (8)$$

$$P(i+1) = \begin{cases} \max(P(i) * p, P_{\min}), & if((R_{LM}(i) - R_{LM}(i-1))/R_{LM}(i-1) > thr2) \\ P(i), & if((R_{LM}(i) - R_{LM}(i-1))/R_{LM}(i-1) > thr2) \\ \min(P(i)/p, 1), & else \end{cases} \quad (9)$$

Where c is a constant value and recommended as 10 packets/s by US Department of Transportation (DOT) [11]. u is the current velocity of the vehicle and $u_{\max}$ stands for the maximum velocity on highway. Here we take $u_{\max}$ as 120 km/h. P is an iteration function of R_{LM}'s change ratio. $P_{\min}$ is the minimum constant value of P function. If a vehicle has not received any LTE traffic safety multicast messages, P function will keep its primitive and maximum constant value, which equals 1. Like the calculation of LTE uplink update rate, p is the decrease factor, and $thr2$ is the increase threshold of R_{LM}. When R_{LM} increases by a certain percentage, we should lower the beacon rate, which means is less than 1. We recommend the decrease factor $p = 2/3$, $thr2 = 1/3$, since we find that the mechanism can achieve stable performance under the circumstance by simulation.

3 Simulation and Performance

As we mentioned above, ABRL helps prevent and alleviate channel congestion in VANETs, and thus can improve the reliability for emergency messages. In this section, the performance is discussed.

3.1 Simulation Tools and Setup

We evaluate the performance of the proposed adaptive beacon rate scheme in highway scenarios with different traffic density. We build a realistic traffic scenario simulation via VISSIM and realize inter-vehicle communication of IEEE 802.11p network via NS-3. Our simulation work can be divided into two parts. First, VISSIM provides modeling of real-world traffic scenarios in a highly detailed manner. We use it to generate realistic vehicle movement patterns of different scenarios, convert them to trace files that NS-3 supported and import the files into NS-3. Second, we take Nakagami propagation loss model as small scale fading model in addition to path loss model. The Nakagami fading model has been performed in real-world tests on highways. It suggests that it is a suitable model for vehicular scenarios when estimating the mobile communication channels [12].

In the simulation of VISSIM, we build a traffic flow model of a 2 km bidirectional way with four lanes. Vehicle types include cars, trucks and buses, and the velocity varies from 60 km/h to 120 km/h. The vehicle normalized density varies from 0.1 to 1, and the max density is 10000 vehicles per hour. For each scenario, the simulation runs more than 120 s. In order to simplify the simulation, we suppose that when vehicles receive the LTE multicast emergency messages, all of them will decrease their velocity to about 80 km/h and lower their 802.11p beacon rates based on ARBL. We assume that each vehicle node would not start to send packets until the traffic flow is steady. Besides, packet interval between every vehicle node is fixed in 802.11p network. We use 1 eNB and 1 UE to simulate the LTE eMBMS. The related parameters for simulation are listed in Table 1.

Table 1. Related simulation parameters

Parameter	Value
Communication technology	IEEE802.11p + LTE
MBMS service area	1 eNB and 1 cell
Cell covering radius	700 m
LTE Propagation model	EVA 60 km/h
Uplink message size	400 bytes
Beacon radius	250 m
Propagation model	Nakagami
Nakagami m	m0 = 5, m1 = 3, m2 = 1
Central frequency	5.890 [GHz]

(Continued)

Table 1. (*Continued*)

Parameter	Value
CCH Channel bandwidth	10 MHz
date rate	6 Mbps
Beacon size	300 bytes
Emergency message size	500 bytes
Emergency message interval	50 ms
Emergency message contention window	3-7
Beacon message contention window	15-1023
Number of Lanes	2*direction
Vehicle density(normalized)	0.1 to 0.9
Vehicle speed	60 to 120 [km/h]

3.2 Performance Analysis

We use two metrics to analyze the performance of the proposed adaptive beacon rate scheme. The first metric is the average packet delivery ratio (PDR), which means the copies of a packet received by a sender's neighbors. Typical schemes with fixed beacon rate (rate = 10 and 3.3 packets/s) are also simulated for comparison. As it shown in Fig. 3, with the increase of vehicle density, all the average PDRs decrease. But the average PDR of ABRL is higher than that with fixed beacon rate under any circumstance. Free channel condition has good average PDR performance because channel is not occupied by other competitors. Congested channel leads to low average PDR due to the fact that multiple senders compete for limited channel resource, thus packets are likely to collide.

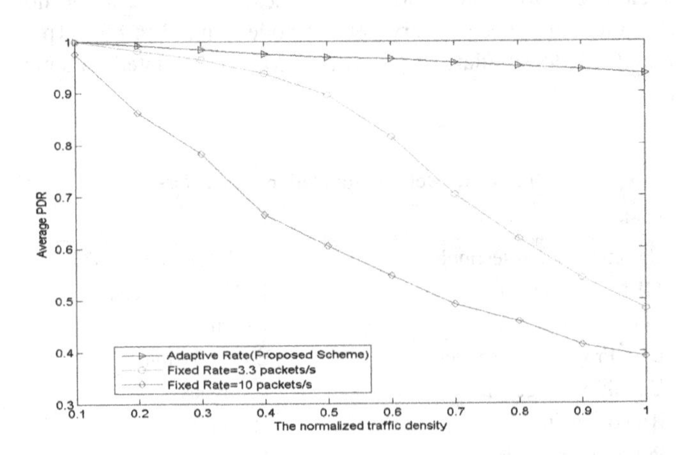

Fig. 3. Average PDR in various scenarios

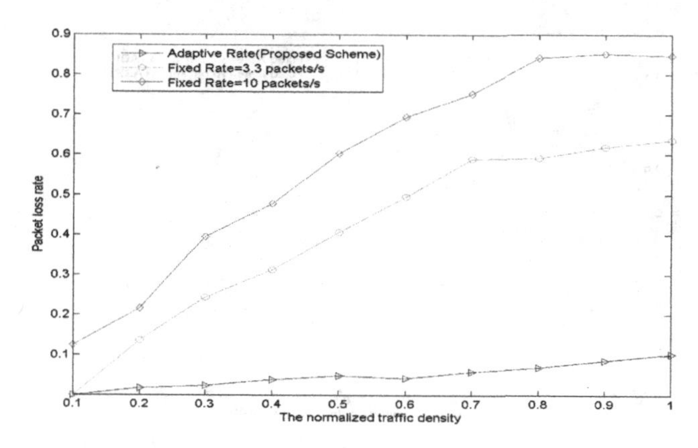

Fig. 4. Average packet loss rate in various scenarios

Figure 4 depicts the results of the second metric: packet loss rate that reflects system reliability. The average packet loss rate increases when the vehicle density increases. Without ABRL, the average packet loss rate rises drastically with the vehicle density. That is because in dense areas, the number of collisions among the broadcast beacons increases. In contrast, our scheme can perform well in average packet loss rate even under the worst circumstance.

Given that not all the vehicle owners will subscribe the traffic safety multicast service, so we conduct a research to reflect the impact of the vehicle proportion applied with ABRL when the vehicle density is 10000 per hour. Figures 5 and 6 show the average PDR and packet loss rate of various vehicle proportions applied with ABRL respectively. With the increase of the proportion increase, the average PDR improves and the packet loss rate decreases dramatically. When the proportion is close to 0.7, the performance of the network can be improved significantly. Our proposed scheme turns out to have better performance than the fixed beacon rate scheme.

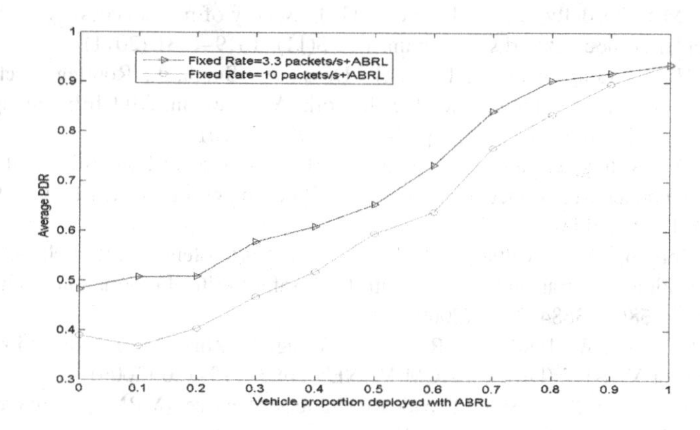

Fig. 5. Average PDR of various vehicle proportion applied with ABRL

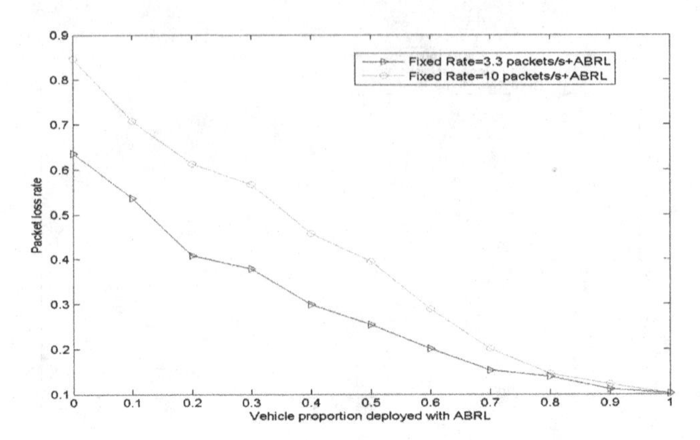

Fig. 6. Packet loss rate of various vehicle proportion applied with ABRL

4 Conclusion

In this paper, we demonstrate an effective beacon rate control scheme based on cellular-VANET heterogeneous networks to solve the problem of congestion in IEEE 802.11p network. The adaption of beacon rates in our system relates to vehicle type, velocity and LTE multicast traffic safety messages update rate. In our scheme, we take advantage of the lager coverage radius of LTE network, which can provide safety messages to more vehicles without 802.11p devices. Simulation based on real traffic models reveals better performance than fixed beacon rate schemes especially in crowed traffic situations.

References

1. Booysen, M.J., Zeadally, S., van Rooyenb, G.-J.: Survey of media access control protocols for vehicular ad hoc networks. Commun. IET **5**(11), 1619–1631 (2011)
2. Trivedi, H., Veeraraghavan, P., Loke, S., Desai, A., Singh, J.: Routing mechanismand cross-layer design for Vehicular Ad Hoc Network: A survey. In: 2011 IEEE Symposium on Computer and Informatics (ISCI), pp. 243–248, 20–23 (2011)
3. Sepulcre, M., Mittag, J., Santi, P., Hartenstein, H., Gozalvez, J.: Congestion and awareness control in cooperative vehicular systems. In: Proceedings of the IEEE, vol. 99, no. 7, pp. 1260–1279 (2011)
4. Torrent-Moreno, M., Mittag, J., Santi, P., Hartenstein, H.: Vehicle-to-Vehicle communication: fair transmit power control for safety criticalinformation. Veh. Technol. IEEE Trans. **58**(7), 3684–3703 (2009)
5. Drigo, M., Zhang, W., Baldessari, R., Le, L., Festag, A., Zorzi, M.: Distributed rate control algorithm for VANETs(DRCV). ACM VANET **2009**, 119–120 (2009)
6. 3GPP TS 23.246: Multimedia Broadcast/Multicast Service (MBMS); Architecture and functional description (Release 9) (2010)
7. Taleb, T., Benslimane, A.: Design guidelines for a network architecture integrating VANET with 3G and beyond networks. In: IEEE Globecom 2010 Proceedings (2010)

8. Yang, Y., Wang, P., Wang, C., Liu, F.: An eMBMS based congestion control scheme in cellular-VANET heterogeneous networks. In: 17th International Conference on Intelligent Transportation Systems(ITSC), pp. 1–5. Qingdao (2014)
9. Dahlman, E., Parkvall, S., Skold, J.: 4G LTE/LTE-Advanced for Mobile Broadband (2012)
10. Shao, C., Wei, L., Jia, B.: Taffic Flow Theory. Publishing House of Electronics Industry, 10–12 (2012)
11. Kandarpa, R., Chenzaie, M., et al.: Final Report: Vehicle Infrastructure Integration Proof-of-Concept Results and Findings-Infrastructure. U.S. Department of Transportation, Washington, DC, USA (2009)
12. Sen, I., Matolak, D.W.: Vehicle-Vehicle channel models for the 5-GHz band. Intell. Trans. Syst. IEEE Trans. **9**(2), 235–245 (2008)

Modeling and Simulations

Link Delay Modeling for Two-Way Traffic Road Segment in Vehicular Networks

Jinho Lee[1], Jaehoon (Paul) Jeong[2(✉)], and David H.C. Du[3]

[1] Department of Computer Science and Engineering, Sungkyunkwan University, Suwon 440-746, Republic of Korea
jinholee@skku.edu

[2] Department of Interaction Science, Sungkyunkwan University, Suwon 440-746, Republic of Korea
pauljeong@skku.edu

[3] Department of Computer Science and Engineering, University of Minnesota, Minneapolis, MN 55455, USA
du@cs.umn.edu

Abstract. This paper proposes expected link delay (i.e., data delivery delay) on a two-way road segment for carry-and-forward data delivery schemes in vehicular networks. Recently, a lot of vehicles are able to communicate with each other by dedicate short-range communications (DSRC) for vehicular networking. In the near future, more vehicles will be equipped with DSRC devices because of governmental policies for driving safety. In this paper, we derive link delay model on a two-way road segment. This link delay model is essential to support multihop infrastructure-to-vehicle or vehicle-to-vehicle data delivery in vehicular networks as disruption tolerant networks. Through simulation, it is shown that our two-way link delay model is more accurate than the legacy two-way link delay model.

Keywords: Vehicular networks · VANET · Link delay · Two-way · Expectation

1 Introduction

Vehicular Ad Hoc Networks (VANETs) have been researched widely recently. The importance of VANET is getting higher as the demand on vehicular networks increases for the communication among vehicles for the driving safety and the Internet connectivity [1,2]. For example, a vehicle in the blind spot can be detected by inter-vehicle communications and a smartphone can give a pedestrian an alarm message when a vehicle is approaching from behind. This communications is achieved by Dedicated Short Range Communications (DSRC) devices [3]. As U.S. Department of Transportation tries to mandate to equip DSRC devices to all light vehicles [4] for driving safety, a lot of vehicles will be equipped with DSRC devices in the near future. These technologies will be more important as autonomous vehicles are under development by major

C.-H. Hsu et al. (Eds.): IOV 2015, LNCS 9502, pp. 339–350, 2015.
DOI: 10.1007/978-3-319-27293-1_30

automotive vendors, such as Audi [5], Ford [6], and Mercedes-Benz [7]. Furthermore, inter-vehicle communications can facilitate internet connectivity of vehicles. These communications can reduce the dependence on 4G-LTE networks for cost effectiveness.

In multihop infrastructure-to-vehicle data delivery, accurate link delay is required for reliable unicast [2] or multicast [8]. Many data forwarding scheme [2,8] are based on one-way link delay model (i.e., the expected data delivery on a road segment with one-way road traffic). However, two-way roads are dominant over one-way roads in real road traffic environments. Precise two-way link delay is necessary to offer better services and connectivity to vehicles. This paper proposes a formulation of expected link delay for two-way road segment, assuming that the length of road, average arrival rate, vehicle speed are available.

Our intellectual contributions are as follows:

- Two-way Link Delay Model. We propose a two-way link delay model utilizing road statistics such as average arrival rate and vehicle speed.
- Validation of our model. We validate our model with extensive simulations.

The remainder of this paper is structured as follows. Section 2 is the literature review of link delay modeling. Section 3 formulates our two-way link delay model. Section 4 describes the modeling of link delay in a two-way road segment. Section 5 evaluates two-way link delay model with simulation results. Section 6 concludes the paper along with future work.

2 Related Work

Much research has been done on multihop Vehicle-to-Infrastructure (V2I) and Vehicle-to-Vehicle (V2V) data forwarding for safety and driving efficiency [1,2]. VANETs have distinctive characteristics from conventional Mobile Ad hoc Networks (MANETs) such as vehicles' restricted moving area and predictable mobility in a short period. Due to these characteristics, we can expect vehicles' partitioning and merging on a road segment. There are several research activities [1,9] to formulate expected link delay on a road segment with these characteristics of VANETs. TBD [1] proposes link delay for a road segment with one-way traffic and [9] suggests average message delivery delay with a bidirectional traffic model.

Link delay for one-way traffic is modeled by TBD [1]. It models link delay for one-way traffic assuming that inter-arrival times between vehicles are distributed exponentially. First, a source vehicle can transmit its packets in a neglectably short time through vehicles constructing a network component. Then, the next carrier carries the packets through the rest of the road segment. We refer to the length of the rest of the road as *carry distance* (l_c). In this scenario, the main portion of link delay is the carry delay which is $\frac{l_c}{v}$ where the average vehicle speed is v. Since this model assumes that the link delay is approximately the same as the carry delay, the link delay is $\frac{l_c}{v}$.

In order to derive average carry distance, [1] formulates the average distance between the source vehicle and the next carrier. It is modeled as a sum of inter-distances between adjacent vehicles. Since it assumes that the inter-arrival time is distributed exponentially, the inter-distance is also exponentially distributed. If the inter-distance is shorter than the communication range of vehicles, we can say that they are connected. This model suggests the average number of hops between the source vehicle and the next carrier and the average distance of two connected vehicles. With this information, [1] derives average carry distance and carry delay. However, in reality, two-way roads are dominant over one-way roads. Therefore, We need link delay information of two-way road traffic situation to make decisions about delay over real roads.

Expected link delay for two-way traffic [9] is formulated, assuming that the two-way traffic is a combination of two Poisson point processes. If two vehicles are moving toward each other with constant speed v, one vehicle can see that the other is approaching with $2v$. In the sense of relative speed, it is identical to one stationary vehicle and another vehicle moving toward the stationary vehicle with $2v$. This model assumes that vehicles on one side of road is stationary. On the other hand, vehicles on the other side drive two times faster than the average speed of the road segment.

In this case, the only way to deliver a packet forward is constructing a network component containing stationary vehicles. The source vehicle transmits its packets immediately if it belongs to a network component. Otherwise, the source vehicle waits until a network component arrives. If the length of the network component is long enough, the source vehicle forwards its packets toward the next carrier. This forward and wait process is repeated until the packets are delivered to the end of the road segment. In this model, we can get the average number of stationary vehicles through the assumption of a Poisson distribution. The number of vehicles is the same with the number of hops to forward packets to the end of the road segment. This model suggests that the link delay is the sum of per-hop delays. The probability to construct a network component long enough to connect stationary vehicles decreases more quickly as the average distance between stationary vehicles becomes larger. The expected delay for each hop becomes very long in the case of a sparse road situation.

3 Problem Formulation

In this section, we describe our goal, assumptions, and high-level design of our model. Given the road statistics, our goal is to model link delay for a two-way road segment. This link delay information is necessary to guarantee on VANETs with two-way traffic road situation. Our assumptions and high-level idea are as follows:

Assumptions

- Vehicles are equipped with DSRC [3] devices.
- RSUs collect road statistics such as speeds and arrival rates from vehicles through DSRC devices.

- Relay nodes are deployed for each intersections. These relay nodes receive and deliver packets as a media.

High-level Idea. We define link delay as the elapsed time to deliver a packet form an intersection (I_i) to another intersection (I_j). When a packet arrives at I_i, a relay node holds the packet until a proper packet carrier arrives. The packet carrier carries the packet and forwards it as soon as it encounters a network component. Then, the next carrier carries until it encounters another network component. This carry and forwarding are repeated while the packet is being delivered. We model two-way link delay as we derive the average forwarding distance and average carry time of the packet carriers.

4 Delay Model

In this section, we model the link delay, considering road statistics such as average speed and average vehicle arrival rate. We assume that relay nodes are installed at both ends of a road segment. When packets arrive at a relay node, it holds packets until a vehicle passes by the relay node.

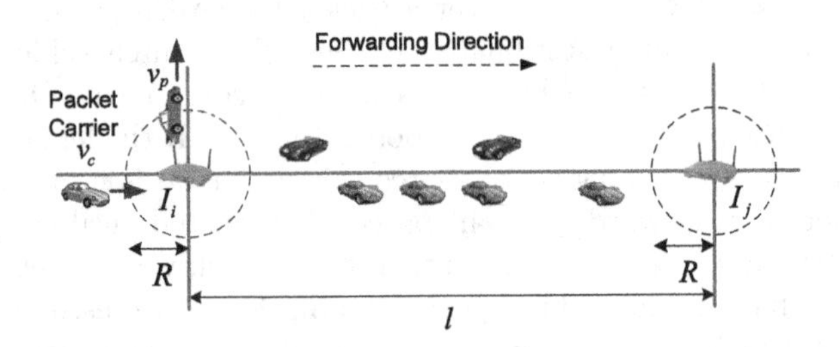

Fig. 1. Network environment

In Fig. 1, the previous packet carrier (v_p) passed by intersection I_i and forwarded its packets to the relay node installed at I_i. Then, the relay node holds them until a vehicle moving in the Forwarding Direction arrives. Once the new packet carrier (v_c) toward the intended direction arrives, the relay node transmits packets to v_c. The packets are delivered to I_j by repetitive carry and forwarding process. We define the *link delay* as the time between the packet arrival time instants at I_i and I_j.

Let's consider a road segment with length l, vehicle arrival rates λ_f and λ_b, average vehicle speed v, and the communication range R. λ_f is the arrival rate of vehicles moving forward (from I_i to I_j). λ_b is that of vehicles moving backward (from I_j to I_i).

Note that the forwarding delay is ignorable compared to the carry delay. It only takes only a few microseconds to forward packets under VANETs conditions. Thus, for simplicity, we consider the link delay is the same as the carry delay.

Our goal in this paper is: *Given the road statistics such as vehicle arrival rates and the average vehicle speed, how can we formulate the expected link delay on a two-way road segment?* For the *link delay*, we assume that packets are delivered by the cycles of carry and forwarding. In order to derive the expected link delay, we need to derive the average lengths of the *carry distance* and the *forwarding distance*. We define the following terms to derive the *link delay*.

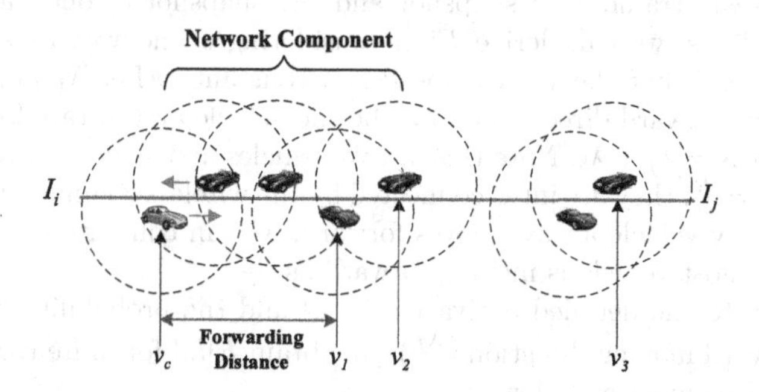

Fig. 2. A road segment with relay nodes at the both ends

Definition 1 (Network Component). *Let Network Component be a group of vehicles that can communicate with each other via either one-hop or multi-hop communication. Figure 1 shows a network component consisting of vehicles* $v_c, \ldots, v_2$.

Definition 2 (Component Length). *Let Component Length (denoted as* l_n*) be the length of a Network Component.*

Definition 3 (Forwarding Distance). *Let Forwarding Distance (denoted as* l_f*) be the physical distance which a packet travels through forwarding within a Network Component from the packet carrier (*v_c*). When the packet carrier (*v_c*) encounters a Network Component, it immediately forwards its packets to the farthest vehicle moving to the same direction with* v_c *in the Network Component. In Fig. 2,* v_c *forwards packets to* v_1*. In this case, the Forwarding Distance is the distance between* v_c *and* v_1.

Definition 4 (Carry Distance). *Let Carry Distance (denoted as* l_c*) be the physical distance where a packet is carried by a packet carrier (*v_1*) until it encounters another vehicle (*v_3*) moving backward, belonging to another network component.*

Definition 5 (Carry Delay). *Let Carry Delay (denoted as* d_c*) be the delay that a packet is carried by a packet carrier (*v_1*) for carry distance* l_c *such that* $d_c = l_c/v$ *for vehicle speed* v.

4.1 Average Component Length for Finite Road Length

In this subsection, we formulate *average component length* ($E[l_n]$) for a finite road. $E[l_n]$ can be computed as the expected sum of the inter-distances of adjacent vehicles (D_h) within a network component. For the simplicity, we consider a road snapshot to calculate $E[l_n]$. Let's suppose that the vehicles on the road have identical shapes of front side and rear side. Then, one cannot tell the difference between two-way traffic road snapshot and the snapshot of one-way, two-lane traffic road. Thus, we can derive $E[l_n]$ considering a one-way, two-lane traffic road. We assume that the vehicle speed is a constant v. Let λ_f be the vehicle arrival rate for forward direction and λ_b be the vehicle arrival rate for backward direction. Let $\lambda = \lambda_f + \lambda_b$. Note that if two vehicles arrive at a certain intersection within $a = \frac{R}{v}$, they are inter-connected by the wireless communication range R. Since a carry vehicle always moves forward, we can compute the probability that the backmost vehicle is moving forward as $\frac{\lambda_f}{\lambda}$.

According to the detailed derivation in [1] and the probability of the carry vehicle's forward moving direction ($\frac{\lambda_f}{\lambda}$), we obtain $E[l_n]$ for finite road length in two-way road segment as follows:

$$E[l_n] = \frac{\lambda_f}{\lambda}\Big(\frac{\alpha((N-1)\beta^N - N\beta^{N-1} + 1)}{(1-\beta)^2} + l\beta^N\Big), \tag{1}$$

where $\alpha = ve^{-\lambda a}(\frac{1}{\lambda} - (a + \frac{1}{\lambda})e^{-\lambda a})$, $\beta = 1 - e^{-\lambda a}$, and $N = \lceil \frac{\beta(1-\beta)}{\alpha} l \rceil$.

4.2 Average Forwarding Distance for Finite Road

Now, we derive the *expected forwarding distance* ($E[l_f]$) considering the directions of vehicles on a finite road. In Fig. 2, the forwarding distance is the distance between v_c and v_1. According to (1) and (2), we can formulate $E[l_f]$ as we derive $E[l_n - l_f]$.

$$\begin{aligned} E[l_f] =& E[l_n - (l_n - l_f)] \\ =& E[l_n] - E[l_n - l_f]. \end{aligned} \tag{2}$$

Since l_n is formulated as the expected sum of the inter-distances of adjacent vehicles (D_h), a network component consists of $\lfloor \frac{E[l_n]}{E[D_h|D_h \le R]} \rfloor$ vehicles in average where $E[D_h \mid D_h \le R]$ is the average vehicle inter-distance within a network component. Let $m = \lfloor \frac{E[l_n]}{E[D_h|D_h \le R]} \rfloor$ where $\lfloor x \rfloor$ is the largest integer less than or equal to x. If we choose a vehicle on the road snapshot, it is either moving forward or moving backward. Considering the ratio of forward-moving vehicles to total vehicles, it is moving forward with probability λ_f/λ or moving backward with probability λ_b/λ. The direction of the vehicle is determined by Bernoulli trials. $l_n - l_f$ is determined by the number of successive vehicles moving backward from the head vehicle in a network component. For example, in Fig. 2, the head vehicle (v_2) is moving backward and the next one (v_1) is moving forward. Considering the probability mass function of Geometric distribution, the probability of this case is $\frac{\lambda_f}{\lambda_f+\lambda_b}\frac{\lambda_b}{\lambda_f+\lambda_b}$. In the same way, $l_n - l_f = k \times E[D_h \mid D_h \leq R]$ with probability $\frac{\lambda_f}{\lambda_f+\lambda_b}(\frac{\lambda_b}{\lambda_f+\lambda_b})^k$ where k is the successive number of backward-moving vehicles from the head vehicle in the network component. Thus,

$$E[l_n - l_f] = \sum_{k=0}^{m} kE[D_h \mid D_h \leq R]\frac{\lambda_f}{\lambda_f+\lambda_b}(\frac{\lambda_b}{\lambda_f+\lambda_b})^k, \tag{3}$$

where $E[D_h \mid D_h \leq R] = v\frac{1/\lambda-(a+1/\lambda)e^{-\lambda a}}{1-e^{-\lambda a}}$, according to [1].

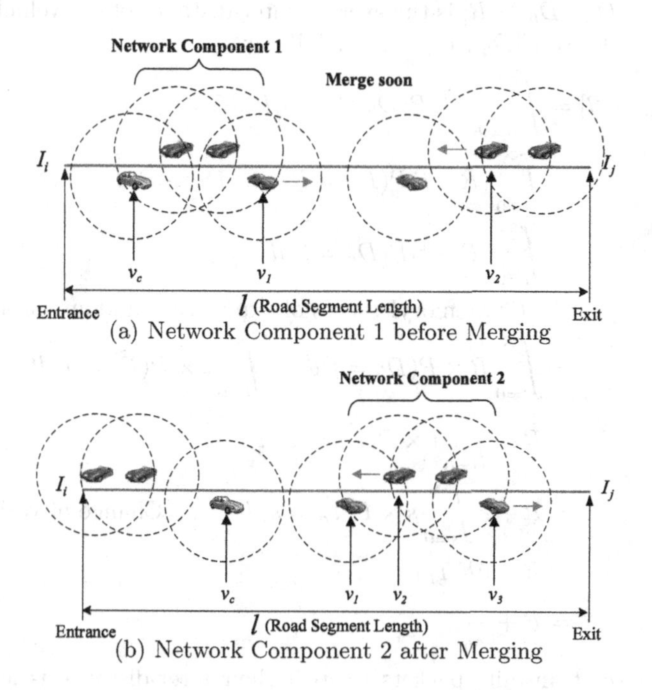

(a) Network Component 1 before Merging

(b) Network Component 2 after Merging

Fig. 3. Renewal process scenario

4.3 Link Delay Derivation with Renewal Process

A packet carrier forwards its packets when it encounters a new network component. As shown in Fig. 3, the current packet carrier (v_c) forwards its packets to the next carrier (v_1) immediately. Then, v_1 carries packets until it comes to the communication range of v_2. When v_1 encounters v_2, v_1 forwards its packets to v_3 as shown in Fig. 3(b). Then, v_3 carries the packets until it encounters another vehicle moving backward belonging to another network component. In this way, the packets are delivered over a road segment. From this example, we can generalize the packet delivery process as a *renewal process* where each transaction consists of forwarding and carry process.

It can be seen that the renewal process consists of repetitive *cycles* of forwarding and carry. Since the forwarding delay is neglectably short compared to the carry delay, the link delay for a cycle is approximately the same with carry delay. In order to derive the expected carry delay ($E[d_c]$) for a cycle, we need to formulate the expected distance between the packet carrier (v_1) and the approaching vehicle from the opposite side (v_2) in Fig. 3(a).

Since we assume that the inter-arrival time of backward-moving vehicles ($\tilde{T_h}$) is exponentially distributed with the arrival rate λ_b, the inter-distance between backward-moving vehicles ($\tilde{D_h}$) is also exponentially distributed. As shown in Fig. 2, the expected distance between v_1 and v_2 is $E[l_n - l_f] + E[\tilde{D_h} \mid \tilde{D_h} > R]$ where $E[l_n - l_f]$ is the expected carry distance within the current network component and $E[\tilde{D_h} \mid \tilde{D_h} > R]$ is the average inter-distance of the vehicles moving backward. We derive $E[\tilde{D_h} \mid \tilde{D_h} > R]$ as follows:

$$\begin{aligned}
E[\tilde{D_h} \mid \tilde{D_h} > R] &= \int_{\tilde{D_h}=R}^{\infty} \tilde{D_h} P(\tilde{D_h} \mid \tilde{D_h} > R) d\tilde{D_h} \\
&= \int_{t=0}^{\infty} (R+t) P(\tilde{D_h} = R+t \mid \tilde{D_h} > R) dt \\
&= \int_{t=0}^{\infty} (R+t) P(\tilde{D_h} = t) dt \\
&\quad (\because \text{ Memorylessness of expoential random variable}) \\
&= \int_{t=0}^{\infty} R \times P(\tilde{D_h} = t) dt + \int_{t=0}^{\infty} t \times P(\tilde{D_h} = t) dt \\
&= R + \int_{t=0}^{\infty} t \times P(\tilde{D_h} = t) dt \\
&= R + v \int_{s=0}^{\infty} s \times P(\tilde{T_h} = s) ds \quad (\because \text{ Change of variable }) \\
&= R + vE[\tilde{T_h}] \\
&= R + \frac{v}{\lambda_b}.
\end{aligned} \tag{4}$$

Note that v_1 transmits packets to v_2 if their inter-distance is less than or equal to R and their relative speed is $2v$. Then, according to (4),

$$\begin{aligned}
E[d_c] &= (E[l_n - l_f] + E[\tilde{D_h} \mid \tilde{D_h} > R] - R)/2v \\
&= (E[l_n - l_f] + \frac{v}{\lambda_b})/2v.
\end{aligned} \tag{5}$$

Then, the *carry distance* is:

$$\begin{aligned}
E[l_c] &= vE[d_c] \\
&= (E[l_n - l_f] + \frac{v}{\lambda_b})/2.
\end{aligned} \tag{6}$$

The packets moves by $E[l_c]$ after a carry phase, hence the expected length of a cycle is $E[l_f] + E[l_c]$. Based on renewal process, this process is repeated for $\frac{l-R}{E[l_f]+E[l_c]}$ times, since a relay node is installed on I_j along with the communication range R. When packets arrive at I_i, there are two cases to deliver packets to I_i.

- **Case 1: Immediate Forward:** Assume that T^* is the inter-arrival time between the vehicles moving forward. If there is a next packet carrier in the communication range of the relay node at I_i, such a probability and the conditional expectation of link delay are:

$$\begin{aligned} P(\text{Case 1}) &= P(T_h^* < a) \\ &= 1 - e^{-\lambda_f a}, \\ E[\text{Link Delay}|\text{Case 1}] &= \frac{l - R}{E[l_f] + E[l_c]} \times E[d_c]. \end{aligned} \tag{7}$$

- **Case 2: Wait and Carry:** If there is no vehicle moving forward in the communication range of the relay node at I_i, such a probability and the conditional expectation of link delay are:

$$\begin{aligned} P(\text{Case 2}) &= P(T_h^* \geq a) \\ &= e^{-\lambda_f a}, \\ E[\text{Link Delay}|\text{Case 2}] &= \frac{1}{\lambda_f} + \frac{l - R}{E[l_f] + E[l_c]} \times E[d_c]. \end{aligned} \tag{8}$$

Considering both cases, the average link delay with relay nodes on each intersection is:

$$\begin{aligned} E[\text{Link Delay}] &= P(\text{Case 1})E[\text{Link Delay}|\text{Case 1}] + P(\text{Case 2})E[\text{Link Delay}|\text{Case 2}] \\ &= \frac{1}{\lambda_f} e^{-\lambda_f a} + \frac{l - R}{E[l_f] + E[l_c]} \times E[d_c]. \end{aligned} \tag{9}$$

5 Performance Evaluation

We validate of our model by comparing its expectation with simulation result. As shown in Table 1, vehicles travel over path length $1000m$ from both ends of the straight road. They move with speed $v \sim N(40, 5)$ MPH. The communication range of DRRC deivces is $200m$. At both the ends, relay nodes are installed.

- **Performance Metric**: We compare *average link delay* with the expected link delay.
- **Parameters**: We investgate the impacts of *average arrival rate* $\lambda (= \lambda_f + \lambda_b)$, *average vehicle speed* μ_v, and *vehicle speed deviation* σ_v.

Our model is compared with simulation result which approximates the ground truth. Furthermore, we also compare the simulation result of two-way traffic with that of one-way traffic.

5.1 Impact of Vehicle Arrival Rate λ

At first, we show how the link delay changes as the vehicle inter-arrival time varies. Note that the vehicle inter-arrival time is the reciprocal of vehicle arrival rate. One-way simulation result has longer delay than two-way simulation result. Vehicles deliver packets faster by using two-way traffic. Thus, we can deliver packets with shorter delay if we utilize the direction and location information from the GPS based navigation system. As shown in Fig. 4, our model accurately expects the link delay. Since [9] does not consider the mobility of vehicles

Table 1. Simulation Configuration

Parameter	Description
Road condition	The road is straight and $1km$ long.
Communication range	$R = 200$ meters (i.e., 656 feet).
Arrival rates	$\lambda_f = \lambda_b = 1/10$ where λ_f is the arrival rate for forward vehicular traffic and λ_b is the arrival rate for backward vehicular traffic.
Vehicle speed	$v \sim N(40, 5)$ MPH.

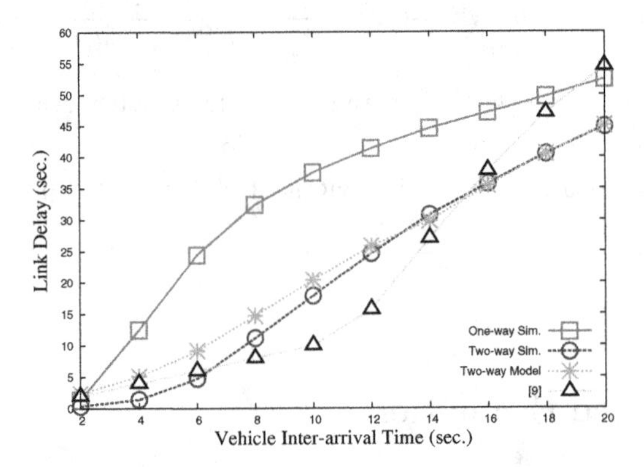

Fig. 4. Link delay versus vehicle inter-arrival time

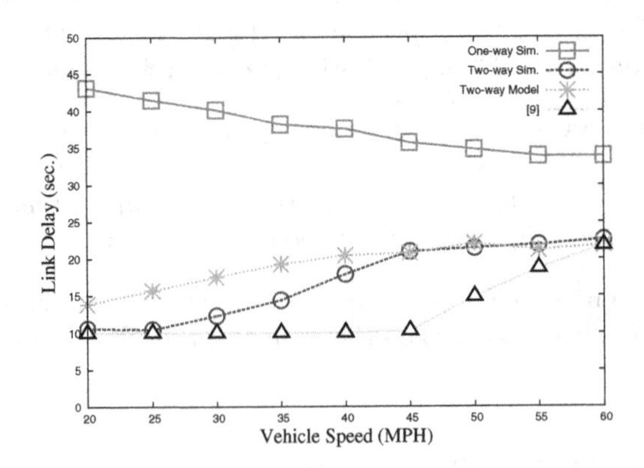

Fig. 5. Link delay versus average vehicle speed

toward forwarding direction, the expectation diverges exponentially. In comparison with [9], our model provides relatively closer result to the simulation result (Fig. 5).

5.2 The Impact of Vehicle Speed μ_v

Here, we investigate the impact of vehicle speed to the link delay. The higher vehicular speed results in the longer delivery delay for the dense traffic case. This is because a higher speed causes a longer inter-distance between vehicles. Moreover, the probability to construct a network component becomes low. Thus, high speed results in the longer delay for even heavy traffic case (Fig. 6).

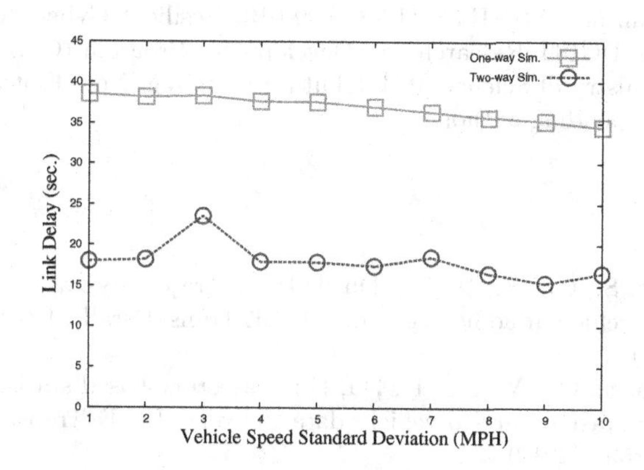

Fig. 6. Link delay versus vehicle speed standard deviation

5.3 The Impact of Vehicle Speed Deviation σ_v

We observe the impact of vehicle speed deviation to the link delay. We increase vehicle speed standard deviation from 1 to 10 MPH. Link delay becomes shorter as the speed standard deviation becomes larger. If two vehicles move with the same speed to the same direction, and they are out of communication range, there is no chance to make a network component. However, in case of different speed case, a faster vehicle can catch up a slower vehicle and they can construct a network component. This phenomenon happens more often in case of the large standard deviation than the low standard deviation.

6 Conclusion

In this paper, we propose average link delay for a road segment with two-way road traffic. In order to derive the expectation, we introduce the concept *renewal process.* This assumes that the forwarding and carry phases alternate repeatedly. We formulate link delay for a cycle because the carry delay is dominant, assuming that the inter-arrival time is distributed exponentially. This link delay modeling can be used for multihop infrastructure-to-vehicle data delivery in road networks. Two-way roads are dominant over one-way roads in real road networks. As future work, we will investigate link delay model, considering more realistic environments such as road networks with traffic signals. Also, we will evaluate data forwarding schemes for the multihop infrastructure-to-vehicle data delivery with our link delay model.

Acknowledgment. This research was supported by Basic Science Research Program through the National Research Foundation of Korea (NRF) funded by the Ministry of Science, ICT &Future Planning (2014006438). This research was supported in part by Global Research Laboratory Program (2013K1A1A2A02078326) through NRF, and the ICT R&D program of MSIP/IITP (14-824-09-013, Resilient Cyber-Physical Systems Research) and the DGIST Research and Development Program (CPS Global Center) funded by the Ministry of Science, ICT &Future Planning. Note that Jaehoon (Paul) Jeong is the corresponding author.

References

1. Jeong, J., Guo, S., Gu, Y., He, T., Du, D.H.C.: Trajectory-based data forwarding for light-traffic vehicular ad hoc networks. IEEE Trans. Parallel Distrib. Syst. **22**(5), 743–757 (2011)
2. Jeong, J., Guo, S., Gu, Y., He, T., Du, D.: Trajectory-based statistical forwarding for multihop infrastructure-to-vehicle data delivery. IEEE Trans. Mob. Comput. **11**(10), 1523–1597 (2012)
3. Carter, A.: The Status of Vehicle-to-Vehicle Communication as a Means of Improving Crash Prevention Performance. Technical report 05–0264 (2005). http://www-nrd.nhtsa.dot.gov/pdf/nrd-01/esv/esv19/05-0264-W.pdf
4. U.S. Department of Transportation. Federal Motor Vehicle Safety Standards: Vehicle-to-Vehicle (V2V) Communications. http://www.gpo.gov/fdsys/pkg/FR-2014-08-20/pdf/2014-19746.pdf
5. Rauwald, C., Nichols, H.: Audi Tests Driverless-Car Technology at 190MPH, October 2014
6. Ramsey, M., Fowler, G.: Ford chief says Mass-Market autonomous vehicle is priority, January 2015
7. LeBeau, P.: Mercedes' futuristic view: Autonomy meets connectivity, January 2015
8. Jeong, J., He, T., Du, D.H.C.: TMA: Trajectory-based Multi-Anycast for Multicast Data Delivery in Vehicular Networks. Technical Report 11–013. http://www.cs.umn.edu/tech_reports_upload/tr2011/11-013.pdf
9. Liu, Y., Niu, J., Ma, J., Shu, L., Hara, T., Wang, W.: The insights of message delivery delay in VANETs with a bidirectional traffic model. J. Netw. Comput. Appl. **36**, 1287–1294 (2013)

Biological Prototype Model for Road Abnormal Vehicle Behavior Simulation

Lingqiu Zeng[1], Xueying He[2(✉)], Qingwen Han[2,3], Xiaoying Liu[2], Chong Peng[2], Bin Yang[2], and Lei Ye[2]

[1] Chongqing Key Laboratory of Software Theory & Technology, College of Computer Science, Chongqing University, Chongqing, China
zenglq@cqu.edu.cn

[2] College of Communication Engineering, Chongqing University, Chongqing, China
{hxy,hqw,Yelei}@cqu.edu.cn, m18725810426_1@163.com, pengchong333@126.com, billyyong@live.cn

[3] Chongqing Automotive Collaborative Innovation Center, Chongqing University, Chongqing, China

Abstract. A striking feature of abnormal vehicle (AV) is the considerable infectiousness to normal vehicle, which greatly influence the collision avoidance strategy for driving model of autonomous vehicle. However, the overall transmission of vehicle's abnormal behavior has not been well understood although many factors affecting this process have been found out. In view of social and human factors in driving, we propose an idea which considers the road vehicle as 'living creature' to explore the essence of vehicle's abnormal status transmission. Here we synthesize what is currently known about the natural history of AV by developing a biological prototype model to mimic the transmission process of abnormal behavior in a road system. The current model can also be extended to include human activities and thus be used to investigate the influences of driver style on collision avoidance.

Keywords: Accident detection · Traffic simulation · Model predictive control · Autonomous vehicles · Biological system modeling

1 Introduction

Road transportation safety is a 'grand challenge' problem for a modern industrial society. With the rapid development of intelligent vehicle technologies, road vehicles are able to communicate with each other for safety commercial applications, and furthermore, exhibit decision-making abilities. In this case, the vehicle

X. He—This research are supported by National Nature Science Foundation of China, Project No. 61272194, Fundamental Research Funds for the Central Universities, Project No. CDJZR13180043, CDJPY12160002 and Chongqing university graduate student research innovation project fund, 2015. Thanks for the open research fund of Chongqing Key Laboratory of Emergency Communications.

C.-H. Hsu et al. (Eds.): IOV 2015, LNCS 9502, pp. 351–362, 2015.
DOI: 10.1007/978-3-319-27293-1_31

is not only a physical entity but also a 'living creature' and should be considered as an intelligence agent. Therefore, we believe that road accident are the result of interactions among road vehicles.

Most transportation researchers consider that traffic models are one of the most effective tools to investigate road traffic status. Existing studies have tried to identify the overall relationship between road traffic conditions, but ignore the influence of driver behavior. Meanwhile, Subir believed that driver behavior was the leading cause of more than 90 percent of all accidents [1]. Taking into account driver behavior, Dixit proposed a two-fluid model whose parameters are influenced not only by the roadway characteristics but also by the behavioral aspects of the driver population, e.g., aggressiveness [2].

With the advent of autonomous vehicles, a question that has been raised is what type of traffic model is suitable as a collision avoidance strategy for autonomous driving and an incentive for good driving [3,4]. In general, an accident case involves more than one vehicle, a multi-car chain accident creates a serious traffic jam and leads to a very dangerous situation. In the opinion of Mather and Romo, both social and human factors should be considered in road safety research [5]. A road traffic model covering both social and human factors is a complex system in which various factors should be taken into account and different levels of analysis are needed to achieve multi-objective optimization. Biological networks have been honed by several cycles of evolutionary selection pressure and are likely to yield reasonable solutions to combinatorial optimization problems [6]. Assuming road vehicles to be 'living creatures', we combine the human factor and social factor in driving as biological features, and use them to construct a road traffic model. In the proposed model, an infection circle, which is similar to that in parasites, is assumed to illustrate the progression of road hazard diffusion.

Based on the biological features assumed in the proposed model, we divided abnormal road vehicles into two categories: infector and infectee. When a vehicle performs abnormally, it is considered to be an abnormal vehicle (AV, primary infector) [7], which could then lead to it infecting any given region on the road. A normal vehicle is an infectee if it is infected as a member of an infected region in which there is another AV. If that vehicle becomes an infectee, it acts as an infector in the next infecting stage. We can then consider AVs to be parasite carriers for both infectors and infectees, which should lead to regional diffusion. Notably, the primary infector is generated by the combined effect of road factors and driver behavior, whereas the infectee is generated as a result of the infection progress.

In biological terms, the infection progress is decided by three key factors, which are the toxicity of the initial parasite, the immunity of a road vehicle, the distance between the infector and the infectee. Most biological infection models are constructed to illustrate a hybrid effect of these three factors on parasite diffusion. Using this method, we believe that we can build a model to investigate the transmission features of abnormal road vehicle behavior. However, although the only problem is what type of biological model would be suitable for a vehicle model, we should consider not only the human behavior of the driver but also the social behavior of the object region.

Numerous biological infection models have been proposed in the past few years. The results of recent studies have led to the opinion that T.gondii may affect human behavior [8] and even human culture [9]. To some degree, an abnormal road vehicle event is a type of experience event, which should affect driver behavior and his social attributes in a similar manner as T.gondii does. Of course, until recently, studies just assumed that T.gondii may affect human behavior and culture, but did not know how. On the other hand, we have also conducted this study to investigate the influence of an AV on an object driver. Perhaps additional research on T.gondii would benefit the study of driver behavior. In this view, we selected the T.gondii transmission biological model as the prototype of the AV infection model.

In view of epidemiological principles, this paper employs a T.gondii model to illustrate the interactions among road vehicles. Based on the AV infection behavior, we exploited the dynamics of parasite infection features to develop a biological model of abnormal vehicle behavior.

The rest of the paper is organized as follows. In Sect. 2, we develop a prototype model to describe the transmission of an AV particle on a road segment. In Sect. 3, an agent-based modeling simulation platform called AnyLogic is employed to determine the dynamics and discrete events for an AV infection procedure. Finally, the conclusion and discussion are given in Sect. 4.

2 Prototype Model for Abnormal Behavior Transmission

AV is the main body of the proposed model. To construct the model, the following definitions are given first [10].

Definition 1: AV Behavior. AV behavior is classified by corresponding to abnormal vehicle events, such as sudden braking, erratic large steering angle and hence, any dramatic change in the moving direction of a moving vehicle. In-car sensors gather the AV information for which the process of information generating can be described by a function f_c, which represents the AV features. Importantly, the information concerning the hazard level of an AV j, can be determined by

$$H(j) =< f_c, Density_{veh}, Road_j > \tag{1}$$

where the hazard level is H, j is the AV to be considered, $Density_{veh}$ is the vehicle density in the area influenced by an AV, and $Road_j$ is the traffic information in the vicinity of an AV event.

Definition 2: AV Particle. The first emergence of an AV in a given region is defined as the primary AV particle, whereas an infection AV particle is a vehicle that reacts with the primary one and becomes an AV. A primary AV particle occurs earlier than an infected one, while the hazard level is not comparable. The distribution characteristics of vehicles depend on region parameters, such as vehicle density and speed.

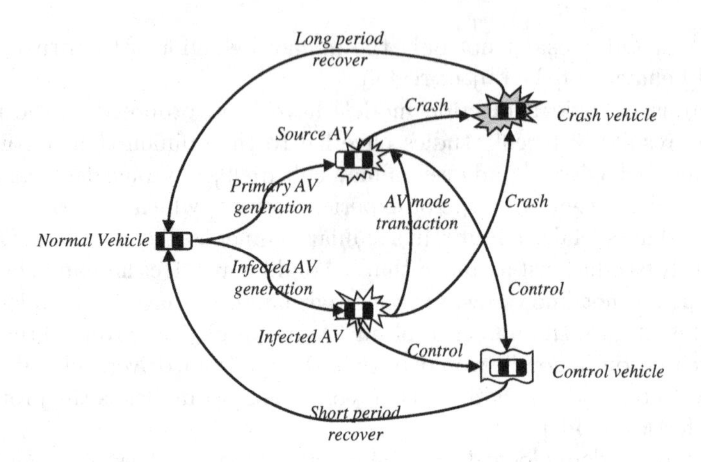

Fig. 1. Life cycle of AV particle

To describe the complete life cycle of AV particles and to quantify their effects on surrounding vehicles, a schematic is used and is given in Fig. 1.

As shown in Fig. 1, based on its status, a vehicle can be compartmentalized into one of three divisions, which are susceptible to, infected with, or recovered from an AV particle. Susceptible vehicles are those that have not been previously infected and can be infected by an AV particle. After a primary AV particle is generated, it becomes the first infection source for a series of subsequent infection events. After infection, a vehicle will undergo a latent period before recovering from its abnormal status. Here, an infected AV may influence other surrounding vehicles that would then become new infection sources. After a period of time, an AV gradually relieves its abnormal status. This recovery period varies from several seconds to several hours depending on the AV damage level. Exposure to infection should increase vehicular immunity to AV particles, which can be explained as the acquisition of driver experience.

2.1 Environment

To explain the AV course of infection, we use a 180 m road with 4 lanes in a single direction, which is divided into 4×60 cells. The surface area of each cell is 9 m^2. We generated vehicles with a random distribution in a given region and assumed that the radiation characteristic of an AV is omni-directionality.

Based on a study by Wen Jiang [11], the T.gondii infection status is quite decided by the prey distance. We then suppose that the AV influence exponentially decreases with the distance between an AV and a victim vehicle, while the distance probability of infection is described as follows,

$$p_{ij} = \frac{e^{\frac{r_{ij}}{r_0}} - e}{1 - e} \tag{2}$$

where r_{ij} is the distance between the infector AV i and the infectee vehicle j, r_0 is the maximum influence distance whose unit is the cell. Note that r is computed at the centers of the cells and that its value is discrete.

2.2 Infection

The AV infection procedure is influenced by three factors.

Immune Capability of Infecting Vehicle. A vehicle with a novice driver lack antibodies to an AV particle and can be easily infected, whereas an experienced driver should maintain its status, which presents a much lower infection probability.

We assume parameter a_i to denote the immune level of vehicle i as follows:

$$a_i(n) = 1 - A^n \tag{3}$$

where n is the number of infected times. A is the convergence coefficient that decides the convergence rate of a_i and should be selected reasonably.

Taking into account driver experience, a vehicle with a novice driver exhibits a high infection probability, whose growth slope cannot be caught up with immune level a_i. As a supplement, we define the novice driver losing rate as follows:

$$l_i = \begin{cases} e^{2-a_i}, a_i \leq th_n \\ 1, \quad else \end{cases} \tag{4}$$

where th_n is the novice threshold.

Then, the vehicle immune factor is denoted as follows:

$$I_i = \frac{l_i}{a_i} \tag{5}$$

Hazard Parameter of an Abnormal Vehicle. The rate at which vehicles are infected by an AV particle depends on the AV hazard degree. Infection with an AV may cause damage whose severity is determined by the surrounding AV quantity as well as the hazard characteristics of the source AV.

T.gondii with intermediate virulence possesses a high infecting capability. For convenience, Jiang et al. assumed that the relationship between virulence and transmission is a normal distribution [11]. Therefore, in this paper, according to Jiangs hypothesis, a normal distribution is used to represent the diffusion ability of AV j.

$$D_j = e^{-(v-\mu)^2/2\sigma^2} \tag{6}$$

where D_j stands for the diffusion ability of different AV hazard levels, v represents the AV hazard level, and μ and σ are the mean and standard deviation of the distribution, respectively. In this paper, a 5 level criterion is employed to illustrate v.

The infection feature of a normal vehicle is related to the diffusion ability of the source AV and the distance probability, which define the hazard parameter of a normal vehicle i to AV j as,

$$h_{ij} = D_j p_{ij} \tag{7}$$

where D_j is the diffusion ability of the source AV j.

Infection Risk. We define the infection risk of a normal vehicle i to AV j as follows:

$$f_{ij} = 1 - e^{-kI_i h_{ij}} \tag{8}$$

where k is the infection constant.

One normal vehicle can be infected by multiple AVs, and then the accumulating infection risk is as follows:

$$f_i = \begin{cases} \sum_{j=1}^{m} f_{ij}, f_i \leq 1 \\ 1 \quad , else \end{cases} \tag{9}$$

where m is the number of source AVs.

To evaluate the status of the infecting vehicle, we set the infection threshold th_{infect}. If $f_i > th_{infect}$, then the infected vehicle is considered to be an AV.

2.3 Recovery

After primary infection, a vehicle starts a recovery procedure. The recovery time may vary from several seconds to several hours depending on the degree of AV damage.

The degree of AV damage i at the infection time is defined as follows:

$$t_{ini} = \rho v_i l_i \tag{10}$$

where v_i is the hazard level of AV i, ρ is the vehicle density and l_i is the novice driver degradation rate.

As shown in Fig. 1, there are three vehicle states in the AV recovery procedure: normal vehicle, abnormal vehicle and crash vehicle. The occurrence of an accident or crash eventually depends on the degree of AV damage. In this work, we assume that an accident is considered to have happened if t_{ini} is higher than the crash threshold th_{crash}.

If an accident happened, the recovery time should be much longer than that when there is no crash. Here, we define the residual degree of AV i damage as decaying exponentially with time as follows:

$$t_c(t) = t_{ini} e^{-t/k_{rm}} \tag{11}$$

where k_{rm} is the recovery constant and is set based on the result of the AV event, which should be under control for slight crashes and serious crashes. Moreover, we assume that an AV changes to a normal vehicle if $t_c(t)$ is lower than the recovery threshold $th_{recover}$.

2.4 Secondary Infection

After primary infection, an un-crashed AV shall enter the recovery period whose duration corresponds to its residual degree of damage. If AV i is infected by another source AV during its immunity period as a secondary infectee, its degree of damage should change due to the accumulation of a hazard level.

Here, we define the secondary infection coefficient AV j as follows:

$$n_{CI} = 1 + f_j \tag{12}$$

where f_j is the accumulating probability for secondary infection.

Then, the new hazard level of AV j is as follows:

$$v_j^* = \begin{cases} ceiling(v_j \cdot v_{CI}), v_j^* \leq 5 \\ 5 \quad , else \end{cases} \tag{13}$$

If repeated infection occurs, the duration of the recovery period should be recalculated according to the new source AV behavior, which can be obtained by Eq. 10.

3 Simulation for Biological Model of AV Infection

3.1 Model Parameters Selection

Statistic mathematical models benefit biological research particularly because they allow for the examination of variables that are difficult to control under complicated conditions. A road accident is a complex and unpredictable event that encompasses various properties, such as driver experience, vehicle density, traffic conditions, and these properties are often intrinsically stochastic. We then believe that the road traffic statistic results are suitable for selection of parameters for the proposed biological AV model.

As reported by the traffic management office of China [12], 3906164 accidents happened in 2011, while the total number of vehicles was 90859439. For an individual vehicle, the probability of an accident in one year is approximately 4.3 %. We assume that the accident probability is 5 % of all AV events. Then, 17 AV events should happen to an individual vehicle in a 1-year period. It is general believe that a driver with 36 month driving experience is considered as a skilled driver while a driver with less than 12 months driving experience is regarded as a novice driver [13]. Also reported by the traffic management office of China was that 3686652 road accidents out of 3906164 total met the simple processing requirement, which means that 94 % of road accidents should recover within a relatively short period of time.

Based on the facts above and the characters of these equations we proposed, we obtained default values of the parameters for proposed AV model, which are given in Fig. 2.

Definition		Value
Number of X grid		60
Number of Y grid		4
Maximum influence distance r_0		4 cells
Vehicle density ρ		0.5
Convergence coefficient parameter A		0.9
Novice threshold th_n		0.8
Distribution means of diffusion ability μ		4
Standard deviation of diffusion ability σ		0.8
Infection threshold th_{infect}		0.35
Crash threshold for initial damage th_{crash}		0.6
Serious accident threshold for initial damage		0.9
Recovery threshold for initial damage		0.1
Recovery constant k_{rm}	Control AV	5
	Slightly crashed AV	85
	Serious crashed AV	820

Fig. 2. Default values of parameters in the model

3.2 Simulation Platform

In this paper, an agent-based modeling simulation platform called AnyLogic is employed to fully analyze the AV clustering procedure. Based on the AnyLogic 6.0 platform, a formal specification framework is coupled with an agent-based model for modeling and simulation of road vehicles in a complex environment.

Two agents are defined in our model. One is the main object, which is used to control the main flow, and the other is the vehicle object, which has four statuses: normal vehicle, infected vehicle, control vehicle and crash vehicle. The sequence flow model is shown in Fig. 3.

At the first step, initialization of the model is implemented by a startup code in the main agent. Two tasks are involved during the initialization processing; one is the generation of the source AV, which is randomly selected from normal vehicles, and the other task is the initialization of vehicle sets. A timer is set for the main agent, which in turn periodically triggers vehicle agent functions. When the functions are called, they can scan all of the vehicles to calculate the AV parameter and at last fulfill any vehicle status change.

3.3 Experimental Setup

Experimental runs are conducted through the initial setup as follows:

- Each individual experiment generates only one primary AV particle;
- All vehicles in the object region randomly scatter in a grid cell manner;
- AV behavior should diffuse into the surrounding area, which is the 'diffusion environment'. The AV particle transmission factor follows the infection feature definition given in Fig. 2;
- A 240 meter road with 4 one-direction lanes is then used to simulate the road AV infection procedure.

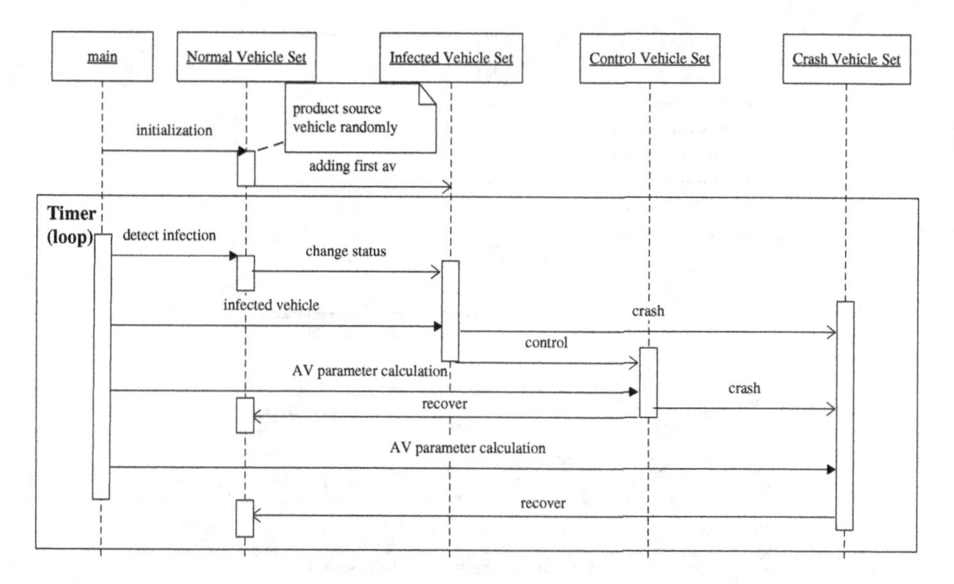

Fig. 3. Models sequence flow

3.4 Experiment Sets

Experiment Set I: In the first experiment set, the density of vehicles is set to 33.3 %, and the hazard level of primary AV particles is level 5.

Simulation results are shown in Fig. 4.

In Fig. 4, a whole infection and recover procedure is illustrated. Here, when primary AV particle emerges, an infection progress is undergoing.

At time $t_1 = 1$ s, the first infection step goes on. Normal vehicles infecting probability is obtained according to the definition of Eq. 7. A vehicle near the infector perhaps maintains its status due to high immunity.

With continuing progress of infection, at time $t_2 = 3$ s, new infectees emerge, while some of the infectees at time t_1 recover to normal.

At time $t_3 = 7$ s, the infection progress produces the largest infected area, whose AV is at a peak value.

At time $t_4 = 8$ s, the amount of AV decreases while the infected area holds.

Finally, at time $t_5 = 16$ s, the infected area shrinks while the amount of AV decreases.

The final status presents only one serious crash vehicle and stays stable. At this time, the whole infection and recovery process finishes.

Notably, most of the infected AVs do not crash, and they undergo a short recovery period. In general, the chain reaction accident always happens in a high vehicle density area. To investigate the relationship between the AV infection probability and vehicle density, experiment set II is performed.

Experiment Set II: The second experiment set is used to illustrate the relationship between the AV infection probability and vehicle density.

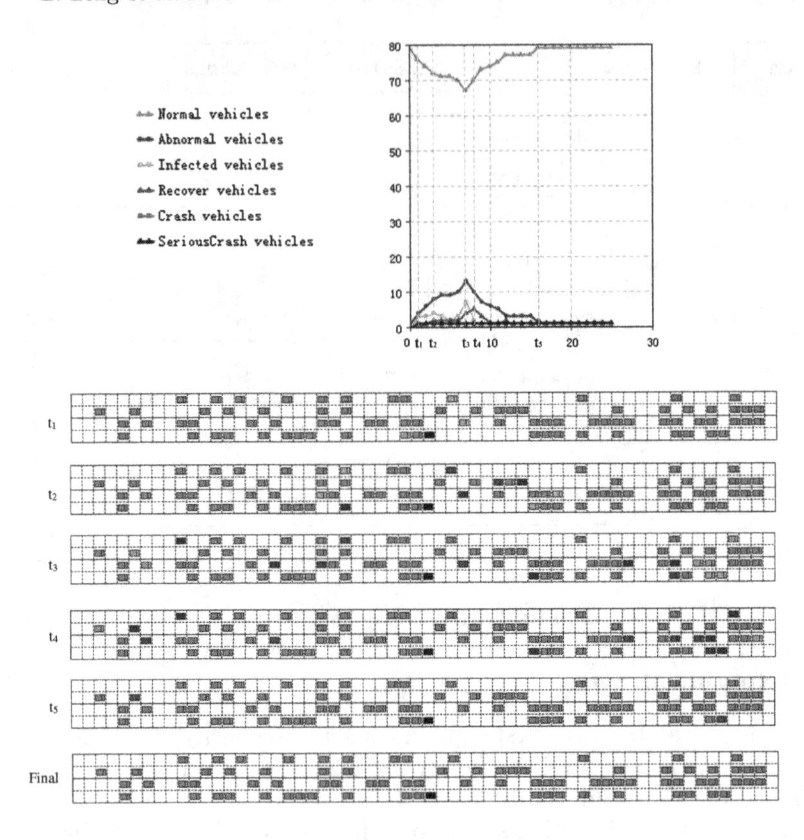

Fig. 4. A whole procedure of infected and recovery

We repeated the experiment set I simulation 100 times and plotted the averaged results of those 100 simulations in Fig. 5.

In Fig. 5(a), the primary AV is a crashed AV whose hazard level is set to 4. It is not surprising that an AV on a road with higher vehicle density infects more vehicles than on a road with lower vehicle density. Furthermore, we can see that if the vehicle density is below 40 %, there should be only one crashed vehicle, which corresponds to a primary AV particle with a hazard level 4 setting. With an increase in vehicle density, more crashed vehicles emerge. In Fig. 5(b), we change the primary status of the AV to un-crashed and the hazard level to 3. The probability of an AV slightly decreases, and one crashed vehicle status corresponds to a 40 % vehicle density.

4 Discussion and Conclusion

In view of social and human factors while driving, an idea that considers the road vehicle as a 'living creature' has been proposed, and a prototype model has been developed to study the transmission dynamics of AV particles in a vehicle population and the influences of various factors. The model consists of modeling

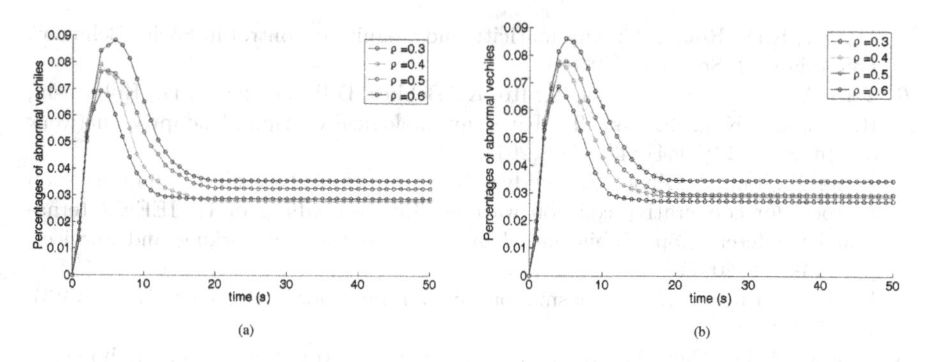

Fig. 5. Interaction between AV infection probability and vehicle density

the interactions between normal vehicles and abnormal individuals and explicitly describes the complete transmission cycle of AV particles. We believe that such a model may provide a useful start to improving road traffic safety and topology control for self-organized networks, such as remote sensor arrays, mobile ad hoc networks, or wireless mesh networks.

Model parameters are examined to catch up with the real-world road accident characteristics, which are set based on stochastic results of Chinese traffic accident reports and a series of analytical calculations.

The proposed AV prototype model just considers a static vehicle state and lacks adaptive decision-making ability based on internal states and local interactions. Future research will consider vehicles moving status and deeply consider driver characteristics, especially keeping track of driver ages and driving style. A more realistic AV model can be implemented to account for structured territories and various environmental influences. The model can be evaluated with new data and simulation studies to elucidate AV particle transmission. Moreover, we hope that the research achievement of T.gondii will provide a good reference to improve the AV model.

References

1. Biswas, S., Tatchikou, R., Dion, F.: Vehicle-to-vehicle wireless communication protocols for enhancing highway traffic safety. IEEE Commun. Mag. **44**(1), 74–82 (2006)
2. Dixita, V.V., Pande, A., Abdel-Atyc, M., Dasc, A., Radwanc, E.: Quality of traffic flow on urban arterial streets and its relationship with safety. Accid. Anal. Prev. **43**(5), 1610–1616 (2011)
3. Chen, W., Chen, L., Chen, Z., Tu, S.: WITS.: A wireless sensor network for intelligent transportation system. In: IEEE International Multi-Symposiums on Computer and Computational Sciences, vol. 2, pp. 635–641 (2006)
4. Kim, D., Moon, S., Park, J., Kim, H.J., Yi, K.: Design of an adaptive cruise control/collision avoidance with lane change support for vehicle autonomous driving. In: ICCAS-SICE 2009, pp. 2938–2943. IEEE Fukuoka (2009)

5. Mather, R.D., Romo, A.: Automaticity and Cognitive Control in Social Behavior. Fountainhead, Southlake (2007)
6. Tero, A., Takagi, S., Saigusa, T., Ito, K., Bebber, D.P., Fricker, M.D., Kobayashi, R., Yumiki, K., Nakagaki, T.: Rules for biologically inspired adaptive network design. Sci. **327**(5964), 439–442 (2011)
7. Yang, X., Lui, J., Zhao, F., Vaidya, N.H.: A vehicle-to-vehicle communication protocol for cooperative collision warning. In: Proceedings of 1st IEEE International Conference on Mobile and Ubiquitous System Networking and Services, pp. 114–123 (2004)
8. Flegr, J.: Effects of toxoplasma on human behavior. Schizophr. Bull. **33**(3), 757–760 (2007)
9. Lafferty, K.D.: Can the common brain parasite, Toxoplasma gondii, influence human culture? Proc. Roy. Soc. B: Biol. Sci. **273**(1602), 2749–2755 (2006)
10. Han, Q., Zeng, L., Liu, Y., Liu, Y.: An adaptive clustering algorithm for road abnormal region analysis. Trans. Inst. Measur. Control **36**, 1–11 (2013)
11. Jiang, W., et al.: An agent-based model for the transmission dynamics of Toxoplasma gondii. J. Theor. Biol. **293**, 15–26 (2012)
12. China, T.m.o.o. http://www.tranbbs.com/news/cnnews/Construction/news_82586.shtml
13. China, T.m.o.o. http://society.nen.com.cn/society/56/3594056.shtml

Public Bicycle Prediction Based on Generalized Regression Neural Network

Min Yang, Yingnan Guang(✉), and Xuedan Zhang

Electrical Engineering Department of Tsinghua University, Beijing, China
{m-yang13, gyn13}@mails.tsinghua.edu.cn,
zhangxuedan@sz.tsinghua.edu.cn

Abstract. Nowadays building a green and efficient public transportation system for the expanding urban population is undoubtedly a big challenge. In recent years, public bicycle system has been widely appreciated and researched worldwide. Unlike traditional public transportation system, public bicycle system doesn't need to follow fixed schedule. This flexibility brings high efficiency as well as uncertainty-we don't know whether there are available bikes or bike stands when they are indeed needed. This paper aims to predict the number of available bikes at given future time point so as to optimize the user's travel choices. In this article, we propose a new prediction model and use generalized regression neural network as our prediction algorithm to optimize prediction accuracy. Experimental results show that in this way we can properly handle this nonlinear problem.

Keywords: Public bicycle system · Prediction model · Generalized Regression Neural Network

1 Introduction

Public bicycle system as a green and efficient public transportation system has caught a lot of attention in the past few years worldwide. Globally the development of public bicycle system has gone through three generations [1]. The concept of public bicycle first originated in Europe. As early as 1965, an anarchist organization in Amsterdam provided free access to several white and unlocked bikes for the citizens. This project is the so called "White Bicycle Plan". Unfortunately, most of these bikes were missing just after several days. That is to say, the project failed just from the beginning. However, this attempt is considered to be a meaningful trial and treated as the origin of the first generation of public bicycle system. The second generation of public bicycle system appeared in Denmark. In this system, you should hypothecate some coins when you checked out a bike and the coins would be returned to you after you restored the bike. But this system still didn't function well because of poor management. Since the 90 s of the 20th century, with the use of modern advanced electronic, wireless communication and Internet technology, the third generation of public bicycle system began to arise in Europe. Such as Vélib' system in Paris, Vélo'v system in Lyon, SmartBikeDC system in Washington, DC and Bicing system in Barcelona. In these systems, Internet Service Provider(ISP) can monitor who is using a certain bike, the

C.-H. Hsu et al. (Eds.): IOV 2015, LNCS 9502, pp. 363–373, 2015.
DOI: 10.1007/978-3-319-27293-1_32

real-time bicycle number of each bike station and the state of each bike station (whether there exist broken bikes or bike slots and whether the station is in service) [2].

Unlike traditional public transportation system, public bicycle system doesn't need to follow fixed schedule. This flexibility brings high efficiency. Users can go to proper bike station for available bikes or bike stands. But this flexibility also brings about uncertainty [3]. We don't know whether there are available bikes or bike stands when they are indeed needed. So to us the biggest challenge is to accurately predict the number of available bike or bike stands at given future time point. Many scholars have done meaningful research in this field. Froehlich et al. [4] use Bayesian Networks for short-term and long-term (5-minute and 2-hour-ahead) predictions of bike availability in Barcelona's Bicing system. Their prediction model incorporates the time of the day, currently available bikes and prediction window. Kaltenbrunner et al. [5] use another quiet different algorithm. Considering the time sequence of available bikes, they use Autoregressive Moving Average (ARMA) time series model. Based on the instability of the number of available bikes, Yoon et al. [6] use Autoregressive Integrated Moving Average (ARIMA) time series model to promote prediction accuracy. Chen et al. [3] not only consider internal, but also the impact of external factors (such as weather) on the prediction results. By applying Generalized Additive Model (GAM) [7], they can predict the number of available bikes as well as provide estimates of the waiting time distribution when there are not available bikes.

To promote the efficiency of the whole system, we also need to consider the locations and number of bike stations. And the allocation of bikes in each station is also very important [8, 12]. Lin et al. have worked on strategic design of public bicycle system based on service level [8]. To better design public bicycle system, Girardin et al. [10] also analyze the mobility of tourist with user-generated content.

In particular, the main contributions of this paper are: (1) taking the impact of surrounding stations on prediction result into consideration. In our experiment we get to know that surrounding stations have great impact on prediction result in some situations; (2) using Generalized Regression Neural Network (GRNN) [9] as our prediction algorithm. Since GRNN can almost fit all nonlinear functions, we can greatly improve the prediction accuracy with it; (3) analyzing factors that may affect prediction result.

2 Data Collection and Definition

We got the data used in this paper by crawling the website of Bicing system. The website offer realtime data of 420 bicycle stations (Fig. 1). Our program collects station ID, available bikes and available bike slots of each of the 420 bicycle stations in Barcelona every 5 min. We separate the collected data into 2 parts. The first part is the data we collected from June 1st to July 17th, and we use it as our training set; the second part is the data we collected from September 1st to September 30th, and we use it as test set. Some of the data we collected are useless and they may do harm to our prediction accuracy. So we must reject the useless part to get better prediction result. Next we will show some important parameters in this paper:

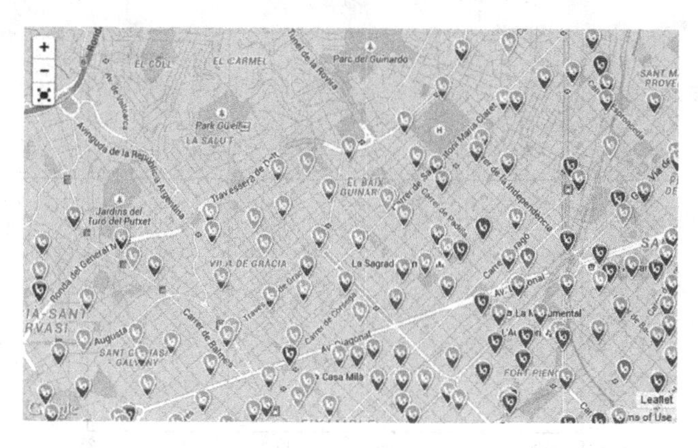

Fig. 1. The 420 bicycle stations distributed across the city of Barcelona, Spain

Normalized available bikes (NAB):

$$\mathrm{NAB} = \alpha/(\alpha + \beta) \tag{1}$$

In this equation, α stands for the number of available bikes at time t, β stands for the number of available bike stands at time t. Since the sum of available bikes and available bike slots is not a constant (some bikes or bike slots may be damaged, so we cannot use them), NAB can reflect the percentage of available bikes.

Normalized activity score (NAS):

$$NAS = |\gamma - \alpha|/(\alpha + \beta) \tag{2}$$

In this equation, γ stands for the number of available bikes at time t-1.NAS can effectively indicate how active a station is at a given time t.

Checked-out bikes (CB):

$$CB = \eta \tag{3}$$

In this equation, η stands for the number of available bikes of all stations. This parameter can reflect the number of bikes that have been checked out from the whole system.

3 Analyse of Time Pattern

In order to optimize our prediction accuracy, we should first analyse the time pattern of Bicing system.

Time pattern can be divided into 2 categories: single-station time pattern and overall time pattern. Single-station time pattern has 2 aspects: one is NAB's changing

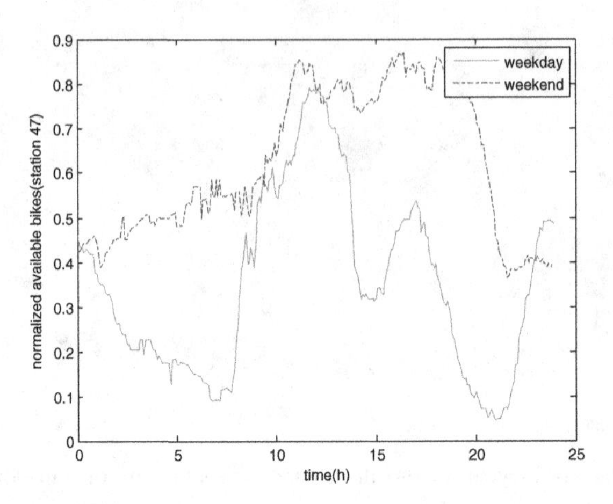

Fig. 2. Comparison of weekday and weekend normalized available bikes over 24 h at Station 47

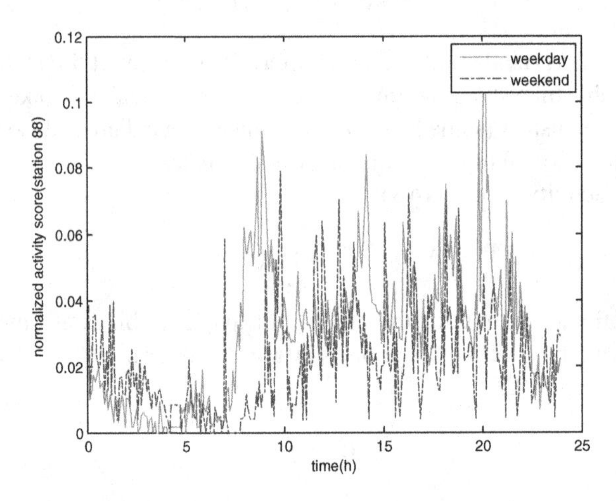

Fig. 3. Comparison of weekday and weekend normalized activity scores over 24 h at Station 47

over time on weekdays and weekends; the other one is NAS's changing overtime on weekdays and weekends. Overall time pattern is CB's changing over time on weekdays and weekends. Figure 2 compares weekday NAB with weekend NAB over 24 h at Station47; Fig. 3 compares weekday NAS with weekend NAS over 24 h at Station47; Fig. 4 compares weekday CB with weekend CB over 24 h.

According to these three pictures, both single-station and overall time patterns differ on weekdays and weekends significantly. So we must discuss these two situations separately. In the following part of this article we only discuss weekday situation and we can handle weekend situation with the same method.

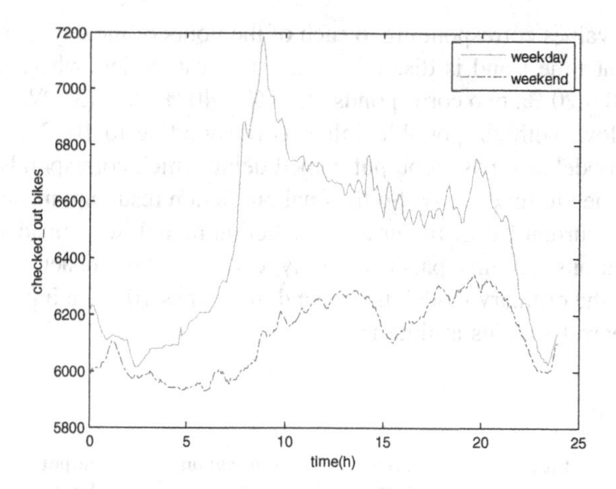

Fig. 4. Comparison of weekday and weekend checked_out bikes over 24 h

4 Numerial Prediction

4.1 Introduction to Generalized Regression Neural Network

Generalized Regression Neural Network (GRNN) was first proposed by American Scholar Donald F.Specht [9] in 1991. Generally it is a kind of Radial Basis Function (RBF) [11] neural network. Its structure has a lot in common with that of RBF neural network. To be brief, its structure is made up of four layers: the first layer is the input layer. We use this layer to accept input data (the input data may have several dimensions). The second layer is the pattern layer. In this layer we can get the Euclidean distance between input data and testing sample. The third layer is the summation layer. From this layer we can get the weighted sum of the output of pattern layer. And the last layer is the output layer. Its output is our prediction result. We can see the structure chart of GRNN in Fig. 5. Besides, GRNN has strong nonlinear mapping ability, flexible network structure and a high degree of fault tolerance and robustness. Compared with RBF, GRNN has more advantages on learning speed and approximation ability. Apart from this, GRNN can also deal with unstable data. Therefore, GRNN has been widely used in various fields such as structural analysis, financial engineering and signal processing. In our paper, we use GRNN as our default prediction algorithm.

4.2 Prediction Models

we propose a new prediction model which considers the effect of surrounding bicycle stations (Fig. 7). In contrast to ordinary prediction model (Fig. 6), the bicycle number of surrounding stations can be explicitly factored into prediction model, resulting in significant gains in terms of prediction accuracy.

In the ordinary model, we don't consider the impact of surrounding stations on the prediction result. This model has 3 inputs: (1) time. It represents time of the day and has 24 possible values corresponding to each of the hours of the day; (2) bikes. It is the value of NAB at time t and is discretized into five categories, where a value of one corresponds to 0 ~ 20 %, two corresponds to 20 % ~ 40 %, etc.; (3) PW. The size of the prediction window, with six possible values corresponding to 10, 20, 30, 60, 90 and 120 min. This model also has an output named delta, which corresponds to the change of NAB from time t to time t+PW. So the final prediction result are made by adding the value of delta to current NAB. In our new prediction model, we consider the impact of surrounding stations (in this paper we only consider the impact of their NABs). Compared with the ordinary model, the second model has 10 more inputs. They are the NABs of 10 nearest stations at time t.

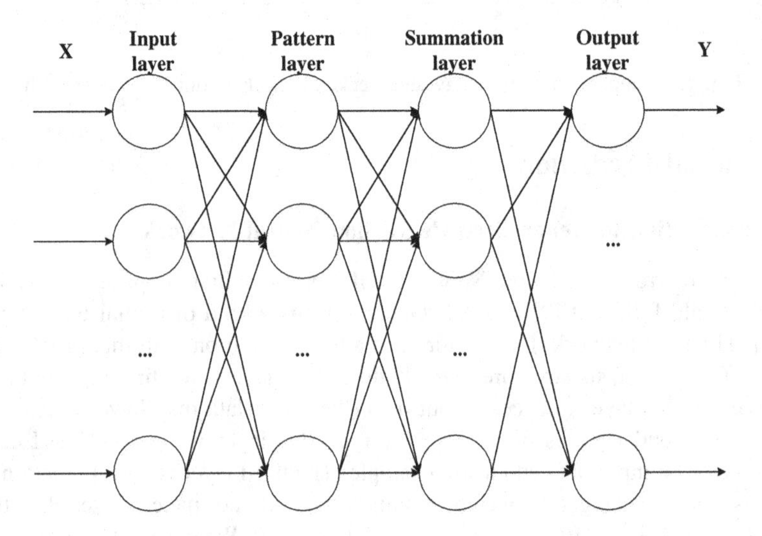

Fig. 5. The structure chart of GRNN

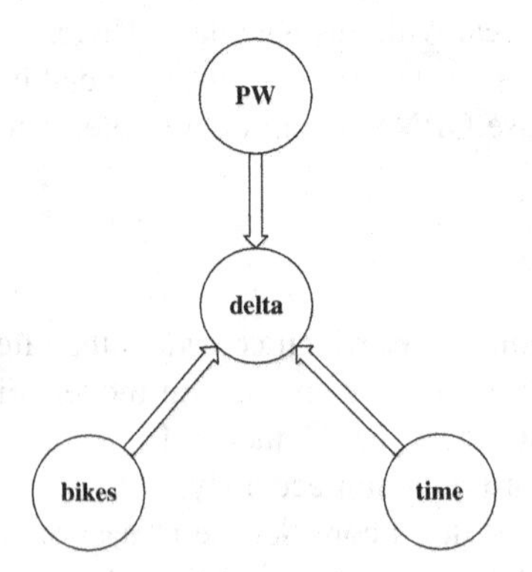

Fig. 6. The structure of ordinary model

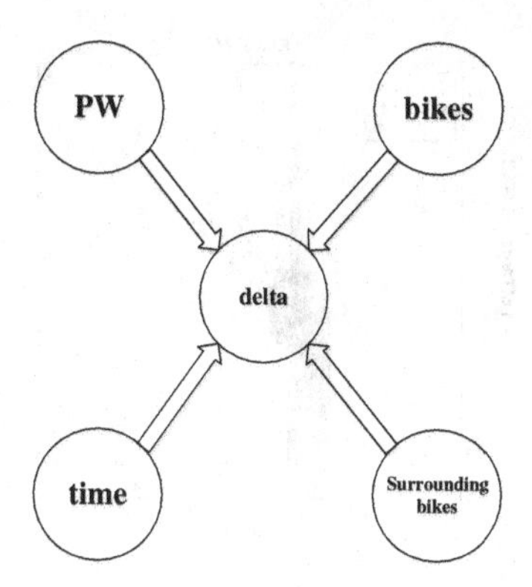

Fig. 7. The structure of our new model

4.3 Prediction Result and Evaluation

In this section, first we will compare the performance of ordinary model and new model. Then we will see whether generalized regression nerual network has a better performance than prevalent algorithm(autoregressive moving average algorithm).

Figures 8 and 9 show the fitted curves of GRNN algorithm and ARMA algorithm in ordinary model. Figures 10 and 11 show the fitted curves of GRNN algorithm and ARMA algorithm in our new model.

According to the four pictures, we can get the comparison table (Table 1). In Table 1 we report the error comparison between ordinary model and new model for both short(20 min ahead) and long(2 h ahead) term predictions using different prediction algorithm. We notice that our new model outperforms the ordinary model and GRNN algorithm outperforms ARMA algorithm.

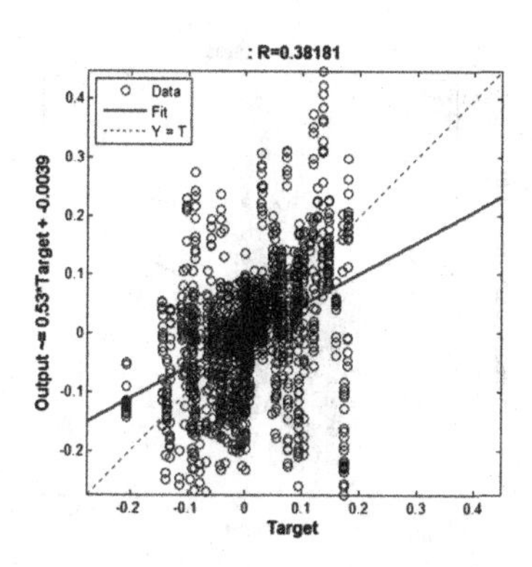

Fig. 8. Fitted curve of GRNN algorithm in ordinary model (target represents prediction value and output represents real value)

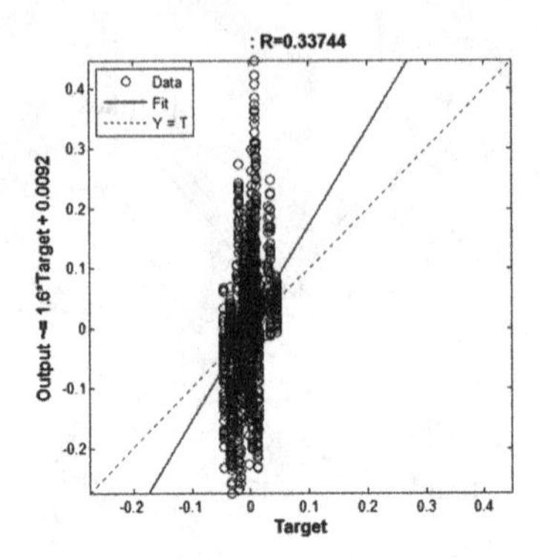

Fig. 9. Fitted curve of ARMA algorithm in ordinary model (target represents prediction value and output represents real value)

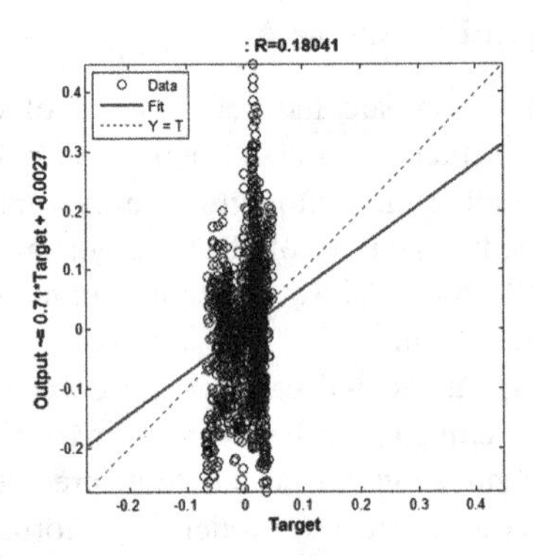

Fig. 10. Fitted curve of GRNN algorithm in the new model(target represents prediction value and output represents real value)

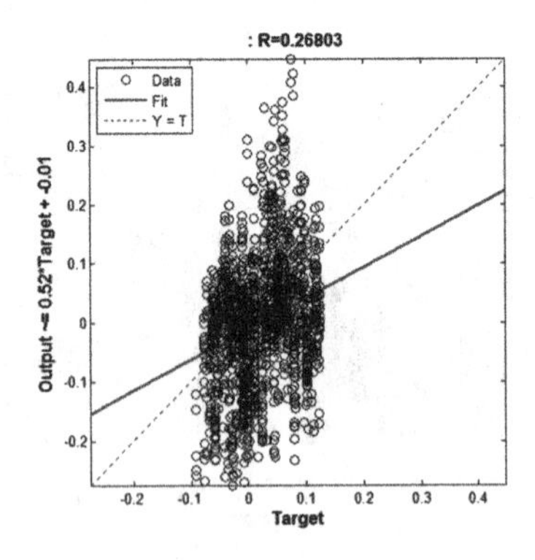

Fig. 11. Fitted curve of ARMA algorithm in the new model(target represents prediction value and output represents real value)

Table 1. Error comparison of prediction models and prediction algorithms

	ARMA	GRNN
ordinary model	3.12(PW=20 min) 3.89(PW=2 h)	1.95(PW=20 min) 2.04(PW=2 h)
new model	2.10(PW=20 min) 2.32(PW=2 h)	0.72(PW=20 min) 0.92(PW=2 h)

4.4 Factors that Impact Prediction Result

In this part, we choose ordinary model and GRNN algorithm to illustrate factors that impact prediction result.

Prediction Window. We can see in Fig. 8 that real value is about 0.53 times of prediction value. So the error is about 0.47 times of prediction value. If prediction window is big, prediction value will be big, then the error will be big. On the other hand, if prediction window is small, prediction value will be small, then the error will be small. In a word, bigger prediction window will contribute to bigger error.

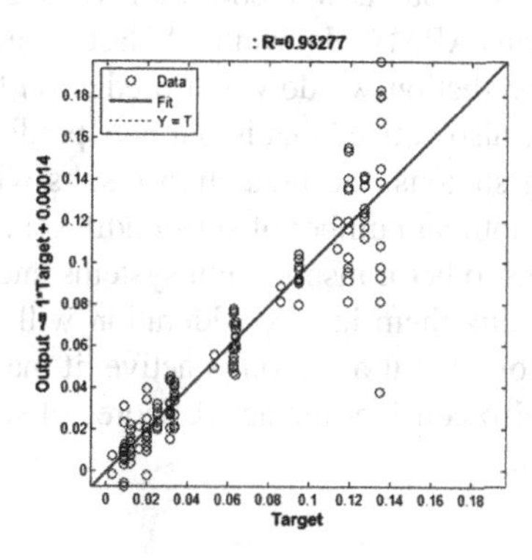

Fig. 12. Fitted curve on 5:00 a.m. (target represents prediction value and output represents real value)

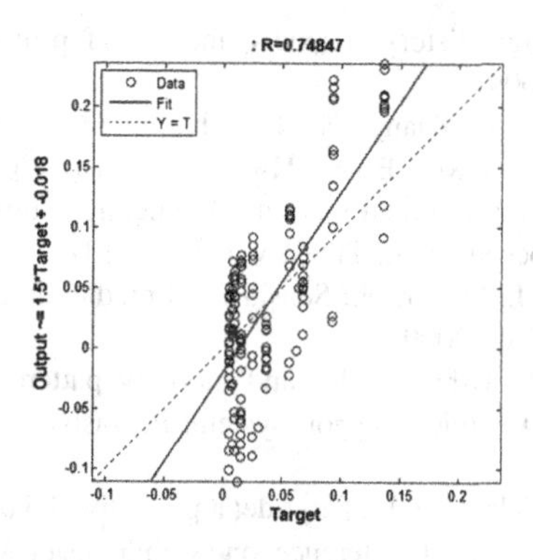

Fig. 13. Fitted curve on 9:00 a.m. (target represents prediction value and output represents real value)

Time Point of Prediction. We can see from Fig. 3 that NAS gets its first peak on 9:00 a.m. and first valley on 5:00 a.m. This tells us that the bicycle station is active around 9:00 a.m. and inactive around 5:00 a.m. So we can make our prediction on 9:00 a.m. and 5:00 a.m. and compare their prediction results. Figures 12 and 13 show the prediction results in these two situations. From these two pictures, we know that prediction result will be more accurate when the station is not so active.

5 Conclusion and Future Work

With the rapid expanding of urban population, public bicycle system has drawn worldwide attention. Many scholars have done meaningful research in this field. In this paper, we first introduce the development history and research status of public bicycle system. Then we bring general regression neural network algorithm into the prediction of bike availability and compare it with autoregression moving average algorithm. Based on whether considering the impact of surrounding stations we propose our new prediction model and compare its performance with that of ordinary prediction model. Experimental results show that our new model outperforms the ordinary model and GRNN algorithm outperforms ARMA algorithm. At last we analyze factors that impact prediction result (such as prediction window and prediction time point).

During our research, we also noticed that in our new prediction model, if we change the number of surrounding stations, the prediction results will be different. So in the future we can work on the optimal number of surrounding stations we should consider. Besides surrounding stations, other transportation systems and weather can also impact the prediction result. So taking them into consideration will also be one of our future work. What's more, we know that if a station is active, it may has a greater impact on its neighbors. So we may also consider the active scores of surrounding stations in our prediction.

References

1. DeMaio, P.: Bike-sharing: History, impacts, models of provision, and future. J. Public Transp. **12**(4), 41–56 (2009)
2. Shaheen, S.A., Guzman, S., Zhang, H.: Bikesharing in europe, the americas, and asia. Transp. Res. Rec. J. Transp. Res. Board **2143**(1), 159–167 (2010)
3. Chen, B., et al.: Uncertainty in urban mobility: Predicting waiting times for shared bicycles and parking lots. In: Proceedings of ITSC, vol. 13 (2013)
4. Froehlich, J., Neumann, J., Oliver, N.: Sensing and predicting the pulse of the city through shared bicycling. In: IJCAI (2009)
5. Kaltenbrunner, A., et al.: Urban cycles and mobility patterns: Exploring and predicting trends in a bicycle-based public transport system. Pervasive Mob. Comput. **6**(4), 455–466 (2010)
6. Yoon, J.W., Pinelli, F., Calabrese, F.: Cityride: a predictive bike sharing journey advisor. In: 2012 IEEE 13th International Conference on Mobile Data Management (MDM), IEEE (2012)

7. Hastie, T.J., Tibshirani, R.J.: Generalized additive models. CRC Press, Boca Raton, Florida (1990)
8. Lin, J.-R., Yang, T.-H.: Strategic design of public bicycle sharing systems with service level constraints. Transp. Res. Part E: Logistics Transp. Rev. **47**(2), 284–294 (2011)
9. Specht, D.F.: A general regression neural network. Neural Netw. IEEE Trans. **2**(6), 568–576 (1991)
10. Girardin, F., et al.: Digital footprinting: Uncovering tourists with user-generated content. Pervasive Comput. IEEE **7**(4), 36–43 (2008)
11. Chen, S., Colin, F.N.C., Grant, P.M.: Orthogonal least squares learning algorithm for radial basis function networks. Trans. Neural Netw. IEEE **2**(2), 302–309 (1991)
12. Krykewycz, G.R., et al.: Defining a primary market and estimating demand for major bicycle-sharing program in Philadelphia, Pennsylvania. Transp. Res. Rec. J. Transp. Res. Board **2143**(1), 117–124 (2010)

Evaluation of Routing Protocols in VANET Based on a Synergetic Simulation Using SUMO and ns-3

Siyuan Hao(✉), Yue Yang, and Haibin Cai

Shanghai Key Laboratory of Trustworthy Computing, East China Normal University, No.3663, North Zhongshan Rd, Shanghai, China
siyuanhao@foxmail.com, yangyueecnu@163.com, hbcai@sei.ecnu.edu.cn

Abstract. Vehicular Ad-hoc Network (VANET) is an emerging autonomous dynamic topology network. It is a special kind of Mobile Ad-hoc Network in which the vehicles exchange their information with each other. VANET turns every car in it into a mobile node and use these nodes to create a mobile dynamic network. The purpose of VANET is to supply a wireless connectivity and deploy various applications such as collision avoidance, safety and improving the traffic efficiency as provisioned by the Intelligent Transportation System (ITS). The vehicles are constrained by the realistic traffic environment, a nd now the simulations are mainly network simulations which cannot simulate the real trace of the vehicle. In this paper we propose a synergic simulation with sumo and ns-3 to evaluate the performance of the Ad-hoc routing protocols. We set ratio of packets delivered, delay and throughput as our performance metrics.

Keywords: Protocol · Simulation · Ad-hoc · Performance · VANET

1 Introduction

Vehicle Ad-hoc network [1] (VANET) is a special form of the mobile Ad-hoc networks (MANET).As the vehicle and traffic industry play an important role in the modern society, Governments embark on attaching importance to applying this new technology to enhance the efficiency of the traffic systems. An increasing number of the researches of VANET emerge. Researchers are applying new generation of wireless technology to the emerging Ad-hoc network. The purpose of VANET is to supply a wireless connectivity for the vehicles in a limited area to exchange information of each other and deploy various applications for collision avoidance, safety and the traffic efficiency as provisioned by the Intelligent Transportation System (ITS) [8] community. Building such a system in the realistic traffic environment needs plenty of human and material resources. In order to avoid squandering these resources and verify the effectiveness of this system, we are inclined to build a simulating traffic environment to research the performance of the vehicles in VANET. Network simulations and traffic simulations

C.-H. Hsu et al. (Eds.): IOV 2015, LNCS 9502, pp. 374–384, 2015.
DOI: 10.1007/978-3-319-27293-1_33

come to play important and effective roles for the researchers study. While many network simulators cannot generate the traces which is similar to the real world and the network simulation and traffic simulation are in two different areas. To solve this problem, we propose a synergetic simulation using a network simulator and a traffic simulator. In this paper we propose a coordinating simulation with a network simulator and a traffic simulator, then we evaluate the capabilities of three Ad-hoc routing protocols in the city roads. Several simulators have been widely used in the other researchers studies, we give some simulation tools in the following parts:

1.1 Network Simulator

Ns-2 [11], OMNeT++[9] and JiST [2] are widely used open source network simulators, they perform well in the general wireless network simulation, however, ns-3, the discrete event network simulator provides more flexible modularity developing environment. It balances the performance and the recompilations in a relatively perfect way. Modules in ns-3 are easier to be extended with C++ as users will.

1.2 Traffic Simulator

In this work we choose ns-3 as the network simulator and SUMO as the road traffic simulator. In order to create a connection between those two simulators, we use an interface to implement the communication. TraCI is short for Traffic Control Interface. It has the architecture of C/S to make SUMO accessible and allows network simulator to manipulate the movements of the vehicles dynamically. TraCI works as a server which is started by the execution of the command-line: –remote-port INT, note that INT is the number of the port which SUMO will keep listening on for the connection request from the network simulators. Once started by the command-line, SUMO waits for an external application which will control the simulation. It will not terminate when SUMO acts as a TraCI server until client asks the simulation to be an end. Figure 1 shows how the TraCI work between ns-3 and SUMO.

1.3 TraCI [10]

In this work we choose ns-3 as the network simulator and SUMO as the road traffic simulator. In order to create a connection between those two simulators, we use an interface to implement the communication. TraCI is short for Traffic Control Interface. It has the architecture of C/S to make SUMO accessible and allows network simulator to manipulate the movements of the vehicles dynamically. TraCI works as a server which is started by the execution of the command-line: –remote-port INT, note that INT is the number of the port which SUMO will keep listening on for the connection request from the network simulators. Once started by the command-line, SUMO waits for an external application

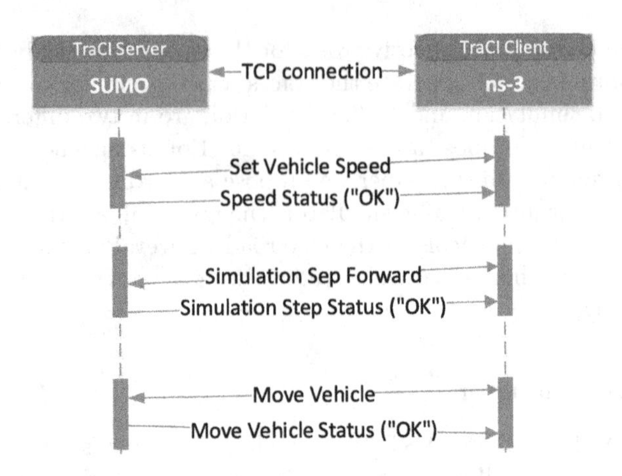

Fig. 1. Interaction between SUMO and ns-3

which will control the simulation. It will not terminate when SUMO acts as a TraCI server until client asks the simulation to be an end. Figure 1 shows how the TraCI work between ns-3 and SUMO works.

The rest of this paper is structured as follows. In Sect. 2, we review the related work, Sect. 3 represents the brief content of three mainstream Ad-hoc routing protocols, we show our models of simulation in Sect. 4, in Sect. 5, we describe the simulation and provide the evaluation of the simulation results. Finally, the conclusions and the future work will be presented in Sect. 6.

2 Related Work

Building a scenario in the realistic traffic environment will take much time, manpower and material resources. In recent years, the techniques of simulation provide an effective way to do the research of the VANETs. In this paper we propose a synergic simulation with sumo and ns-3 to evaluate the performance of the Ad-hoc routing protocols There are plenty of other attempts for the network simulation and the evaluation of the performance for different Ad-hoc routing protocols. [3] presents the implementation of DSDV, and describe the detailed explanation of the components and how each class interacts with one other is also provided the attributes that can be modified in the protocol. They analyzed their implementation and its performance. [7] gives a comparison of the Performance of the Ad-hoc routing protocols at the Grid Environment, it propose a method to evaluate the performance of certain kind of protocols under the specific circumstance. [10] has presented an interface called TraCI to control the traffic simulators via a flexible and extendable request-reply protocol. They demonstrated they can use the TraCI interfaces to evaluate complex scenarios. Such as accidents or hazardous road conditions. [5] presents a concept and design of an integrated realistic simulation scenario for VANETs named TraNS. Their main purpose is to make detailed and realistic evaluation of VANETs at network-centric possible, as well as application-centric levels. In [4], They compared performance of routing protocols for mobile ad-hoc networks considering TCP as the transport protocol and make FTP as the traffic generator. TCP was mainly developed to be deployed within wired networks. It is observed that TCP is not appropriate transport protocol for highly mobile multi-hop wireless networks because TCP protocol is unable to manage efficiently the effects of mobility. Nodes in Ad-hoc network may cause route failures. [6] compared three Ad-hoc routing protocols in a VANETs safety application scenario. They make the analysis of the performance and give a conclusion that DSDV is more encouraged than AODV. The usage of OLSR is not recommended in any circum-stances in safety application. The current researches does not give much synergetic simulations combined network and traffic simulators.

3 Overview of Routing Protocols

3.1 AODV

There have been several routing protocols for VANETs, and Ad-hoc On-demand Distance Vector Routing Protocol (AODV) is one of the protocols that have satisfying performances. AODV does not need to retain the routes all the time. To find a route, the source node broadcasts a Route Request packets and these packets create route entries for the reverse path through all the nodes it passes provisionally. There will be a feedback, Route Reply, being sent back through the path that Route Request propagated. Note that the route will not last long to complete packets transmission, particularly in the fast changing network topology.

3.2 DSR

Dynamic Source Routing protocol (DSR) has two phases which are route discovery phase and the route maintenance phase. It is a kind of source routing protocols which means the source node must know the complete path to the destination node. It is designed for multi-hop Ad-hoc especially for the mobile nodes. Every node has a cache to store the routes to every other nodes. A source node will start to discover the path to the destination node when it does not have a route in its own route cache. When the node detects that the route to a certain node works, the node will maintain its route. While the number of the nodes or the topology changes, the route will no longer work, route maintenance indicates a broken route and try to use another. Source node invokes routing discovery to find a new route when there is no other paths to the destination node.

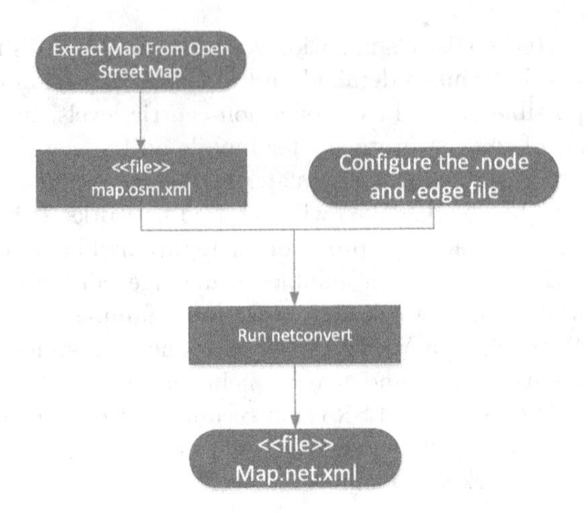

Fig. 2. The configuration of SUMO file

3.3 DSDV

Destination Sequenced Distance Vector is a table-driven routing protocol for Ad-hoc network. It was developed by C. Perkins and P.Bhagwat in 1994 and solved the routing loop problems. DSDV routing protocol maintains a routing table which records the routes to all destination nodes. When a route receives new information, it will use the path which is most recently used. A route with a better metric will be used when the sequence number is same to the one that has already existed in the route table. The method of incremental updates with sequence number labels, marks the existing wired network protocols adaptable to Ad-hoc wireless networks. Consequently, all available wired network protocol can be used to Ad-hoc networks without much modification. However, DSDV updates its routing tables regularly, and this behavior will cost much power even when the networks are in the free time. Besides, it need a new sequence number whenever the topology of the network changes. Obviously, DSDR is not suitable for the network that are always changing.

4 Methodology

In this part we show our model of the simulation, and the overview of the metrics which we use to evaluate the performance of different routing protocols.

4.1 Road Topology

There are several ways to generate the .net.xml file to run with SUMO. One is to use a free edible map, Open Street Map, to generate .osm file. Another way to obtain the scenario file is to configure the .node file and .edge file manually.

Fig. 3. Road topology

After doing these, we use the command to transform our source files to object .net.xml file. Figure 2 indicates the process how the process is to obtain the map we apply in our simulation scenarios. We produce a map.net.xml that has the topology as Fig. 3. It was extracted from the Open Street Map by the real small partial city of the highway. We add some traffic lights at the intersections to make our simulation closer to the realistic scenario.

4.2 Performance Metrics of Routing Protocols

Our simulation focuses on the performance of different protocols, thus we choose different metrics to observe and evaluate the performance of AODV, DSR and DSDV in the simulation scenario.

Ratio of Packets Delivered(RPD). This value indicates the ratio of the packets that delivered from the source node to the destination node. The value is higher, the more successfully the routing protocol performs. The formula of RPD is given by:

$$RPD = \frac{\sum_{i=1}^{n} PketRecv}{\sum_{i=1}^{n} PketSend} \times 100 \quad (1)$$

where PketRecv represents the number of the sent packets for each source nodes. PketSend presents the number of packets that the destination node receives.

Throughput. Throughput is the total number of the packets that have been delivered in the net-work in a unit time. To a certain extent, the density of the nodes will affect the throughput. The formula of the throughput is given by:

$$Throughput = \frac{pkt(i) - pkt(j)}{recvT(i) - recvT(j)} \quad (2)$$

Table 1. Simulation parameter

Parameter	Value
Area(m*m)	100, 300, 500, 700, 900
Simulation time(s)	200
Speed(m/s)	20
MAC type	802.11
Traffic type	TCP
Packet size	512 bytes
Protocols	AODV, DSR, DSDV
Mobility model	Constant Velocity Mobility Model

where pkt(n) is the representative of the total number of the packets when the packet n is received by the destination node. recvT(n) represents the time point the packet n is received by the destination node.

Packet Delay. Delay indicates that how long a certain packet will take from the sending time point to the receiving time point. Apparently, the small value would reveal a better performance. The formula is given by:

$$PacketDelay = RecvTime - RecvTime \tag{3}$$

where RecvTime and the SendTime is representative of the sending time point and the receiving time point respectively of a certain packet.

4.3 Simulation Model

The mobility of the vehicles in the scenario is the constant velocity mobility model, the vehicles in the scenario move with a constant velocity from a certain point, and select a random destination point in the scenario. To simplify the mobility model, we define that a destination point must be an end point of a lane. In our simulation, the evaluation of the performance is carried out by varying two parameters, the number of the vehicles and the geographic size of the scenario. Three Ad-hoc routing protocols we describe briefly in Sect. 3 are taken into consideration for a comparison purpose in the performance analysis. The other simulation parameters are reported in Table 1.

5 Evaluation

Each simulation runs ten times to obtain the average value to plot the performance of three protocols with the varying number of the vehicles, another variable parameter is the size of the network.

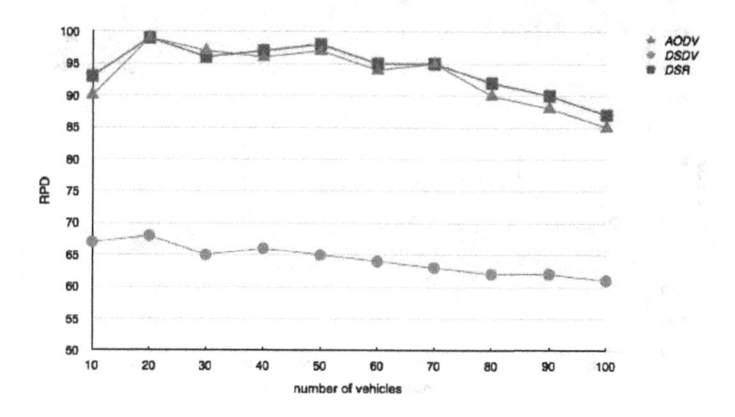

Fig. 4. Performance of RPD

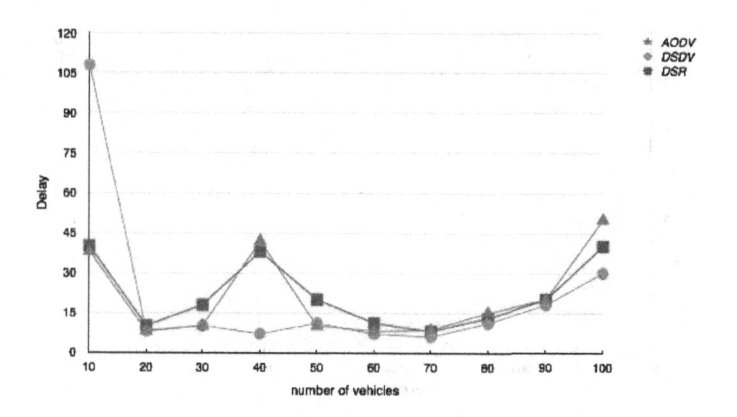

Fig. 5. Performance of delay

5.1 Varying the Number of the Vehicles

During the simulation, the number of the vehicles varies from 10 to 100 with an increment of 10 and set the size of the scenario to be 500*500 (m2). The number of vehicles vs RPD, delay and throughput are respectively shown in Figs. 4, 5, and 6. We can observe from Fig. 4 that DSR has a slightly lager value of RPD than AODV, but outperforms DSDV much by varying the number of the vehicles in the scenario. In terms of the value of the delay DSR gives a relatively stable performance compared with AODV and DSDV which are presented in Fig. 5. In Fig. 6, AODV has the lowest throughput in comparison with other two protocols. The throughput of DSDV is highest. DSR still has an average value which is higher than AODV but lower than AODV.

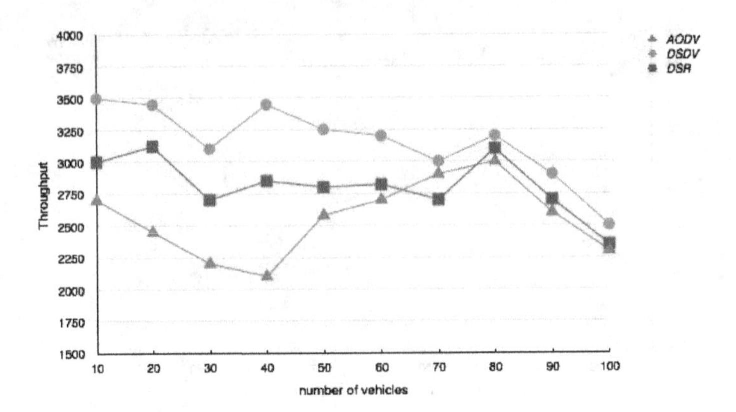

Fig. 6. Performance of throughput

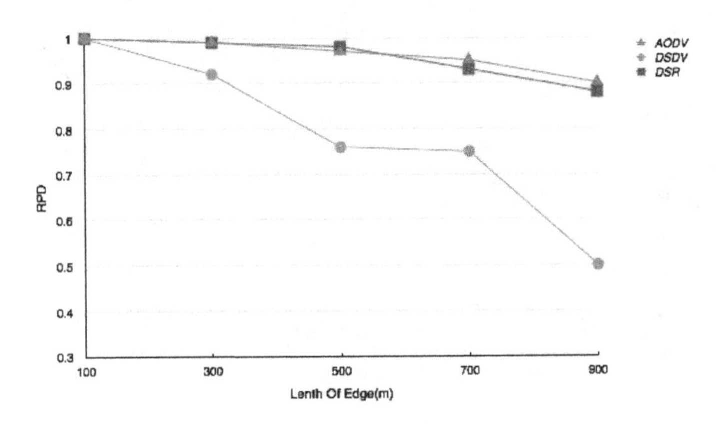

Fig. 7. Performance of RPD

5.2 Varying the Size of the Scenario

The geographic size of the scenario varies from 100*100 (m2) to 900*900 (m2). Size of the scenario vs RPD, delay and throughput are presented with Figs. 7, 8, and 9. In this scenario, we set the number of vehicles to be 50. The performance of DSDV is still poorer than the other protocols. AODV and DSR have a similar value again whereas AODV is little higher than DSR when the size of scenario becomes larger as shown in Fig. 7. In Fig. 8, All protocols have a little delay when the size of scenario is relatively small. However, DSDV still keeps a preeminent performance when the size increases up to 500*500 (m2). The delay of DSR and AODV increase due to the larger size of the scenario and the network topology changing fast. Figure 9 shows that DSDV outcomes AODV and DSR again, it has the largest throughput and AODV performs worse when the size of scenario rise up. DSR is close to AODV whereas the through-put of it falls with the variation of the size of the scenario.

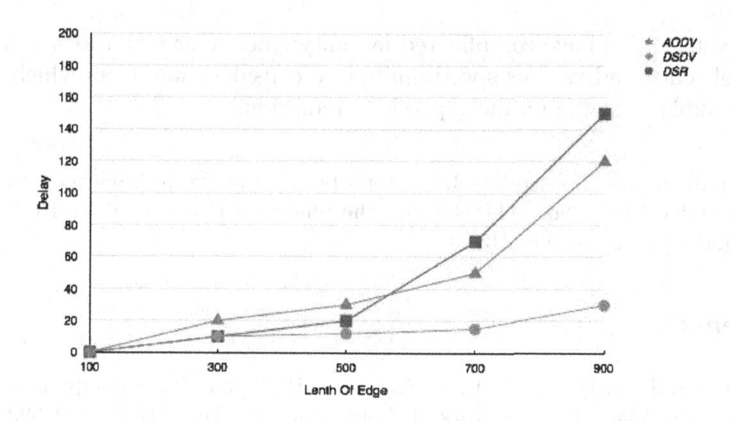

Fig. 8. Performance of delay

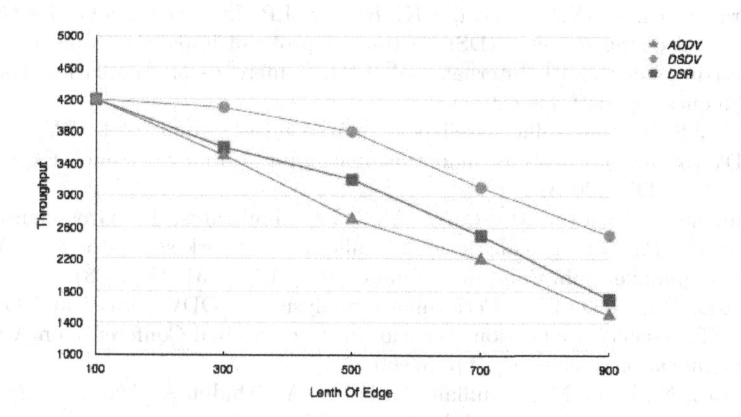

Fig. 9. Performance of throughput

6 Conclusion and Future Work

In our work we create a scenario to simulate a simple traffic environment. Our model realizes the dynamic evaluation of different protocols in VANET. We have evaluated three performance metrics of AODV, DSR and DSDV with varying number of vehicles, size of scenario. The simulation reported clearly DSDV has the lowest value of RPD, although it has the highest value of throughput. DSR has an average performance compared with the other protocols when the density of the vehicles changes. In another scenario, we set the number of the vehicles to be 50, from the result we can imply that all protocols have a similar performance when the size of the network is small. With the size growing lager, we observe DSDV as inferior to DSR and AODV in terms of the value of RPD. On the contrary, DSDV outperforms AODV and DSR in the value of delay and throughput when the size of network grows up. In our further research, we will pay attention to the various mobility model in terms of the realistic traffic environment since the real vehicle will have complicated mobility trace. Our scenario will take the density of roads and various speed limits into consideration, from which we can obtain reliable data from a more precise simulation.

Acknowledgments. This research is supported by the Natural Science Foundation of China under Grant No. 91118008 and the Shanghai Knowledge Service Platform Project under Grant No. ZF1213.

References

1. Al-Sultan, S., Al-Doori, M.M., Al-Bayatti, A.H., Zedan, H.: A comprehensive survey on vehicular ad hoc network. J. Netw. Comput. Appl. **37**(1), 380–392 (2014)
2. Barr, R., Haas, Z.J., Van Renesse, R.: JiST: an efficient approach to simulation using virtual machines. Softw. Pract. Exp. **35**(6), 539–576 (2005)
3. Narra, H., Cheng, Y., etinkaya, E.K., Rohrer, J.P., Sterbenz, J.P.G.: Destination-sequenced distance vector (DSDV) routing protocol implementation in ns-3. In: Proceedings of the 4th International ICST Conference on Simulation Tools and Techniques, pp. 439–446 (2011)
4. Payal, J.S.K.: Tcp traffic based performance investigations of DSDV, DSR and AODV routing protocols for manet using ns2. Int. J. Innov. Technol. Explor. Eng. **3**(2), 2278–3075 (2013)
5. Pirkowski, M., Raya, M., Lugo, A.L., Papadimitratos, P., Grossglauser, M., Hubaux, J.P.: TraNS: realistic joint traffic and network simulator for VANETs. ACM Sigmobile Mob. Comput. Commun. Rev. **12**(1), 31–33 (2008)
6. Santoso, G.Z., Kang, M.: Performance analysis of AODV, DSDV and OLSR in a VANETs safety application scenario. In: International Conference on Advanced Communication Technology, pp. 57–60 (2012)
7. Surayati, N., Usop, M., Abdullah, A., Faisal, A., Abidin, A., Darul, U., Malaysia, I.: Performance evaluation of AODV, DSDV and DSR routing protocol in grid environment. Int. J. Comput. Sci. Netw. Secur. **9**, 261–268 (2009)
8. Taniguchi, E., Shimamoto, H.: Intelligent transportation system based dynamic vehicle routing and scheduling with variable travel times. Transp. Res. Part C Emerg. Technol. **12**, 235–250 (2004)
9. Varga, A.: The OMNeT++ discrete event simulation system. In: ESM 2001 (2001)
10. Wegener, A., Pirkowski, M., Raya, M., Hellbrck, H., Fischer, S., Hubaux, J.P.: TraCI: an interface for coupling road traffic and network simulators. In: Communications and Networking Simulation Symposium (2008)
11. Xue, Y., Lee, H.S., Yang, M., Kumarawadu, P., Ghenniwa, H.H., Shen, W.: Performance evaluation of ns-2 simulator for wireless sensor networks. In: Canadian Conference on Electrical and Computer Engineering, CCECE 2007, pp. 1372–1375 (2007)

Real-Time Traffic Light Scheduling Algorithm Based on Genetic Algorithm and Machine Learning

Biao Zhao[1(✉)], Chi Zhang[2], and Lichen Zhang[1]

[1] Shanghai Key Laboratory of Trustworthy Computing,
National Trusted Embedded Software Engineering Technology Research Center,
East China Normal University, Shanghai 200062, China
{zbhdsfdx,zhanglichen1962}@163.com

[2] The Institute for Data Science and Engineering, East China Normal University,
Shanghai 200062, China
chizhang@ecnu.cn

Abstract. Traffic signals are essential to provide safe driving that allows all traffic flows to share road intersection. However, they decrease the traffic flow fluency because of the queuing delay at each road intersection. In order to improve the traffic efficiency all over the road network, Intelligent Traffic Light Scheduling (ITLS) algorithm has been proposed. In this work, we introduce an ITLS algorithm based on Genetic Algorithm (GA) merging with Machine Learning (ML) algorithm. This algorithm schedules the time phases of each traffic light according to each real-time traffic flow that intends to cross the road intersection, whilst considering next time phases of traffic flow at each intersection by ML. In order to get each next time phases of traffic flow, we use Linear Regression (LR) algorithm as ML algorithm. The introduced algorithm aims to increase traffic fluency by decreasing the total waiting delay of all traveling vehicles at each road intersection in the road network. We compare the performance of our algorithm with the unimproved one for different simulated data. Results shows that, our algorithm increases the traffic fluency and decreases the waiting delay by 21.5 % compared with the unimproved one.

1 Introduction

Nowadays, the existing traffic lights system could satisfy the scheduling of traffic flows at each traffic intersection partly. But as the increasing number of urban vehicles in the past few years have resulted in an awkward situation that urban congestion is becoming increasingly severe. Sometimes, the situation that the horizontal driveway has already jammed whereas the vertical lane has few cars would be occurred. The traffic light cannot adjust light cycle schedules considering the change of traffic flow at each traffic crossing. Then the queuing delay at each road intersection decreases the traffic fluency and decreases the traffic efficiency all over the network. In order to improve the traffic efficiency, over past years researchers have actively studied intelligent traffic light scheduling algorithms [1, 5]. Many researchers have considered that the phases of the traffic light are set according to the real-time traffic flows [2–4]. The optimal scheduling algorithm minimizes the traffic flow delay at each road intersection. The less the average

C.-H. Hsu et al. (Eds.): IOV 2015, LNCS 9502, pp. 385–398, 2015.
DOI: 10.1007/978-3-319-27293-1_34

queuing delay at each road intersection is, the more efficient the scheduling algorithm becomes [6].

In this paper, we propose an Intelligent Traffic Light Scheduling (ITLS) algorithm based on GA merging LR (ITLSGMLR). The introduced algorithm aims at decreasing the total waiting delay at the signalized road intersections. At each light cycle schedules, we compute a total waiting delay of all traveling vehicles at each intersection. The all traveling vehicles at each road intersection with the least waiting delay is the best schedule, without exceeding the maximum allowable color switching time for that phase(APPENDIX A.7). We predict next time phases of traffic flow at each intersection by LR. The prediction aims to guarantee the queuing delay is not only the least at this time cycle phase, but also the least at next time cycle phase. During the period in the simulated scenario, the result with the least sum of the total waiting delay at every light cycle schedules is what we want. We compare the performance of our proposed algorithm with the Intelligent Traffic Light Scheduling algorithm based on GA (ITLSG). From the experimental results, we infer that our algorithm ITLSGMLR decreases the total waiting delay by 21.5 % compared with the ITLSG. Moreover, the less the average queuing delay and the larger average throughput at each road intersection is obtained by using ITLSGMLR. Our simulated data derives from the Big challenge of Intelligent transportation system algorithm of 2014 China Guizhou cloud, Big Data business model contest.

The remainder of this paper is organized as follows: In Sect. 2, we introduce several algorithms that have been proposed in this field. Then, we illustrate our algorithm model and investigate the characteristics of GA and LR which we used in Sect. 3. The phases of ITLSGMLR are detailed in Sect. 4. After that, Second V shows the performance evaluations of ITLSGMLR compared to the unimproved one. At last, Sect. 6 concludes the paper.

2 Related Work

The intelligent traffic light takes the traffic characteristics of each traffic flow that intend to cross the road intersection into consideration [1]. In this section, we discuss some algorithms that have been proposed in this field.

Bani Younes et al. proposed an Intelligent Traffic Light Controlling (ITLC) algorithm through Vehicular Ad-hoc Networks (VANETs) [6]. It utilizes VANETs to gather the real-time traffic characteristics of each surrounding traffic flow. The introduced algorithm aims to decrease the waiting delay time at each road intersection and to increase traffic fluency. They defined a ready area that around the signalized road intersection where vehicles are ready to cross the intersection. The ready area is proposed to guarantee fair sharing of the road intersection without exceeding the maximum allowable green time. The traffic flow with the largest traffic density is scheduled first, that is implemented to set the phases of each traffic light cycle. Bani Younes et al. [6] concluded that ITLC algorithm minimizes the queuing delay at each traffic light and increases the throughput of each road intersection.

In [8], it introduced a traffic light control algorithm that is embedded with SUMO simulation [7]. This algorithm measures the jam lengths at each intersection and uses the length of the jam in front of the traffic light as input. Each traffic light attempts to solve the detected jams by this algorithm. If jam becomes longer at one of these intersections, the intersection is allocated a longer time of green light to reduce the jam.

Maslekar et al. reduces the waiting time and the queue length at each intersection based on car-to-car communication in [2]. Gradinescu et al. [3] improves the traffic fluency with the less cost by developed adaptive traffic signal control systems. Pandit et al. [4] proposes a real-time traffic light control algorithm. It utilizes VANETs to collect speed and position information for each vehicle to optimize the traffic signals control at each road intersection. In [5], the Vehicle-to-infrastructure (V2I) communications are considered. At each road intersection, an RSU is installed to gather the basic information of traveling vehicles. Each RSU uses the phase-based strategy algorithm [9] to optimize the traffic fluency and to minimize the waiting delay time [5].

As early introduced, the less the average queuing delay at each road intersection is, the more efficient the scheduling algorithm becomes. In this work, we aim to introduce an ITLSGMLR algorithm that schedules the time phases of each traffic light according to each real-time traffic flow that intends to cross the road intersection, whilst considering next time phases of traffic flow at each intersection by LR. At each light cycle schedules, we compute a total waiting delay of all traveling vehicles at each intersection. During the period in the simulated scenario, the result with the least sum of the total waiting delay at every light cycle schedules is what we want. The detail of the proposed ITLSGMLR algorithm and the performance evaluations are introduced in Sects. 4 and 5.

3 ITLSGMLR Model, Genetic Algorithm

3.1 ITLSGMLR Model

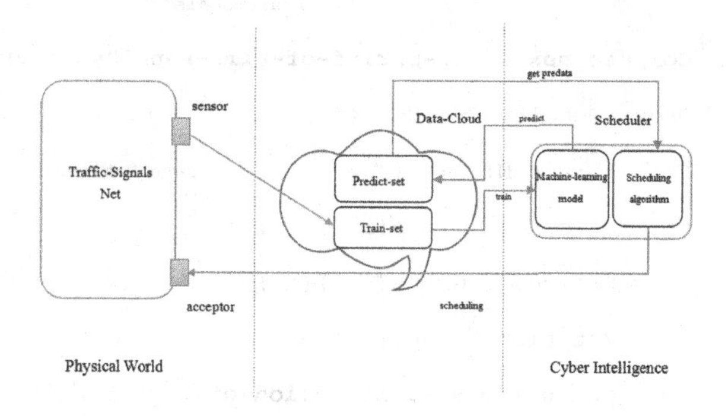

Fig. 1. Details of ITLSGMLR model

ITLSGMLR model consists of two main parts. One is Traffic-Signals Net that represents Physical World and acts continuously. The other is the Scheduler which stands for Cyber Intelligence, that schedules the traffic lights when it gets prediction result data from a Data-Cloud. Figure 1 shows the details of ITLSGMLR model. In this paragraph, the model will be mainly illustrated. The model is constituted by three parts: Traffic-Signals Net, Data-Cloud and Scheduler. The Traffic-Signals Net monitors vehicles volume of each traffic intersection. Then, the sensor collects such kind of information and transfers the data to the Train-set in Data-Cloud. Besides, the acceptor gets scheduling strategy from the scheduler. The Data-Cloud consists of two sets. One is a Train-set that gets raw data from sensor and provides training data to Machine-learning model in scheduler, and the other is Predict-set whose data is predicted by LR model and supplies prediction result data for scheduling algorithm in the scheduler. For Scheduler, there are two main portions which are LR model and Scheduling algorithm.

3.2 Genetic Algorithm (GA)

In order to schedule traffic lights, we use GA as brain of ITLSGMLR algorithm. GA [10] is one of the best-known techniques for solving optimization problems. In general, GA has five operators which are Initialization, Evaluation, Selection, Crossover and Mutation [11–15] in their computational framework.

4 The Phases of ITLSGMLR Algorithm

This algorithm based on GA merging LR algorithm. It give back a schedule with the least waiting delay at each light cycle schedules. The essential difference between ITLSGMLR with ITLSG is that: ITLSGMLR algorithm considers next time phases of traffic flow at each intersection by LR when it schedules the time phases of each traffic light according to each real-time traffic flow that intends to cross the road intersection.

The ITLSGMLR algorithm:

```
BEGIN:
    Initialize Group(0)(m);  /**Group(0)(m): the 0th generation has m
                               individuals**/
    Compute-next-unit-period-of-time-Penalty Group(0)(m);
    Order-Penalty Group(0)(m);
    For(i=1; i<=N; i++){          //Iterate N times
        Selection Group(i-1)(m);
        Crossover Group(i-1)(m);
        Mutation Group(i-1)(m);
        Compute-next-unit-period-of-time-Penalty Group(i)(m);
/** who can make the next unit period of time penalty also small, whose fitness will higher**/
        Order-Penalty Group(i)(m);}
    Transform-To-Light-Status Group(N)(0);
    //Group(N)(0) is the optimal solution
END;
```

4.1 Gene-Coding and Initialize Group

4.1.1 Gene-Coding

We can get a road network that include 45 intersections from traffic_light_table (APPENDIX B. Table 4). There are two different types of road intersection. Each road intersection connects three or four legs. Firstly, in order to locate each intersection, we give each road intersection a label number. Then, for the purpose of scheduling the lights by GA, we code 45 road intersections used binary code at each label number on the sequence. '0' denotes horizontal traffic flow while '1' denotes vertical traffic flow as shown in Fig. 2.

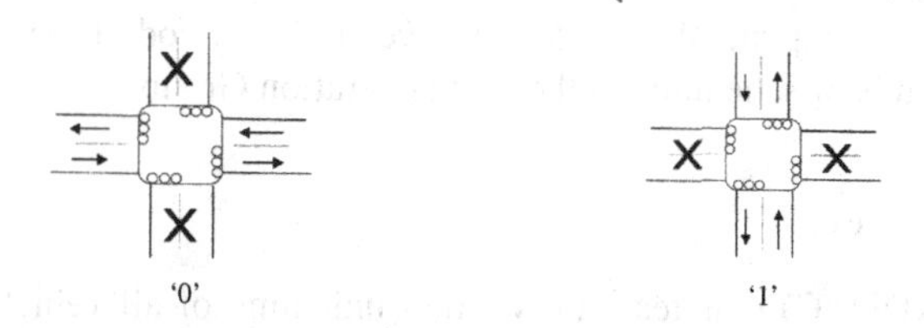

Fig. 2. '0': horizontal traffic flow, '1': vertical traffic flow

"010101010001111100000111000010111100001111000"

There is an initial binary string. In this string, it can be regarded as a scheduling for a traffic cycle phases. Each binary code in string denotes an intersection. The value of the code [4] is '1' as shown in Fig. 3a. It states that its traffic flow is vertical. Its location in road network is tl_4 and its type is 4-leg seen in Fig. 3b. Having gotten these information, we can set traffic_light_status_table(t_x) (APPENDIX B. Table 5) as Table 1. Through this string, we can set traffic_light_status_table(t_x) of each road in the network.

Table 1.

Traffic-LLightID	From-ID	leftLight-Status	rightLight-Status	straightLight-Status
tl_4	tl_1	0	1	0
tl_4	tl_2	1	1	1
tl_4	tl_5	0	1	0
tl_4	tl_{10}	1	1	1

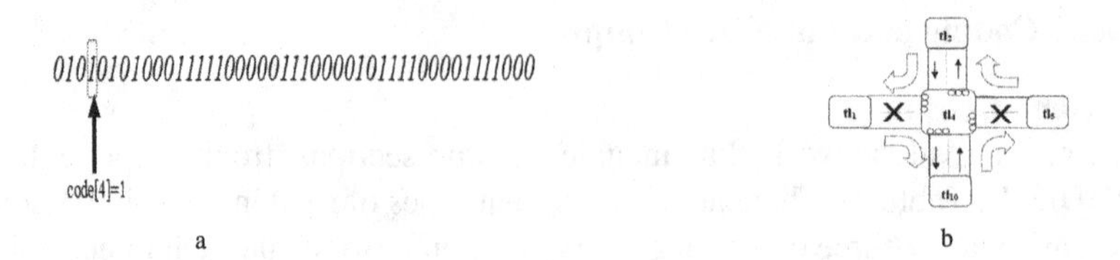

Fig. 3. a. code [4] = 1; b. The intersection whose label number is 4

4.1.2 Initialize Group

The number of the Group member is 150, thus we randomly code 150 individuals which represents 150 kinds of schedules as the first generation Group.

4.2 Evaluate Fitness

Penalty (APPENDIX C) denotes total waiting unit time of all vehicles at each road intersection in the road network. We use penalty to evaluate fitness.

The ITLSG Algorithm objective function is:

$$\min_{Group}\{\sum_{EachIntersection} penalty(t_i)\} \quad (1)$$

The ITLSGMLR objective function is:

$$\min_{Group}\{\sum_{EachIntersection} Max\,(VerticalPenalty)\,(t_{i+1}), HorizontalPenalty(t_{i+1}))\} \quad (2)$$

In order to compute VerticalPenalty(t_{i+1}) and HorizontalPenalty(t_{i+1}), we must obtain predict flow. And we use LR to get prediction traffic flow. Shown in Fig. 4, we could compute road S to K traffic flow of the next unit period of time, the computational formula is:

$Flow_{(s\ to\ k)}(t_{i+1}) = stayFlow(t_i) + rightInFlow(t_i)$
$+ straightInFlow(t_i) + leftInFlow(t_i)$
$+ predictFlow_{(s\ to\ k)}(t_{i+1})$

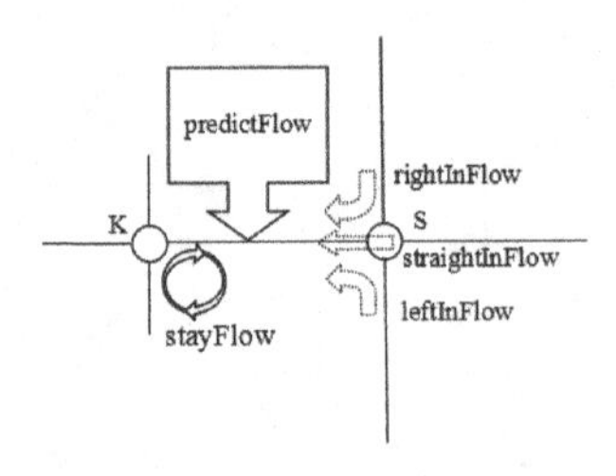

Fig. 4. Prediction traffic flow of next period

4.3 Genetic Operations

The genetic operations has three operators which are Selection, Crossover and Mutation.

4.3.1 Selection

To further improve the convergence performance, the elitist selection scheme is also used to ensure that the best chromosome is always taken to the next generation [11, 16, 17]. In nature, an individual who has high fitness can reproduce offspring the most possible. First, we order each member penalty in group by ascending. Second, we choose individuals which satisfies our request.

4.3.2 Crossover

Crossover is the main search operator in GA which performs the exchange of information among chromosomes through a combination and disruption of schemata [11, 13, 18]. We choose single-point crossover (show in Fig. 5). Firstly, we choose a point of code randomly. Secondly, we exchange the part that is selected of string.

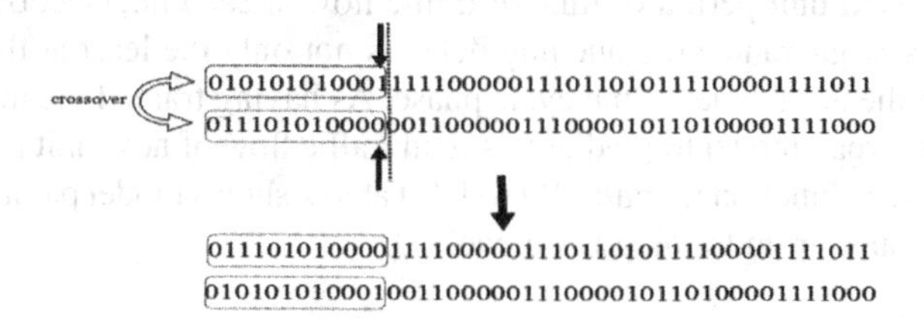

Fig. 5. Single-point crossover

4.3.3 Mutation

Mutation serves as a background operator to restore genetic materials in GA more commonly [11, 13, 18]. If a mutation occurs, algorithm choose a point of code randomly. We change the point selected like this: '1' change to '0' or '0' change to '1' (seen in Fig. 6).

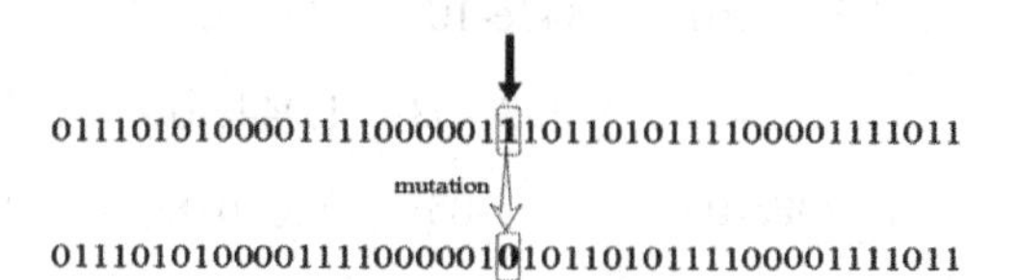

Fig. 6. Mutation

In each traffic light cycle, ITLSGMLR gives back a schedule with the least penalty. It is expected to increase the throughput of the road intersection and decreasing the waiting delay.

5 Performance Evaluation

In this section, we evaluate the performance of our proposed algorithm (ITLSGMLR). The simulated scenario is that traveling vehicles aim to pass through a road intersection that is controlled by an intelligent traffic light. In our experiments, we first compute each road LR model. Then, we can use ITLSGMLR to schedule the time phases of each traffic light according to each real-time traffic flow that intends to cross the road intersection, whilst considering next time phases of traffic flow at each intersection by LR model.

We have seven days traffic flow data as simulated data derived from the big challenge of Intelligent transportation system algorithm of 2014 China GuiZhou cloud, Big Data business model contest. We save the simulated data into traffic_flow_table(APPENDIX B. Table 6). In order to improve GA objective function, we used LR to predict traffic flow. As the only training feature is time, the hypothesis function for every road is:

$$H_\theta(x) = \theta_0 x^0 + \theta_1 x^1 + \theta_2 x^2 + \ldots + \theta_8 x^8 \quad (3)$$

(x: Time; θ_i: i^{th} parameter; $H_\theta(x)$: traffic flow)

We predict next unit period of time of traffic flow at each intersection by LR. The prediction aims to guarantee the queuing delay is not only the least at this time cycle phase, but also the least at next time cycle phase. As having trained the model for each road, we can use road model to predict this road traffic flow of next unit period of time. We use Hypothesis function to train LR model. Table 2 shows model parameters of four roads that we obtain from training LR model:

Table 2. Model parameters

	tl_9-tl_2	tl_9-tl_{10}	tl_9-tl_{16}	tl_9-tl_8
θ_0	-2.728e-23	4.273e-24	-2.427e-22	-1.559e-22
θ_1	-3.320e-16	-3.325e-20	1.501e-18	9.149e-19
θ_2	-3.320e-16	1.081e-16	-3.703e-15	-2.107e-15
θ_3	3.312e-13	-1.884e-13	4.625e-12	2.387e-12
θ_4	-1.329e-10	1.842e-10	-3.066e-09	-1.354e-09
θ_5	-6.360e-09	-9.388e-08	1.044e-06	3.580e-07
θ_6	1.4238e-05	1.777e-05	-0.0001806	-7.035e-05
θ_7	0.0001251	0.0016805	0.0274157	0.0367806
θ_8	0.2640605	0.0226113	0.1287508	0.0389523

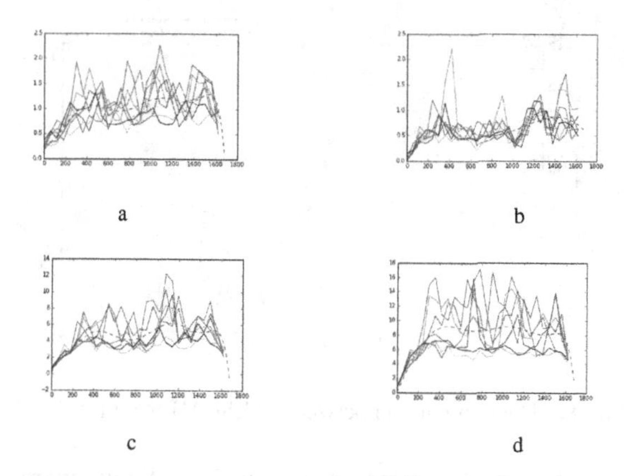

Fig. 7. x: time unit during the period in the simulated scenario(1680T); y: number of traveling vehicles

a. $tl_9 - tl_2$**: LR model** **b.** $tl_9 - tl_{10}$**: LR model**
c. $tl_9 - tl_{16}$**: LR model** **d.** $tl_9 - tl_8$**: LR model**

The LR model that simulates the real traffic flow can be seen in Fig. 7. The solid line represents seven days simulated data of this road. Dashed line represents LR model fitting curve. It can express how this road LR model can fit the real traffic flow. From the experimental results, we infer that these LR models can fit real traffic flow approximately.

Table 3. Genetic algorithm parameters

Parameter	Value
Iteration Times	500
Group	150
Probability of Crossover (Pc)	0.85
Probability of Mutation (Pm)	0.05

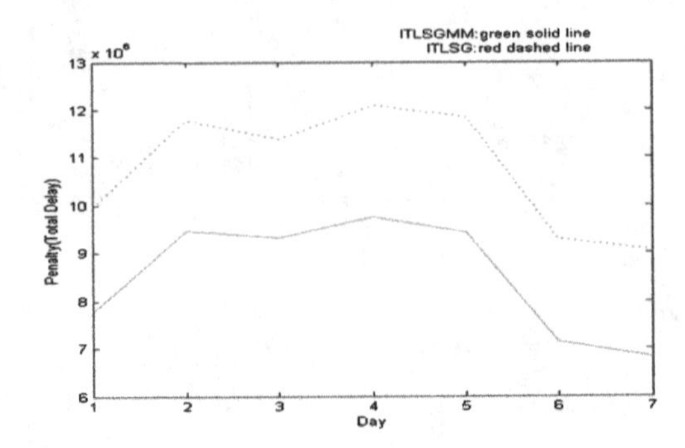

Fig. 8. The comparison between ITLSGMLR with ITLSG

Table 3 illustrates the main parameters used in GA. We compare the performance of our algorithm with the unimproved one for different simulated data. As we can see from Fig. 8, the penalty using ITLSGMLR is 21.5 % less than the penalty using ITLSG. This is due to ITLSGMLR algorithm considers next time phases of traffic flow at each intersection by LR model. At each light cycle schedules, Our algorithm gives back a schedule with the least penalty (waiting delay). ITLSGMLR can guarantee penalty not only being the least at this time cycle phase, but also the least at next time cycle phase. During the period in the simulated scenario, ITLSGMLR can get the less total penalty of all traveling vehicles at each road intersection in the road network compared with ITLSG. Our algorithm also increases throughput of the road intersection, because less time is wasting for vehicles to pass the intersection.

6 Conclusion

In this paper, we introduce an Intelligent Traffic Light Scheduling algorithm based on Genetic Algorithm (GA) merging Linear Regression (LR) algorithm (ITLSGMLR). This algorithm considers next time phases of traffic flow at each intersection by LR, whilst scheduling the time phases of each traffic light according to each real-time traffic flow that intends to cross the road intersection. The all traveling vehicles at each road intersection with the least waiting delay is the best schedule. The maximum allowable time for that phase should not exceed the maximum color switching time (APPENDIX A.7). During the period in the simulated scenario, the result with the least sum of the total waiting delay at each light cycle schedules is what we want. From the experimental results, we can infer that ITLSGMLR achieves a better performance compared to ITLSG. ITLSGMLR decreases traffic queuing delay and increases the traffic fluency by 21.5 % compared with unimproved one. Meanwhile, ITLSGMLR minimizes the average queuing delay and increases average throughput of each road intersection.

Our proposed algorithm just simulates in this special problem scenario. While, In the real world, the scenarios are more complicated. In future, our work is to make our algorithm could be applied in real world, and obtain a more accurate prediction model.

Acknowledgment. This work is supported by the national natural science foundation of China under grant (No.61370082, No.61173046, No.91318301)), natural science foundation of Guangdong province under grant (No.S2011010004905). This work is also supported by Shanghai Knowledge Service Platform Project (No.ZF1213) and NSFC Creative Team 61321064 and Shanghai Project 012FU125X15.

Appendix: ITLSGMLR Scenario

A Scene Simplification

In order to expedite scheduling algorithm modeling, we have to simplify the problem as follows:

(1) Only red light and green light are taken into consideration, regardless of yellow illumination;
(2) Driving directions have three conditions which are going straight, turning left or right, without turning around;
(3) At each intersection, the probability of going straight, turning left or right respectively are $\alpha = 0.8$, $\beta = 0.1$, $\gamma = 0.1$, where $\alpha + \beta + \gamma = 1$;
(4) On Traffic signal location network, we just care about 4-leg intersection or 3-leg intersection;
(5) T is unit time and the initial time of red and green light is T;
(6) Traffic light cycle time must be set in multiples of T;
(7) At each light cycle scheduling, light must switch to another color without exceeding maximum allowable time (4T) for that phase;
(8) The distance between any two traffic lights is equal, meanwhile, any vehicle just cost T time from one traffic light to another adjacent traffic light.

B Data structure schema

Table 4. Traffic_light_table

Traffic-LightID	From-ID	leftTarget-ID	rightTarget-ID	straightTarget-ID
tl_4	tl_1	tl_2	tl_{10}	tl_5
tl_4	tl_5	tl_{10}	tl_2	#

It illustrates the location of traffic light in the road network which can be expressed as table as above. All type of table field are Integer.

Table 5. Traffic_light_status_table(t_x)

Traffic-LightID	From-ID	leftLight-Status	rightLight-Status	straightLight-Status
tl_4	tl_1	0	0	1
tl_4	tl_5	-1	0	1

This table is a dynamic time dependent table. It was shown as above at the moment of t_x. All type of table field are Integer. '1' as green light on, '0' as red light on, '-1' as no light in this direction.

Table 6. Traffic_flow_table

TrafficLightID	FromID	traffic_flow
tl_4	tl_1	[10,8,5,0,0,23,13,…,23]

It says traffic flow between two traffic signals during the period in the simulated scenario. The type of table field, TrafficLightID and FromID are Integer. trafficFlow's type is Array (Table 7).

Table 7. Vehicle_through_rate_table

Traffic-Light-ID	From-ID	leftTh-rough-Rate	right-Throug-Rate	straight-Through-Rate
tl_4	tl_1	2	2	16

It illustrates how many number of vehicles can pass the green light during unit time T.

C Penalty

Penalty denotes total waiting unit time of all traveling vehicles at each road intersection in the road network.

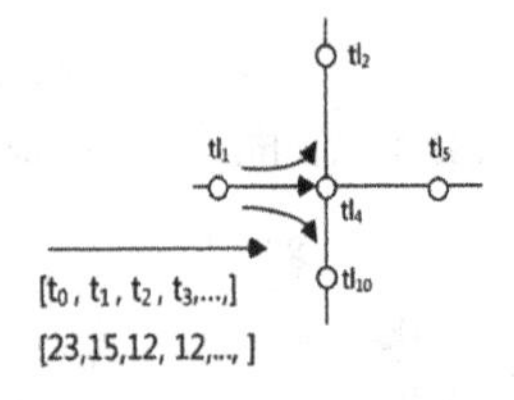

Fig. 9. Part of road network

Table 8. Traffic_light_status_table(t_0)

Traffic-LightID	From-ID	leftLight-Status	rightLight-Status	straight-Light-Status
tl_4	tl_1	0	0	1

Table 9. Traffic_light_status_table(t_1)

Traffic-LightID	From-ID	leftLight-Status	rightLight-Status	straight-Light-Status
tl_4	tl_1	0	1	1

At moment t_0, from Fig. 9, we can get traffic_flow_table (tl_4, tl_1, t_0) = 23. This variable states that 23 vehicles will arrive at tl_4 from tl_1 at moment t_0. According to $\alpha = 0.8$, $\beta = 0.1$, $\gamma = 0.1$(APPENDIX A.3), the number of cars going straightly, turning left and turning right are 18, 2 and 3 respectively. Then, the traffic_light_status_table(t_0) (APPENDIX B.Table 8) presents that the straightLight of road tl_1-tl_4 is green, with the red leftLight and rightLight. Since straightThroughRate is 16, only 16 cars which go straightly can pass tl_4, 2 cars stay. Cars turning left or turning right stay. At last,

$$\text{Penalty}\,(tl_4,\ tl_1,\ t_0) = 2 + 2 + 3 = 7$$

Now, we get penalty of intersection tl_4 at time t_0. Using this method, we can get each road penalty at time t_0. At moment t_1, traffic_flow_table (tl_4, tl_1, t_1) = 15. According to $\alpha = 0.8$, $\beta = 0.1$, $\gamma = 0.1$(APPENDIX A.3), the number of cars going straightly, turning left and turning right are 12, 2 and 1 respectively. The traffic_light_status_table(t_1) (APPENDIX B.Table 9) illustrates that the straightLight and rightLight color of road tl_1-tl_4 are green, while the leftLight is red. So the 12 going straight cars with 2 cars which was stayed during last unit time(t_0), as well as the straightThroughRate is 16, 14 cars can all pass the intersection tl_4 straightly. The number of turning left and turning right cars are both 4(already add the number of stayed cars during last unit time(t_0)). While rightThroughRate is 2, so 2 turning right cars have to stay. And, the 4 turning left cars also have to stay because of the red leftLight. Eventually, the penalty at t_1 of the road tl_1-tl_4 is:

$$\text{Penalty}\,(tl_4, tl_1, t_1) = 2 + 4 = 6$$

The total penalty:

$$\text{Penalty}\,(tl_4, tl_1, t_0 + t_1) = 7 + 6 = 13$$

Now, we get penalty of road tl_1-tl_4 during t_0 and t_1.

The key point is: the less the penalty, the less the waiting delay.

References

1. Krajzewicz, D., Erdmann, J., Behrisch, M., et al.: Recent development and applications of SUMO–simulation of urban mobility. Int. J. Adv. Syst. Meas. **5**(3,4), 128–138 (2012)
2. Maslekar, N., Boussedjra, M., Mouzna, J., et al.: VANET based adaptive traffic signal control. In: 2011 IEEE 73rd Vehicular Technology Conference (VTC Spring), IEEE, pp. 1–5 (2011)
3. Gradinescu, V., Gorgorin, C., Diaconescu, R., et al.: Adaptive traffic lights using car-to-car communication. In: IEEE 65th Vehicular Technology Conference, 2007 VTC2007-Spring IEEE, pp. 21–25 (2007)

4. Pandit, K., Ghosal, D., Zhang, H.M., et al.: Adaptive traffic signal control with vehicular ad hoc networks. IEEE T. Veh. Technol. **62**(4), 1459–1471 (2013)
5. Priemer, C., Friedrich, B.: A decentralized adaptive traffic signal control using V2I communication data. In: 12th International IEEE Conference on Intelligent Transportation Systems, ITSC 2009. IEEE, pp. 1–6 (2009)
6. Bani, Y.M., Boukerche, A.: An intelligent traffic light scheduling algorithm through VANETs. In: 2014 IEEE 39th Conference on Local Computer Networks Workshops (LCN Workshops), IEEE, pp. 637–642 (2014)
7. Behrisch, M., Bieker, L., Erdmann, J., et al.: SUMO– simulation of urban mobility: an overview. In: Simul the Third International Conference on Advances in System Simulation, pp. 63–68 (2011)
8. Krajzewicz, D., Brockfeld, E., Mikat, J., et al.: Simulation of modern traffic lights control systems using the open source traffic simulation SUMO. In: Proceedings of Industrial Simulation Conference (2005)
9. Gordon, R.L., Tighe, W.: Traffic control systems handbook Publication FHWA-HOP-06-006. In: Road Engineering Association of Asia & Australasia Conference Kuala Lumpur Malaysia, September 2005
10. Goldberg, D.E.: Genetic algorithms in search, optimization, and machine Learning. Addion wesley, Boston (1989)
11. Gen M, Cheng R. Genetic algorithms and engineering optimization. John Wiley & Sons, (2000) [20]
12. Eiben, A.E., Schippers, C.A.: On evolutionary exploration and exploitation. Fundam. Informaticae **35**(1), 35–50 (1998)
13. Golberg, D.E.: Genetic algorithms in search, optimization, and machine learning. Addion wesley, Boston (1989)
14. Goldberg, D.E.: Real-coded genetic algorithms, virtual alphabets, and blocking. Urbana **51**, 61801 (1990)
15. Holland, J.H.: Adaptation in natural and artificial systems: an introductory analysis with applications to biology, control, and artificial intelligence. In: Taniguchi, E., Thompson, R.G., Yamada, T. et al. U Michigan Press, 1975. City Logistics. Network modeling and intelligent transport systems (2001)
16. Carroll, D.L.: Genetic algorithms and optimizing chemical oxygen-iodine lasers. Dev. Theor. Appl. Mech. **18**(3), 411–424 (1996)
17. Xu, Y.G., Li, G.R., Wu, Z.P.: A novel hybrid genetic algorithm using local optimizer based on heuristic pattern move. Appl. Artif. Intell. **15**(7), 601–631 (2001)
18. Chan, C.H.: Actuator Hysteresis Modelling and Compensation with an Adaptive Search Space Based Genetic Algorithm (2003)

Estimation of External Cost and Security Governance on Traffic Accidents

Chunchao Chu(✉) and Yang Liu

China Academy of Transportation Science, Beijing 100029, China
chucc79@163.com

Abstract. This paper focuses on the issues about road traffic accident losses' estimation and security governance, which is paid close attention to in theoretical research and practical application analysis. Firstly, from the external economics view dividing into road traffic accident losses, the external cost was defined as the difference between total social cost and private cost, which were caused by road traffic accidents. Secondly, estimation models were presented about road traffic accident external cost, which included not only total social cost calculation method but also private cost one. Finally, based on existing statistics of China, both road traffic accident losses and external cost were calculated and analyzed in China since 2008, and a series of measures were put forward about strengthening the road traffic security governance. Overall, the research results can provide a reference for the theory research on externality, external cost estimation and the safety management about road traffic accidents.

Keywords: Traffic accident · External cost · Loss · Security governance · Private cost

1 Introduction

Traffic accidents generally refer to the event of human casualties and property damage during transportation processes. Since reform and opening-up, along with the road scale expansion and highway mileage growth, both the motor vehicle amount and freight volume were increasing rapidly, and which induced obviously accident casualties and property loss. As a result, both road traffic accidents and safety management were becoming issues. By the end of 2013, the length of highway was 4.35 million kilometres, and the quantity of automobile broke through 0.25 billion vehicles, motor vehicle drivers reached 0.28 billion people. At the same year, road traffic accidents occurred 198.4 thousand times, killed 58.5 thousand people, injured 213.7 thousand people, and lost directly property 1.04 billion Yuan [1]. Compared with developed countries, road traffic safety situation was very serious in China, and the loss caused by road traffic accidents was relatively high. According to statistics, the number of deaths around the world was about 0.5-0.6 million in the road traffic accidents every year, accounting for 10-15 % in China. In other words, the number of China's motor vehicles accounted for 2 % of the world, but the number of traffic accident deaths was 10 % [2].

Because the road traffic accidents caused severe casualties and property losses, all governments around the world paid more attention to governance "civilized social

C.-H. Hsu et al. (Eds.): IOV 2015, LNCS 9502, pp. 399–408, 2015.
DOI: 10.1007/978-3-319-27293-1_35

disease", and invested tremendous manpower, material and financial resources. At the same time, road traffic accidents were also researched widely about externality and inherent laws in the academic circles, such as about the accident losses' division, estimation model of accident casualty losses, loss index system of accident evaluation, and so on [3–5]. However, among these studies, traffic accident losses were discussed only at the level of theoretical analysis or mathematical model [6–10], and it didn't consider indirect losses of social public service caused by the accidents, all of which can't meet the need of practical application. In this paper, it combined with theoretical research and practical application analysis. Firstly, dividing into road traffic accident losses from the external economics view, and the external cost was defined as the difference between total social cost and private cost, which were caused by road traffic accidents. Secondly, estimation models were presented about road traffic accident external cost, which included both total social cost calculation method and private cost one. Finally, based on existing statistics of China, both road traffic accident losses and external cost were calculated and analyzed in China since 2009, and a series of measures were put forward about strengthening the road traffic security governance.

2 Externality and Loss Composition of Road Traffic Accidents

2.1 Road Traffic Accident Externality

The theory of external economics came from the welfare economics, which mainly included three doctrines of Marshall's external economy theory, Pigou's Pigou tax theory and Coase's Coase Theorem. External economics of road traffic accidents refers to the part social cost which was not responsible for traffic accident parties self, and resulted in the loss of social welfare [11], including mainly traffic congestion, traffic accidents, air pollution, noise pollution and negative global climate changes [12]. So external economic losses of road traffic accidents includes mainly loss of life, human capital loss, property loss, medical care cost, social management cost, relatives' pain etc., all of which are caused by the road traffic accidents.

2.2 Dividing into Road Traffic Accident Losses

According to the different types of accounting method, accident responsibility and accident parties, the losses of road traffic accidents had different classification. In this paper, from the view of externality, the losses caused by road traffic accidents were divided into: direct property losses, direct losses of deaths and injuries which included the social labour value loss, losses of relatives' spirit and pain, medical expenditures and funeral expenses etc., social service agency's cost which referred to spending medical care, the courts and traffic police etc.

Direct Property Losses. It refers to the direct economic losses of infrastructure and vehicle which were caused by the traffic accidents at the accident scene, including vehicle damage, goods damage, road damage and facilities damage etc.

Labour Value Losses of Casualties. It refers to the losses of which casualties can't continue to create the social labour value due to death or injury in the remaining period of labour ability.

Losses of the Spirit. It refers to that traffic accidents cause casualty families' sadness and other emotions. In fact, it will not only make the accident victims lose social labour value, but also result in the loss of the spirit after the traffic accidents.

Social Service Agency Cost. It refers to that social service organizations pay all of human, material and financial resources to deal with traffic accidents, including all expense of emergency centre, department of transportation, fire control institutions, insurance companies, the public security organs etc.

3 Estimation Model of Road Traffic Accident External Cost

Estimation of traffic accident external cost means mainly to monetize negative external economics of traffic accidents, and calculates the proportion which road traffic accident losses account for the gross domestic product (GDP). According to the previous definition of externality, the external cost of road traffic accidents (*EC*) is defined as the different part between total social cost caused by traffic accidents (*TC*) and the private cost of road traffic accidents (*PC*). The total cost of society has two components: direct property losses (*PL*) and labour value losses of casualties (*LL*), where direct property losses (*PL*) can be obtained by the statistics of China Statistical Yearbook and labour value losses of casualties (*LL*) can be calculated by total yield method. The private cost, the losses of related accident parties' direct undertaking, can be obtained by a court ruling method, which includes life loss, spirit loss, social services agency expenditure etc. which has been internalized by means of insurance. In conclusion, estimation model of external cost can be shown in road traffic accidents in formula 1.

$$EC = PL + LL - PC \quad (1)$$

Where *EC*—the external cost of road traffic accidents;
PL—direct property losses;
LL—Labour value losses of casualties;
PC—the private cost of road traffic accidents.

3.1 Calculation Method of Labour Value Losses of Casualties

The total output method is also named the human capital approach, and reflects the losses that accident casualties don't devote to society from the accident taking place to the future expected life. Estimates labour value losses of casualties (*LL*) need three elements: the quantity of casualties (*QC*), lost working time (*LT*) and the per capita annual output value (*PV*). Calculation method of labour value losses of casualties can be shown in formula 2.

$$LL = \sum_{i=1}^{n} QC_i \times LT_i \times PV \tag{2}$$

Where, QC_i—the quantity of casualties which is corresponding to ith age stage;
LT_i—lost working time of casualties which is corresponding to ith age stage;
PV—the per capita annual output value;
i—the division number of casualty age stage.

Lost Working Time of Casualties. Lost working time of a casualty is closely related to his current age, level of injury and expected life value. In the paper, some assumptions were made as followed: if the age of the death was under 16 years old, at the stage of compulsory education, which wouldn't be considered the social labour value losses because of regardless of the ability of social production. Over the age of 16 years old, lost working time of casualties was estimated according to complete labour capacity. Between 60 years old (usually male retirement age) and expected life value (75 years old), lost working time of casualties was computed according to 30 % of complete labour capacity. Above 75 years old, it was assumed approximately that they created no longer social economic value, and ignored the working time loss.

The Per Capita Annual Output Value. The per capita annual output value of casualties can be measured by GDP per employment.

In reality, the injured losses varied greatly with injury level, but subdivision data of injury level weren't provided according to: minor injury, general injury and serious wound in present traffic accident statistics. Referring to the relevant research results, the injured losses were about 1 ~ 3 % the losses of the death [13], so this paper adopted the average level 2 % to calculate the injured losses.

3.2 Calculation Method of Private Cost

According to the court ruling method, it represented the private cost of traffic accidents, which should be paid to participants' losses by the traffic accident crime or negligence party. On this subject, the Supreme People's Court of the People's Republic of China offered a specific explanation and the corresponding compensation standard in "Some explanations on the application of law in the trial of personal injury compensation cases" (2003). Accordingly, this paper constructs the private cost accounting model of the road traffic accident casualties as was shown in formula 3.

$$PC = \begin{cases} 20 \times PCI \times \beta & PA \leq 60 \\ (80 - AC) \times PCI \times \beta & 60 < PA < 75 \\ 5 \times PCI \times \beta & PA \geq 75 \end{cases} \tag{3}$$

Where, PC—the private cost of road traffic accidents;
PCI—urban households per capita disposable income (PCDI) or rural households per capita annual income (PCAI) according to the location of the court;

AC—the age of casualties;

β– compensation coefficient of casualties, where divided into injuries from 1 level to 10 level according to compensation coefficient from 1 to 0.1, and death compensation coefficient was set to 1.

4 Application and Analysis

4.1 Composition of Road Traffic Accident Casualties

At present, the special ages of casualties weren't provided in the relevant road traffic accident statistics. Therefore, it was assumed that the casualty age distribution of road traffic accidents was consistent to the national population age distribution. Based on the assumption, Casualty composition of road traffic accidents can be approximately calculated by virtue of the national population age distribution and the total quantity of road traffic accident casualties. The results were shown in Table 1.

Table 1. Composition of road traffic accident casualties by ages during 2009-2013 in China

Age stage	2009		2010		2011		2012		2013	
	Deaths	Injuries	Deaths	Injuries	Deaths	Injuries	Deaths	Injuries	Deaths	Injuries
0-4	3499	14207	3530	13752	3532	13440	3413	12762	3323	12132
5-9	3664	14878	3509	13668	3338	12704	3271	12229	3268	11933
10-14	4267	17325	3833	14931	3404	12953	3193	11937	3013	11000
15-19	4858	19723	4627	18025	4379	16666	3943	14743	3597	13131
20-24	5097	20696	5545	21600	5914	22508	5428	20294	5098	18614
25-29	4390	17826	4655	18132	4863	18505	4798	17939	4875	17798
30-34	4579	18594	4543	17698	4475	17031	4459	16672	4327	15799
35-39	6168	25043	5728	22312	5279	20088	4751	17763	4414	16116
40-44	6535	26534	6203	24165	5850	22264	5736	21449	5431	19830
45-49	5372	21814	5458	21260	5494	20909	5298	19809	5136	18752
50-54	4905	19917	4132	16097	3389	12896	3303	12350	3639	13287
55-59	4602	18684	4256	16577	3904	14858	3809	14242	3701	13514
60-64	3239	13152	3060	11920	2871	10927	2957	11056	3049	11132
65-69	2333	9473	2150	8374	1964	7475	2005	7496	2053	7494
70-74	1890	7674	1730	6741	1570	5976	1506	5630	1461	5332
75-79	1310	5320	1234	4805	1153	4390	1134	4239	1112	4061
80-84	686	2785	671	2613	652	2480	648	2423	668	2440
85-89	278	1131	271	1054	261	994	255	953	275	1004
90-94	68	277	74	287	78	298	77	287	82	298
95+	18	73	17	66	16	60	15	56	16	58
Total	67759	275125	65225	254075	62387	237421	59997	224327	58539	213724

4.2 Labour Value Losses of Casualties

The Per Capita Annual Output Value (*PV*). The per capita annual output value can be measured by GDP per employment which had been defined in the Sect. 3.1. According to the China Statistical Yearbook, the per capita annual output value can be calculated and shown in Table 2 in China during 2009-2013.

Table 2. The per capita annual output value in China during 2009-2013

Year	GDP (100 millions Yuan)	Employment (10 thousands people)	*PV* (Yuan)
2009	340903	77995	43708
2010	401202	76105	52717
2011	473104	76420	61908
2012	519470	76704	67724
2013	568845	76977	73898

Lost Working Time of Casualties (*LT*). According to the method in the Sect. 3.1, combined with the casualty composition in Table 1, Lost working time of casualties can be estimated and shown in Table 3, where the average values were adopted in each age range.

Calculating Labour Value Losses of Casualties. By virtue of these data of the quantity of casualties(*QC*), the per capita annual output value (*PV*) and lost working time of casualties (*LT*) which were obtained above, labour value losses of casualties (*LL*) can be calculated by formula 2, including two parts: labour value losses of deaths and labour value losses of injuries. The results were shown in Table 3.

Table 3. Labour value losses of casualties (*LL*) in road traffic accidents during 2009-2013

Age stage	*LT* (Years)	*LL* in 2009	*LL* in 2010	*LL* in 2011	*LL* in 2012	*LL* in 2013
0-14	0	0.00	0.00	0.00	0.00	0.00
15-19	47.5	109.04	124.89	138.58	136.33	135.46
20-24	42.5	102.37	133.91	167.46	167.90	171.81
25-29	37.5	77.80	99.19	121.48	130.96	144.95
30-34	32.5	70.33	83.91	96.89	105.48	111.52
35-39	27.5	80.15	89.51	96.70	95.09	96.25
40-44	22.5	69.49	79.31	87.69	93.95	96.90
45-49	17.5	44.43	54.27	64.05	67.49	71.27
50-54	12.5	28.98	29.35	28.22	30.05	36.07
55-59	7.5	16.31	18.14	19.51	20.79	22.01
60-64	3.9	5.97	6.78	7.46	8.39	9.43
65-69	2.4	2.65	2.93	3.14	3.50	3.91
70-74	0.9	0.80	0.89	0.94	0.99	1.04
75+	0	0.00	0.00	0.00	0.00	0.00
Total		608.32	723.08	832.13	860.93	900.63

4.3 Calculating Private Cost

The Proportion and Per Capita Income of Rural and urban Population. Including GDP and the proportion of rural and urban population, which can be obtained from the China Statistical Yearbook, through further analysis, and more information can be

Table 4. The per capita income of urban and rural residents

Year	GDP (100 millions Yuan)	The proportion of urban population (%)	The proportion of rural population (%)	Per capita income of urban residents (Yuan)	Per capita income of rural residents (Yuan)
2009	340903	46.59	53.41	15781	4761
2010	401202	49.95	50.05	19109.4	5919
2011	473104	51.27	48.73	21809.8	6977.3
2012	519470	52.57	47.43	24564.7	7916.6
2013	568845	53.73	46.27	26955.1	8895.9

obtained such as per capita income of urban and rural residents. The results were shown in Table 4. These data were the basis of completing *PC* calculation of road traffic accidents.

The Quantity of Casualties and Lost Working Time. On the basis of method presented in the Sect. 3.2, combined with these data in Table 1, both the special quantity of casualties and lost working time can be obtained to calculate private cost by age stages. The results were shown in Table 5.

Table 5. The quantity of casualties and lost working time by age stages

Year	Status	0-59 years old	60-64 years old	65-69 years old	70-74 years old	Over 75 years old
2009	Death number	57936	3239	2333	1890	2360
	Injury number	235241	13152	9473	7674	9586
2010	Death number	56019	3060	2150	1730	2267
	Injury number	218217	11920	8374	6741	8825
2011	Death number	53821	2871	1964	1570	2160
	Injury number	204822	10927	7475	5976	8222
2012	Death number	51402	2957	2005	1506	2129
	Injury number	192189	11056	7496	5630	7958
2013	Death number	49822	3049	2053	1461	2153
	Injury number	181906	11132	7494	5332	7861
lost working time (Years)		20	18	13	8	5

Table 6. Estimation of the private cost in road traffic accidents

Year	The private cost in road traffic accidents (100 millions Yuan)		
	Private cost of deaths	Private cost of injuries	Private cost of casualties
2009	126.09	281.59	407.68
2010	153.67	329.22	482.89
2011	171.63	359.24	530.87
2012	188.36	387.34	575.7
2013	204.68	411.01	615.69

Calculating Private Cost (*PC*) by Court Ruling Method. Using the results in Tables 4 and 5, private cost in traffic accidents can be obtained during 2009-2013 by formula 3, as were shown in Table 6. Because of no relative statistics about injury levels, the average coefficient 0.55 was adopted which came from compensation standard level 1 to level 10.

4.4 Calculating External Cost

Utilizing these data obtained in Tables 3 and 6, and direct property losses from the China Statistical Yearbook, external cost of road traffic accidents can be computed and shown in Table 7 in China during 2009-2013.

Some development laws can be drawn from above estimated results by virtue of analyzing. From the variation of *TC*/GDP, the proportion of *TC*/GDP presents downtrend basically. It may be approved that the safe situation of road transportation is being improved gradually since 2009, due to government departments strengthening road transportation safety governance. From the estimated results of road traffic accident external cost, about one third of total society cost in road traffic accidents aren't still undertaken by related accident parties, which are known as external cost. So external cost should be internalize more through the judicial system and insurance system in the future, and let parties responsible for the accidents take full responsibility for the social welfare losses.

Table 7. External cost of road traffic accidents during 2009-2013 (Unit: 100 millions Yuan)

Year	Total social cost (TC)				Private cost (PC)	External cost (EC)	EC/GDP
	TC	LL	PL	TC/GDP			
2009	617.46	608.32	9.14	0.18	407.68	209.78	0.06
2010	732.34	723.08	9.26	0.18	482.89	249.45	0.06
2011	842.92	832.13	10.79	0.18	530.87	312.05	0.07
2012	872.68	860.93	11.75	0.17	575.7	296.98	0.06
2013	911.02	900.63	10.39	0.16	615.69	295.33	0.05

5 Security Governance of Road Transportation

Through quantitative calculating road traffic accident external cost, the purpose is for not only the special cost values, but also reference for the government decision and policy making. By virtue of systemic estimation and analysis on road traffic accident external cost, more focused measures need be taken to maximize the transportation positive effect and reduce the negative effect, including infrastructure pricing, tax, market regulation policy etc. Therefore, based on the above calculated results, some suggest as followed.

Implementing Plans of the Road Security and Intelligent Management Optimize Travel Safety Environment. To carry out the highway security engineering, optimizing road design and construction, increasing the safety facilities, enhancing the signs and markings, greatly improves the security environment of people travel and transportation. To implement intelligent supervision and information service engineering, improving the ability of emergency management and rescue, raising the traffic accident disposal ability, obviously improves service quality.

Strengthening the Safety Standards and Market Access Management of Commercial Vehicles Improves the Safety Performance of Vehicles Operation. To enhance the safety manufacturing standards of commercial vehicles, improves vehicle automatic collision prevention and safety early warning intelligence level. To perfect safety operation standards, strengthens commercial vehicle market supervision. To execute annual inspection regulations and market exit mechanism, prevent these vehicles from entrancing market which don't meet safety standards. To strengthen transportation market supervision and crack down on overloading transportation, standardize transportation market order.

Strengthening the Driver Quality-oriented Trainings and National Safety Awareness Natures a Transportation Safety Culture of People-centric. At present, just entering the mobile society, transportation safety culture is still in the developing stage. To strengthen practitioners' behaviour training and safety supervision, norms practitioners especially the drivers working behaviour through technical means and management regulations. To enhance the national transportation safety culture, lets the civilized travel behaviour become a conscious action of all citizens.

6 Conclusion

From the external economics view, the external cost caused by road traffic accidents was defined as the difference between total social cost and private cost, based on dividing into road traffic accident losses. Based on presented estimation models and existing statistics of China, both road traffic accident losses and external cost were quantitatively calculated and analyzed in China since 2009. The research paid close attention to the combination theoretical research with practical application analysis, and the results can provide a reference for the theory research on externality, external cost estimation and the safety management about road traffic accidents.

References

1. National Bureau of Statistics of China. China Statistical Yearbook, PeiKing: China Statistics Press (2013)
2. China Academy of Transportation Sciences. Study on evaluation method of road traffic accident economic losses. Beijing (2008)
3. Trawen, A., Maraster, P.: International comparison of cost of a fatal casualty of road accident in 1990 and 1999. Accid. Anal. Prev. **34**, 323–332 (2002)
4. Zhengyu, L., Ling, J.: Research on several problems about transport external cost. J. Tongji Univ. (Nat. Sci.) **33**, 931–936 (2005)
5. Internalization Measures and Policies for All external Cost of Transport (IMPACT). Handbook on estimation of external costs in the transport sector. Netherlands: Delft (2013)
6. Derriks, H.M., Mak, P.M.: Underreporting of Road Traffic Casualties: IRTAD Special Report. Geneva: IRTAD (2007)
7. Sauerzapf, V., Jones, A.P., Haynes, R.: The problems in determining international roadmortality. Accid. Anal. Prev. **42**, 492–499 (2010)
8. Clarke, D.D., Ward, P., Bartle, C.: Killer crashes: fatal road traffic accidents in the UK. Accid. Anal. Prev. **42**, 764–770 (2010)
9. Aidoo, E.N., Amoh-Gyimah, R.: The effect of road and environmental characteristics on pedestrian hit-and-run accidents in ghana. Accid. Anal. Prev. **53**, 23–27 (2013)
10. Shi, J., Bai, Y., L, T., Atchley, p: A model of beijing drivers' scrambling behaviors. Accid. Anal. Prev. **43**, 1540–1546 (2011)
11. Yafei, W., Qunren, L.: Study on the external cost of China transportation industry. Railway Transport Econ. **28**, 1–3 (2006)
12. Sen, A.K., Tiwar, G., Upadhyay, V.: Estimating marginal external cost of transport in Delhi. Transp. Policy **17**, 27–37 (2010)
13. China Academy of Transportation Sciences. Study on estimation of transportation external cost. Beijing (2009)

Research on Adaptive Wrapper in Deep Web Data Extraction

Donglan Liu(✉), Lei Ma, and Xin Liu

State Grid Shandong Electric Power Research Institute,
2000 Wangyue Road, Jinan 250002,
People's Republic of China
liudonglan2006@126.com

Abstract. As the rapid development of Internet technology, Deep Web has the vast amounts of data information, and in the rapid growth of the Web to become a huge data source. Many documents share common HTML tree structure on script generated websites, allowing users to effectively extract interested information from deep webpages by wrappers. However, since tree structure evolves over time, the wrappers break frequently and need to be re-learned. In this paper, we explore the problem of constructing adaptive wrappers in deep webpages. In order to keep web extraction robust when webpages changes, a minimum cost script edit model based on machine learning techniques is proposed. With the method, three edit operations under structural changes are considered, i.e. inserting nodes, deleting nodes and substituting nodes' labels. By obtaining the extraction model for 51job site and then random sampling pages at zhaopin site using this extraction model for training the new wrapper. Besides, the wrapper has high versatility, realizing the adaptation extraction. Experimental results show that the proposed approach can improve the extraction accuracy of target data and effectively solve the adaptive wrapper for the massive Deep Web data.

Keywords: Deep web data extraction · Wrapper · Minimum cost script · Machine learning

1 Introduction

With the rapid development of Internet technology, Deep Web has the vast amounts of data information, and in the rapid growth of the Web to become a huge data source. The information can be accessed by the query interfaces provided by backend database. Deep Web contains a lot of valuable information, but the Deep Web data has the characteristics of heterogeneity and dynamic, it is a very challenging task to use the abundant information effectively.

Several websites use scripts to generate highly structured HTML from backend databases, such as: recruitment sites, shopping sites, electric power sites, academic sites and form-based websites. The structural similarity of script-generated webpages can help information extraction systems extract information from the webpages using simple rules. These rules are called wrappers. Nowadays, wrappers are becoming a dominant strategy for extracting web information from script-generated pages.

C.-H. Hsu et al. (Eds.): IOV 2015, LNCS 9502, pp. 409–423, 2015.
DOI: 10.1007/978-3-319-27293-1_36

However, the extraction operation of wrappers greatly depends on the structure of webpage. Since the information of webpage changes dynamically, even very slight change may lead to the breakdown of wrappers and require them to re-learn. This is the so-called Wrapper Breakage Problem. Therefore, it is very important for web data integration to effectively improve the adaptive capacity of web data extraction.

To illustrate, Fig. 1 displays an XML document tree of a script-generated job page from 51job site. If we want to extract working place from this page, following XPath expression can be used:

$$W_1 \equiv /\mathrm{html}/\mathrm{body}/\mathrm{div}[2]/\mathrm{table}/\mathrm{td}[2]/\mathrm{text}(). \quad (1)$$

which is an instruction on how to traverse HTML DOM trees. However, there are several small changes which can break this wrapper. For instance, the first div is deleted or merged with the second one, a new table or td is added under the second div, the order of "Position" and "Working place" is changed, and so on.

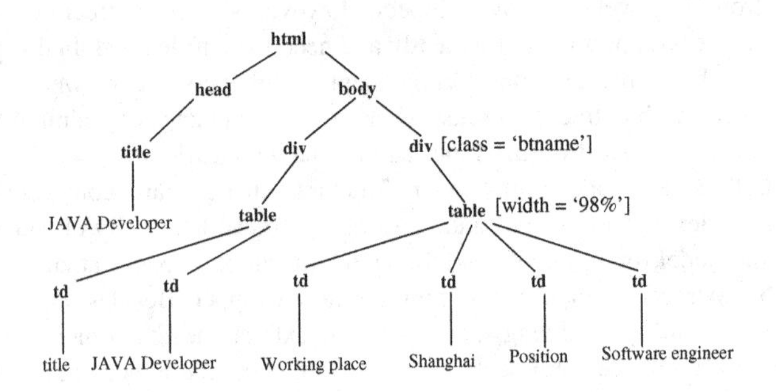

Fig. 1. An HTML webpage of 51job site

In fact, the expression W_1 is one form of a simple wrapper, and the problem of robust web extraction has caught much attention and has been widely researched [1–3]. Jussi Myllymaki and Jared Jackson [1] proposed that certain wrappers are more robust than the others, and the wrappers can have lower breakage. For instance, each of the following XPath expressions can be used as an alternative to W_1 to extract the working place.

$$W_2 \equiv //\mathrm{div}[@\mathrm{class} = \text{'btname'}]/*/\mathrm{td}[2]/\mathrm{text}(). \quad (2)$$

$$W_3 \equiv //\mathrm{table}[@\mathrm{width} = \text{'98\%'}]/\mathrm{td}[2]/\mathrm{text}(). \quad (3)$$

Intuitively, these wrappers may be preferable to W_1 since they have a lesser dependence on the tree structure. For example, if the first div is deleted, W_1 will stop working, while W_2 and W_3 might still work.

In this paper, we aim to design a adaptive wrapper which can achieve substantially higher robustness. Wrapper can complete and extract web data records from deep webpage accurately, which is denoted as "distinguished node" in this paper. Then, wrapper constructs a set of extraction rules according to website template, extracts information from webpage and translates into structured data automatically. In general, when webpage changes beyond the limitation of wrapper script, we can only re-locate the data by modifying wrapper scripts. Otherwise, the information extraction might be failed. When webpage evolves over time, we use the existing information of web data integrated system combining with other techniques to recognize and label the data elements and attribute tags. Then, we can generate an optimal training sample. Finally, we can rebuild a new wrapper by using the existed wrapper induction techniques, and make it possible for wrappers to cope with the changes in websites effectively. In order to keep web extraction robust when webpage changes, we propose a novel and highly efficient approach based on minimum cost script edit model with machine learning techniques for robust web extraction. With the method, we consider three edit operations under structural changes, i.e., inserting nodes, deleting nodes and substituting nodes' labels. Firstly, we obtain the change frequencies of three edit operations for each HTML label according to the frequency of webpage change on real web data with machine learning method. Then, we compute the corresponding edit costs for three edit operations on the basis of change frequencies and minimum cost model. Finally, we choose the most proper data to extract the interested information by applying the minimum cost script.

Our contributions involve three aspects in this work. Firstly, we propose a general framework for constructing robust wrapper to extract interested information from deep webpages. Secondly, we present a bottom-up fashion for enumerating all minimal candidate wrappers to speed up the robustness of the evaluation. Finally, we perform an extensive set of experiments covering over multiple websites, and the experiments are able to achieve very high precision and recall. It demonstrates that our wrapper can improve the extraction accuracy of target data, effectively solve the adaptive wrapper for the massive Deep Web data.

The rest of the paper is organized as follows. In Sect. 2, we briefly define a few related concepts. We introduce our robust web extraction framework and the minimum cost script edit algorithm in Sect. 3, and generate minimal candidate wrappers based on bottom-up method in Sect. 4. Our experimental evaluation is presented in Sect. 5, related work is discussed in Sect. 6 and conclusion is provided in Sect. 7. Finally, acknowledgements and related references are given in the following sections.

2 Problem Definition

Some related concepts about robust web extraction are defined in this section.

Definition 1. Order Labeled Trees: Let w be a webpage. We represent w as an ordered, labeled tree corresponding to the parsed HTML DOM tree of the webpage [3]. To illustrate, we consider Fig. 1 representing the HTML of a recruitment page. The children of every node are ordered and every node has a label from a set of labels L.

Definition 2. Edit Operations: we consider three edit operations under structural changes, i.e., inserting nodes, deleting nodes and substituting labels of nodes. Each change is one of the three edit operations and the tree structures evolve by choosing these edit operations randomly. A sequence S of edit operations is defined to be an *edit script*. $S(w)$ is used to denote the new version of the webpage obtained by applying the operators in S to webpage w in sequence. We use $S(n)$, $n \in w$, to denote the node in $S(w)$ that n maps to when edit script S is applied.

Definition 3. Edit Costs: We describe edit costs for three edit operations: inserting nodes, deleting nodes and substituting labels of nodes. We define a cost function for computing the edit costs corresponding to each of operations.

Definition 4. Isomorphic Trees: Suppose two webpages $w1$ and $w2$ to be isomorphic, written as $w1 \equiv w2$, if they have identical structures and labels [3]. Similarly, the parsed HTML DOM trees corresponding to the two webpages are called isomorphic trees.

Definition 5. Wrapper: A wrapper is a function $f(x)$ from a webpage to a node in the webpage, since the core of data extraction is positioning in the entire document. Supposing w is a webpage with a distinguished node $d(w)$ which includes the interested information. We want to construct a wrapper that extracts from future target versions of w. Let $w' = S(w)$ denote a new version of the webpage, namely w' is obtained by applying the operators in S to webpage w in sequence. We want to find the position of distinguished node $d(w)$ in the new webpage by XPath expression. If wrapper function meets $f(w') = d(w')$, we say that a wrapper $f(x)$ works on a future target version w' of a webpage w. If $f(w') \neq d(w')$, then we say that wrapper has broken or failed.

Definition 6. Confidence of Extraction: Confidence [3] is a measure of how much we trust our extraction of the distinguished node for the given new version.

3 Minimum Cost Script Edit Model

3.1 Robust Web Extraction Framework

In this section, we give an overview of our robust web extraction framework on the basis of the framework which was recently proposed by Nilesh Dalvi et al. [2]. This framework is depicted in Fig. 2. We describe the main components as follows and the functions in italic indicate our new methods.

In the framework, the *Archival Data* component contains various evolutions of webpage. Suppose a webpage w denotes the recruitment page for the Java Developer, undergoing various changes across its time. Let w_0, w_1, ... denote the various future versions of w. The archival data component is a collection of pairs (w_t, w_{t+1}) for the various future versions of w, i.e. it includes $\{(w_0, w_1), (w_1, w_2), \ldots, (w_t, w_{t+1})\}$. The archival data can be obtained by monitoring a set of webpages over time.

The *Model Learner* component is mainly responsible for minimum cost model. Model learner takes the archival data as input and learns a model that best fits the data. It learns the parameter values that minimize the cost of each edit operation, i.e.

$$\arg\min_{x} \prod_{(w_t, w_{t+1}) \in ArchivalData} \{\cos t(insert : x), \cos t(delete : x), \cos t(substitute : x \rightarrow y)\}. \quad (4)$$

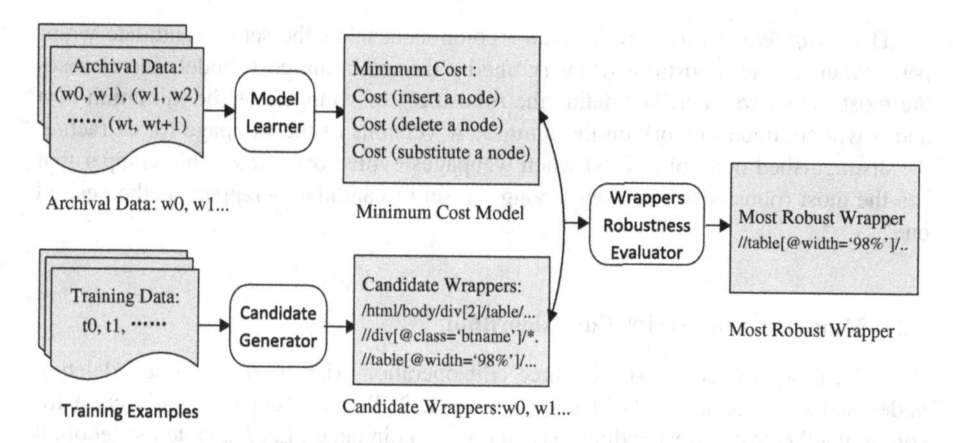

Fig. 2. Robust web extraction framework

where x represents a node of webpage; *cost* (*insert:* x) represents the cost of inserting a node with the label x; *cost* (*delete:* x) represents the cost of deleting a node with the label x; *cost* (*substitute:* $x \rightarrow y$) represents the cost of substituting node x to node y.

The *Minimum Cost Model* component consists of costs of three edit operations: inserting nodes, deleting nodes and substituting labels of nodes. This is the salient component of our framework. The minimum cost model is specified by a set of parameters, computing the minimum cost of each edit operations by means of the parameter obtained by model learner. We intend to compute the change frequencies of three edit operations for each HTML label according to the frequency of webpage change on real web data with machine learning techniques. Then, we derive the cost of edit operations according to the frequency of webpage change combining with model learner. If the operations of webpages are frequent, then we need to make the corresponding cost as low as possible. Since the cost of each edit operation is minimal, the cost sum of a sequence of edit operations for webpage is minimum.

The *Training Examples* component for an extraction task consists of a small subset of the interested webpages that specify some fields. Such as: the working place, release date, education background and position for job information of the webpages from recruitment sites.

The *Candidate Generator* component takes labeled training data and generates a set of candidate alternative wrappers. The problem of learning wrappers from labeled examples has been extensively studied [4–11], and some focus specifically on learning XPath rules [1, 4]. Any of the techniques can be used as part of candidate generator in this paper. In this section, we consider a method that generates wrappers in a bottom-up fashion, by starting from the most general XPath that matches and specializes every node until it matches the target node in each document. We want to generate an XPath expression, which makes both precision and recall equal to 1. Precision reflects the accuracy of the results, and Recall reflects the cover of getting correct results. We can enumerate all the wrappers according to the above idea, but enumerating all the wrappers is relatively time-consuming for evaluating the robustness of wrappers. Consequently, we consider enumerating all "minimal" candidate wrappers to improve our algorithm.

The *Wrappers Robustness Evaluator* component takes the set of candidate wrappers, evaluates the robustness of every one by the minimum cost model, and chooses the most robust wrapper. We define the robustness of wrappers as the minimum cost and it will continue to work on the future new versions of the webpage for extracting the distinguished node of interest when webpages evolve over time. The wrapper that has the most robustness is chosen among the set of candidate wrappers as the desired one.

3.2 Minimum Cost Script Edit Algorithm

We have described edit costs for three edit operations (i.e. inserting nodes, deleting nodes and substituting labels of nodes) in Sect. 2. We now depict cost function for computing the cost corresponding to each operation in detail. Let L denote the set of all labels, $l_i \in L$. We assume that there is a cost function $cost(x)$ for computing the cost for each edit operation, such as $cost\ (\varnothing, l_1)$ represents the cost of inserting a node with label l_1, $cost\ (l_1, \varnothing)$ represents the cost of deleting a node with label l_1, and $cost\ (l_1, l_2)$ represents the cost of substituting a node with label l_1 to another node with label l_2. Note that $cost\ (l_1, l_1) = 0$. In addition, we assume that cost functions are satisfied for the triangle inequality, i.e., $cost(l_1, l_2) + cost(l_2, l_3) \geq cost(l_1, l_3)$. S is an given edit script and the cost denoted as $cost(S)$, which is simply the sum of costs of each of the edit operations in S as given by the cost function $cost(x)$.

It is easy to compute the minimum cost scripts according to the cost model is trained in Sect. 3.1, since each cost of the operations is minimum which are trained in model. We simply introduced the concept of confidence in Sect. 2. Then, the following part provides a detailed explanation of the confidence of extraction. Intuitively, if the page w' differs a lot from w, then our confidence in the extraction should be low. However, if all the changes in w' are in a distinct portion away from the distinguished node, then the confidence should be high despite those changes. Based on this intuition, we define the confidence of extraction on a given new version as follows.

Let S_1 be the smallest cost edit script that takes w to w', namely $S_1(w)$ and w' are isomorphic trees, the node extracted by wrapper is denoted $S_1(d(w))$, and the corresponding cost is $cost(S_1)$. We also look at the smallest cost edit script S_2 that takes w to w' but does not map $d(w)$ to the node corresponding to $S_1(d(w))$, and the corresponding cost is $cost(S_2)$. We define the confidence of extraction as $cost(S_2)$-$cost(S_1)$. Intuitively, if this difference between $cost(S_2)$ and $cost(S_1)$ is large, the extracted node is well separated from the rest of nodes, and the extraction is likely to be correct. If this difference is low, the extracted node is likely to be wrong.

Aditya Parameswaran et al. [3] proposed a method by enumeration method for computing the minimum cost scripts and extracting the distinguished node of interest. The method is simple but inefficient, for the efficiency of enumerative algorithm is very low. Next, we design a more efficient algorithm using dynamic programming to compute the costs of all edit scripts efficiently. The basic idea behind the more efficient algorithm is to pre-compute the costs of all edit scripts, and finally choose the most suited data to extract the interested information. In Table 1, the process is illustrated.

In Algorithm 1, output parameter *New_d(W)'* represents the new location of distinguished node of interest, expressed by XPath expression. Output parameter *Extr_Conf* represents the confidence of extraction. If *Extr_Conf* is large, the extraction is likely to be correct. The confidence of extraction can be used to decide whether to use the extracted results or not.

Table 1. Minimum cost script edit algorithm

Algorithm 1 Minimum Cost Script Edit Algorithm

Input: *W*: A webpage;
  W': the future version of webpage *W*;
  d(*W*): a distinguished node of interest.
Output: *New_d(W)'*: new location of *d*(*W*) in *W'*;
  Extr_Conf: confidence of extraction.

1. Compute the change frequencies of three edit operations for each HTML label.
2. Compute the corresponding edit costs for three edit operations.
3. cost(insert: x) := *cost* (∅, *x*):=Cost(insert a new node);
4. cost(delete: x) :=*cost* (*x*, ∅) := Cost(delete a exist node);
5. cost(substitute: x→y) := *cost*(*x*, *y*):=Cost(substitute a node to another node);
6. Take webpage *W* to *W'* by a set of edit scripts, using the dynamic programming to compute the costs of all edit scripts and save the results with array.
7. Choose the minimum cost script S_1 such that *W'* are obtained by applying S_1 in *W*, namely $S_1(W)$ and *W'* are isomorphic, i.e.
   $S_1(W) \equiv W'$;
   $New\_d(W)' = S_1(d(W))$;
   $Extr\_C1 = \text{cost}(S_1)$;
8. Choose the minimum cost script S_2 such that *W'* are obtained by applying S_2 in *W*, namely $S_2(W)$ and *W'* are isomorphic but does not map *d*(*W*) to the node corresponding to $S_1(d(W))$, i.e.
   $S_2(W) \equiv W'$ and $New\_d(W)' \neq S_1(d(W))$;
   $Extr\_C2 = \text{cost}(S_2)$;
9. $Extr\_Conf = Extr\_C2 - Extr\_C1$;
10. **Return** *New_d(W)'*, *Extr_Conf*

4 Generating Minimal Candidate Wrappers Based on Bottom-up Method

We intend to obtain a set of alternative wrappers for our extraction task to pick the most robust wrapper according to our model. The set of candidate wrappers should contain a variety of potentially robust ones. Previous work [11] on automatically learning XPath rules from labeled webpages works in a top-down fashion, namely, it starts from the specific paths in each webpage and generalizes them to a single XPath. But unfortunately, this results in the most specific XPath which contains all possible predicates across all webpage. For instance, in Fig. 1, a full XPath expression W_1 is obtained by applying a top-down fashion when we extract the information of "Working place".

For W_1, if the first div is deleted or merged with the second one, it will not work, namely the robustness of wrapper W_1 is extremely poor. Therefore, the resulting XPath is complex and not robust, thus, it is not a suitable candidate for a robust wrapper.

Nilesh Dalvi et al. [2] recently proposed an algorithm for enumerating wrappers; however, the algorithm is not complete. Besides, the problem of constructing the most robust wrapper is still unsolved. In this paper, we describe a more complete and effective method based on their algorithm. It generates wrappers in a bottom-up fashion, by starting from the most general XPath that matches and specializing every node until it matches the target node in each document.

Suppose D represents a set of labeled XML documents, i.e. it contains several distinguished nodes of interest in corresponding webpage. For each $d \in D$, we want to extract the target nodes coming from the set of labeled nodes of D, which is written as $T(d)$. For an given XPath expression x, we intend to generate such an XPath expression x which satisfies the following rule: for each d, $x(d) = T(d)$, namely, the result of extraction is exactly equal to the target node. We define evaluation criteria according to the information retrieval as follows.

$$\mathrm{Precision}(x) = \sum_{d} (x(d) \cup T(d))/x(d). \tag{5}$$

$$\mathrm{Recall}(x) = \sum_{d} (x(d) \cup T(d))/T(d). \tag{6}$$

We intend to generate an XPath expression, which makes both precision and recall equal to 1. To illustrate, we use the following XPath expression:

$$x_0 \equiv //\ \mathrm{html/body/div/\ table/} * \mathrm{/td/text()}. \tag{7}$$

We define a one-step specialization of x to be an XPath expression obtained by any of the following four operations on x:

1. Adding a predicate to some node in x. E.g.

$$x_1 \equiv //\ \mathrm{html/body/div/\ table}\left[@width = `\ 98\%'\right]/ * \mathrm{/td/text()}. \tag{8}$$

2. Adding a//* at the top of x. E.g.

$$x_2 \equiv // * \mathrm{/html/body/div/table/} * \mathrm{/td/text()}\,. \tag{9}$$

3. Adding child position information to some node in x. E.g.

$$x_3 \equiv \mathrm{//html/body/div[2]/table/} * \mathrm{/td/text()}\,. \tag{10}$$

4. Converting a * to an HTML label name. E.g.

$$x_4 \equiv \mathrm{//html/body/div/table/tr/td/text()}\,. \tag{11}$$

We represent $x_0 \rightarrow x_1$ as a one-step specialization of x_0, and we identify $x_0 \xrightarrow{*} x_1$ as multi-step specializations, namely, x_1 can be obtained from x_0 using a sequence of specializations. The basic idea behind the proposed algorithm is maintaining a set P of partial wrappers. Each element of set P is an XPath expression which has a recall equal to 1 and a precision less than 1. Initially, set P contains the single XPath expression "//*" that matches every node in each document. The algorithm applies specialization steps to XPaths in P to obtain new ones repeatedly until the precision of XPaths reaches 1. Then, XPaths are removed from P and added to the set of output wrappers. By this method, we can enumerate all XPaths expressions. However, it is really time-consuming to evaluate the robustness of wrappers by enumerating all XPaths, besides, the efficiency is quite low. Thus, we consider enumerating all "minimal" candidate wrappers to improve our method.

For a set of given documents D and an XPath expression x, we say x is minimal if there is no other XPath x_0 fits: $x_0 \xrightarrow{*} x$, $Precision(x_0) = Precision(x)$ and $Recall(x_0) = Recall(x)$.

Suppose x is a wrapper expressed by XPath expression, namely its precision and recall equal to 1, but x does not meet the conditions of minimal wrapper. Then we need to discover a smaller XPath expression x_0 which is also a wrapper. Moreover, smaller XPaths expressions are less likely to break when extracting the information in the webpages.

Obviously, suppose X be any XPath expression and x a wrapper, such that $X \xrightarrow{*} x$. If x is minimal, X is also minimal. Thus, we can obtain all minimal wrappers by enumerating wrappers and discarding non-minimal XPath expressions in the set P after each specialization.

Table 2. Generating minimal candidate wrappers based on bottom-up method

Algorithm 2 Generating minimal candidate wrappers based on bottom-up method

Input: A set of labeled webpages, i.e. webpages include several distinguished nodes.

Output: *ResultSet*: a set of wrappers expressed by XPath expressions.

1. $ResultSet = \varnothing$
2. $P = \{\text{"//*"}\}$
3. **while** $P \neq \varnothing$ **do**
4. Suppose x be any XPath expression in P of partial wrappers, namely, each $x \in P$.
5. $P = P - x$
6. **for all** x_0 s.t. $x \rightarrow x_0$ **do**
7. **if** $isMinimal(x_0)$ = false **then**
8. $P = P - x_0$
9. **end if**
10. **if** $isMinimal(x_0)$ and $Precision(x_0) \lhd$ and $Recall(x_0) = 1$ **then**
11. $P = P \cup x_0$
12. **end if**
13. **if** $isMinimal(x_0)$ and $Precision(x_0) = Recall(x_0) = 1$ **then**
14. $ResultSet = ResultSet \cup x_0$
15. **end if**
16. **end for**
17. **end while**
18. **Return** *ResultSet*

The algorithm includes three cases. Firstly, if x_O in set P is not the minimal wrapper, we will remove it from P. Then, if the wrapper x_O meets the conditions of minimal wrapper, the recall equals to 1 but the precision is less than 1, we will put x_O in set P and continue looping until the precision reaches to 1. Finally, if the wrapper is the minimal one and the precision reaches to 1, then we remove the wrapper x_O from P and add it to the set of ResultSet. In Table 2, the algorithm for enumerating minimal wrappers is described in detail. It can be seen that Algorithm 2 is very reasonable and complete, thus, it generates all minimal candidate wrappers and only minimal ones.

5 Experimental Evaluation

In our experiments, we evaluate the effect of our robust adaptation extraction framework on two datasets of crawled pages on two real-world recruitment sites. They include 51job site and zhaopin site.

5.1 Data Sets

To test the robustness of our techniques, we use archival data from two sites: 51job site and zhaopin site. Each data set consists of a set of webpages from the above websites monitored over last 6 months respectively. We choose about 100 webpages acting as archival versions from each website, and we crawl every version once a week.

In each of our data sets, we manually select distinguished nodes which can be identified. We choose distinguished nodes including "Working place", "Position", "Release date" and "Education background" and so on.

5.2 Evaluation Criterion

In this paper, evaluation criteria from information retrieval [12] are adopted to evaluate the effect of this method.

Suppose A represents the number of data elements extracted by the method, B the number of correct data elements, and C the number of incorrect data elements.

Precision, Recall and F1 measure is:

$$\text{Precision} = \frac{B}{A}. \tag{12}$$

$$\text{Recall} = \frac{B}{B+C}. \tag{13}$$

$$\text{F1} = \frac{2 \times \text{Precision} \times \text{Recall}}{\text{Precision} + \text{Recall}}. \tag{14}$$

Precision reflects the believe level of the results, and Recall reflects the cover of getting correct results, with F1 synthesizing precision and recall.

5.3 Experimental Results and Analysis

(1) The Change Frequencies and Costs of Edit Operations

In this section, firstly, we obtain the change frequencies of edit operations by using machine learning techniques, then edit costs are calculated. Intuitively, the edit cost for an operation should capture the difficulty for a website maintainer to make edit operation. If the operations of webpages are frequent, we need to make the corresponding cost as low as possible. We set the cost of edit operations as 1, same with the frequencies of edit operations.

Table 3 shows the top 10 change frequencies that we implement our approach on two datasets of crawled pages on two real-world recruitment sites. Table 4 shows the edit costs according to the frequency of webpages change combining with our minimum cost model.

Table 3. The change frequencies of edit operations

Insertion frequency		Deletion frequency		Substitution frequency	
Label name	Frequency	Label name	Frequency	Label name	Frequency
a	0.0281	a	0.1031	a	0.0381
span	0.0142	li	0.0659	td	0.0261
div	0.0124	td	0.0395	li	0.0136
li	0.0092	div	0.0261	div	0.0082
td	0.0053	img	0.0182	span	0.0057
b	0.0044	span	0.0143	img	0.0054
img	0.0037	tr	0.0119	b	0.0045
tr	0.0031	b	0.0085	tr	0.0039
br	0.0026	br	0.0064	br	0.0034
ul	0.0009	ul	0.0059	ul	0.0027

Table 4. The edit costs of edit operations

Edit operations	Edit costs
Insertion	0.5236
Deletion	0.1947
Substitution	0.2817

(2) Wrappers for Extracting "Working Place" from 51job Site

In this section, we construct the wrappers from 51job site for data extraction. Many job seekers consider lots of factors, such as working place, educational requirements and salary and so on. We analyze the features of webpages and construct the wrappers with bottom-up method. Table 5 shows the wrappers for extracting "Working place" from 51job site.

Table 5. Wrappers for extracting "Working place" from 51job site

Wrappers Extracting "Working place" from 51job site
//table[@width ='98%']/td[2]/text()
//div[@class ='btnname']/*/td[2]/text()
//*html/body/div[2]/table/td[2]/text()
//*[preceding-sibling::*[position()=1][text()='Working place']]/text()
//*[preceding-sibling::*[position()=1][text()='Working place']]/a/text()
//*html/body/div[@class = 'btnname']/table/td[2]/text()
//* body/div[@class = 'btnname']/table/td[2]/text()
//*[preceding-sibling::*[position()=1][td/text()='Position']]/a/text()
//*[preceding-sibling::*[position()=1][text()='Position']]/a/text()
//div[position()=2]/ div[position()=2]/a/text()

(3) Evaluating the Robustness of Wrappers

In our experiments, we implement two other wrappers for comparison. One uses the full XPath containing the complete sequence of nodes' labels from the root node to the distinguished node in the initial version of the webpage. The other one uses the probabilistic robust XPath wrappers from [2]. We call these wrappers FullXPath and ProRobustXPath respectively, and our wrapper MinCostScript. The input of a wrapper consists of the old and new versions of a page as well as the location of the distinguished node in the page. We plot the results of three methods on 51job site and zhaopin site in Fig. 3. As we can see, our wrapper performs much better.

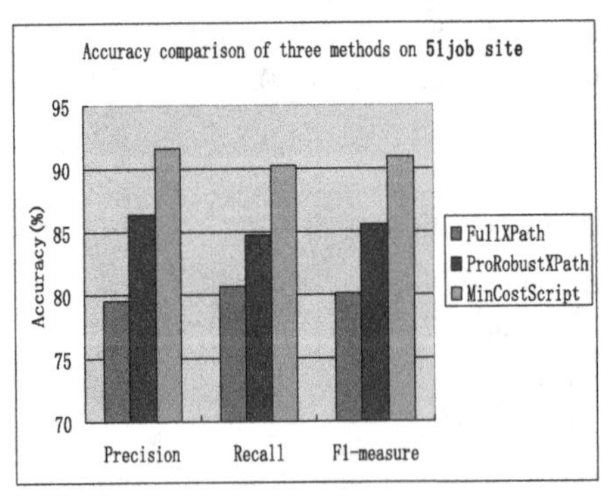

(a). 51job site

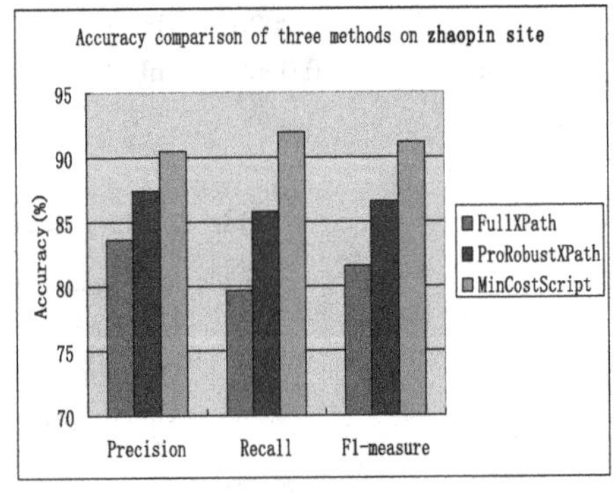

(b). Zhaopin site

Fig. 3. Performance comparisons of three methods

(4) Performance Comparison with RoadRunner [17].

This experiment compares the performance of our approach vs. RoadRunner with the sets of recruitment sites. We select 50 databases from the recruitment sites as testing data and construct wrappers using our approach and RoadRunner. The average result is considered as the evaluation criteria. Table 6 shows the performance comparison for 10 recruitment sites. Figure 4 shows the average performance comparison for these two methods. From Fig. 4 we can find our approach excelled RoadRunner, due to the low precision of training samples with RoadRunner.

Table 6. Performance comparison of MinCostScript and RoadRunner for 10 recruitment sites

Recruitment Websites (URL)	MinCostScript			RoadRunner		
	Precision (%)	Recall (%)	F1 (%)	Precision (%)	Recall (%)	F1 (%)
www.51job.com	91.6	90.2	90.9	79.5	77.4	78.4
www.zhaopin.com	90.4	91.9	91.1	80.8	78.4	79.5
www.chinahr.com	90.9	90.7	90.8	80.8	75.4	78
www.yingjiesheng.com	91.6	91.5	91.5	85.8	79.4	82.5
www.dajie.com	90.2	89.7	89.9	70.5	67.8	69.1
www.ganji.com	90.5	91.2	90.8	76.4	71.9	74.1
www.lietou.com	89.4	88.9	89.1	72.9	69.5	71.2
www.guolairen.com	90.9	90.4	90.6	78.3	80.4	79.3
www.cjol.com	91.2	90.6	90.9	75.2	81.6	78.3
www.baijob.com	89.8	89.6	89.7	74.1	78.7	76.3

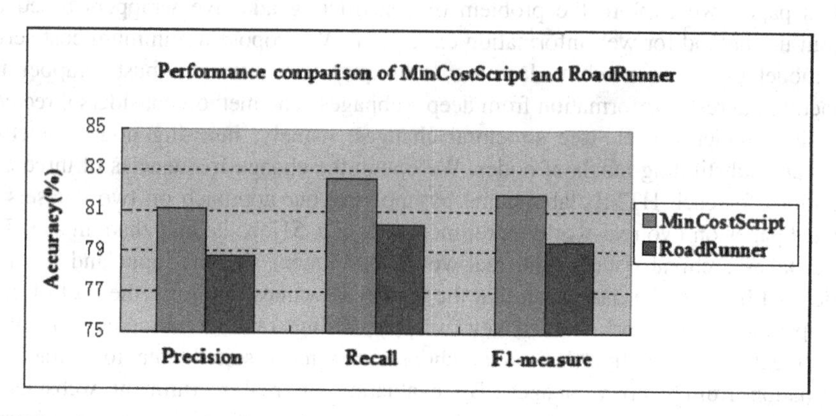

Fig. 4. Performance comparison of our approach(MinCostScript) and RoadRunner

6 Related Work

Jussi Myllymaki and Jared Jackson [1] proposed that certain wrappers are more robust than others, and it can have obviously lower breakage. They constructed robust wrappers manually, and left open the problem of learning such rules automatically. Nilesh Dalvi et al. [2] proposed a probabilistic tree-edit model to capture how webpage evolves over time, and the method can be used to evaluate the robustness of wrappers. They proposed the first formal framework to capture the concept of wrappers' robustness. Aditya Parameswaran et al. [3] considered two models to study for constructing the most robust wrapper, i.e. the adversarial model where look at the worst-case robustness of wrappers, and probabilistic model where look at the expected robustness of wrappers, as web-pages evolve. They presented the adversarial model by enumeration method for computing the minimum cost scripts and extracting the distinguished node of interest.

Evaluating wrapper robustness is supplementary to wrapper repair [13]. The idea here is generally to use content models of the desired data to learn or repair wrappers. Wrapper induction techniques focus on how to find a small number of wrappers from a few training examples [4–11]. Any of these techniques, whether manual or automatic, can benefit from robustness metric on the resulting wrappers.

Lastly, there has been some work recently on discovering robust wrappers. However, most of this work either discovers robust wrappers by human help [1] or discovers them from a fixed wrapper language [2]. Mapping one tree to another or searching the smallest tree edit distance has been used to solve other information extraction problems in the past. Also, it has been used to map similar portions to each other between two trees in order to identify repeating data elements in HTML pages, as well as to identify clusters of pages with similar HTML tree structure [14, 15].

7 Conclusion

In this paper, we explore the problem of constructing adaptive wrappers based on bottom-up method for web information extraction. We propose a minimum cost script edit model based on machine learning techniques to construct robust wrapper for extracting interested information from deep webpages. The method considers three edit operations under HTML tree structural changes, namely, inserting nodes, deleting nodes and substituting labels of nodes. We obtain the change frequencies of three edit operations for each HTML label-name by applying our approach on two datasets of crawled pages on two real-world recruitment sites, i.e. 51job site and zhaopin site. The model takes archival data on the real-world recruitment sites as input and learns a model that best fits the data, such that the parameter values minimize the cost of each edit operation. This work learned new wrappers using training examples from other recruitment sites. Finally, the model chooses the most suited data to extract the information from deep webpages. By evaluating on real recruitment websites, it demonstrates that the proposed approach can improve the extraction accuracy of target data, effectively solve the adaptive wrapper for the massive Deep Web data.

Acknowledgments. Most of this work was done during my working period at State Grid Shandong Electric Power Research Institute. The authors wish to acknowledge all colleagues and the anonymous reviewers for their helpful comments.

References

1. Myllymaki, J., Jackson, J.: Robust web data extraction with xml path expressions. In CiteSeer (2002)
2. Dalvi, N., Bohannon, P., Sha, F.: Robust web extraction: an approach based on a probabilistic tree-edit model. In: SIGMOD (2009)
3. Parameswaran, A., Dalvi, N., Garcia-Molina, H., Rastogi, R.: Optimal schemes for robust web extraction. In: VLDB (2011)
4. Dalvi, N., Kumar, R., Soliman, M.: Automatic wrappers for large scale web extraction. In VLDB (2011)
5. Baumgartner, R., Gottlob, G., Herzog, M.: Scalable web data extraction for online market intelligence. In: VLDB (2009)
6. Gupta, R., Sarawagi, S.: Domain adaptation of information extraction models. SIGMOD Rec. **37**(4), 35–40 (2008)
7. Cafarella, M.J., Madhavan, J., Halevy, A.: Web-scale extraction of structured data. In: SIGMOD (2008)
8. Cafarella, M.J., Halevy, A., Khoussainova, N.: Data integration for the relational web. In: VLDB (2009)
9. Kasneci, G., Ramanath, M., Suchanek, F., Weikum, G.: The YAGO-NAGA approach to knowledge discovery. SIGMOD Record **37**(4), 41–47 (2008)
10. Kim, Y., Park, J., Kim, T., Choi, J.: Web information extraction by HTML tree edit distance matching. In: ICCIT (2007)
11. Anton, T.: Xpath-wrapper induction by generating tree traversal patterns. In: LWA, pp. 126–133 (2005)
12. van Rijsbergen, C.: Information Retrieval. Butterworths, London (1979)
13. Chidlovskii, B., Roustant, B., Brette, M.: Documentum ECI self-repairing wrappers: performance analysis. In: SIGMOD, pp. 708–717 (2006)
14. de Castro Reis, D., Golgher, P.B., da Silve, A.S.: Automatic web news extraction using tree edit distance. In: WWW, pp. 502–511 (2004)
15. Wang, W., Xiao, C., Lin, X., Zhang, C.: Efficient approximate entity extraction with edit distance constraints. In: SIGMOD, pp. 759–770 (2009)
16. Zhai, Y., Liu, B.: Web data extraction based on partial tree alignment, In: Proceedings of 14th International Conference on World Wide Web, pp. 76–85(2005)
17. Crescenzi, V., Mecca, G., Merialdo, P.: RoadRunner: Towards automatic data extraction from large Web sites. In: VLDB, pp. 109–118 (2001)

Miscellaneous Issues

Automatic Recognition of Emotions and Actions in Bi-modal Video Analysis

Chenjian Wu[1], Chengwei Huang[2], and Hong Chen[3](✉)

[1] School of Electronic Information, Soochow University, Suzhou, China
cjwu@suda.edu.cn
[2] College of Physics, Optoelectronics and Energy, Soochow University, Suzhou, China
[3] School of Mathematical Sciences, Soochow University, Suzhou, China
chenhong@suda.edu.cn

Abstract. Driver's psychological state is a major concern in traffic safety and an automatic detection of unusual emotions and actions of motor vehicle driver may help prevent and predict the traffic accidents. In this paper, we first study a number of stimulation methods to induce driver's emotions and actions in a simulated driving experiment. We adopt video games, noise stimulation, sleep deprivation, etc., to induce negative emotions including fidgetiness, anger, tiredness, and anxiety. Potentially dangerous actions, such as looking away, texting with a phone, talking on a phone, etc., is posed by volunteer subjects. Using the recorded video data and speech data we established a basic emotion and action database. Second, we propose an efficient multi-modal emotion and action recognition framework, in which facial expression recognition, speech emotion recognition, and action recognition are combined and fused for improved performance.

Keywords: Emotion recognition · Video analysis · Action recognition

1 Introduction

Emotion and action recognition has many potential applications in human-computer interaction [1–5]. Recently, a number of researches on driver's emotion recognition have been reported [6–9]. In these studies, basic emotions have been modelled, however, practical emotions related to driving have not been studied. Detection special types of actions that may cause traffic accidents have not been paid much attention before.

Driver's emotional states and unusual actions may be detected from audio-visual data collected in a simulated driving experiment. The eliciting experiment is carried out in an in-door environment. The subject, who plays the driver, is required to play a video game of car racing. We then collect multi-modal data under different driving situations, including focusing on the road, looking away, reaching over, using a phone, drinking water, etc. These actions are potentially dangerous and we would like to develop automated system to detect them.

C.-H. Hsu et al. (Eds.): IOV 2015, LNCS 9502, pp. 427–438, 2015.
DOI: 10.1007/978-3-319-27293-1_37

Fig. 1. Examples of the posed actions and induced emotions

In the rest of the paper we will first present a database of stimulated emotions and actions. Although some of the actions and emotions are posed, we select the most believable segments for our recognition study. It is difficult to collect actual driving data when drivers are doing dangerous actions, such as drinking, reaching over, looking away, etc. Because this kind of actions rarely happen and very few drivers are willing to be recorded by a camera. Therefore the simulated driving data might be a good choice for the study (Fig. 1).

2 Multi-modal Emotion and Action Database

2.1 The Simulated Driving

The subjects' ages are between 20 to 30. Most of them have actual driving experience and all of them have played driving video games before. During the video recording, the dialog takes place spontaneously and since all of the subjects are native Chinese speakers, the language used is Mandarin Chinese. The camera is placed approximately at the same height of the subjects head, hence to catch the front face image. The light source is located approximately at 45 degrees front above the subject and providing a good lighting condition that is similar to the actual driving scenario. However we havent considered the night driving situation, in which the lighting condition is totally different. At this stage of study, we aim to provide video data under good lighting condition, which makes the recognition easier.

Computer games can be used for eliciting subject's emotions [10–12]. The subject is required to play the car racing game for several minutes, and the recorded video data may be used for analyzing emotions such as fidgetiness, anger, happiness, and neutral state. When the subject succeeds in the car racing game, he or she tends to show the positive emotions. When the subject failed in the car racing game, he or she tends to show the negative emotions. The recorded video data may also be used for modeling the action focusing on the road, or focused driving.

The subject is also required to act the following actions, using cell phone to make a call, using cell phone to text a friend, drinking from a cup or a bottle, reaching over to the side seat, reaching over to the back seat, looking away to the right, looking away to the left, pushing the horn button, and closing the eyes. Looking away to the right or to the left is a potentially dangerous action. The subject is required to look away to the side for three to five seconds to simulate this situation. When using the cell phone to text a friend, the hands of the driver are sometimes not visible. This is close to the actual situation, since in the in-car environment the lower part of the body is usually not inside of the camera. We suppose the camera is located near the dashboard of a vehicle.

We also require the subject to simulate several emotions, and the quality of the emotion data is largely dependent on the cooperativeness of the subject. The subject is required to recall some past experience to induce the target emotion. The facial expressions collected include neutrality, smile, and anger. The emotional actions collected include getting angry on the phone, and getting sleepy. The facial expression data collected here may be used for basic modeling of expressing recognition. The emotional actions collected here are two typical dangerous situations. When the driver is talking on the phone and getting angry, he or she may not be able to focusing on the driving. We record the emotional speech data and facial expression data in this simulated scenario. Many traffic accidents happen because the drivers get tired and sleepy. We simulated this situation and record the video data of a subject getting sleepy. Detecting sleepiness may contribute greatly to an automatic alarm system. Some of the sleepiness action data collected in this database is not spontaneous, but acted by the subjects. Other sleepiness action data is spontaneous which is induced by sleep deprivation. In the sleep deprivation experiment, the subject is required to stay up and play a car racing game all night. At the time the spontaneous data is recorded, the subject hasnt sleep for about 24 h and the self evaluation is reported as very tired and sleepy.

2.2 Perception Experiment and Annotation

After the collection of the drivers simulated data, we design a perception experiment to evaluate the quality of the action data and emotion data. There is no clear boundary between the action data and the emotion data. The same video segment may present a certain action and at the same time contains a certain emotion. The raw data contains some undesirable frames or shot that need to be removed. The video data is then segmented into short clips which contains single action or emotion. The experimenter who conducts the recording labels each of these clips with the target actions and emotions. In the perception experiment, another group of volunteers, 2 males and 2 females, who has not participated in the recording evaluate these video clips, and decide whether the target actions and emotions are obvious in the clips. All the video clips are scaled from 1 to 10, 10 as the most believable (Table 1).

2.3 The Additional Databases

An additional expression database we use to train our expression recognizer is the JAFFE database [13]. The database contains 213 images of 7 facial expressions (6 basic facial expressions + 1 neutral) posed by 10 Japanese female models. Each image has been rated on 6 emotion adjectives by 60 Japanese subjects.

Table 1. Statistics of the database

Number of subjects	10
Sex of the subjects	5 males and 5 females
Total length of the video data	30 min
Total storage size of the video data	3 G
Average duration of the sample clips	5 s
Shortest duration of the sample clips	1 s
Longest duration of the sample clips	15 s

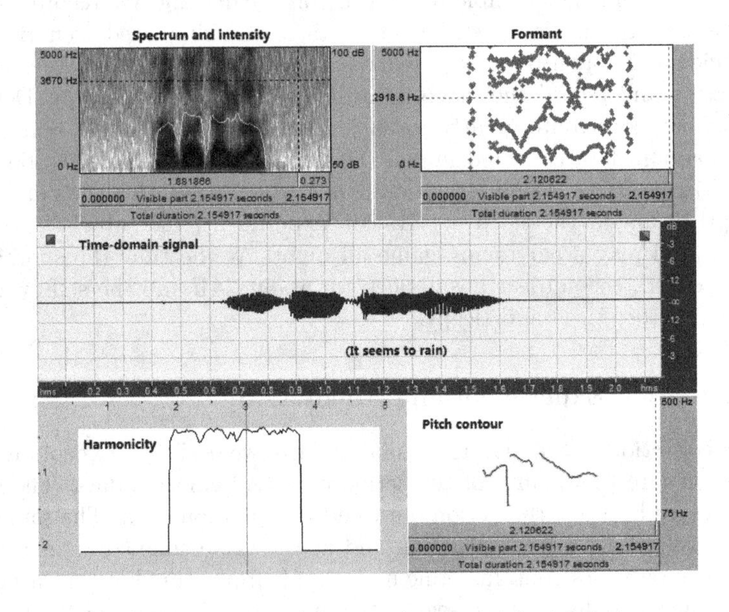

Fig. 2. A depiction of the emotional speech and features

An additional speech emotion database is used for training and testing our automatic speech emotion recognition system. A depiction of emotional utterance and the extracted speech features are shown in Fig. 2.

Harmonic-to-Noise-Ratio (HNR) features are analysed for various emotional states, e.g. happiness, tiredness, and fidgetiness. As shown in Figs. 3, 4, and 5, we

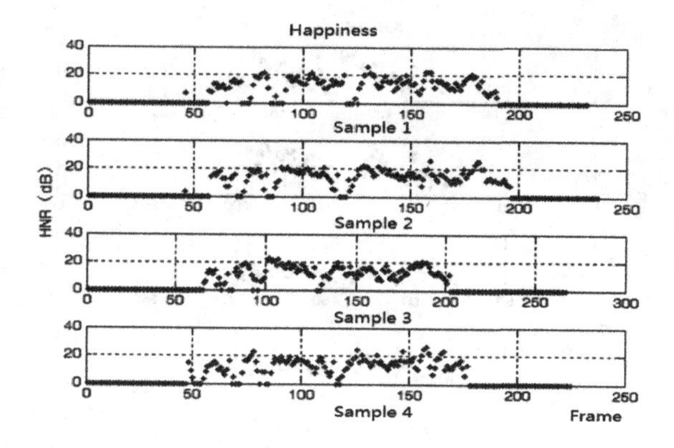

Fig. 3. Harmonic-to-Noise-Ratio as an index of emotional change: happiness.

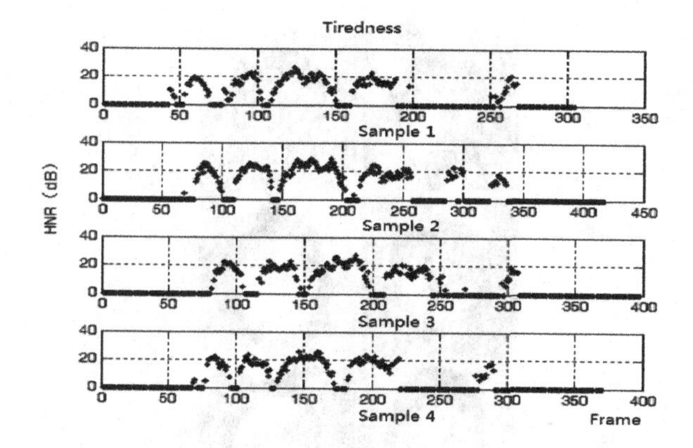

Fig. 4. Harmonic-to-Noise-Ratio as an index of emotional change: tiredness.

can see the contour of the HNR features over time is affected by speaker's emotional changes. Other prosodic feature and voice quality features are adopted for speech emotion feature extraction, include pitch, intensity, formant, MFCC, etc.

3 Recognition Methodology

3.1 Expression Recognition

Facial expression recognition is crucial to many affective and behaviour recognition systems. Extracting proper expression features on a facial image is an important step in our facial expression recognition.

In order to automate the computer system to recognize expressions, we need to construct features that explains what happens to our faces when we express different emotions.

We adopt the geometric features to model the expressions in our simulated driving, since the geometric features are closely related to facial expressions, relatively robust to lighting changes, and simple in implementation.

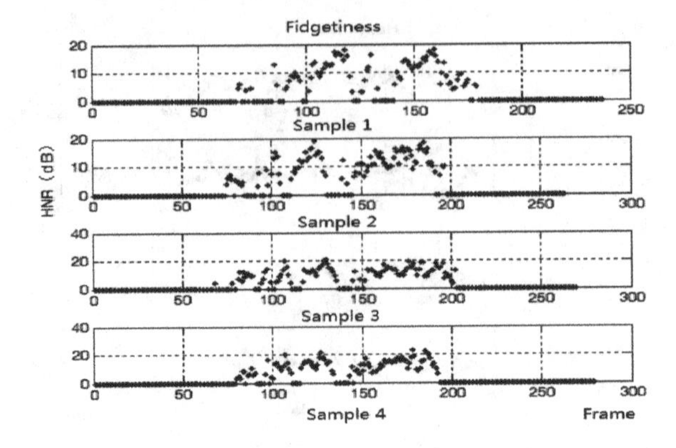

Fig. 5. Harmonic-to-Noise-Ratio as an index of emotional change: fidgetiness.

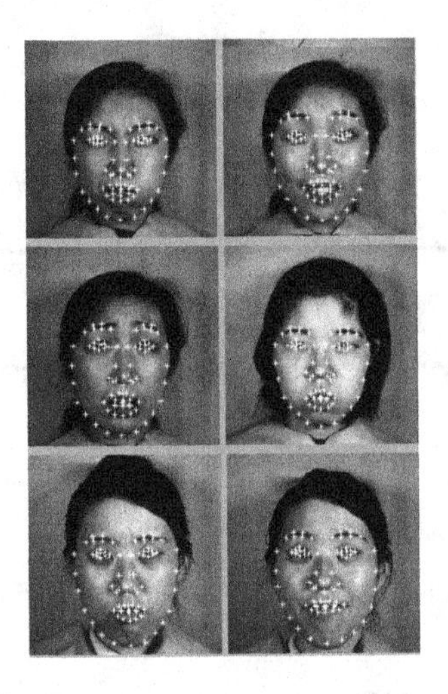

Fig. 6. Examples of landmark detection

Most facial expression recognition system adopts the action units (AU) to describe various expressions, we propose to use the facial landmark points [14,15] to model the changes of AU.

3.2 Facial Feature Extraction

The software tool Stasm which is based on active shape model is used to extract facial fiducial points, as shown in Fig. 6.

Stasm is a C++ software library for finding features in faces. You give it an image of a face and it returns the positions of the facial features. Stasm is designed to work on front views of faces with neutral expressions.

3.3 Speech Emotion Recognition

The definition of Gaussian Mixture Model is:

$$p(\mathbf{X}_t|\lambda) = \sum_{i=1}^{M} a_i b_i(\mathbf{X}_t) \tag{1}$$

where $\mathbf{X}_t$ is a D-dimension random vector, $b_i(\mathbf{X}_t)$ is the ith member of Gaussian distribution, t is the index of utterance, a_i is the mixture weight, and M is the number of Gaussian mixture members. λ denotes the GMM parameter. Each member is a D-dimension variable follows the Gaussian distribution with the mean $\mathbf{U}_i$ and the covariance $\mathbf{\Sigma}_i$:

$$b_i(\mathbf{X}_t) = \frac{1}{(2\pi)^{D/2}|\mathbf{\Sigma}_i|^{1/2}} exp\left\{ -\frac{1}{2}(\mathbf{X}_t - \mathbf{U}_i)^T \mathbf{\Sigma}_i^{-1}(\mathbf{X}_t - \mathbf{U}_i) \right\}. \tag{2}$$

Note that

$$\sum_{i=1}^{M} a_i = 1. \tag{3}$$

Expectation-Maximization (EM) algorithm is then used for the estimation of GMM parameters. Under the Bayesian framework, emotion recognition can be done by maximizing the posterior probability:

$$EmotionLabel = \underset{k}{\operatorname{argmax}}\left(p(\mathbf{X}_t|\lambda_k)\right). \tag{4}$$

3.4 In-vehicle Action Recognition

In order to understand the driver's state for driving safety we study several types of actions that may cause potential danger.

Based on the video signal we may classify the driver's actions using temporal motion features and SVM classifier.

Preprocessing of the video data involves with manual annotation and image resizing. Video data is annotated into short clips for action and emotion labelling. Each sample (clip) is labelled with a unique emotion category, sometimes with an action type for the ground-truth in training the classifiers.

Feature extraction is an essential step in our study of action recognition. Since most of the target actions in vehicle involve with head, face and hands, we focus on the changes over time related to these parts of the body.

We propose two basic types of features that can be used to classify the driver actions, (i) the position of the face, and (ii) the track of any moving subject.

For the first type of features, we localize the facial landmarks. By study the coordinates change over time, we may classify some head movement actions, such as, eye closing, looking away, reaching over.

For the second type of features, they are most likely related to the hand movement. Drinking water or using a phone may be detected by the track of the motion relative to the position of the face.

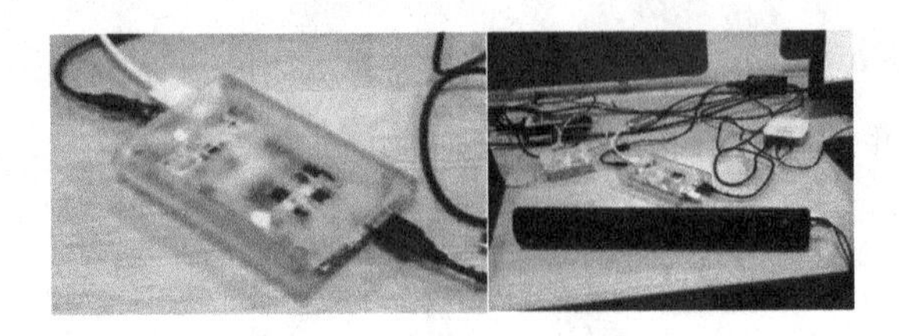

Fig. 7. Hardware implementation of a mobile speech emotion recognition system using pcDuino

A problem with these features is that some of them are effective at the beginning of a certain action. Therefore detecting the beginning and the end of an action is a necessary step.

Using a simple threshold of the background, we may separate the moving subjects from the still background. Considering the in-vehicle environment, we suppose there is no change of camera view point or multiple-person case in the driver's seat area (Fig. 7).

4 Experimental Results

"pcDuino" is a mini PC platform that runs PC like Ubuntu. It has hardware headers interface with microphones and cameras, and it outputs screen to HDMI. During the experiments, we run our core speech emotion recognition algorithms on pcDuino.

We use JAFFE database to train our facial expression recognizer. The database contained 213 images of 10 females in total. Expression recognition algorithm is implemented use Support Vector Machine combined with landmark features. Happiness, angry, and neutrality are modelled and recognized. Positive and negative emotions are also modelled and recognized by grouping the basic six emotions into two category in the valence dimension. Ten images are selected for testing, results are shown in Table 2.

Table 2. Results of expression recognition test in dimensional model

Emotion category	Recognition Rate
Happiness	50 %
Angry	70 %
Sadness	60 %
Surprise	60 %
Disgust	50 %
Fear	40 %
Positive	50 %
Negative	60 %

We use a local speech emotion database to train our speech emotion model. Testing results on are shown in Fig. 8.

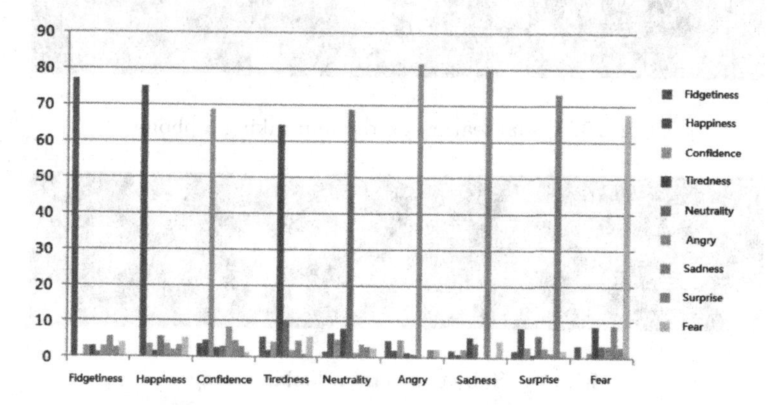

Fig. 8. Speech emotion recognition test

Fig. 9. Subject's action: taking a phone

The simulated driving data contains facial expression data, action data and speech emotion data. We randomly selected half of the facial expression data as an additional training data set, which is added to the JAFFE data for retraining. In our simulated driving database, we collected emotional speech when the driver is talking on the phone. And we randomly selected half of the speech emotion data as an additional training data set, which is added to the previous data set for retraining.

Several key frames of a male subject's action of taking a phone while simulated driving are shown in Fig. 9. After extracting the motion features by computing the difference between frames, we separate the motion from the still background, as shown in Fig. 10. The face area is detected by using Viola-Jones face detector, or by prediction in neighbouring frames when subject's face turned away. The motion detected outside the facial area is considered as the hand movements which corresponding to drinking from a bottle, taking a phone, etc. The motion detected within the facial area is corresponding to the action looking away or covering face with a hand. When the subject is reaching over, the difference computed from neighbouring frames is significantly different.

Figure 11 shows a female subject looking away wearing sun-glasses. By simply computing the difference between two frames, the motion of head moving is shown in Fig. 12. Corresponding facial landmark feature are extracted using Stasm. As we tracking the facial points through frames in time duration, the

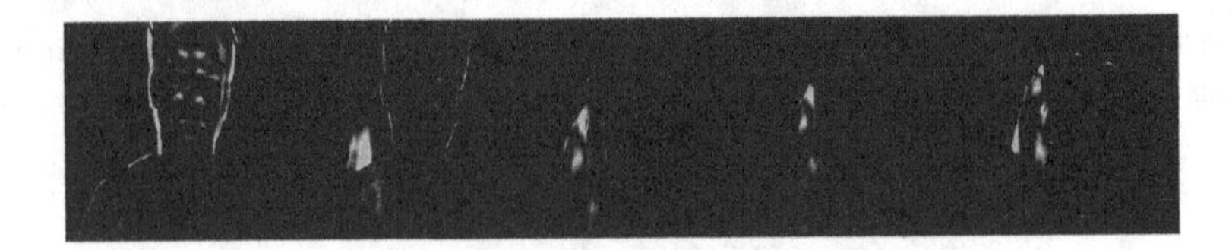

Fig. 10. Visual feature extraction: taking a phone

Fig. 11. Subject's action: looking away

looking away action may be estimated by the lost tracking time, as shown in Fig. 13. We take the time window of frames that missing facial landmark points as the time of looking away. Practically there may be other causes that lead to the lost tracking of face, such as covering facial area with hands, reaching over, or drink water (where the face is not frontal either). Although, the detection accuracy of these actions need to be improved, we can still come to the decision that lost track of frontal face is related to unusual behaviour in driving. When that happens and last for a significant duration of time (e.g. three seconds) we may set off an alarm to remind the driver for driving safety.

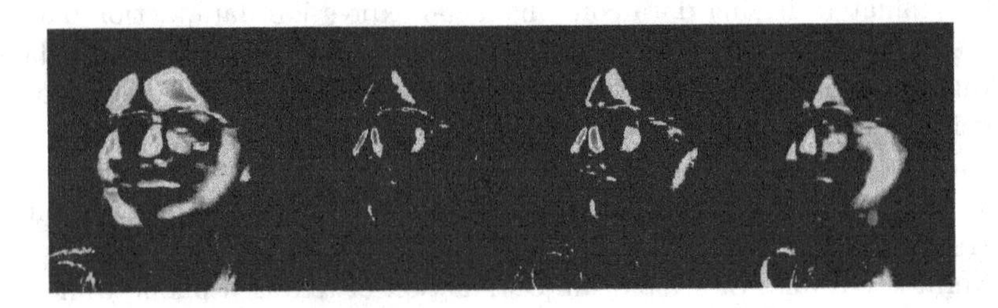

Fig. 12. Visual feature extraction: looking away

A successful tracking of frontal face is shown in Fig. 14. A male subject who is drunk, sleepy, is playing the simulated driving video game while talking on the phone. His emotion changes from happiness to exciting, and finally angry. The facial expressions are not posed deliberately, and the corresponding emotional speech is also spontaneous.

In testing, some of the video clips are recognized through three channels: facial expression, action, and speech, others are recognized based on two of them. Total recognition accuracy is shown in Table 3. A comparison on bi-modal recognition and uni-modal recognition is also shown in Table 3. Ten clips are used for each category.

Fig. 13. Frontal face tracking: calculating the lost tracking window as an index of head movement.

Table 3. Results of emotion and action recognition test with multi-modality

Detection Category	Expression	Speech	Action	Bimodal
Anger	60 %	60 %	N/A	70 %
Fidgetiness	50 %	70 %	N/A	70 %
Happiness	50 %	50 %	N/A	60 %
Drinking from bottle	N/A	N/A	70 %	N/A
Taking a Phone	N/A	N/A	70 %	N/A
Looking away	N/A	N/A	70 %	N/A
Reaching over	N/A	N/A	70 %	N/A
Sleepy	N/A	N/A	40 %	N/A

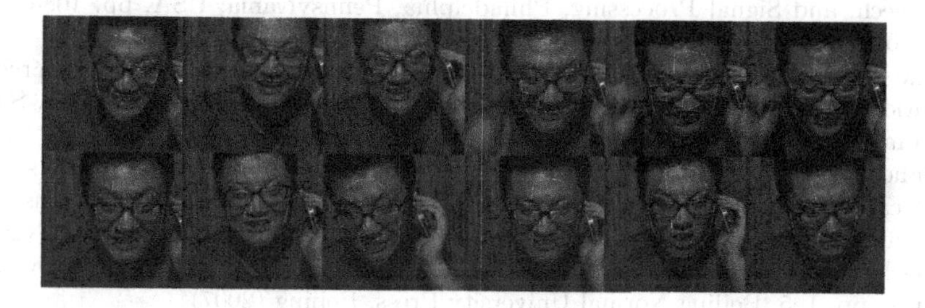

Fig. 14. Frontal face tracking: an example on a sleepy male subject with spontaneous emotions.

5 Conclusions

In this paper, we focus on the recognition of driver's state in order to improve the driving safety. Various behavioural cues are used for the detection of the driver's emotion related states, including facial expression, speech emotion and in-vehicle actions. We use existing databases to train our facial expression and speech emotion classifiers. Support vector machine and Gaussian mixture model are proved to be efficient algorithms in modelling human emotions. Since the in-vehicle action data is difficult to achieve, we induced and collected such data in a simulated driving experiment.

Acknowledgments. This work has been supported by the National Natural Science Foundation of China (No. 11401412) and Natural Science Foundation of Jiangsu Province of China (No. BK20150342).

References

1. Huang, C., Jin, Y., Zhao, Y., Yu, Y., Zhao, L.: Speech emotion recognition based on re-composition of two-class classifiers. In: International Conference on Affective Computing and Intelligent Interaction and Workshops, Amsterdam, Netherlands, pp. 10–12 (2009)
2. Huang, C., Liang, R., Wang, Q., Xi, J., Zha, C., Zhao, L.: Practical speech emotion recognition based on online learning: from acted data to elicited data. Math. Prob. Eng., **2013**, 1–10 (2013)
3. Huang, C., Chen, G., Hua, Y., Yongqiang, B., Zhao, L.: Speech emotion recognition under white noise. Arch. Acoust. **38**, 1–10 (2014)
4. Wang, J., Wu, B., Huang, C., Qin, H., Zha, C., Zhao, L.: Segment-based static feature analysis and recognition of emotional speech for manned space mission. ICIC Express Lett. **8**(6), 11–13 (2014)
5. Huang, C., Han, D., Bao, Y., Yu, H., Zhao, L.: Cross-language speech emotion recognition in German and Chinesel. ICIC Express Lett. **6**(8), 2141–2146 (2012)
6. C. You, et al.: CarSafe App: Alerting drowsy and distracted drivers using dual cameras on smartphones. In: ACM Mobisys, pp. 1–4 (2013)
7. Hoch, S., Althoff, F., McGlaun, G., Rigoll, G.: Bimodal fusion of emotional data in an automotive environment. In: IEEE International Conference on Acoustics, Speech, and Signal Processing, Philadelphia, Pennsylvania, USA, pp. 1085–1088 (2005)
8. Katsis, C.D., Katertsidis, N., Ganiatsas, G., Fotiadis, D.I.: To- ward emotion recognition in car-racing drivers: a biosignal processing approach. IEEE Trans. Syst., Man Cybern., Part A: Syst. Hum. **38**(3), 502–512 (2008)
9. Jones, C.M., Jonsson, I.M.: Automatic recognition of affective cues in the speech of car drivers to allow appropriate responses. In: Proceedings of 17th Australia Conference on Computer-Human Interaction, Canberra, Australia, pp. 1–10 (2005)
10. Barbara, A., Spellman, D., Willingham, T.: Current Directions in Cognitive Science, pp. 1–5. Beijing Normal University Press, Beijing (2007)
11. Johnstone, T.: Emotional speech elicited using computer games. In: Fourth International Conference on Spoken Language, vol. 3, pp. 1985–1988 (1996)
12. Johnstone, T., van Reekum, C.M., Hird, K., Kirsner, K., et al.: Affective speech elicited with a computer game. Emot. **5**(4), 513–518 (2005)
13. Lyons, M.J., Kamachi, M., Gyoba, J.: Japanese female facial expressions (JAFFE). In: Database of digital images (1997)
14. Efraty, B., Huang, C., Shah, S.K., Kakadiaris, I.K.: Facial landmark detection in uncontrolled conditions. In: International Joint Conference on Biometrics, Washington, DC, USA, pp. 11–13, October 2011
15. Huang, C., Efraty, B.A., Kurkure, U., et al.: Facial landmark configuration for improved detection. In: IEEE International Workshop on Information Forensics and Security, Tenerife, Spain (2012)

A Dynamic Programming Algorithm for a Generalized LCS Problem with Multiple Subsequence Inclusion Constraints

Daxin Zhu[1], Yingjie Wu[2](✉), and Xiaodong Wang[3](✉)

[1] Quanzhou Normal University, Quanzhou 362000, China
[2] Fuzhou University, Fuzhou 350002, China
yjwu@fzu.edu.cn
[3] Fujian University of Technology, Fuzhou 350108, China
wangxd135@139.com

Abstract. In this paper, we consider a generalized longest common subsequence problem with multiple subsequence inclusive constraints. For the two input sequences X and Y of lengths n and m, and a set of d constraints $P = \{P_1, \cdots, P_d\}$ of length l_i for each $P_i \in P$, the problem is to find a common subsequence Z of X and Y including each of constraint string in P as a subsequence and the length of Z is maximized. A simple dynamic programming algorithm to this problem is presented in this paper. The correctness of the new algorithm is demonstrated. The time complexities of the new algorithm is $O(nmdt)$, where $t = \prod_{1 \le i \le d} l_i$.

Keywords: Longest common subsequence problem · Dynamic programming · Subsequence inclusion constraints · Time complexity

1 Introduction

The longest common subsequence (LCS) problem is a classic computer science problem, and has applications in bioinformatics. It is further widely applied in diverse areas, such as file comparison, pattern matching and computational biology [1–4]. Given two sequences X and Y, the longest common subsequence problem is to find a subsequence of X and Y whose length is the longest among all common subsequences of the two given sequences. It differs from the problems of finding common substrings: unlike substrings, subsequences are not required to occupy consecutive positions within the original sequences. The most referred algorithm, proposed by Wagner and Fischer [5], solves the LCS problem by using a dynamic programming algorithm in quadratic time. Other advanced algorithms were proposed in the past decades [1,2,6–10]. If the number of input sequences is not fixed, the problem to find the LCS of multiple sequences has been proved to be NP-hard [11]. Some approximate and heuristic algorithms were proposed for these problems [12,13].

For some biological applications some constraints must be applied to the LCS problem. These kinds of variants of the LCS problem are called the constrained

C.-H. Hsu et al. (Eds.): IOV 2015, LNCS 9502, pp. 439–446, 2015.
DOI: 10.1007/978-3-319-27293-1_38

LCS (CLCS) problem. One of the recent variants of the LCS problem, the constrained longest common subsequence (CLCS) which was first addressed by Tsai [14], has received much attention. It generalizes the LCS measure by introducing of a third sequence, which allows to extort that the obtained CLCS has some special properties [15]. For two given input sequences X and Y of lengths m and n, respectively, and a constrained sequence P of length r, the CLCS problem is to find the common subsequences Z of X and Y such that P is a subsequence of Z and the length of Z is the maximum. The most referred algorithms were proposed independently [3,16], which solve the CLCS problem in $O(mnr)$ time and space by using dynamic programming algorithms. Some improved algorithms have also been proposed [17,18]. The LCS and CLCS problems on the indeterminate strings were discussed in [19]. Moreover, the problem was extended to the one with weighted constraints, a more generalized problem [20].

Recently, a new variant of the CLCS problem, the restricted LCS problem, was proposed [21], which excludes the given constraint as a subsequence of the answer. The restricted LCS problem becomes NP-hard when the number of constraints is not fixed. Some more generalized forms of the CLCS problem, the generalized constrained longest common subsequence (GC-LCS) problems, were addressed independently by Chen and Chao [22]. For the two input sequences X and Y of lengths n and m, respectively, and a constraint string P of length r, the GC-LCS problem is a set of four problems which are to find the LCS of X and Y including/excluding P as a subsequence/substring, respectively. The four generalized constrained LCS [22] can be summarized in Table 1.

For the four problems in Table 1, $O(mnr)$ time algorithms were proposed [22]. For all four variants in Table 1, $O(r(m+n)+(m+n)\log(m+n))$ time algorithms were proposed by using the finite automata [23]. Recently, a quadratic algorithm to the STR-IC-LCS problem was proposed [24], and the time complexity of [23] was pointed out not correct.

The four GC-LCS problems can be generalized further to the cases of multiple constraints. In these generalized cases, the single constrained pattern P will be generalized to a set of d constraints $P = \{P_1, \cdots, P_d\}$ of total length r, as shown in Table 2.

Table 1. The GC-LCS problems

Problem	Input	Output
SEQ-IC-LCS	X,Y, and P	The longest common subsequence of X and Y including P as a subsequence
STR-IC-LCS	X,Y, and P	The longest common subsequence of X and Y including P as a substring
SEQ-EC-LCS	X,Y, and P	The longest common subsequence of X and Y excluding P as a subsequence
STR-EC-LCS	X,Y, and P	The longest common subsequence of X and Y excluding P as a substring

Table 2. The Multiple-GC-LCS problems

Problem	Input	Output
M-SEQ-IC-LCS	X,Y, and a set of constraints $P = \{P_1, \cdots, P_d\}$	The LCS of X and Y including $P_i \in P$ as a subsequence
M-STR-IC-LCS	X,Y, and a set of constraints $P = \{P_1, \cdots, P_d\}$	The LCS of X and Y including $P_i \in P$ as a substring
M-SEQ-EC-LCS	X,Y, and a set of constraints $P = \{P_1, \cdots, P_d\}$	The LCS of X and Y excluding $P_i \in P$ as a subsequence
M-STR-EC-LCS	X,Y, and a set of constraints $P = \{P_1, \cdots, P_d\}$	The LCS of X and Y excluding $P_i \in P$ as a substring

The problem M-SEQ-IC-LCS has been proved to be NP-hard in [24]. The problem M-SEQ-EC-LCS has also been proved to be NP-hard in [25,26]. In addition, the problems M-STR-IC-LCS and M-STR-EC-LCS were also declared to be NP-hard in [22], but without a proof. The exponential-time algorithms for solving these two problems were also presented in [22]. We will discuss the problem M-SEQ-IC-LCS in this paper.

2 A Dynamic Programming Algorithm

A sequence is a string of characters over an alphabet $\sum$. A subsequence of a sequence X is obtained by deleting zero or more characters from X (not necessarily contiguous). The symbol $\oplus$ is also used to denote the string concatenation. For example, if $S_1 = aaa$ and $S_2 = bbb$, then it is readily seen that $S_1 \oplus S_2 = aaabbb$.

For the two input sequences $X = x_1x_2 \cdots x_n$ and $Y = y_1y_2 \cdots y_m$ of lengths n and m, respectively, and a set of d constraints $P = \{P_1, \cdots, P_d\}$ of total length r, the problem M-SEQ-IC-LCS is to find an LCS of X and Y including each of constraint $P_i \in P$ as a subsequence.

Definition 1. *Let $Z(i, j, k)$ denote the set of all LCSs of $X[1:i]$ and $Y[1:j]$ with state k, where $1 \leq i \leq n, 1 \leq j \leq m$, and k is a state vector $k = (k_1, \cdots, k_d) \in V$, where $V = \{(k_1, \cdots, k_d) | 0 \leq k_i \leq l_i, 1 \leq i \leq d\}$. It is indicated by the state vector $k \in V$ that for any $z \in Z(i, j, k)$, z includes $P_i[1:k_i], 1 \leq i \leq d$, as its subsequences. The length of an LCS in $Z(i, j, k)$ is denoted as $f(i, j, k)$.*

The constraint state of $z \in Z(i, j, k)$ is defined by its state vector $k = (k_1, \cdots, k_d)$. If we add a character $c \in \Sigma$ to the end of z, then the state vector of $z \bigoplus c$ becomes $\delta(k, c) = (\bar{k}_1, \cdots, \bar{k}_d)$, where the function $\delta(k, c)$ can be defined as follows.

$$\bar{k}_i, = \begin{cases} 1 + k_i \ \texttt{if} \ P_i(1 + k_i) = c \\ k_i \quad \texttt{else} \end{cases} \tag{1}$$

If we can compute $f(i, j, k)$ for any $1 \leq i \leq n, 1 \leq j \leq m$, and $k = (k_1, \cdots, k_d)$, $0 \leq k_i \leq l_i, 1 \leq i \leq d$ efficiently, then the length of an LCS of X and Y including P must be $f(n, m, k^*)$, where $k^* = (l_1, \cdots, l_d)$.

We can give a recursive formula for computing $f(i,j,k)$ by the following Theorem.

Theorem 1. *For the two input sequences $X = x_1x_2\cdots x_n$ and $Y = y_1y_2\cdots y_m$ of lengths n and m, respectively, and a set of d constraints $P = \{P_1, \cdots, P_d\}$ of total length r, let $Z(i,j,k)$ and $f(i,j,k)$ be defined as in Definition 1. Then, for any $1 \le i \le n, 1 \le j \le m$, and $k \in V$, $f(i,j,k)$ can be computed by the following recursive formula.*

$$f(i,j,k) = \begin{cases} \max\{f(i-1,j,k), f(i,j-1,k)\} & \text{if } x_i \neq y_j, \\ \max\left\{f(i-1,j-1,k), 1 + \max\limits_{\bar{k}\in S(k,x_i)} \{f(i-1,j-1,\bar{k})\}\right\} & \text{if } x_i = y_j. \end{cases} \tag{2}$$

where,

$$S(k,x_i) = \{\bar{k} \in V | \delta(\bar{k},x_i) = k\} \tag{3}$$

The boundary conditions of this recursive formula are $f(i,0,0) = f(0,j,0) = 0$ for any $0 \le i \le n, 0 \le j \le m$.

Proof. For any $0 \le i \le n, 0 \le j \le m$, and $k \in V$, suppose $f(i,j,k) = l$ and $z = z_1 \cdots z_l \in Z(i,j,k)$.
First of all, we notice that for each pair $(i',j'), 1 \le i' \le n, 1 \le j' \le m$, such that $i' \le i$ and $j' \le j$, we have $f(i',j',k) \le f(i,j,k)$, since a common subsequence z of $X[1:i']$ and $Y[1:j']$ with state k is also a common subsequence of $X[1:i]$ and $Y[1:j]$ with state k.

(1) In the case of $x_i \neq y_j$, we have $x_i \neq z_l$ or $y_j \neq z_l$.

(1.1) If $x_i \neq z_l$, then $z = z_1 \cdots z_l$ is a common subsequence of $X[1:i-1]$ and $Y[1:j]$ with state k, and so $f(i-1,j,k) \ge l$. On the other hand, $f(i-1,j,k) \le f(i,j,k) = l$. Therefore, in this case we have $f(i,j,k) = f(i-1,j,k)$.

(1.2) If $y_j \neq z_l$, then we can prove similarly that in this case, $f(i,j,k) = f(i,j-1,k)$.

Combining the two subcases we conclude that in the case of $x_i \neq y_j$, we have

$$f(i,j,k) = \max\{f(i-1,j,k), f(i,j-1,k)\}.$$

(2) In the case of $x_i = y_j$, there are also two cases to be distinguished.

(2.1) If $x_i = y_j \neq z_l$, then $z = z_1 \cdots z_l$ is also a common subsequence of $X[1:i-1]$ and $Y[1:j-1]$ with state k, and so $f(i-1,j-1,k) \ge l$. On the other hand, $f(i-1,j-1,k) \le f(i,j,k) = l$. Therefore, in this case we have $f(i,j,k) = f(i-1,j-1,k)$.

(2.2) If $x_i = y_j = z_l$, then $f(i,j,k) = l > 0$ and $z = z_1 \cdots z_l$ is an LCS of $X[1:i]$ and $Y[1:j]$ with state k.

Let the state of $(z_1, \cdots, z_{l-1})$ be $\bar{k}$, then we have $\bar{k} \in S(k,x_i)$, since $z_l = x_i$. It follows that $z_1 \cdots z_{l-1}$ is a common subsequence of $X[1:i-1]$ and $Y[1:j-1]$ with state $\bar{k}$. Therefore, we have

$$f(i-1,j-1,\bar{k}) \ge l-1$$

Furthermore, we have

$$\max_{\bar{k}\in S(k,x_i)} \{f(i-1,j-1,\bar{k})\} \geq l-1$$

In other words,

$$f(i,j,k) \leq 1 + \max_{\bar{k}\in S(k,x_i)} \{f(i-1,j-1,\bar{k})\} \tag{4}$$

On the other hand, for any $\bar{k} \in S(k,x_i)$, and $v = v_1 \cdots v_h \in Z(i-1,j-1,\bar{k})$, $v \oplus x_i$ is a common subsequence of $X[1:i]$ and $Y[1:j]$ with state k. Therefore, $f(i,j,k) = l \geq 1 + h = 1 + f(i-1,j-1,\bar{k})$, and so we conclude that,

$$f(i,j,k) \geq 1 + \max_{\bar{k}\in S(k,x_i)} \{f(i-1,j-1,\bar{k})\} \tag{5}$$

Combining (4) and (5) we have, in this case,

$$f(i,j,k) = 1 + \max_{\bar{k}\in S(k,x_i)} \{f(i-1,j-1,\bar{k})\} \tag{6}$$

Combining the two subcases in the case of $x_i = y_j$, we conclude that the recursive formula (2) is correct for the case $x_i = y_j$.

The proof is complete. ■

3 The Implementation of the Algorithm

According to Theorem 1, our algorithm for computing $f(i,j,k)$ is a standard 3-dimensional dynamic programming algorithm. By the recursive formula (2), the dynamic programming algorithm for computing $f(i,j,k)$ can be implemented as the following Algorithm 1.

In Algorithm 1, the variable S is used to record the current states created. When a state is first created, the new state is added to S. Therefore, in Algorithm 1, the current state set S is extended gradually while the for loop processed. In the worst case, the set S will have a size of $t = |V| = \prod_{1\leq i\leq d} l_i$. The body of the triple for loops can be computed in $O(d)$ time in the worst case. Therefor, the total time of Algorithm 1 is $O(nmdt)$ in the worst case. The space used by Algorithm 1 is also $O(nmdt)$.

The number of constraints is an influent factor in the time and space complexities of our new algorithm. If a string P_i in the constraint set P is a proper substring of another string P_j in P, then an LCS of X and Y including P_j must also exclude P_i. For this reason, the constraint string P_i can be removed from constraint set P without changing the solution of the problem. Without loss of generality, we can put forward the following two assumptions on the constraint set P.

Assumption 1. *There are not any duplicated strings in the constraint set P.*

Algorithm 1. M-SEQ-IC-LCS

Input: Strings $X = x_1 \cdots x_n$, $Y = y_1 \cdots y_m$ of lengths n and m, respectively, and a set of d constraints $P = \{P_1, \cdots, P_d\}$

Output: The length of an LCS of X and Y including $P_i \in P$ as a subsequence

```
1: for all i, j, 0 ≤ i ≤ n, 0 ≤ j ≤ m do
2:    f(i,0,0) ← 0, f(0,j,0) ← 0 {boundary condition}
3: end for
4: S ← {0} {current set of states}
5: for i = 1 to n do
6:    for j = 1 to m do
7:       for each k ∈ S do
8:          if x_i ≠ y_j then
9:             f(i,j,k) ← max{f(i−1,j,k), f(i,j−1,k)}
10:         else
11:            k̄ ← δ(k, x_i)
12:            f(i,j,k̄) ← max{f(i−1,j−1,k̄), 1 + f(i−1,j−1,k)}
13:            S ← S ⋃ {k̄}
14:         end if
15:      end for
16:   end for
17: end for
18: return f(n,m,k*)
```

Assumption 2. *No string in the constraint set P is a proper substring of any other string in P.*

If Assumption 1 is violated, then there must be some duplicated strings in the constraint set P. In this case, we can first sort the strings in the constraint set P, then duplicated strings can be removed from P easily and then Assumption 1 on the constraint set P is satisfied. It is clear that removed strings will not change the solution of the problem. If Assumption 2 is violated, we can remove all proper substrings from the constraint set P in $O(|P|)$ time. It is clear that the removed proper substrings will not change the solution of the problem.

If we want to compute the longest common subsequence of X and Y including P, but not just its length, we can also present a simple recursive backtracking algorithm for this purpose as the following Algorithm 2. A function call $back(n, m, k^*)$ will produce the answer LCS accordingly.

Since the cost of $\delta(k, x_i)$ is $O(d)$ in the worst case, the time complexity of the algorithm $back(i, j, k)$ is $O(ndr)$.

Finally we summarize our results in the following Theorem.

Theorem 2. *For the two input sequences $X = x_1x_2 \cdots x_n$ and $Y = y_1y_2 \cdots y_m$ of lengths n and m, respectively, and a set of d constraints $P = \{P_1, \cdots, P_d\}$, the Algorithms 1 and 2 solve the M-SEQ-IC-LCS problem correctly in $O(nmdt)$ time and $O(nmdt)$ space in the worst case, where $t = \prod_{1 \le i \le d} l_i$.*

Algorithm 2. *back*(i, j, k)

Comments: A recursive back tracing algorithm to construct the answer LCS

```
 1: if i = 0 or j = 0 then
 2:    return
 3: end if
 4: if x_i = y_j then
 5:    if f(i, j, k) = f(i − 1, j − 1, k) then
 6:       back(i − 1, j − 1, k)
 7:    else
 8:       for each k̄ ∈ S do
 9:          if k = δ(k̄, x_i) and f(i, j, k) = 1 + f(i − 1, j − 1, k̄) then
10:             back(i − 1, j − 1, k̄)
11:             print x_i
12:          end if
13:       end for
14:    end if
15: else if f(i − 1, j, k) > f(i, j − 1, k) then
16:    back(i − 1, j, k)
17: else
18:    back(i, j − 1, k)
19: end if
```

Acknowledgments. This work was supported by the Science and Technology Foundation of Quanzhou under Grant No. 2013Z38, Fujian Provincial Key Laboratory of Data-Intensive Computing and Fujian University Laboratory of Intelligent Computing and Information Processing.

References

1. Ann, H.Y., Yang, C.B., Peng, Y.H., Liaw, B.C.: Efficient algorithms for the block edit problems. Inf. Comput. **208**(3), 221–229 (2010)
2. Apostolico, A., Guerra, C.: The longest common subsequences problem revisited. Algorithmica **2**(1), 315–336 (1987)
3. Chin, F.Y.L., Santis, A.D., Ferrara, A.L., Ho, N.L., Kim, S.K.: A simple algorithm for the constrained sequence problems. Inform. Process. Lett. **90**(4), 175–179 (2004)
4. Crochemore, M., Hancart, C., Lecroq, T.: Algorithms on Strings. Cambridge University Press, Cambridge (2007)
5. Wagner, R., Fischer, M.: The string-to-string correction problem. J. ACM **21**(1), 168–173 (1974)
6. Ann, H.Y., Yang, C.B., Tseng, C.T., Hor, C.Y.: A fast and simple algorithm for computing the longest common subsequence of run-length encoded strings. Inform. Process. Lett. **108**(11), 360–364 (2008)
7. Hirschberg, D.S.: Algorithms for the longest common subsequence problem. J. ACM **24**(4), 664–675 (1977)
8. Hunt, J.W., Szymanski, T.G.: A fast algorithm for computing longest common subsequences. Commun. ACM **20**(5), 350–353 (1977)

9. Iliopoulos, C.S., Rahman, M.S.: A new efficient algorithm for computing the longest common subsequence. Theor. Comput. Sci. **45**(2), 355–371 (2009)
10. Iliopoulos, C.S., Rahman, M.S., Vorcek, M., Vagner, L.: Finite automata based algorithms on subsequences and supersequences of degenerate strings. J. Discrete Algorithm **8**(2), 117–130 (2010)
11. Maier, D.: The complexity of some problems on subsequences and supersequences. J. ACM **25**, 322–336 (1978)
12. Blum, C., Blesa, M.J., Lnez, M.: Beam search for the longest common subsequence problem. Comput. Oper. Res. **36**(12), 3178–3186 (2009)
13. Shyu, S.J., Tsai, C.Y.: Finding the longest common subsequence for multiple biological sequences by ant colony optimization. Comput. Oper. Res. **36**(1), 73–91 (2009)
14. Tsai, Y.T.: The constrained longest common subsequence problem. Inform. Process. Lett. **88**(4), 173–176 (2003)
15. Tang, C.Y., Lu, C.L.: Constrained multiple sequence alignment tool development and its application to RNase family alignment. J. Bioinform. Comput. Biol. **1**, 267–287 (2003)
16. Arslan, A.N., Egecioglu, O.: Algorithms for the constrained longest common subsequence problems. Int. J. Found. Comput. Sci. **16**(6), 1099–1109 (2005)
17. Deorowicz, S., Obstoj, J.: Constrained longest common subsequence computing algorithms in practice. Comput. Inform. **29**(3), 427–445 (2010)
18. Iliopoulos, C.S., Rahman, M.S.: New efficient algorithms for the LCS and constrained LCS problems. Inform. Process. Lett. **106**(1), 13–18 (2008)
19. Iliopoulos, C.S., Rahman, M.S., Rytter, W.: Algorithms for two versions of LCS problem for indeterminate strings. J. Comb. Math. Comb. Comput. **71**, 155–172 (2009)
20. Peng, Y.H., Yang, C.B., Huang, K.S., Tseng, K.T.: An algorithm and applications to sequence alignment with weighted constraints. Int. J. Found. Comput. Sci. **21**(1), 51–59 (2010)
21. Deorowicz, S.: Quadratic-time algorithm for a string constrained LCS problem. Inform. Process. Lett. **112**(11), 423–426 (2012)
22. Chen, Y.C., Chao, K.M.: On the generalized constrained longest common subsequence problems. J. Comb. Optim. **21**(3), 383–392 (2011)
23. Farhana, E., Ferdous, J., Moosa, T., Rahman, M.S.: Finite automata based algorithms for the generalized constrained longest common subsequence problems. In: Chavez, E., Lonardi, S. (eds.) SPIRE 2010. LNCS, vol. 6393, pp. 243–249. Springer, Heidelberg (2010)
24. Gotthilf, Z., Hermelin, D., Lewenstein, M.: Constrained LCS: hardness and approximation. In: Ferragina, P., Landau, G.M. (eds.) CPM 2008. LNCS, vol. 5029, pp. 255–262. Springer, Heidelberg (2008)
25. Gotthilf, Z., Hermelin, D., Landau, G.M., Lewenstein, M.: Restricted LCS. In: Chavez, E., Lonardi, S. (eds.) SPIRE 2010. LNCS, vol. 6393, pp. 250–257. Springer, Heidelberg (2010)
26. Tseng, C.T., Yang, C.B., Ann, H.Y.: Efficient algorithms for the longest common subsequence problem with sequential substring constraints. J. Complexity **29**, 44–52 (2013)

Financial System Networks Modeling Based on Complex Networks Theory

Lingling Zhang(✉) and Guoliang Cai

Nonlinear Scientific Research Center, Jiangsu University, Zhenjiang 212013, China
852745370@qq.com

Abstract. Complex networks theory has been widely used in many areas. Many financial scholars regard it as a new method to study financial system. In this paper, the principle, general path and method of constructing financial networks model are given. Then a delay nonlinear complex financial networks model is constructed by using differential equation method. Finally, the practical significance of the new financial networks is given.

Keywords: Complex networks · Modeling principle · Modeling method · Delay financial networks

1 Introduction

Complex networks theory has been widely applied in many areas, such as computer science, communications, engineering, economics, epidemiology, biology, sociology [1]. Due to financial system has many important characteristics similar with complex networks, it has begun to apply complex networks theory to research on the financial field in recent years. Especially, some western scholars and policy makers consider that the complex networks theory is the important and new tool to study financial systemic risk after the 2008 financial crisis.

In recent years, more and more scholars research on complex networks and use it to study the financial markets [2–4]. Financial markets are typically complex system with interaction units. It attracts experts on economics, mathematics, even physics to understand the dynamical evolution of such a complex system. Constructing networks is a key method for understanding the interaction between the individual systems, which makes the phenomenon in the field of biological research including diversity organization, computer networks, bibliometrics, electronic shopping networks that exist to be unified. In order to study the structure and the properties of financial networks, we need to build the networks at first.

Some domestic and foreign economists began to try new perspective to study the financial system by using complex networks theory. Hairong Cui used complex networks theory to study the systemic risk of bank [5]. Yanbo Zhou established a complex networks model based on the stock price volatility sequence. He found the networks model whose node degree distribution of financial market satisfies scale-free feature [6]. Zhaohan Sheng used the computational experimental method to provide a

C.-H. Hsu et al. (Eds.): IOV 2015, LNCS 9502, pp. 447–457, 2015.
DOI: 10.1007/978-3-319-27293-1_39

new tool for the study of complex management system on the interaction between dynamic evolution and objective and micro level [7]. Fang Liu discussed the cause of financial system risk and how to manage it based on complex networks theory [8]. Jasemi M presented a new model to do stock market timing on the basis of a supervised feed-forward neural networks and the technical analysis of Japanese Candlestick [9]. Hao Li constructed finance networks model to study relation based on minority game and complex [10]. Sanford, AD explored the structure of the Bayesian networks modeling operational risk in structured finance business by the designing of Bayesian networks structure [11]. Hua Han related the random matrix theory to the networks construction to study the financial networks model in terms of the random matrix [12]. Ningning Song took financial networks as research object, proposed a new method to determine the threshold depending on the cross-correlation matrix and the intrinsic attribute of the financial networks based on the threshold method of edge correlation coefficient when building the networks in complex system [13].

Access to various properties of financial market is a basic point to understand the nature of financial market. Based on complex networks theory, networks nodes may represent various companies, stock and so on. The edge from networks nodes to nodes can be represented as relationships of these companies and the stocks [14]. Complex networks exists in every corner of society, such as communications networks, social networks, biological networks, metabolic networks, the Internet and it also exists in financial system. How to describe and characterize the complex Internet banking systems by complex networks model and how to construct complex networks of nonlinear financial system are great meaningful.

This paper is organized as follows. The basic conceptions of complex networks are described in Sect. 2. Section 3 presents the general ways and methods of complex networks modeling. In Sect. 4, we describe the construction of financial networks. In Sect. 5, we present the meaning of constructing financial networks. In Sect. 6, a conclusion is made.

2 Basic Conception of Complex Networks Theory

2.1 Complex Networks

Xuesen Qian defined complex networks to be a networks with part or all properties of self-organizing, self-similar, attractor, small world, scale-free [15]. Complex networks is an abstract description of complex system, it highlights topological characteristics of system structure. In principle, any material that contains a large number of constituent units (or subsystem) is a complex system. When we abstract unit into a node abstracted, the relationship between the elements into edge, then a complex networks is made. Complex networks can be used to describe the social relations between humans, prey relationship between species, the semantic relation between words, the networks links between computers, communication feedback between neurons, interaction relations between proteins and so on. A specific networks can be abstracted as figure $G = (V, E)$ that consists of a point set V and edge set E. The number of nodes is denoted as $N = |V|$ and the number of edges is denoted as $M = |E|$. Each edge in E corresponds to a couple of point in V. If at any point (i, j) and (j, i) correspond to the same edge, the networks is called undirected networks, otherwise it is known as a directed networks [16].

Three typical models of complex networks are small world networks, scale-free networks, hierarchical networks, and the latter is true subset of the former.

Small Word Networks. A small-world networks is a type of mathematical graph in which most nodes can be reached from every other by a small number of hops or steps rather than neighbors. Specifically, a small-world networks is defined to be a networks where the typical distance L between two randomly chosen nodes (the number of steps required) grows proportionally to the logarithm of the number of nodes N in the networks, that is $L \propto \log N$.

A certain category of small-world networks were identified as a class of random graphs by Duncan Watts and Steven Strogatz in 1998 [17]. They noted that graphs could be classified according to two independent structural features: the clustering coefficient, and average shortest path length. Purely random graphs built according to the Erdős–Rényi (ER) model, exhibit a small average shortest path length (varying typically as the logarithm of the number of nodes) along with a small clustering coefficient. Watts and Strogatz measured that in fact many real-world networks have a small average shortest path length, but also a clustering coefficient significantly higher than expected by random chance. Watts and Strogatz then proposed a novel graph model, currently named the Watts and Strogatz model, with (i) a small average shortest path length, and (ii) a large clustering coefficient. The crossover in the Watts-Strogatz model between a "large world" (such as a lattice) and a small world was first described by Barthelemy and Amaral in 1999 [18].

Scale-Free Networks. A scale-free networks' degree follows a power law, at least asymptotically. That is, the fraction $P(k)$ of nodes in the networks having k connections to other nodes goes for large values of k as $P(k) \approx k^{-\gamma}$, where γ is a parameter whose value is typically in the range $2 < \gamma < 3$, although occasionally it may lie outside these bounds [19, 20].

Many networks have been reported to be scale-free, although statistical analysis has refuted many of these claims and seriously questioned others. Preferential attachment and the fitness model have been proposed as mechanisms to explain conjectured power law degree distributions in real networks [21].

Hierarchical Networks. The hierarchical networks model is part of the scale-free model family sharing their main property of having proportionally more hubs among the nodes than by random generation. However, it significantly differs from the other similar models (Barabási Albert, Watts Strogatz) in the distribution of the node's clustering coefficients: as other models would predict a constant clustering coefficient as the function of the degree of the node, in hierarchical models nodes with more links are expected to have a lower clustering coefficient. Moreover, while the Barabási-Albert model predicts a decreasing average clustering coefficient as the number of nodes increases, in the case of the hierarchical models there is no relationship between the size of the networks and its average clustering coefficient [22].

The development of hierarchical networks models was mainly motivated by the failure of the other scale-free models in incorporating the scale-free topology and high clustering into one single model. Since several real-life networks (metabolic networks, the protein interaction networks, the World Wide Web or some social networks) exhibit such properties, different hierarchical topologies were introduced in order to account for these various characteristics.

2.2 The Basic Properties of Complex Networks

Several of the basic properties of networks are great essential to understand the networks topology. Basic geometric properties of complex networks are average path length, cluster, degree and degree distribution, betweenness centrality etc. [23].

Average Path Length. The average path length of the networks is used to indicate node dependency. The average path length can be denoted as $L = \frac{1}{\frac{1}{2}N(N+1)} \sum_{i \neq j} d_{ij}$, where N is the number of nodes, d_{ij} represents the shortest distance between nodes i and j. Notice that distance in here is not physical distance, but the minimum number of connections between any two nodes in the networks.

Clustering. The so-called "average cluster coefficient" refers to the average probability of interconnection between two nodes in a networks connected to the same node. This factor is often used to characterize the structural properties of the local area networks, which can be defined as follows: Suppose any node i of the networks is connected to other k_i nodes, then the maximum number of connections that may exist in this node is k_i (k_i -1)/2. However, the actual number of connections is E_i, we define $C_i = 2\ E_i/k_i\ (k_i\ -1)$ as the clustering coefficient of node i. Then the clusters coefficient averaged for the entire networks can be obtained as $C = \frac{1}{N} \sum_i C_i$. That is, the average clustering coefficient of the entire networks.

Degree and Degree Distribution. Degree is an important concept which descripts node properties and it is denoted as k_i. Degree of node i represents the number of nodes connected to the node i. Most networks, including the World Wide Web, the Internet, and Metabolic Networks, etc., which of them has power-law degree distribution properties: $P\ (k) \sim k^{-\gamma}$. Table 1 shows main topological properties of several networks, we can clearly find the similarities and differences between various networks models and actual networks.

Betweenness Centrality. Betweenness centrality of a node reflects the importance of information transmission through the networks. It is defined as $B_n = \sum_{i,j} \frac{g_{inj}}{g_{ij}}$, $i,\ j = n,\ i \neq j$, where g_{ij} represents the number of the shortest path between nodes i and j, g_{inj} is the number of the shortest path between nodes i and j though node n.

Table 1. Main topology properties of some networks.

	Average path length	Clustering	Degree distribution
Regular networks	big	big	δ function
Random networks	small	small	Poisson distribution
Small Word networks	small	big	Exponential distribution
BA Scale-free networks	small	small	Power-law distribution
Some actual networks	small	big	Approximate power-law distribution

The two most important statistical properties of complex networks are small-world effect and scale-free property.

Small World Effect. Small world effect refers to the large scale complex networks between any two points in the structure that exist a relatively shortest connection path. Small world phenomenon reveals the objective things in motion some of the most efficient way of transmitting information and pathways that can be used to describe the many interesting life events occurring in a period. The small world effect is usually used to quantitatively describe the characteristic path length in complex networks. The definition of small-world effects are still being discussed, at present a more reasonable definition in follows. If the average shortest distance between two networks nodes as the networks size increasing shows logarithmic growth(number of nodes in the networks is N), namely $L \sim lnN$, when the size of the growth of the networks system is fast and an average growth relative shortest distance slow, thus it is called the networks a small-world effect.

Scale-Free Property. Scale-free of scale-free networks is an intrinsic property to describe the serious unbalance distribution on a whole of many complex systems.

3 General Approach and Method of Complex Networks Modeling

3.1 The Modeling Method of Complex Networks

In essence, the networks model is an abstract "image" that departure from the networks concept on a small part or a few aspects of the real world. Therefore, establishing the networks model needs to be established abstract as follows: a function of input, output, and relationships between state variables. This process is called an abstract model construction. You must contact the real target systems and modeling, which describe-variable plays a very important role in abstract, regardless it is observable or unobservable. Variables that can be observed from the outside of the system to apply sound or interference are called input variables. Variables that responses to networks are called output variables. Set of input and output variables characters "input– output" traits (relationship) of the real system. To sum up, a real networks can be regarded as a certain trait data sources, models have the same traits and real networks data set of instructions. The abstract played a role of media [24]. Mathematical modeling is process of the real system abstracted into mathematical expressions. Usually, there are several ways to build a networks model as follows.

The Graphic Modeling. It is made of points and lines. It uses graphics to describe system model. Graphical model of a structural model can be used to describe the relationship between a lot of things in nature and human society. Graph theory as a tool in modeling. Analyzing figure character provides an effective approach for the study of the various systems especially in complex system. Example of the common model of urban public transport networks use directed shows relationship between nodes (station).

The Analytic Hierarchy Process. It is a new system analysis method put forward by T.L. Saaty a famous professor at the University of Pittsburgh in the 1970s. It is a practical method for making multi-criteria decision. The basic idea of studying complex systems is to break down complex problems into factors that formed orderly hierarchical structure according to the dominating relationship and weigh the various aspects. Then the order is decided by the importance of the merits and comprehensive personal judgment.

The Probability Statistics Method. It is an inductive method from the specific to the general that base on probability theory. It observes objective phenomena in part of information, collecting and collating analysis, and then inference whole phenomena based on the sample. Probability statistics method has been widely applied to various fields, such as biostatistics, medicine statistics, project statistics, and business statistics.

Differential Equation Modeling Method. Continuous system refers to a system which state variable changes with changing time continuous. Its main features can be described by ordinary differential equations or partial differential equations. Common continuous system modeling methods are differential equation modeling method, state space modeling method and the variation principle modeling method. Discrete event system is the system which the state of the system occurs only at discrete points in time changes, and these discrete time points are generally uncertain. In this system, the event causes state change and usually one to one state change with events. Occurrence of an event of no continuity and it can be viewed instantaneously on the same point of time. The event occurrence time is discrete, thus such systems are called discrete event systems. It should be noted that a real system being discrete or continuous, essentially refers to a descriptive model of the system being discrete or continuous. Discrete event system modeling methods are physical flow diagram, activity cycle diagram method and Petri net method [25].

3.2 The General Approach to Complex Networks Modeling

For the so-called white-box (most of engineering systems), that is, the internal structure and properties of system are clear, we can use some basic laws of the known, through analysis and deduction derived system model. For the so-called gray box and black box, that is, the internal structural characteristic of system are not known or not very clear, if allowed direct experimental observation, we can assume that model, then experiment verification and correct it. For those systems that belong to a black box but not allow directly experiment observation (most non-engineering systems), we can collect data and induct statistical to assume model [26, 27].

4 Financial System Networks Modeling

In this section, we mainly use ordinary differential equations method to build a delay differential equations financial networks model. To construct a financial system networks, the following basic principles must be obeyed in this paper.

- **Ensure that the System Dynamics without Distortion**
  The purpose of the financial system simulation is to study the dynamic characteristic of system, to determine the level of stability of the system, to establish the development style of the system, to analyze the development plan of system and to research on the security and stability system technical measures. Financial system simulation accuracy or not have a significant impact on the stability of the financial system operating, planning and designing. And the accuracy simulation results are determined by the accuracy of the simulation models and parameters. Therefore, when studying the financial system networks construction model, we must first ensure that the financial networks model is established without distortion through dynamic simulation.
- **Dynamic Characteristic of the System Should Maintain Greater Impact Financial Dynamics Properties**
  As the financial constitutions with complexity, diversity, distribution, time-varying and stochastic characteristic, it is very difficult to construct financial system model precisely. Therefore, the financial system networks model should maintain great impact on the dynamic characteristic of the financial dynamics to ensure maintain financial dynamic system characteristic simulation without distorted.
- **Ensure that the Financial Model has Meaningful and Strong Adaptability**
  Financial system modeling is not just for a single financial institution, it mainly reflects characteristic of financial institutions constituting to the entire financial market. However, due to the complexity, diversity, time-varying, random and distribution of the financial constitution, the establishment financial networks model system can not be static. It requires financial networks model established with practical clearly, strong adaptability and easy analyze to provide proper guidance for operating financial markets stably.

References [28] have reported a dynamic model of finance composed of four sub-blocks: production, money, stock and labor force, and expressed by three first-order differential equations. The model describes the time variations of three state variables: the interest rate x, the investment demand y, and the price exponent z. The chaotic finance system is given as follows:

$$\begin{cases} \dot{x} = z + (y - a)x \\ \dot{y} = 1 - by - x^2 \\ \dot{z} = -x - cz \end{cases}$$

The parameter a is the saving, b is the per investment cost, c is the elasticity of demands of commercials, and they are positive constants. When parameters $a = 0.9, b = 0.2, c = 1.2$, with the initial state $(x(0), y(0), z(0)) = (1, 3, 2)$, system shows chaotic behavior.

Based on the chaotic finance system above, we found that the factors which affect the interest rates not only in addition to investment demand and price index, but also related with the average profit margin. And the average profit margin and interest rate is proportional. Therefore, the following improved chaotic finance system is constructed.

The novel financial hyperchaotic system [29] which is judged by two positive Lyapunov exponents can be generated by adding an additional state variable u, and u is the average profit margin. The system of four-dimensional autonomous differential equations has the following form:

$$\begin{cases} \dot{x} = z + (y - a)x + w, \\ \dot{y} = 1 - by - x^2, \\ \dot{z} = -x - cz, \\ \dot{w} = -dxy - kw, \end{cases}$$

where the model describes the time variations of four state variables: the interest rate x, the investment demand y, the price exponent z, and the average profit margin w, a, b, c, d, k are the parameters of the system (1), and they are positive constants. When set up parameters $a = 0.9$, $b = 0.2$, $c = 1.5$, $d = 0.\ 2$ and $k = 0.17$, the four Lyapunov exponents of the system (1) calculated with Wolf algorithm are $L_1 = 0.034432$, $L_2 = 0.018041$, $L_3 = 0$ and $L_4 = -1.1499$. The system is constructed in the background of the global economic crisis which occurred in 2007 and still existence nowadays. As this global economic crisis did not cause the great depression like in 1920s-1930s, so the system is a weak hyperchaotic finance system. The largest Lyapunov exponent of the system is a little small, but this hyperchaotic finance system exactly reflects this finance phenomenon. By calculation, the hyperchaotic system has three equilibrium points: P_0 (0, $1/b$, 0, 0), $P_{1,2}(\pm\theta, \frac{k+ack}{c(k-d)}, \mp\frac{\theta}{c}, \frac{d\theta(1+ac)}{cd-ck})$ when the parameters a, b, c, d, k satisfies $\frac{kb+dc+abck-ck}{c(d-k)} > 0$ and $\theta = \sqrt{\frac{kb+abck}{c(d-k)}} + 1$. As the eigenvalues of the Jacobian Matrix at the equilibrium points P_0 are 3.6433, 0.1278, -0.200.-1.3411. Hence, equilibrium P_0 is an unstable saddle point in this nonlinear four-dimensional autonomous system. Similarly, system undergoes a Hopf bifurcation at the equilibrium $P_{1,2}$. Therefore, the equilibrium $P_{1,2}$ are unstable saddle points. According to the hyperchaotic parameter values given above, the equilibrium points are calculated as: P_1(0, 5, 0, 0), P_2(1.66, -8.87, -1.11, 17.4) and P_3(-1.66, -8.87, 1.11, -17.4). More dynamic behaviors analyze about system in reference [29–32].

The financial system has many characteristics of complex networks. We regard financial institutions as networks nodes such as banks, stock markets, all kinds of companies. Let edge represents the relationship between these companies, stocks, banks. Such as credit relations, risk exposure, the interbank payment system financial flows with each other. However, financial market information spreads usually with time delay phenomenon. The time delay phenomenon will affect the whole financial market, so it is necessary and sensible to discuss this phenomenon in financial markets. Therefore, we will present a delay complex financial networks differential equation model in this paper. The model is described as follows.

$$\dot{x}_i(t) = Ax_i(t) + Bf\left(x_i(t)\right) + \sum_{j=1}^{N} c_{ij}\Gamma_1 x_j(t) + \sum_{j=1}^{N} d_{ij}\Gamma_2 x_j(t-\tau) + I, \quad t > 0,$$

where A, B are coefficient matrixes of linear and nonlinear terms respectively, x_i is financial institution, $f(x_i(t))$ is nonlinear function, Ax_i represent linear relationship between some financial institutions and some financial institutions exists non-linear relationship signed as $Bf(x_i(t))$, c_{ij} and d_{ij} are internal coupling coefficient matrix without time delay and with time delay between different financial institutions x_i and x_j.. Γ_1, Γ_2 are financial institutions own internal connection weights without time delay and with time delay, τ is a time lag of information spread in financial institutions and financial institution affected by external interference denotes as I.

5 Significance of the Financial System Networks Model

In this paper, complex networks theory is applied to the financial system and a delayed financial system networks model is established. Using the complex networks theory to analyze model and evaluate the new financial system. Through detailed explanation and analysis of the financial and economic system, the results can be applied to the government and the enterprise's economic management for economic decision-making reference, making corresponding countermeasures and suggestions. On this basis, further study on control of financial system about dynamics provides a reference for the normal operation of the financial system. There is no doubt that this paper's study is useful for Chinese economic development, governors to make economic decisions accurately and scientifically. And it has important reference value and practical significance. The paper is to serve Chinese economic construction. The method of modeling has higher application value in economic evaluation, economic decision-making and economic management and especially in the financial field which has a broad application prospect.

6 Conclusions

This paper studies the construction of complex financial networks. Through the relevant concepts' understanding of the complex networks, we gives principles in financial system networks modeling by using complex networks modeling method and general way. Finally, we construct a delayed differential equation complex networks model of financial system and give the establishment meaning of financial system networks. The method for constructing financial networks model is given. Actually, in the financial networks, the financial institution's coupling strength is related to distance and the external environment. In addition, this paper only considers the coupling between nodes from the static aspect, time varying delay is not considered in the financial networks model. The characteristics, synchronization and control of time-varying delay of complex financial networks will be considered in next content.

Acknowledgment. This work was supported by the National Nature Science foundation of China (Grants 51276081, 11171135), the Society Science Foundation from Ministry of Education of China (Grants 12YJAZH002, 15YJAZH002), The Priority Academic Program Development of Jiangsu Higher Education Institutions, the Advanced Talents' Foundation of Jiangsu University (Grants 07JDG054,10JDG140), the Postgraduate research and innovation projects of Jiangsu University (Grant KYXX-0038), and the Students' Research Foundation of Jiangsu University (Grants Y13A126, 13A293). Especially, thanks for the support of Jiangsu University.

References

1. Zheng, X., Chen, J.P., Shao, J.L., Bie, L.D.: Analysis on topological properties of Beijing urban public transit based on complex network theory. Acta Phys. Sin. **61**, ID 190510 (2012)
2. Qiu, T., Zheng, B., Chen, G.: Financial networks with static and dynamic thresholds. New J. Phys. **12**, ID 043057 (2010)
3. Zhuang, X.T., Huang, X.Y., Nie, H.M.: The model and optimization of financial networks. Chin. J. Mana. Sci. **11**, 7–10 (2003)
4. Li, H., Chen, T.Q., He, J.M.: The interbank market network structure evolution model based on the complex network theory. J. Beijing Instit. Technol. (Social Sci. Edition) **14**, 71–76 (2012)
5. Cui, H.R., He, J.M.: Systemic risk in banking based on complex network theory. J. Xidian Univ. (Social Sci. Edition) **19**, 12–18 (2009)
6. Zhou, Y.B., Cai, S.M., Zhou, P.L.: Scale-free properties of financial markets. J. Univ. Sci. Technol. China **39**, 880–884 (2009)
7. Sheng, Z.H.: The research of large scale complex engineering integrated management mode–Sutong Bridge project management theory. Construct. Economy. **5**, 20–22 (2009)
8. Liu, F.: The causes and governance of financial system risk based on complex network theory. Spec. Zone Econ. **12**, 91–93 (2011)
9. Jasemi, M.: A modern neural network model to do stock market timing on the basis of the ancient investment technique of Japanese Candlestick. Expert Syst. Appl. **38**, 3884–3890 (2011)
10. Li, H., Cao, H.D., Xing, H.K.: Modeling and simulation of complex finance networks based on minority game. Syst. Engin. Theor. Pract. **32**, 1882–1890 (2012)
11. Sanford, A.D.: A Bayesian network structure for operational risk modeling in structured finance operations. J. Operat. Res. Soci. **63**, 431–444 (2012)
12. Han, H., Wu, L.Y., Song, N.N.: Financial networks model based on random matrix. Acta Phys. Sin. **63**, ID 138901 (2014)

13. Song, N.N., Han, H., Wu, L.Y.: New method to construct financial network based on threshold. Comp. Eng. Appl. **51**, 249–253 (2015)
14. Guo, Y.L., Xue, Y.W., Zhang, P.Z.: Research on model construction of complex financial network. J. Taiyuang Univ. Sci. Technol. **30**, 113–116 (2009)
15. Wang, X.F., Li, X., Chen, G.R.: Complex network theory and application, pp. 8–9. University Qinghua Press, Beijing (2005)
16. Ren, X.Y., Guo, Li., Zhou, P.L.: Research on financial market modeling based on complex networks. Univ. Sci. Technol, China 7–8 (2013)
17. Watts, D.J., Strogatz, S.H.: Collective dynamics of 'small-world' networks. Nature **393**, 440–442 (1998)
18. Barthelemy, M., Amaral, L.A.N.: Small-world networks: Evidence for a crossover picture. Phys. Rev. Lett. 82, ID 3180 (1999)
19. Onnela, J.P., Saramaki, J., Hyvonen, J., Szabo, G., Lazer, D., Kaski, K., Kertesz, J., Barabasi, A.L.: Structure and tie strengths in mobile communication networks. Proc. Nat. Acad. Sci. **104**, 7332–7336 (2007)
20. Choromański, K., Matuszak, M., MiKisz, J.: Scale-Free graph with preferential attachment and evolving internal vertex structure. J. Statist. Phys. **151**, 1175–1178 (2013)
21. Clauset, A., Cosma, R.S., Newman, M.E.J.: Power-law distributions in empirical data. Phys. Data-an. **6**, ID 00361445 (2007)
22. Ravasz, E., Somera, A.L., Mongru, D.A.: Hierarchical organization of modularity in metabolic networks. Science **297**, 1551–1555 (2002)
23. Tang, C., Tang, Y.: Research on complex networks model and its statistical properties, pp. 23–26. University Xiangtan, Xiangtan (2005)
24. Qi, H., Wang, X.P.: System modeling and simulation, pp. 26–53. University Qinghua Press, Beijing (2013)
25. Guo, Q.S.: System modeling theory and method, pp. 41–83. National University of Defense Technology Press, Changsha (2003)
26. Peng, F., Yang, R.Y., Xiao, H.J., He, Y.M.: Mathematics modeling method, pp. 4–5. Science Press, Beijing (2012)
27. Ma, L., Zhang, M.Z.: Research Progress on War Complex System of Systems Modeling Based on Complex Network. J. Syst. Simulat. **27**, 217–225 (2015)
28. Jian, J., Deng, X., Wang, J.: Globally exponentially attractive set and synchronization of a class of chaotic finance system. In: Yu, W., He, H., Zhang, N. (eds.) ISNN 2009, Part I. LNCS, vol. 5551, pp. 253–261. Springer, Heidelberg (2009)
29. Yu, H.J., Cai, G.L., Li, Y.X.: Dynamic analysis and control of a new hyperchaotic finance system. Nonlinear Dyn. **67**, 2171–2182 (2012)
30. Cai, G.L., Hu, P., Li, Y.X.: Modified function lag projective synchronization of a financial hyperchaotic system. Nonlinear Dyn. **69**, 1457–1464 (2012)
31. Cai, G.L., Yao, L., Hu, P., Fang, X.L.: Adaptive full state hybrid function projective synchronization of financial hyperchaotic systems with uncertain parameters. Disc. Cont. Dyn. B **18**, 2019–2028 (2013)
32. Cai, G.L., Zhang, L.L., Yao, L., Fang, X.L.: Modified function projective synchronization of financial hyperchaotic systems via adaptive impulsive controller with unknown parameters. Disc. Dyn. Natur. Soc. **2015**, 572–735 (2015)

Bi-directional Coverage in the Connectivity of Sensor Networks

Bin Xia[1,2] and Zuoming Yu[3](✉)

[1] State Grid Suzhou Power Supply Company, Suzhou 215000, People's Republic of China
[2] Dongwu Business School of Suzhou University, Suzhou 215000, People's Republic of China
[3] School of Zhangjiagang, Jiangsu University of Science and Technology, Zhangjiagang 215600, People's Republic of China
yuzuoming1981@gmail.com

Abstract. Coverage theory plays an important role not only in Topology but also in information science. In this paper, we study a novel coverage model, named bi-directional coverage. We characterize it with matrix. As an application, we prove that connectivity of sensor network is a special kind of bi-directional coverage by digraph theory.

Keywords: Bi-directional coverage · Sensor network · Connectivity · Digraph

1 Introduction

Coverage is a classic problem in both topology theory [4,5] and sensor network studies [1,2]. It is a hot topic in recent years to find varies of optimal sensor network deployments [3,7,9]. In these works, an object A covers B means that B lies in the some geometry region derived from A. For example, one sensor covers an point implies that the point lies in the sensing area of this sensor. We can see that in this case, if we divide the whole things into two parts, sensors and points, sensors cover points "actively" and points are covered "negatively". It is an one-way coverage. However, if we look into another important issue, connectivity, in sensor network [6,11–13], we can see that one-way "coverage" is not enough [10,14]. Let us describe it in detail. Two sensors, named sensor A and sensor B are connected if and only if A and b are able to communicate with each other. From the coverage perspective, A should cover B while B has to cover A at the same time. Inspired by this fact, we study a novel kind of coverage, called bi-directional coverage.

Zuoming Yu—This paper is supported by Natural Science Foundation of China (Nos: 11401262, 11401263, 61472469), Natural Science Foundation of Jiangsu Province (No. BK20140503) and Undergraduate innovation plan of Jiangsu University of Science and Technology (No. 126031084).

C.-H. Hsu et al. (Eds.): IOV 2015, LNCS 9502, pp. 458–468, 2015.
DOI: 10.1007/978-3-319-27293-1_40

The remainder of this paper is structured as follows. Section 2 describes concepts and some properties of bi-directional coverage. Section 3 mainly presents the characterization of connectivity of sensor networks with bi-directional coverage. Section 4 concludes.

2 Concepts and Properties of Bi-directional Coverage Systems

2.1 Definitions of Bi-directional Coverage Systems

Let system $\mathcal{S}(\mathcal{A}, \mathcal{B})$ consisting of subsystem $\mathcal{A}$ and $\mathcal{B}$.

Bi-directional coverage, BDC for short:
(1) $\forall\ a \in \mathcal{A}$, $\exists\ b \in \mathcal{B}$, such that b covers a;
(2) $\forall\ b \in \mathcal{B}$, $\exists\ a \in \mathcal{A}$, such that a covers b.

Strongly Bi-directional coverage, SBDC for short:
(1) $\forall\ a \in \mathcal{A}$, $\exists\ b \in \mathcal{B}$, such that b covers a and a covers b;
(2) $\forall\ b \in \mathcal{B}$, $\exists\ a \in \mathcal{A}$, such that a covers b and b covers a.

Weakly Bi-directional coverage, WBDC for short:
(1) $\forall\ a \in \mathcal{A}$, $\exists\ b \in \mathcal{B}$, such that a covers b;
(2) $\forall\ b \in \mathcal{B}$, $\exists\ a \in \mathcal{A}$, such that b covers a.

Locally Bi-directional coverage, LBDC for short:
(1) $\exists\ a \in \mathcal{A}$, $\exists\ b \in \mathcal{B}$, such that b covers a;
(2) $\exists\ b \in \mathcal{B}$, $\exists\ a \in \mathcal{A}$, such that a covers b.

Locally Strongly Bi-directional coverage, LSBDC for short:
(1) $\exists\ a \in \mathcal{A}$, $\exists\ b \in \mathcal{B}$, such that b covers a and a covers b;
(2) $\exists\ b \in \mathcal{B}$, $\exists\ a \in \mathcal{A}$, such that a covers b and b covers a.

Locally weakly Bi-directional coverage, LWBDC for short:
(1) $\exists\ a \in \mathcal{A}$, $\exists\ b \in \mathcal{B}$, such that a covers b;
(2) $\exists\ b \in \mathcal{B}$, $\exists\ a \in \mathcal{A}$, such that b covers a.

Example 2.1. BDC and LBDC. See Fig. 1.

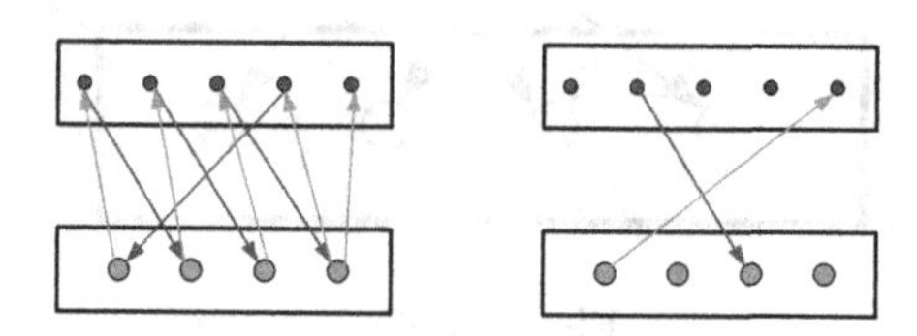

Fig. 1. BDC and LBDC

Remark 2.1. "(1)" and "(2)" are equivalent in the definition of LSBDC.

Example 2.2. WBDC not BDC. See Fig. 2.

Example 2.3. BDC not SBDC. See Fig. 3.

Example 2.4. SBDC. See Fig. 4.

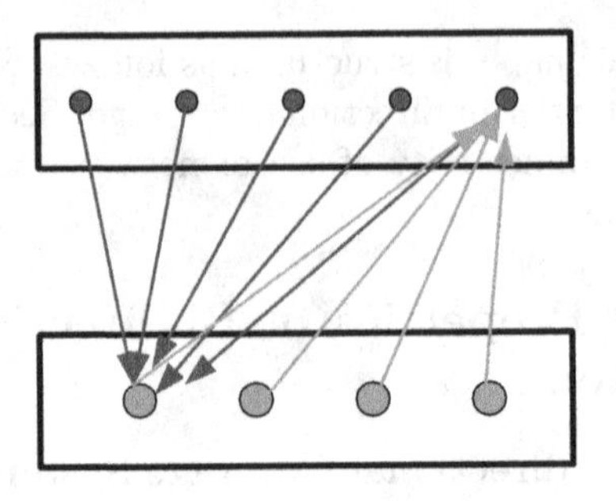

Fig. 2. WBDC not BDC

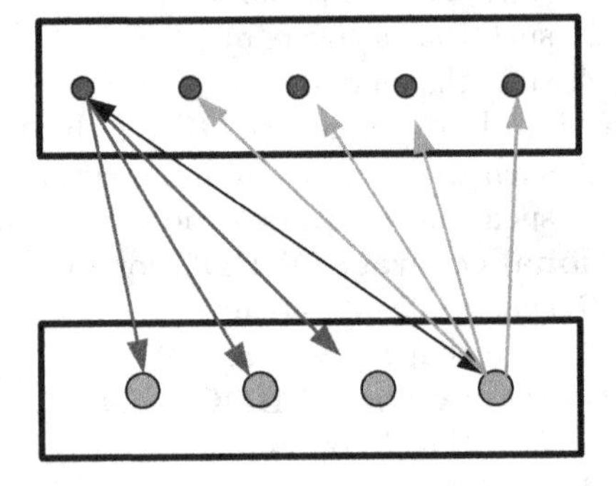

Fig. 3. BDC not SBDC

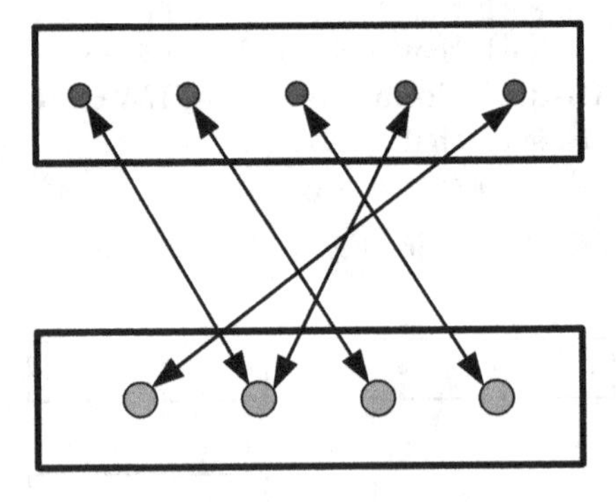

Fig. 4. SBDC

2.2 Basic Properties of Matrixes of Bi-directional Coverage Systems

Matrixes of Bi-directional Coverage Systems. Given a bi-directional coverage system $\mathcal{S}(\mathcal{A}, \mathcal{B})$, where $\mathcal{A} = \{a_1, a_2, ..., a_m\}$ and $\mathcal{B} = \{b_1, b_2, ..., b_n\}$, then the matrix $M_{\mathcal{A},\mathcal{B}}$ of this bi-directional coverage system is defined as following:

$M_{\mathcal{A},\mathcal{B}}=(x_{kl})_{(m+n)\times(m+n)}$.
$x_{i(m+j)} = 1$ if and only if a_i covers b_j, $i = 1, 2, ..., m$, $j = 1, 2, ..., n$;
$x_{(m+j)i} = 1$ if and only if b_j covers a_i, $i = 1, 2, ..., m$, $j = 1, 2, ..., n$;

Else, $x_{i(m+j)} = 0$, $x_{(m+j)i} = 0$.
$M_{\mathcal{A}} = (x_{i(m+j)})_{i \times j}$, $i = 1, 2, ..., m$, $j = 1, 2, ..., n$.
$M_{\mathcal{B}} = (x_{(m+j)i})_{j \times i}$, $i = 1, 2, ..., m$, $j = 1, 2, ..., n$.
$M_{\mathcal{A},\mathcal{B}}$ of Fig. 1 BDC and LBDC: See Fig. 5.

	x1	x2	x3	x4	x5	x6	x7	x8	x9
x1	0	0	0	0	0	0	1	0	0
x2	0	0	0	0	0	0	0	1	0
x3	0	0	0	0	0	0	0	0	1
x4	0	0	0	0	0	1	0	0	0
x5	0	0	0	0	0	0	0	0	0
x6	1	0	0	0	0	0	0	0	0
x7	0	1	0	0	0	0	0	0	0
x8	0	0	1	0	0	0	0	0	0
x9	0	0	0	1	1	0	0	0	0

	x1	x2	x3	x4	x5	x6	x7	x8	x9
x1	0	0	0	0	0	0	0	0	0
x2	0	0	0	0	0	0	0	1	0
x3	0	0	0	0	0	0	0	0	0
x4	0	0	0	0	0	0	0	0	0
x5	0	0	0	0	0	0	0	0	0
x6	0	0	0	0	1	0	0	0	0
x7	0	0	0	0	0	0	0	0	0
x8	0	0	0	0	0	0	0	0	0
x9	0	0	0	0	0	0	0	0	0

Fig. 5. $M_{\mathcal{A},\mathcal{B}}$ of Fig. 1 BDC and LBDC

$M_{\mathcal{A},\mathcal{B}}$ of Fig. 2 WBDC: See Fig. 6.

	x1	x2	x3	x4	x5	x6	x7	x8	x9
x1	0	0	0	0	0	1	0	0	0
x2	0	0	0	0	0	1	0	0	0
x3	0	0	0	0	0	1	0	0	0
x4	0	0	0	0	0	1	0	0	0
x5	0	0	0	0	0	1	0	0	0
x6	0	0	0	0	1	0	0	0	0
x7	0	0	0	0	1	0	0	0	0
x8	0	0	0	0	1	0	0	0	0
x9	0	0	0	0	1	0	0	0	0

Fig. 6. $M_{\mathcal{A},\mathcal{B}}$ of Fig. 2 WBDC

$M_{\mathcal{A},\mathcal{B}}$ of Fig. 3 BDC: See Fig. 7.
$M_{\mathcal{A},\mathcal{B}}$ of Fig. 4 SBDC: See Fig. 8.

Proposition 2.1. *A system $\mathcal{S}(\mathcal{A}, \mathcal{B})$, where $\mathcal{A} = \{a_1, a_2, ..., a_m\}$ and $\mathcal{B} = \{b_1, b_2, ..., b_n\}$, is BDC if and only if $\sum_{k=m+1}^{m+n} x_{kl} \geq 1$ for each $l = 1, ..., m$ and $\sum_{k=1}^{m} x_{kl} \geq 1$ for each $l = m+1, ..., m+n$.*

Proof. "Necessity": Pick $a_l \in \mathcal{A}$, $l \in \{1, 2, ..., m\}$, by the definition of BDC, there is some $i \in \{1, ..., n\}$ such that b_i covers a_l. Therefore, $x_{(m+i)l} = 1$ in $M_{\mathcal{B}}$. Hence, $\sum_{k=m+1}^{m+n} x_{kl} \geq 1$ for l. By the same argument, $\sum_{k=1}^{m} x_{kl} \geq 1$ for each $l = m+1, ..., m+n$.

"Sufficiency": Suppose that $\sum_{k=m+1}^{m+n} x_{kl} \geq 1$ for each $l = 1, ..., m$. For every $l \in \{1, ..., m\}$, there is some $x_{(m+i)l} = 1$ for some $i \in \{1, ..., n\}$ since

	x_1	x_2	x_3	x_4	x_5	x_6	x_7	x_8	x_9
x_1	0	0	0	0	0	1	1	1	1
x_2	0	0	0	0	0	0	0	0	0
x_3	0	0	0	0	0	0	0	0	0
x_4	0	0	0	0	0	0	0	0	0
x_5	0	0	0	0	0	0	0	0	0
x_6	0	0	0	0	0	0	0	0	0
x_7	0	0	0	0	0	0	0	0	0
x_8	0	0	0	0	0	0	0	0	0
x_9	1	1	1	1	1	0	0	0	0

Fig. 7. $M_{\mathcal{A},\mathcal{B}}$ of Fig. 3 BDC

	x_1	x_2	x_3	x_4	x_5	x_6	x_7	x_8	x_9
x_1	0	0	0	0	0	0	1	0	0
x_2	0	0	0	0	0	0	0	1	0
x_3	0	0	0	0	0	0	0	0	1
x_4	0	0	0	0	0	0	1	0	0
x_5	0	0	0	0	0	1	0	0	0
x_6	0	0	0	0	1	0	0	0	0
x_7	1	0	0	1	0	0	0	0	0
x_8	0	1	0	0	0	0	0	0	0
x_9	0	0	1	0	0	0	0	0	0

Fig. 8. $M_{\mathcal{A},\mathcal{B}}$ of Fig. 4 SBDC

$\sum_{k=m+1}^{m+n} x_{kl} \geq 1$. It means that b_i covers a_l by the definition of $M_{\mathcal{A},\mathcal{B}}$. So, the condition (1) of the definition of BDC holds. Meanwhile, we can prove the condition (1) of the definition of BDC is true in this case by the same way.

Proposition 2.2. *A system $\mathcal{S}(\mathcal{A},\mathcal{B})$, where $\mathcal{A} = \{a_1, a_2, ..., a_m\}$ and $\mathcal{B} = \{b_1, b_2, ..., b_n\}$, is WBDC if and only if $\sum_{k=m+1}^{m+n} x_{lk} \geq 1$ for each $l = 1, ..., m$ and $\sum_{k=1}^{m} x_{lk} \geq 1$ for each $l = m+1, ..., m+n$.*

Proof. "Necessity": Given $a_l \in \mathcal{A}$, there is some $b_i \in \mathcal{B}$ such that a_l covers b_i. So according to the definition of $M_{\mathcal{A},\mathcal{B}}$, we have $x_{l(m+i)} = 1$. Obviously, $\sum_{k=m+1}^{m+n} x_{lk} \geq 1$ for l. Similarly, we can show that $\sum_{k=1}^{m} x_{lk} \geq 1$ for each $l = m+1, ..., m+n$.

"Sufficiency": Suppose that $l \in \{1, ..., m\}$ and $\sum_{k=m+1}^{m+n} x_{lk} \geq 1$. Then there is some $x_{lk} = 1$ for some $k \in \{m+1, ..., m+n\}$. By the definition of $M_{\mathcal{A},\mathcal{B}}$, we know that a_l covers b_i, where $i \in \{1, ..., n\}$. This finishes the proof of the condition (1) of the definition of WBDC. By the same way, we can prove that the condition (1) of the definition of WBDC also holds under the assumption.

Proposition 2.3. *A system $\mathcal{S}(\mathcal{A}, \mathcal{B})$, where $\mathcal{A} = \{a_1, a_2, ..., a_m\}$ and $\mathcal{B} = \{b_1, b_2, ..., b_n\}$, is SBDC if and only if*

(1) $\prod_{l=1}^{l=m} (\sum_{k=m+1}^{m+n} x_{lk} + \sum_{k=m+1}^{m+n} x_{kl}$
$- \sum_{k=m+1}^{m+n} |x_{kl} - x_{lk}|)$
> 0
for each $l \in \{1, 2, ..., m\}$ and
(2) $\prod_{j=1}^{l=n} (\sum_{k=1}^{m} x_{(m+j)k} + \sum_{k=1}^{m} x_{k(m+j)}$
$- \sum_{k=1}^{m} |x_{jk} - x_{kj}|)$
> 0
for each $j \in \{1, 2, ..., n\}$.

Proof. "Necessity": As $\mathcal{S}(\mathcal{A}, \mathcal{B})$ is SBDC, for each $a_l \in \mathcal{A}$, there is at least one element $b_i \in \mathcal{B}$ such that a_l and b_i cover each other at the same time. That is to say, $x_{l(m+i)} = x_{(m+i)l} = 1$ by the definition of $M_{\mathcal{A},\mathcal{B}}$. Therefore, $|x_{l(m+i)} - x_{(m+i)l}| = 0$. Notice that $|x_{h(m+j)} - x_{(m+j)h}|$ is still equal to "1" if a_h and b_j do not cover each other at the same time. Thus,

$\prod_{l=1}^{l=m} (\sum_{k=m+1}^{m+n} x_{lk} + \sum_{k=m+1}^{m+n} x_{kl}$
$- \sum_{k=m+1}^{m+n} |x_{kl} - x_{lk}|)$
$> 0.$

"Sufficiency": Suppose that $\prod_{l=1}^{l=m} (\sum_{k=m+1}^{m+n} x_{lk} + \sum_{k=m+1}^{m+n} x_{kl}$
$- \sum_{k=m+1}^{m+n} |x_{kl} - x_{lk}|)$
> 0
for each $l \in \{1, 2, ..., m\}$. Then it is easy to see that $\sum_{k=m+1}^{m+n} x_{lk} + \sum_{k=m+1}^{m+n} x_{kl}$
$- \sum_{k=m+1}^{m+n} |x_{kl} - x_{lk}|$
> 0
for each $l \in \{1, 2, ..., m\}$.
Pick $a_l \in \mathcal{A}$, where $l \in \{1, ..., m\}$. Since $\sum_{k=m+1}^{m+n} x_{lk} + \sum_{k=m+1}^{m+n} x_{kl}$
$- \sum_{k=m+1}^{m+n} |x_{kl} - x_{lk}|$
$> 0,$

by the definition of $M_{\mathcal{A}}$ and $M_{\mathcal{B}}$, there exists $k \in \{m+1, 2, ..., m+n\}$ such that $x_{kl} = x_{lk} = 1$. Let $k = m + j$. It follows that b_j covers a_l and a_l covers b_j. So, condition (1) of SBDC holds.

By the same way, we can use (2) to show that condition (2) of SBDC also holds.

Remark 2.2. In SBDC, $M_{\mathcal{A}}$ and $M_{\mathcal{B}}$ are not necessarily symmetric.

Example 2.5. In Proposition 2.3, neither (1) nor (2) can be omitted. see Figs. 9 and 10.

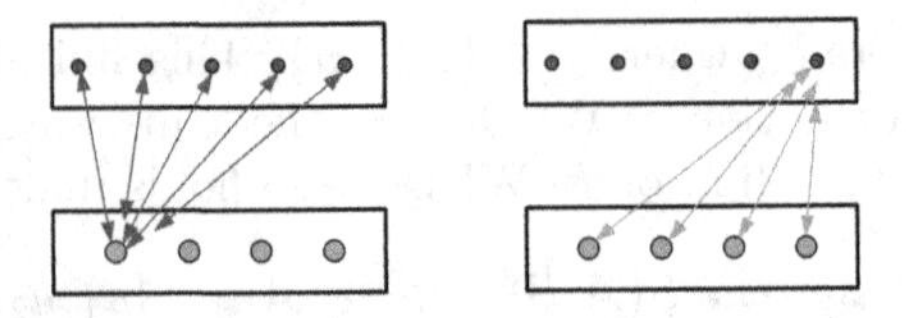

Fig. 9. A matrixes satisfies condition (1) of Proposition 2.3.

	x_1	x_2	x_3	x_4	x_5	x_6	x_7	x_8	x_9
x_1	0	0	0	0	0	1	0	0	0
x_2	0	0	0	0	0	1	0	0	0
x_3	0	0	0	0	0	1	0	0	0
x_4	0	0	0	0	0	1	0	0	0
x_5	0	0	0	0	0	1	0	0	0
x_6	1	1	1	1	1	0	0	0	0
x_7	0	1	0	0	0	0	0	0	0
x_8	0	0	0	0	0	0	0	0	0
x_9	0	0	0	0	0	0	0	0	0

	x_1	x_2	x_3	x_4	x_5	x_6	x_7	x_8	x_9
x_1	0	0	0	0	0	0	0	0	0
x_2	0	0	0	0	0	0	0	0	0
x_3	0	0	0	0	0	0	0	0	0
x_4	0	0	0	0	0	0	0	0	0
x_5	0	0	0	0	0	1	1	1	1
x_6	0	0	0	0	1	0	0	0	0
x_7	0	0	0	0	1	0	0	0	0
x_8	0	0	0	0	1	0	0	0	0
x_9	0	0	0	0	1	0	0	0	0

Fig. 10. A matrixes satisfies condition (2) of Proposition 2.3.

It is not hard to check that the matrixes above satisfies conditions (1) and (2) of Proposition 2.3. But neither of these systems is SBDC system.

Proposition 2.4. *A system $\mathcal{S}(\mathcal{A}, \mathcal{B})$, where $\mathcal{A} = \{a_1, a_2, ..., a_m\}$ and $\mathcal{B} = \{b_1, b_2, ..., b_n\}$ is LBDC if and only if $\sum_{k=m+1}^{m+n} x_{kl} \geq 1$ for some $l = 1, ..., m$ and $\sum_{k=1}^{m} x_{kl} \geq 1$ for some $l = m+1, ..., m+n$.*

Proof. Similar to the proof of Proposition 2.1.

Proposition 2.5. *A system $\mathcal{S}(\mathcal{A}, \mathcal{B})$, where $\mathcal{A} = \{a_1, a_2, ..., a_m\}$ and $\mathcal{B} = \{b_1, b_2, ..., b_n\}$, is LWBDC if and only if $\sum_{k=m+1}^{m+n} x_{lk} \geq 1$ for some $l = 1, ..., m$ and $\sum_{k=1}^{m} x_{lk} \geq 1$ for each $l = m+1, ..., m+n$.*

Proof. Similar to the proof of Proposition 2.2.

Proposition 2.6. *A system $\mathcal{S}(\mathcal{A}, \mathcal{B})$, where $\mathcal{A} = \{a_1, a_2, ..., a_m\}$ and $\mathcal{B} = \{b_1, b_2, ..., b_n\}$, is LSBDC if and only if*

$$\prod_{l=1}^{l=m} (\sum_{k=m+1}^{m+n} x_{lk} + \sum_{k=m+1}^{m+n} x_{kl}$$
$$- \sum_{k=m+1}^{m+n} |x_{kl} - x_{lk}|)$$
$$> 0$$
for some $l \in \{1, 2, ..., m\}$, or
$$\prod_{j=1}^{l=n} (\sum_{k=1}^{m} x_{(m+j)k} + \sum_{k=1}^{m} x_{k(m+j)}$$
$$- \sum_{k=1}^{m} |x_{jk} - x_{kj}|)$$
$$> 0$$
for some $j \in \{1, 2, ..., n\}$.

Proof. Similar to the proof of Proposition 2.3.

3 Applications of Bi-directional Coverage Systems

3.1 Some Related Concepts in Digraph Theory

Digraph: A digraph D consists of a non-empty finite set $V(D)$ of elements called **vertices** and a finite set $A(D)$ of ordered pairs of distinct vertices called **arcs**. For an arc (u, v) the first vertex u is its **tail** and the second vertex v is its **head**. When parallel arcs (pairs of arcs with the same tail and the same head) and loops (arcs whose head and tail coincide) are admissible we speak of **directed pseudographs**.

Walk, Trace, Path, cycle: For a directed pseudograph. A **walk** in D is an alternating sequence $W = x_1a_1x_2a_2x_3 ... x_{k-1}a_{k-1}x_k$ of vertices x_i and arcs a_j from D such that the tail of a_i is x_i and the head of a_i is x_{i+1} for every $i \in [k-1]$. A walk W is closed if $x_1 = x_k$. A **trail** is a walk W which all arcs are distinct. If the vertices of W are distinct, W is a path. If the vertices x_1, x_2,..., x_{k-1} are distinct, $k \geq 3$ and $x_1 = x_k$, W is a cycle.

A walk (path, cycle) W is a **Hamilton** walk (path, cycle) if $V(W) = V(D)$.

Underlying Graph of Digraph: The underlying graph of a digraph D is the unique graph G such that D is a biorientation of G.

Strongly Connected Digraph: A digraph is **strongly connected** if for every pair x, y of distinct vertices in D, there exists an (x, y)-walk and a (y, x)-walk.

Connected Graph: An undirected graph G is connected if for each pair x, y of $V(G)$ there is a (x, y)path

(A, B) of Digraph: (A, B) of digraph D is the collection of arcs in $A(D)$ with tails in A and heads in B.

More related definitions, see Chapter one of [8].

3.2 Connectivity of Sensor Networks

Let $\mathcal{N}$ is a system of sensors. For two sensors x, y of $\mathcal{N}$, we say that x and y is connected if:

(1) there is a trace $x_0x_1...x_n$ between x and y with $x = x_0$, $x_n = y$ and x_i can transmit message to x_{i+1};

(2) there is a trace $y_0x_1...y_n$ between x and y with $y = y_0$, $x = y_n$ and y_i can send information to y_{i+1}.

Digraph $G(D, A(D))$ deduced from connectivity of sensor network, DGC for short. $D = \mathcal{N}$, and $\overrightarrow{n_in_j} \in A(D)$ if and only if sensor n_i is able to send message to sensor n_j.

Lemma 3.1 [8]. For a digraph G and x_1, $x_n \in V(G)$. If there is a walk $W_{x_1x_n}$ from x_1 to x_n, then there is a path $P_{x_1x_n}$ from x_1 to x_n.

Proposition 3.1. *For a sensor network consisting of sensor system $\mathcal{N} = \{n_1, ..., n_k\}$, the following are equivalent:*

(1) $\mathcal{N}$ is connected;
(2) $DGC_{\mathcal{N}}$ is strongly connected;
(3) $DGC_{\mathcal{N}}$ contains a closed Hamilton walk;
(4) $DGC_{\mathcal{N}}$ is connected and each arc of $DGC_{\mathcal{N}}$ is on at least one of directed circle;
(5) $(\mathcal{A}, \overline{\mathcal{A}}) \neq \emptyset$, $(\overline{\mathcal{A}}, \mathcal{A}) \neq \emptyset$ for each $\mathcal{A} \subset \mathcal{N}$;
(6) For each $\mathcal{A} \subset \mathcal{N}$, $\mathcal{S}(\mathcal{A}, \overline{\mathcal{A}})$ is LBDC.

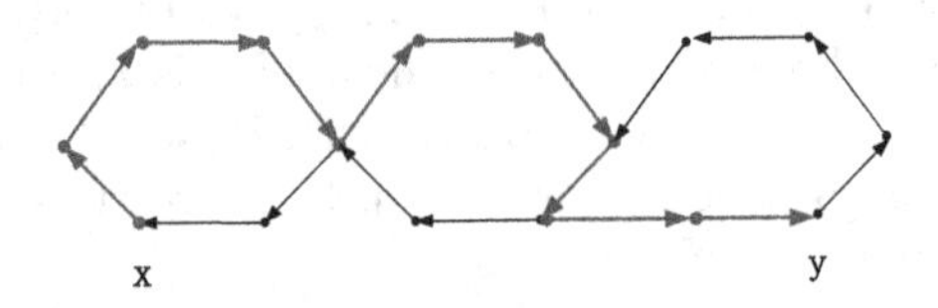

Fig. 11. An example on condition (4) of Proposition 3.1

Proof. (1) $\Leftrightarrow$ (2) is trivial by the definition of connectivity of sensor networks and strongly connectivity of $DGC_{\mathcal{N}}$.

(2) $\Rightarrow$ (3): For each n_i and n_{i+1}, there is a walk $w_{i(i+1)}$ from n_i to n_{i+1} by the strongly connected $DGC_{\mathcal{N}}$. At the same time, there is a walk $w_{(i+1)i}$ from n_{i+1} to n_i. It is easy to see that $W_{12} \cup W_{23} ... \cup W_{(k-1)k} \cup W_{k(k-1)} \cup ... \cup W_{21}$ is a closed Hamilton walk.

(3) $\Rightarrow$ (2): It is obviously from the definition of the strongly connectivity.

(2) $\Rightarrow$ (4): For each pair x, y, there is a walk W_{xy} from x to y. By Lemma 3.1, there is a path from x to y. It means that the underlying graph of $DGC_{\mathcal{N}}$ is connected.

Let $\overrightarrow{n_in_j} \in A(\mathcal{N})$. There is a walk W_{ji} from n_j to n_i by the strongly connected$DGC_{\mathcal{N}}$. There is a path P_{ji} from n_j to n_i by Lemma 3.1. $n_i\overrightarrow{n_in_j}n_j \cup P_{ji}$ is the circle we need.

(4) $\Rightarrow$ (2) For each pair x, $y \in V(G)$, there is a path $x_0e_0x_1e_1...e_{n-1}x_n$ from x to y in the underlying graph of $DGC_{\mathcal{N}}$, where $x_0 = x$ and $x_n = y$. Let C_0,...,C_{n-1} be the directed circles such that e_i is on the circle C_i. Pick a (x_0, x_1)-path P_0 on circle C_0. Since e_1 is on directed circle C_1, we can get a (x_1, x_2)-path P_1 on circle C_1 with the same direction of P_0 (See Fig. 11). Repeat this operation, for each $i < n$, we can get path P_i starting from x_i to x_{i+1} and with the same direction as P_{i-1}. Put $P = P_0 \cup P_1 \cup ... \cup P_{n-1}$. We can see that P is a walk from x to y. By the same way, we can get a walk from y to x. This fact shows that $DGC_{\mathcal{N}}$ is strongly connected.

(2) $\Rightarrow$ (5) For each $\mathcal{A} \subset \mathcal{N}$ with $\mathcal{A} \neq \emptyset$. Pick $x \in \mathcal{A}$ and $y \in \overline{\mathcal{A}}$. By the definition of strongly connectivity, there is a walk $W_{xy} = x_0\overrightarrow{e_0}...\overrightarrow{e_{n-1}}x_n$ from x to y with $x_0 = x$ and $x_n = y$. Let $i_0 = min\{i : x_i \notin \overline{\mathcal{A}}\}$. Then $x_{i_0-1} \in \mathcal{A}$. Hence, $\overrightarrow{e_{i_0-1}} \in (\mathcal{A}, \overline{\mathcal{A}})$.

(5) $\Rightarrow$ (4) Suppose that the underlying graph of $DGC_{\mathcal{N}}$ is not connected. Let $\{\mathcal{A}_i : i = 1, ...k$ be all the collection of all the disjoint connected components. Obviously, $(\mathcal{A}_1, \overline{\mathcal{A}_1}) = \emptyset$, which contradicts with (5).

Let $\overrightarrow{x_0x_1}$ be an arc of $DGC_{\mathcal{N}}$. Let $V_1' = \{v \in V(G) :$ there is an arc $\overrightarrow{x_1v}$ in $DGC_{\mathcal{N}}\}$; $V_1 = \{x_1\} \cup V_1'$. We know that V_1 is not empty since $(\{x_1\}, \overline{\{x_1\}}) \neq \emptyset$. If $x_0 \in V_1$, then $x_0\overrightarrow{e_0}x_1\overrightarrow{x_1x_0}x_0$ is a closed walk, in fact, a circle, containing $\overrightarrow{e_0}$. If $x_0 \notin V_1$ and $(V(G) \setminus (\{x_0\} \cup V_1)) \neq \emptyset$, then put $V_2 = \{v \in V(G) :$ there is an arc $\overrightarrow{x_2v}$ in $DGC_{\mathcal{N}}$, where $x_2 \in V_1\}$. If $x_0 \in V_2$, then there exists $x_2 \in V_2$ such that we can get a closed walk $x_0\overrightarrow{x_0x_1}x_1\overrightarrow{x_1x_2}x_2\overrightarrow{x_2x_0}x_0$. If $x_0 \notin V_2$ and $(V(G) \setminus (\{x_0\} \cup V_2)) \neq \emptyset$, then put $V_3 = \{v \in V(G) :$ there is an arc $\overrightarrow{x_3v}$ in $DGC_{\mathcal{N}}$, where $x_3 \in V_2\}$. So, we can see that, for i, if $x_0 \in V_i$, we can get a closed walk $x_0\overrightarrow{x_0x_1}x_1...\overrightarrow{x_ix_0}x_0$; if $x_0 \notin V_i$, we can get a subset V_{i+1} of $V(G)$ such that $|V_{i+1}| > |V_i|$. For the second case, there is some i_0 such that $V_{i_0} = V(G) \setminus \{x_0\}$. By the assumption, $(V_{i_0}, \overline{V_{i_0}}) \neq \emptyset$, which means that $(V_{i_0}, \{x_0\}) \neq \emptyset$. Therefore, there exists $x_{i_0} \in V_{i_0}$ such that $\overrightarrow{x_{i_0}x_0} \in E(G)$. Hence, we get a closed walk $x_0\overrightarrow{x_0x_1}x_1...\overrightarrow{x_{i_0}x_0}x_0$. By Lemma 3.1, there is a path $P_{x_1x_{i_0}}$ from x_1 to x_{i_0}. We can see that $x_0\overrightarrow{x_0x_1}x_1 \cup P_{x_1x_{i_0}} \cup x_{i_0}\overrightarrow{x_{i_0}x_0}x_0$ is a circle containing arc $\overrightarrow{x_0x_1}$.

(5) $\Leftrightarrow$ (6) If sensor a covers sensor b, then we put $\overrightarrow{ab}$ in $DGC_{\mathcal{N}}$. (5) $\Leftrightarrow$ (6) directly follows from the definition of LBDC.

Remark 3.1. "(5)" of Proposition 3.1 can not be weaken to "$(\{x\}, \overline{\{x\}}) \neq \emptyset$, $(\overline{\{x\}}, \{x\}) \neq \emptyset$ for each $x \in \mathcal{N}$". See Fig. 12.

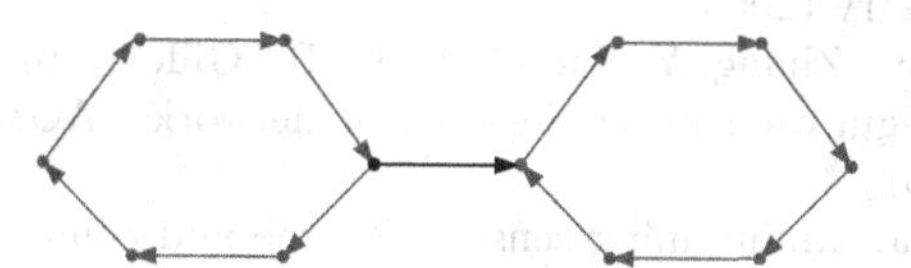

Fig. 12. A counter example on condition (5) of Proposition 3.1

Remark 3.2. A method to analysis wether a sensor network is connected by the matrixes of LBDC. For each sensor in the network, input those sensors it can receive messages, construct the matrix of LBDC to estimate the connectivity of the sensor network.

4 Conclusion

This paper presented the bi-directional coverage concept due to connectivity of sensor networks. We described bi-directional coverage models with matrix. Using digraph theory, we provided some characterizations of connectivity of sensor networks. These results will support a potential role for algorithms to connectivity of sensor networks.

References

1. Alam, S., Haas, Z.: Coverage and connectivity in three-dimensional networks. In: ACM MobiCom (2006)
2. Bai, X., Kumar, S., Xuan, D., Yun, Z., Lai, T.H.: Deploying wireless sensors to achieve both coverage and connectivity. In: ACM MobiHoc (2006)
3. Balister, P., Bollobás, B., Sarkar, A., Kumar, S.: Reliable density estimates for coverage and connectivity in thin strips of finite length. In: ACM MobiCom (2007)
4. Burke, D.K.: Covering Properties. Handbook of Set-theoretic Topology. Elsevier Science Publishers, North-Holland (1984)
5. Engelking, R.: General Topology. Heldermann Verlag, Berlin (1989)
6. Gupta, H., Das, S., Gu, Q.: Connected sensor cover: self organization of sensor networks for efficient query execution. In: ACM MobiHoc (2003)
7. Huang, C., Tseng, Y.: The coverage problem in a wireless sensor network. In: ACM WSNA (2003)
8. Jorgen, B.J., Gregory, Z.G.: Digraphs, Theory, Algorithms and Applications. Springer, Heidelberg (2009)
9. Kumar, N., Gunopulos, D., Kalogeraki, V.: Sensor network coverage restoration. In: Prasanna, V.K., Iyengar, S.S., Spirakis, P.G., Welsh, M. (eds.) DCOSS 2005. LNCS, vol. 3560, pp. 409–409. Springer, Heidelberg (2005)
10. Tung-Shou, C., Jeanne, C., Yun-Ru, T.: A study of bidirectional antenna for indoor localization using zigbee wireless sensor network. Inf. Technol. J. **10**(9), 1836–1841 (2011)
11. Wang, Y.C., Hu, C.C., Tseng, Y.C.: Efficient deployment algorithms for ensuring coverage and connectivity of wireless sensor networks. In: Wireless International Conference, Hungary (2005)
12. Xing, G., Wang, X., Zhang, Y., Lu, C., Pless, R., Gill, C.: Integrated coverage and connectivity configuration in wireless sensor networks. ACM Trans. Sens. Netw. **1**(1), 36–72 (2005)
13. Zhang, H., Hou, J.: Maintaining sensing coverge and connectivity in large sensor networks. In: NSF International Workshop on Theoretical and Algorithmic Aspects of Sensors Ad hoc Wireless, and Peer-to-Peer Networks. (2004)
14. Zorbas, D.: Efficient target coverage in WSNs with bidirectional communication. In: Computer Aided Modeling, Analysis and Design of Communication Links and Networks (CAMAD), pp. 106–110 (2010)

Hemispherical Lens Featured Beehive Structure Receiver on Vehicular Massive MIMO Visible Light Communication System

Tian Lang, Zening Li, Albert Wang, and Gang Chen(✉)

Department of Electrical Engineering, University of California, Riverside, CA 92521, USA
gachen@ee.ucr.edu

Abstract. With the development of internet of vehicles, Visible Light Communication (VLC) is the missing piece of this inevitable technology. However, current VLC systems lack the adequate date rate and reliability. By utilizing the proposed semispherical lens on Beehive Structure Receiver (BSR), we can achieve the significant diversity channel gain which is require to enable massive Multiple Input Multiple Output (MIMO) technology in any VLC systems. Via classical optics, we are able to derive a model to determine the channel gain of the receiver based on the number of transmitter. The simulation result shows images from transmitters to the receiver are distinguishable based on different angle of incidence which confirms that there is a significant spatial diversity for the proposed VLC system.

1 Introduction

1.1 Background

Visible Light Communication (VLC) has been widely investigated as an alternative communication solution for very high speed data transmission in internet of vehicles application while white lighting light emitting diode (LED) emerges as an important energy efficient illumination technology. Compare to its radio frequency (RF) counterpart, VLC possesses various advantages such as impassable via opaque obstacles which cause the visible light band to be reused in different room with no interferences, low-cost frontends, and relative high received signal-to-noise ratio (SNR). Also, there is no health concerns so long as the LEDs' intensity are within the regulation of eye-and-skin safety limit [1].

White light emitted from LEDs can be formed by using either a balanced intensity ratio of red-green-blue (RGB) LED or blue LED light combined with yellow phosphor. Due to their costs, latter method has been widely accepted as the mainstream illumination technology. These white LEDs can be modulated at 25 MHz [2] which can be utilized as the basis of novel data communication system [3–7].

Intensity modulation and direct detection (IM/DD) is well accept modulation and demodulation method for VLC. As information is carried on the intensity of the emitted light, all transmitted signals are non-negatives. Yet, in IM/DD optical systems, the channel gain is given by the ratio of the received optical power to the transmitted optical

C.-H. Hsu et al. (Eds.): IOV 2015, LNCS 9502, pp. 469–477, 2015.
DOI: 10.1007/978-3-319-27293-1_41

power [8]. More than ten transmitters and receivers within a multiple-in-multiple-out (MIMO) system are considered as Massive MIMO. In RF Massive MIMO, its channel matrix are generally complex. However, due to its IM/DD limitation the channel matrix for optical wireless is a real matrix with its elements denoting the power gains of all the LED-photodetector pairs.

1.2 Challenge

In background, it shows that theoretically optical carrier can be considered as having an 'unlimited bandwidth', but photodetector area and channel capacity in the system limit the amount of bandwidth that is practically available for a distortion-free communication system. For any type of communication system, transmission reliability should be one of the highest priority when designing a modulation technique. VLC system's modulation scheme should be able to offer a minimum acceptable error rate in adverse conditions as well as show resistance to the multipath-induced ISI and variations in the data signal DC component. Furthermore, the fundamental issue of designing a VLC system is that the system is located in a confined space. One of the major issue which limits the performance of a VLC system is inter-symbol interference (ISI). ISI is produce from multiple path effect which signals arrive at the receiver at different time slots by light's reflection from vehicles, buildings and road signs etc.

1.3 Motivation and Contribution

Many state-of-the-art techniques have been proposed for high speed transmissions with VLC systems e.g. a spread spectrum technique to combat ISI is given by [9] with the expense of reduced SE. Nonlinear equalization with guard slots were presented to reduce multipath effect in [10]. Also, [11] introduced multiple transmitter beams with narrow multi-beam field of view receiver to mitigate multipath dispersion. Nevertheless, the complexity of equalization in traditional single carrier transmission schemes increases rapidly with the data rates, which motivates the necessity of the research in optical multi-carrier transmission techniques [12]. Many types of optical multi-carrier transmission techniques have been proposed to support high data rates transmission such as orthogonal frequency division multiplexing technique [13, 14]. It has been shown that combining OFDM and multiple-input multiple-output (MIMO) can significantly reduce fading and yields frequency flat MIMO channel by spatial diversity [15].

To the best of our knowledge, massive MIMO Visible Light Vehicular Communication (VLVC) system have not been studied so far. The main contributions of this paper are summarized as follows: Proposed hemispherical lens featured beehive structure (SBS) receiver on Massive MIMO system; verified the spatial diversity of the MIMO technique and channel hardening advantage for the proposed massive MIMO VLC systems.

The remaining of this paper is organized as follows. In Sect. 2, the optical and MIMO optical channel model are explained. The proposed hemispherical lens featured beehive structure (SBS) receiver is described in Sect. 3. Section 4 shows simulation results. Section 5 concludes the paper.

2 Channel Model

2.1 Optical Channel

The characteristics of the channel is essential to the designing, implementation, and operation of any optical communication system. Intensity modulation with direct detection (IM/DD) is used as the method of implementation VLC system. This is method uses the instantaneous optical power of the transmitter as the transmitted waveform $x(t)$ and the received waveform $y(t)$ is the instantaneous current in the receiving photodetector. Therefore, the optical channel with IM/DD is descried as baseband linear system with impulse response $h(t)$. The equivalent baseband model of an IM/DD optical wireless link can be summarized by the following equations:

$$y(t) = Rx(t) \otimes h(t) + n(t) \tag{1}$$

where $n(t)$ is modelled as Gaussian and signal-independent noise, the symbol ⊗ denotes the convolution and R is the detector responsivity (Amperes/Watts). The information of $h(t)$ allows us to determine the multipath penalty, which limits the maximum symbol rate. Also, the second term is related to the signal-to-noise ratio (SNR), which determines the performance of the systems. This impulse response $h(t)$ can be used to analyze or simulate the effects of multipath dispersion in VLC channel. In [16], Gfeller and Bapst modelled the optical channel response as follows:

$$h(t) = f(x) = \begin{cases} \dfrac{2t_0}{t^3 \sin^2(FOV)}, & t_0 \le t \le \dfrac{t_0}{\cos(FOV)} \\ 0, & elsewhere \end{cases} \tag{2}$$

where t_0 is the minimum time delay. The VLC transfer function is defined by

$$H_{\mathrm{VLC}}(f) = H_{\mathrm{los}} + H_{\mathrm{diff}}(f) \tag{3}$$

where H_{los} is the contribution of the LOS, which is mostly independent on the modulation frequency, and H_{diff} is the contribution of the diffuse part of the optical channel.

2.2 Line-of-Sight Propagation Model

A distance d between transmitter T and receiver R in a LOS link configuration as shown in Fig. 1 has the following description:

$$P_R = \frac{1}{d^2} R_T(\phi, n) A_{\mathrm{eff}}(\varphi) \tag{4}$$

The transmitter is modelled using a generalized Lambertian radiation pattern $R_T(\phi, n)$, and $A_{\mathrm{eff}}(\varphi)$ represents the effective signal collection area of the receiver

where n in Fig. 1 represents the mode number of the radiation lobe that specifies the directionality of the transmitter

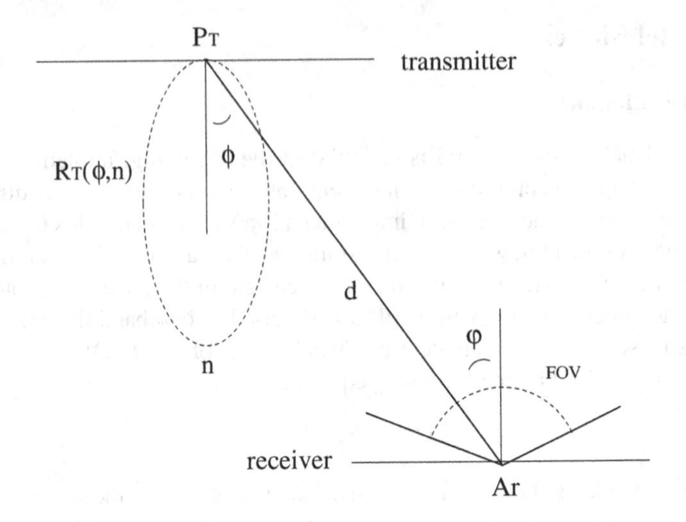

Fig. 1. Basic LOS link configuration of the VLC system

$$R_T(\phi, n) = \frac{n+1}{2\pi} P_T \cos^n(\phi), 0 \le \phi \le \frac{\pi}{2} \tag{5}$$

$$A_{\text{eff}} = A_r \cos \varphi \,\text{rect}(\frac{\varphi}{\text{FOV}}) \tag{6}$$

where P_T is the power radiated by the transmitter, A_r is the physical area of the receiver, and FOV is the receiver field of view (half-angle from the surface normal).

3 System Model

The imaging system of BSR is described in Fig. 2 (left). It is a typical room of size $5 \times 5 \times 2.5$ meters as length, width, and height respectively. There are 16 white light LEDs as transmitters located on the ceiling of the room. The BSR, Fig. 2 (middle) is can be found at the (2.5, 2.5, 0) meter position. The photodetector array can be easily

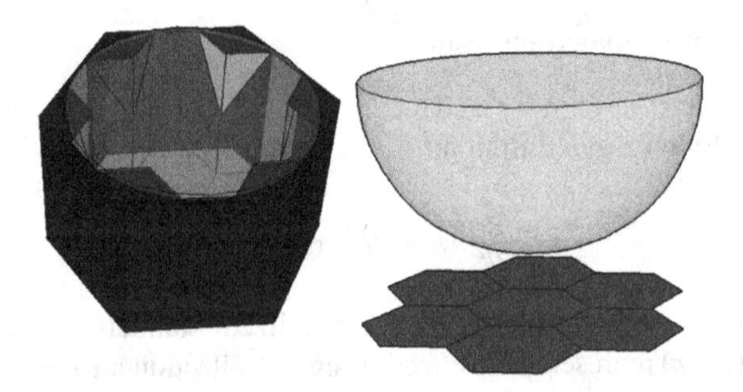

Fig. 2. The configuration of the BSR with opaque casing (left) and BSR components with a hemispherical lens and honeycomb organized photodetector array (right)

expanded due to its maximized photo-detecting area formation which is enabled by the honeycomb structure. In Fig. 3, we are showing some possible the expansion of the BSR photo-detecting area configuration.

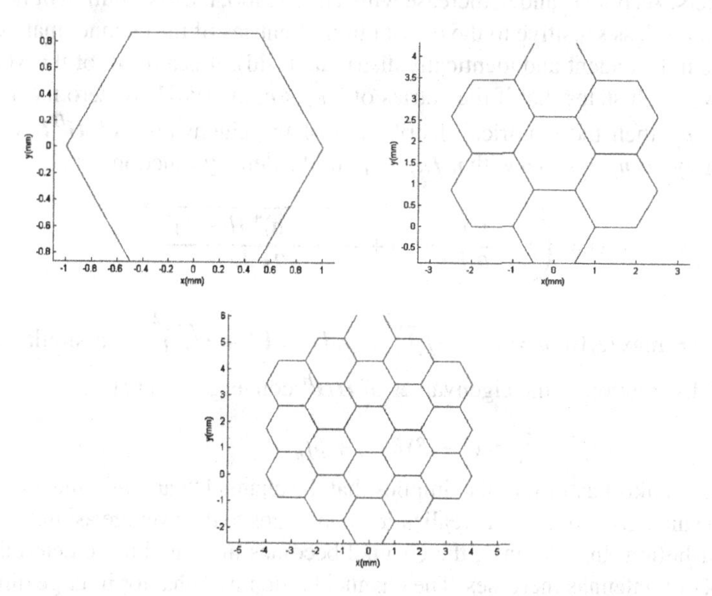

Fig. 3. Photo-detecting array expansion for BSR from one (left), to seven (middle), to 13 (right)

3.1 Massive MIMO

By utilizing a large number of LEDs as transmitters and photodetector as receivers, massive MIMO OFDM VLC systems creates more degrees of freedom in spatial domain which can bring many values such as increased link reliability and data rate without increasing bandwidth [17]. For example, the probability of link outage, P_{outage} in a point-to-point MIMO system with n_t transmit and n_r receive antennas can be described as

$$P_{outage} \propto SNR^{-(n_t n_r)} \tag{7}$$

This relationship shows the system has a potential of reaching a diversity order of $n_r n_t$. Moreover, with increase in SNR, the massive MIMO link error rate can be approximate by an exponential fall. Therefore, the achievable rate is described as

$$\min\left(n_t, n_r\right) \log_2(1 + SNR) \tag{8}$$

Without increasing the bandwidth, (11) shows that the system can achieve a very high data rates using large n_r, n_t. In addition, the rate gain in multiuser massive MIMO with large number of transceivers can also be significant [19]. A large number of transmitting LED in multiuser downlink can allow the use of simple precoding methods and flexible user 4selection and scheduling.

3.2 Channel Hardening in Large Dimensions

Increased data rate and diversity gain are not the only advantages a massive MIMO system offers. As both n_t and n_r increase with a fixed ratio, the distribution of its singular value becomes less sensitive to the distribution of entries of the channel matrix as long as they are independent and identically distributed (iid). It is a result of the Marcenko-Pastur law which states that if the entries of a $n_r \times n_t$ matrix H are zero mean iid with variance $1/n_r$, then the empirical distribution of the eigenvalues of $\boldsymbol{H}^H\boldsymbol{H}$ converges almost surely, as $n_t, n_r \to \infty$ with $n_t/n_r \to \beta$, to the density function [18]

$$f_\beta(x) = \left(1 - \frac{1}{\beta}\right)^+ \delta(x) + \frac{\sqrt{(x-a)^+ (b-x)^+}}{2\pi\beta x} \tag{9}$$

where $(z)^+ = \max(z, 0)$, $a = \left(1 - \sqrt{\beta}\right)^2$, *and* $b = \left(1 + \sqrt{\beta}\right)^2$. In a similar way, the empirical distribution of the eigenvalues of $\boldsymbol{H}\boldsymbol{H}^H$ converges to [17]

$$\tilde{f}_\beta = (1 - \beta)\,\delta(x) + \beta f_\beta(x) \tag{10}$$

The Marcenko-Pastur law also implies that the channel "hardens", meaning that the eigenvalue histogram of a single realization converges to the average asymptotic eigenvalue distribution. In this sense, the channel becomes more and more deterministic as the number of antennas increases. The channel hardening behavior in large dimensions can be seen clearly as $n_t$andn_r increases. Channel hardening results in several advantages in large dimensional signal processing. For example, linear detectors like zero forcing (ZF) and minimum mean square error (MMSE) detectors need to perform matrix inversions. Inversion of large random matrices can be done quickly, using series expansion techniques [19]. Because of channel hardening, approximate matrix inversions using series expansion and deterministic approximations from the limiting distribution become effective in large dimensions. Also, channel hardening can allow simple detection methods/algorithms to achieve a very good performance in large dimensions.

4 Simulation

In simulation configuration, 4 transmitters are placed 1meters apart to simulate the vehicle's tail lights and beehive structure receiver (BSR) which composed of 4 photodiodes is placed 2.5 meters away from the center of the transmitter plane. The photodetectors are evenly distributed on the bottom of the receiver as shown in Fig. 4. The lens' diameter is 5 mm. Its index of refraction is 1.5, and angle of incidence is less than 70 degree.

Assumption for a tractable simulation and analysis can be summarized as the following. Only LOS (without considering diffuse) transmission between LED and the flat surface of the lens and each LED is a point source. The rays from a given LED are approximately parallel when they reach the flat surface of the lens. Power is lost only at the surface of the lens through reflection (not within the lens through dissipation).

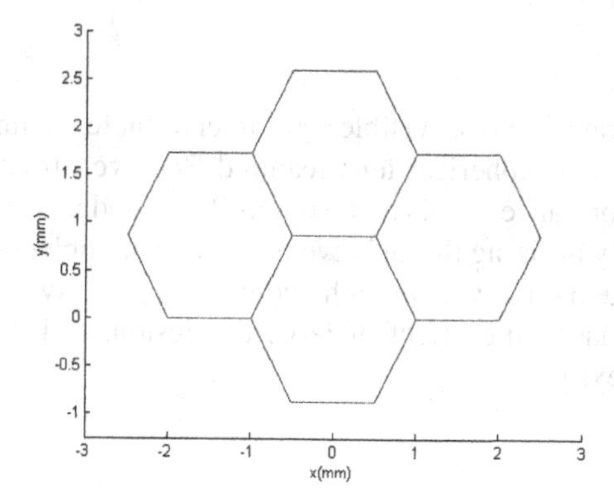

Fig. 4. BSR with 4 photodetector array structure

Light reflected internally at the curved surface of the lens is lost from the system and does not reach the photodetector array after multiple reflections. All of the light reaching the photo-detecting array is detected.

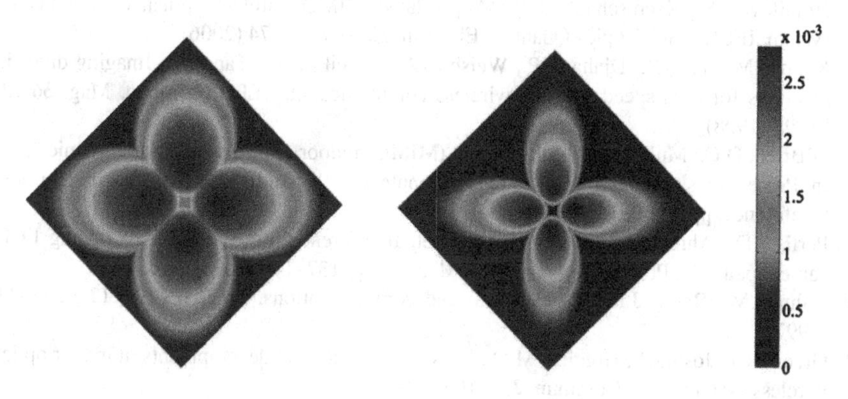

Fig. 5. Power density with symmetrical arrangement on MIMO configuration: 40 degree of the angle of incidence (left), LEDs placements further apart (middle), power density index (right)

The optical power density generated on the photo-detecting array with the angle of incidence at 40 degree is shown on Fig. 5 (left). As we can see, the signals from all four LEDs are clearly separated on the image which means that most of the power emitted from each corresponding LED are received. However, we later increase the LED placement further apart. The power density, shown on Fig. 5 (middle), is almost identical which results show an orthogonal channel matrix. The simulation result for LOS channel power density of the proposed system shows a clear presentation of the spatial diversity which is required to enable the massive MIMO technique.

5 Conclusion

This paper demonstrated a novel visible light inter vehicle communication receiver design which is the hemispherical lens featured Beehive Structure Receiver. The analysis for the performance of this receiver has illustrated its capability of enabling a spatial diversity. By utilizing this achievement, we can implement Massive MIMO technology to any existing visible light communication systems to receive the benefit of channel hardening effect of large dimension, and therefore reduce the computation complexity.

References

1. O'Brien, D., Minh, H.L., Zeng, L., Faulkner, G., Lee, K., Jung, D., Oh, Y., Won, E.T.: Indoor visible light communications: challenges and prospects. In: Proceedings of the SPIE 7091, 709 106–1–709 106–9 (2008)
2. Yuanquan, W., Nan, C.: A high-speed bi-directional visible light communication system based on RGB-LED. In: Communications, China, vol. 11, no.3, pp. 40–44, March 2014
3. Dambul, K.D., O'Brien, D.C., Faulkner, G.: Indoor optical wireless MIMO system with an imaging receiver. IEEE Photon. Technol. Lett. **23**(2), 97–99 (2011)
4. Hranilovic, S., Kschischang, F.R.: A pixelated MIMO wireless optical communication system. IEEE J. Sel. Topics Quantum Electron. **12**(4), 859–874 (2006)
5. Kahn, J.M., You, R., Djahani, P., Weisbin, A.G., Teik, B.K., Tang, A.: Imaging diversity receivers for high-speed infrared wireless communication. IEEE Commun. Mag. **36**(12), 88–94 (1998)
6. O'Brien, D.C.: Multi-Input Multi-Output (MIMO) indoor optical wireless communications. In: Proceedings of the Signals, Systems Computers, 2009 Conference Record 43rd Asilomar Conference, pp. 1636–1639, November 2009
7. Perli, S.D., Ahmed, N., Katabi, D.: PixNet: Interference-free wireless links using LCD-camera pairs. In: Proceedings MOBICOM 2010, pp. 137–148 (2010)
8. Kahn, J.M., Barry, J.R.: Wireless infrared communications. Proc. IEEE **85**(2), 265–298 (1997)
9. Green, R.J., Joshi, H., Higgins, M.D., Leeson, M.S.: Recent developments in indoor optical wireless systems. IET Commun. **2**, 3–10 (2008)
10. Ghassemlooy, Z., Hayes, A., Wilson, B.: Reducing the effects of intersymbol interference in diffuse DPIM optical wireless communications. In: Proceedings of the Institution of Electrical Eng.-Optoelectron, vol. 150, no. 5, pp. 445–452, October 2003
11. Carruther, J., Kahn, J.: Angle diversity for nondirected wireless infrared communication. IEEE Trans. Commun. **48**(6), 960–969 (2000)
12. Armstrong, J.: OFDM for optical communications. Lightwave Technol. J. **27**(3), 189–204 (2009)
13. Zhu, H., Wang, J.: Chunk-based resource allocation in OFDMA systems–Part I: Chunk allocation. IEEE Trans. Commun. **57**(9), 2734–2744 (2009)
14. Zhu, H., Wang, J.: Chunk-based resource allocation in OFDMA systems–Part II: Joint chunk, power and bit allocation. IEEE Trans. Commun. **60**(2), 499–509 (2012)
15. Sharma, M., Chadha, D., Chandra, V.: Capacity evaluation of MIMO-OFDM Free Space Optical communication system. In: 2013 Annual IEEE India Conference (INDICON), pp. 1–4, 13–15 December 2013

16. Gfeller, F.R., Bapst, U.: Wireless in-house data communication via diffuse infrared radiation. Proc. IEEE **67**, 1474–1486 (1979)
17. Tse, D., Viswanath, P.: Fundamentals of Wireless Communication. Cambridge University, Cambridge, UK (2005)
18. Tulino, A., Verdu, S.: Rnadom Matrix Theory and Wireless Communications Foundations and Trends in Communications and Information Theory. Now Publishers Inc, Delft, The Netherlands (2004)

Author Index

Printed in the United States
By Bookmasters

Printed in the United States
By Bookmasters